Lecture Notes in Computer Science   11077

Commenced Publication in 1973
Founding and Former Series Editors:
Gerhard Goos, Juris Hartmanis, and Jan van Leeuwen

More information about this series at http://www.springer.com/series/7407

Vladimir P. Gerdt · Wolfram Koepf
Werner M. Seiler · Evgenii V. Vorozhtsov (Eds.)

Computer Algebra in Scientific Computing

20th International Workshop, CASC 2018
Lille, France, September 17–21, 2018
Proceedings

 Springer

Editors
Vladimir P. Gerdt
Laboratory of Information Technologies
Joint Institute of Nuclear Research
Dubna
Russia

Wolfram Koepf
Institut für Mathematik
Universität Kassel
Kassel
Germany

Werner M. Seiler (iD)
Institut für Mathematik
Universität Kassel
Kassel
Germany

Evgenii V. Vorozhtsov
Institute of Theoretical and Applied
  Mechanics
Russian Academy of Sciences
Novosibirsk
Russia

ISSN 0302-9743   ISSN 1611-3349 (electronic)
Lecture Notes in Computer Science
ISBN 978-3-319-99638-7   ISBN 978-3-319-99639-4 (eBook)
https://doi.org/10.1007/978-3-319-99639-4

Library of Congress Control Number: 2018952056

LNCS Sublibrary: SL1 – Theoretical Computer Science and General Issues

This Springer imprint is published by the registered company Springer Nature Switzerland AG
The registered company address is: Gewerbestrasse 11, 6330 Cham, Switzerland

Preface

The International Workshop on Computer Algebra in Scientific Computing (CASC) is an annual conference that brings together researchers and scientists working in the field of computer algebra and researchers from various application areas that apply pioneering methods of computer algebra in sciences such as physics, chemistry, life sciences, and engineering, to discuss problems and solutions in the area, to identify new issues, and to shape future directions for research.

This year, the 20th CASC conference was held in Lille (France). The computer algebra group of Lille is one of the few French computer algebra groups hosted in a computer science laboratory (initially the LIFL then CRIStAL, since 2015). This context always pushed the group to develop symbolic methods with a focus on applications and software development. Control theory has provided a major focus since the group was founded in the 1980s by Gérard Jacob. The development of symbolic methods dedicated to biological modeling has been continuously growing since 2002, when Michel Petitot became group leader. Software development has always been a major concern for the group, with a particular effort under the leadership of François Boulier, group leader since 2011. In particular, at CRIStAL foundation, the computer algebra group merged with a high-performance computing group, yielding the "algebraic and high-performance computing" (CFHP) team, with the broader "scientific computing" domain. In 2018, control theory received renewed interest from the group, with the creation of GAIA ("geometry, algebra, informatics and applications"), both an Inria team and a subgroup of CFHP, led by Alban Quadrat.

From a computer algebra point of view, the research activity of the group was much influenced by the works of Michel Fliess in control theory. The initial focus was on the application of noncommutative algebra to the problem of expanding series from dynamical systems by means of iterated integrals. Later, the group experience in noncommutative algebra led to results in the theory of poly-logarithms (Hoang Ngoc Minh and Joris Van Der Hoeven were group members at that time), which are related to Riemann zeta function, and the investigation of chemical reaction networks (encoding biological models) endowed with stochastic determination. In the 1990s, the simplification theory of systems of differential equations became a major domain, with two approaches: (1) Ritt and Kolchin differential algebra and its elimination theory; (2) differential geometry and Cartan's equivalence method. The group was involved in research on Ritt's characteristic sets and made an important contribution to the theory of regular chains both in the differential and in the polynomial case (Marc Moreno Maza, now at ORCCA, was a group member for a few years). In terms of software, the group developed an important relationship with Maplesoft. It released the *diffalg* (1996) and the *DifferentialAlgebra* (2008) packages (itself relying on the open source *BLAD* libraries). It contributed to the first version of the *RegularChains* library. All these software packages have been shipped with the MAPLE standard library.

The aforementioned events influenced the choice of Lille as a venue for the CASC 2018 workshop.

This volume contains 24 full papers submitted to the workshop by the participants and accepted by the Program Committee after a thorough reviewing process with usually three independent referee reports. Additionally, the volume includes two contributions corresponding to the invited talks.

Polynomial algebra, which is at the core of computer algebra, is represented by contributions devoted to the computation of Pommaret bases using syzygies, factorization of multivariate polynomials, tropical Newton–Puiseux polynomials, positive solutions of systems of signed parametric polynomial inequalities, sparse polynomial arithmetic with the BPAS library, blackbox polynomial system solver on parallel shared memory computers, investigation of analytic complexity of a bivariate holomorphic function by means of computer algebra tools, localization of polynomial ideals by a new "local primary algorithm," computation of the sparse multivariate polynomial remainder sequence, splitting permutation representations of finite groups by polynomial algebra methods, tropicalization of linear subspaces, and efficient implementation of the algorithms for the computation of Gröbner basis with the aid of the Haskell compiler.

In his invited talk, Jean-Guillaume Dumas promotes the idea that the proof-of-work certificates can be efficiently computed in the cloud. When there is such a cloud-based service, demanding computations are outsourced in order to limit infrastructure costs. The idea of verifiable computing is to associate a data structure, a proof-of-work certificate, to the result of the outsourced computation. This allows a verification algorithm to prove the validity of the result faster than by recomputing it. The problem-specific procedures in computer algebra are also presented for exact linear algebra computations that are Prover-optimal, that is, that have much less financial overhead.

The tutorial of Marc Moreno Maza is devoted to the problem of the symbolic computation of the limits of multivariate functions. Although the calculation of such limits is supported, with some limitations, in general-purpose computer algebra systems such as Maple and Mathematica, Maple is not capable of computing limits of rational functions in more than two variables. In this tutorial, it was shown how various types of limits can be computed by means of algebraic calculations. Examples cover the Zariski closure of a constructible set, the tangent cone of an algebraic set at one of its singular points, and the limit of a real multivariate rational function at one of its poles.

Four papers deal with applications of symbolic and symbolic-numeric computations for investigating and solving partial differential equations (PDEs) and ODEs in mathematical physics and fluid mechanics: a new finite difference strongly consistent scheme for steady Stokes flow, solution of elliptic boundary-value problems using multivariate simplex Lagrange elements, and invertibility of difference operators arising at the approximation of ODEs.

Applications of CASs in mechanics, physics, and biology are represented by the following themes: satellite dynamics with aerodynamic attitude control systems, dynamical systems with irrational first integrals, and modeling of the evolution of a staphylococcus population with the aid of nonlinear integro-differential equations.

The remaining topics include the application of the Bargmann–Moshinsky basis in molecular and nuclear physics, the visualization of planar real algebraic curves with singularities with the aid of a continuation method, signal processing with the aid of Padé approximants, finding multiple solutions in nonlinear integer programming, and investigation of noncommutative evolution equations with singularities.

The CASC 2018 workshop was supported financially by the University of Lille, the Research Center in Computer Science, Signal and Automatics of Lille (CRIStAL - UMR 9189), the National Center for Scientific Research (CNRS), the Inria Lille–Nord Europe research center, and the Maplesoft company.

Our particular thanks are due to the members of the CASC 2018 local Organizing Committee at the University of Lille, i.e., François Boulier, François Lemaire, and Adrien Poteaux, who ably handled all the local arrangements in Lille. In addition, Prof. F. Boulier provided us with the information about the computer algebra activities at the University of Lille.

Furthermore, we want to thank all the members of the Program Committee for their thorough work. We are grateful to Matthias Orth (Universität Kassel) for his technical help in the preparation of the camera-ready manuscript for this volume. Finally, we are grateful to the CASC publicity chair, Andreas Weber (Rheinische Friedrich-Wilhelms-Universität Bonn), and his assistant Hassan Errami for the design of the conference poster and the management of the conference website (http://www. casc.cs.uni-bonn.de).

July 2018   Vladimir P. Gerdt
                                                             Wolfram Koepf
                                                             Werner M. Seiler
                                                         Evgenii V. Vorozhtsov

Organization

CASC 2018 was organized jointly by the Institute of Mathematics at Kassel University and the Université de Lille, Lille, France.

Workshop General Chairs

François Boulier, Lille, France
Vladimir P. Gerdt, Dubna, Russia
Werner M. Seiler, Kassel, Germany

Program Committee Chairs

Wolfram Koepf, Kassel, Germany
Evgenii V. Vorozhtsov, Novosibirsk, Russia

Program Committee

Moulay Barkatou, Limoges, France
François Boulier, Lille, France
Jin-San Cheng, Beijing, China
Victor F. Edneral, Moscow, Russia
Matthew England, Coventry, UK
Jaime Gutierrez, Santander, Spain
Sergey A. Gutnik, Moscow, Russia
Thomas Hahn, Munich, Germany
Jeremy Johnson, Philadelphia, USA
Victor Levandovskyy, Aachen, Germany
Dominik Michels, Jeddah, Saudi Arabia
Marc Moreno Maza, London, Canada

Veronika Pillwein, Linz, Austria
Alexander Prokopenya, Warsaw, Poland
Georg Regensburger, Linz, Austria
Eugenio Roanes-Lozano, Madrid, Spain
Valery Romanovski, Maribor, Slovenia
Timur Sadykov, Moscow, Russia
Doru Stefanescu, Bucharest, Romania
Thomas Sturm, Nancy, France
Akira Terui, Tsukuba, Japan
Elias Tsigaridas, Paris, France
Jan Verschelde, Chicago, USA
Stephen M. Watt, Waterloo, Canada

Local Organization

François Boulier (Chair), Lille, France
François Lemaire, Lille, France
Adrien Poteaux, Lille, France

Publicity Chair

Andreas Weber, Bonn, Germany

Website

http://www.casc.cs.uni-bonn.de/2018 (Webmaster: Hassan Errami)

Contents

Proof-of-Work Certificates that Can Be Efficiently Computed in the Cloud (*Invited Talk*)

Jean-Guillaume Dumas(✉) (iD)

Université Grenoble Alpes, Laboratoire Jean Kuntzmann, CNRS, UMR 5224, 700 avenue centrale, IMAG - CS 40700, 38058 Grenoble, Cedex 9, France
`Jean-Guillaume.Dumas@univ-grenoble-alpes.fr`

Abstract. In an emerging computing paradigm, computational capabilities, from processing power to storage capacities, are offered to users over communication networks as a cloud-based service. There, demanding computations are outsourced in order to limit infrastructure costs.

The idea of verifiable computing is to associate a data structure, a *proof-of-work certificate*, to the result of the outsourced computation. This allows a verification algorithm to prove the validity of the result, faster than by recomputing it. We talk about a Prover (the server performing the computations) and a Verifier.

Goldwasser, Kalai and Rothblum gave in 2008 a generic method to verify any parallelizable computation, in almost linear time in the size of the, potentially structured, inputs and the result. However, the extra cost of the computations for the Prover (and therefore the extra cost to the customer), although only almost a constant factor of the overall work, is nonetheless prohibitive in practice.

Differently, we will here present problem-specific procedures in computer algebra, e.g. for exact linear algebra computations, that are Prover-optimal, that is that have much less financial overhead.

1 Introduction

In an emerging computing paradigm, computational capabilities, from processing power to storage capacities, are offered to users over communication networks as a service.

Many such outsourcing platforms are now well established, as Amazon web services (through the Elastic Compute Cloud), Microsoft Azure, IBM Platform Computing or Google cloud platform (via Google Compute Engine), as shown in Fig. 1. None of these platforms, however, offer any guarantee whatsoever on the calculation: no guarantee that the result is correct, nor even that the computation has even effectively been done.

1.1 Verifiable Computing

This new paradigm holds enormous promise for increasing the utility of computationally weak devices. A natural approach is for weak devices to delegate

© Springer Nature Switzerland AG 2018
V. P. Gerdt et al. (Eds.): CASC 2018, LNCS 11077, pp. 1–17, 2018.
https://doi.org/10.1007/978-3-319-99639-4_1

Fig. 1. Some outsourced computing services

expensive tasks, such as storing a large file or running a complex computation, to more powerful entities (say servers) connected to the same network. While the delegation approach seems promising, it raises an immediate concern: when and how can a weak device verify that a computational task was completed correctly? This practically motivated question touches on foundational questions in cryptography, coding theory, complexity theory, proofs, and algorithms.

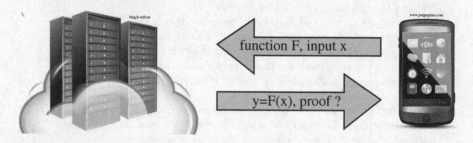

Fig. 2. Verifying the computation should take less time than computing it

More generally, the question of verifying a result at a lower cost (time, memory) than that of recomputing it, as shown on Fig. 2, is of paramount importance. Another example of application is for the extension of the trust about results computed via probabilistic or approximate algorithms. There the idea is to gain confidence into the correctness, but only at a cost negligible when compared to that of the computation.

1.2   Linear Algebra, Global Optimization

For instance, GL7d19 is an $1\,911\,130 \times 1\,955\,309$ matrix whose rank $1\,033\,568$ was computed once in 2007 with a Monte-Carlo randomized algorithm [19]. This required 1050 CPU days of computation. We thus need publicly verifiable certificates to improve our confidence in computational results.

In linear algebra our original motivation is also related to sum-of-squares. By Artin's solution to Hilbert 17th Problem, any polynomial inequality $\forall \xi_1, \ldots, \xi_n \in \mathbb{R}, f(\xi_1, \ldots, \xi_n) \geq g(\xi_1, \ldots, \xi_n)$ can be proved by a fraction of sum-of-squares:

$$\exists u_i, v_j \in \mathbb{R}[x_1, \ldots, x_n], f - g = \left(\sum_{i=1}^{\ell} u_i^2 \right) \Big/ \left(\sum_{j=1}^{m} v_j^2 \right) \tag{1}$$

Such proofs can be used to establish global minimality for $g = \inf_{\xi_v \in \mathbb{R}} f(\xi_1, \ldots, \xi_n)$ and constitute certificates in non-linear global optimization. A symmetric integer matrix $W \in \mathbb{SZ}^{n \times n}$ is positive semidefinite, denoted by $W \succeq 0$, if all its eigenvalues, which then must be real numbers, are non-negative. Then, a certificate for positive semidefiniteness of rational matrices constitutes, by its Cholesky factorizability, the final computer algebra step in an exact rational sum-of-squares proof, namely

$$\exists e \geq 0, \ W^{[1]} \succeq 0, \ W^{[2]} \succeq 0, \ W^{[2]} \neq \mathbf{0}:$$
$$(f - g)(x_1, \ldots, x_n) \cdot (m_e(x_1, \ldots, x_n)^T W^{[2]} m_e(x_1, \ldots, x_n)) =$$
$$m_d(x_1, \ldots, x_n)^T W^{[1]} m_d(x_1, \ldots, x_n), \tag{2}$$

where the entries in the vectors m_d, m_e are the terms occurring in u_i, v_j in (1). In fact, (2) is the semidefinite program that one solves [43]. Then, the client can verify the positiveness by checking Descartes' rule of signs on the *certified* characteristic polynomial of $W^{[1]}$ and $W^{[2]}$. Thus arose the question how to give possibly probabilistically checkable certificates for linear algebra problems.

1.3 Techniques

The tools used to provide such efficient *proof-of-work certificates* stem from programs that check their work [12], to proof of knowledge protocols [7], via error-correcting codes [35, 42] complexity theory [1] or secure multiparty protocols [17], and the interaction of these different methodologies is crucial.

Here we will thus follow this road map:

- We recalled that global optimization can be reduced to linear algebra. Thereupon we will focus on certificates for linear algebra problems [43] in computer algebra, which extend the randomized algorithms of Freivalds [32].
- We combine those with probabilistic interactive proofs of Babai [5] and Goldwasser et al. [39],
- as well as Fiat-Shamir heuristic [9, 29] turning interactive certificates into non-interactive heuristics subject to computational hardness.
- Overall, we obtain problem-specific efficient certificates for dense, sparse, structured matrices with coefficients in fields or integral domains.

2 Interactive Protocols, the PCP Theorem and Homomorphic Encryption

2.1 Arthur-Merlin Interactive Proof Systems

A proof system usually has two parts, a theorem T and a proof Π, and the validity of the proof can be checked by a verifier V. Now, an *interactive proof*, or a $\sum$-*protocol*, is a dialogue between a prover P (or *Peggy* in the following) and a verifier V (or *Victor* in the following), where V can ask a series of questions, or challenges, q_1, q_2, ... and P can respond alternatively, in successive *rounds*, with a series of strings $\pi_1, \pi_2, \ldots$, the responses, in order to prove the theorem T. The theorem is sometimes decomposed into two parts, the hypothesis, or input, H, and the commitment, C. Then the verifier can accept or reject the proof: $V(H, C, q_1, \pi_1, q_2, \pi_2, \ldots) \in \{\text{accept}, \text{reject}\}$.

To be useful, such proof systems should satisfy **completeness** (the prover can convince the verifier that a true statement is indeed true) and **soundness** (the prover cannot convince the verifier that a false statement is true). More precisely, the protocol is *complete* if the probability that a true statement is rejected by the verifier can be made arbitrarily small. Similarly, the protocol is *sound* if the probability that a false statement is accepted by the verifier can be made arbitrarily small. The completeness (resp. soundness) is *perfect* if accepted (resp. rejected) statements are always true (resp. false).

It turns out that interactive proofs with perfect completeness are as powerful as interactive proofs [33]. Thus in the following, as we want to prove correctness of a result more than proving knowledge of it, we will mainly show interactive proofs with perfect completeness.

The class of problems solvable by an interactive proof system (IP) is equal to the class PSPACE [55] and a probabilistically checkable proof, $\text{PCP}[r(n), \pi(n)]$, for an input of length n, is a type of proof that can be checked by a randomized algorithm using a bounded amount of randomness $r(n)$ and reading a bounded number of bits of the proof $\pi(n)$. For instance, $\text{PCP}[O(\log n), O(1)] = \text{NP}$ [3,6].

In general, interactive protocols encompass many kinds of proofs and Prover and Verifier settings. One can think of the difficulty of integer factorization versus that of re-multiplying found factors. Another example could be satisfiability checking, where the solver has to explore the state space, while verifying a variable assignment or a conflict clause could be much simpler [2]. In computer algebra, the Prover can be a probabilistic algorithm or a symbolic-numeric program, where the Verifier would perform the checks exactly or symbolically; further, computer algebra systems could perform complex calculations where an interactive theorem prover (or proof assistant like Isabel-HOL or Coq) only has to a posteriori formally verify the certificate [15,16].

Table 1 gives more examples of such settings.

Table 1. Examples of prover/verifier settings

Prover	Verifier
Computer Scientist	Mathematician
Computer Algebra system	Formal proof assistant
Cloud	User
Server	Client
Cellphone	Trusted platform module

2.2   Goldwasser et al. Prover Efficient Interactive Certificates

Now, efficient protocols (interactive proofs between a *Prover*, responsible for the computation, and a *Verifier*, to be convinced) can be designed for delegating computational tasks.

Recently, generic protocols, mixing zero-sum checks [45] and probabilistically checkable proofs, have been designed by teams around Shafi Goldwasser at the MIT or Harvard, for circuits with polylogarithmic depth [38,57], namely for problems that can be efficiently solved on a parallel computer (in the NC or AC complexity class). These results have also been extended to any structured inputs (any polynomial-time-uniform polylog-depth Boolean circuits in the sense of Beame's et al. [8], division circuits) [23].

The resulting protocols are interactive and there is a trade-off between the number of interactive rounds, the volume of communication and the computational cost [50]; the cost for the verifier being usually only roughly proportional to the input size.

These protocols can, e.g., certify that two supersparse polynomials are relatively prime in verifier cost which is polylog time (and rounds) in the degree. The produced certificates, in analogy to processor-efficient parallel algorithms, are Prover-efficient: if the cost to compute the result by the best known algorithm is $T(n)$ for a size n problem, then the cost to produce the result together with the verifiable certificate is $T(n)^{1+o(1)}$.

Those techniques can however produce a non negligible practical overhead for the Prover and are restricted to certain classes of circuits.

2.3   Parno et al. Homomorphic Solutions

Another approach as been developed by Gentry et al., at Microsoft and IBM research, it is Pinocchio. It solves a broader range of problems, to the cost of using relatively inefficient homomorphic routines [48] in an amortized way.

The idea is that the Prover should run the program (or at least part of the program twice), once normally on the input, and once homomorphically on an encrypted version of the input. The Verifier will then verify the consistency between the normal output and the encrypted one. Usually the Verifier is required to run the algorithm at least once for a given size or structure of the input but can reuse this for multiple inputs.

This generic procedure can be applied on specific linear algebra or polynomial problems [25, 28, 31, 60], or on generic quadratic arithmetic programs [48]. There, fully homomorphic encryption can be used [36] or sometimes just pairings [48] and/or cryptographic hashes [30].

Here also the Prover can be efficient, but subject in practice to the overhead of homomorphic computations.

2.4 Public Verification, Delegatability, and Zero-Knowledge

Interactive certificates require some exchanges between the Prover and the Verifier. With such a protocol, the Verifier can be *privately* convinced that the computation of the Prover produced the correct answer. This does not mean that other people would be convinced by the transcript of their exchange: the Prover and Verifier could be in cahoots and the supposedly random challenges carefully crafted.

Fiat-Shamir heuristic [9, 29] can thus turn interactive certificates into non-interactive heuristics subject to computational hardness: the random challenges are replaced by cryptographic hashes of all previous data and exchanges. Crafting such values would then reduce to being able to forge cryptographic fingerprints [20, Sect. 4.5].

Further, more properties could be sought for such protocols, such *privacy* of data and/or computations. In this setting, a publicly verifiable computation scheme can also be four algorithms (*KeyGen, ProbGen, Compute, Verify*), where *KeyGen* is some (amortized) preparation of the data, *ProbGen* is the preparation of the input, *Compute* is the work of the *Prover* and *Verify* is the work of the *Verifier* [49]. Usually the Verifier also executes *KeyGen* and *ProbGen* but in a more general setting these can be performed by different entities (respectively called a *Preparator* and a *Trustee*).

This allows to define several adversary models but usually the protocols are secure against a *malicious Prover only* (that is the Client must trust both the Preparator and the Trustee).

One can also further impose that there is no interaction between the Client and the Trustee after the Client has sent his input to the Server. Publicly verifiable protocols with this property are said to be *publicly delegatable* [25, 28, 60].

Then, some different properties of the protocol could be desirable, such as not disclosing the result but instead just providing a *proof-of-work*. This results in general in *zero-knowledge* protocols over confidential data, such as cryptocurrency transactions, as in, e.g., [39], with recent efficient implementations [10, 11, 13, 14].

2.5 Problem-Specific Efficient Certificates

Differently, dedicated certificates (data structures and algorithms that are verifiable a posteriori, without interaction) have also been developed, e.g., in computer algebra for exact linear algebra [20, 22, 24, 32, 43], even for problems that are not structured. There the certificate constitute a proof of correctness of a result, not of a computation, and can thus also **detect bugs** in the implementations.

Moreover, problem-specific certificates can gain crucial logarithmic factors for the verifier and allow for optimal prover computational time, see Fig. 3.

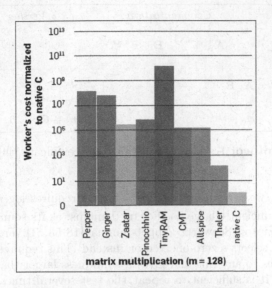

2048×2048	Thaler[57]	Ad-hoc [32]
Server time	18.23s	0.65s
Certificate overhead	0.13s	0.00s
Client time	2.89s	0.01s

Fig. 3. Generic protocols [58] versus dedicated protocols for matrix multiplication

For this, the main difficulty is to be able to design verification algorithms for a problem that are completely orthogonal to the computational algorithms solving it, while remaining checkable in time and space not much larger than the input.

3   Prover-Optimal Certificates in Linear Algebra

We show in this section, that such problem-specific certificates are attainable in linear algebra, where we allow certificates that are validated by Monte Carlo randomized algorithms.

3.1   Freivalds Zero Equivalence of Matrix Expressions

The seminal certificate in linear algebra is due to Rūsiņš Freivalds [32]: quadratic time is feasible because a matrix multiplication AB can be certified by the resulting product matrix C via the probabilistic projection to matrix-vector products (see also [44] who reduced the requirements to only $O(\log(n))$ random bits), shown in Protocol 1.

Prover	Communication	Verifier
	$\mathbf{A} \in \mathbb{F}^{m \times k}, \mathbf{B} \in \mathbb{F}^{k \times n}$	
Compute $\mathbf{C} = \mathbf{A} \cdot \mathbf{B}$	$\xrightarrow{\quad \mathbf{C} \quad}$	$r \xleftarrow{\$} S \subseteq \mathbb{F}$ Form $v = [1, r, r^2, \ldots, r^{n-1}]^T$ $\mathbf{A}(\mathbf{B}v) - \mathbf{C}v \overset{?}{=} 0$

Protocol 1. Matrix multiplication certificate [44].

In Protocol 1, we give the variant of [44] that requires $\log(n)$ random bits, but works over sufficiently large coefficient domains, as its soundness is $1 - \frac{|S|}{n}$ by the DeMillo-Lipton/Schwartz/Zippel lemma [18,53,61]. Freivalds' original version randomly selects a zero-one vector instead. This requires n random bits instead but applies to any ring and gives a soundness larger than $\frac{1}{2}$.

In both cases it is sufficient to repeat the test several times to achieve any desired probability.

3.2 Reductions to Matrix Multiplication

With a certificate for matrix expressions, then one can certify **any algorithm that reduces to matrix multiplication**: the Prover records all the intermediate matrix products and sends them to the Verifier who reruns the same algorithm but checks the matrix products instead of computing them [43], as shown in Protocol 2.

Prover	Communication	Verifier
Runs the algorithm	All intermediate matrix products $\xrightarrow{\hspace{3cm}}$	Runs the algorithm but replace each matrix products by Freivalds' checks

Protocol 2. Certificates with reduction to matrix multiplication [43, Sect. 5].

Overall, the communications and Verifier computational cost are given by taking $\omega = 2$ in the Prover's complexity bounds (with potential additional logarithmic factors due to summations). Further, *the production of the certificate has no computational overhead* for the Prover: it only adds the communication of the intermediate matrix products.

For instance, Storjohann's Las Vegas rank algorithm of integer matrices [56] becomes a non-interactive/non-cryptographic Monte Carlo checkable proof-of-work certificate that has soft-linear time communication and verifier bit complexity in the number of input bits!

3.3   Sparse or Structured Matrices

When the matrices are sparse or present some structure, quadratic run time and/or quadratic communications might be overkill for the Verifier. There it is better if his communications and computational cost is of the form $\mu(A)+n^{1+o(1)}$ where $\mu(A)$ is the number of operations to perform a matrix-vector product. This scheme is thus also interesting if the considered matrix is only given as a black box [40].

In that vein, we now have certificates for:

- **non-singularity**, Protocol 3;
- **an upper bound to the rank**, Protocol 4 (if elimination on the input matrix is possible for the Prover then a variant without preconditioners can be used [24, 26]);
- the **rank**, combining Protocols 3 and 4;
- the **minimal polynomial**, using Protocol 5 (where $f_u^{A,v}$ is the monic scalar minimal generating polynomial of the sequence $u^T v, \ldots, u^T A^i v$, $\rho_u^{A,v}$ is such that $\rho_u^{A,v} = f_u^{A,v} \cdot G$ with G the generating function of the latter sequence, for random vectors u and v, chosen by the Verifier [41, Theorem 5]);
- the **determinant**, Protocol 6, which randomness could be reduced from $O(n)$ to a constant number of field elements [21, Sect. 7].

Additionally, properties of the given matrices can also sometimes be discovered at low cost: whether the blackbox is a **band matrix**, has a **low displacement rank**, has a few or many **nilpotent blocks** or **invariant factors** [27]. Similarly, the existence of a **triangular one sided equivalence**, as well as the **rank profiles** can also be certified without sending an explicit factorization to the Verifier [24]. For the latter, the price to pay is to require a linear number of rounds.

3.4   Integer or Polynomial Matrices

Over an integral domain, the verification procedure can be performed via a randomly chosen modular projection. If there are sufficiently many *small* maximal ideals, then one can uniformly choose one at random and then ask for a certification of the result in the associated quotient field as shown in Protocol 7.

For instance this gives very efficient certificates for polynomial or integer/rational matrices, provided that one has a bound on the degree or the magnitude of the coefficients:

Prover		Verifier
Input	$\mathbf{A} \in \mathbb{F}^{n \times n}$	

Commitment	$\xrightarrow{\text{1 : non-singular}}$	
Challenge	$\xleftarrow{\text{2 : } \boldsymbol{b}}$	$\boldsymbol{b} \xleftarrow{\$} S^n \subset \mathbb{F}^n$
Response $\boldsymbol{w} \in \mathbb{F}^n$	$\xrightarrow{\text{3 : } \boldsymbol{w}}$	$\mathbf{A}\boldsymbol{w} \overset{?}{=} \boldsymbol{b}$

Protocol 3. Blackbox interactive certificate of non-singularity [20]

Prover		Verifier
	$A \in \mathbb{F}^{m \times n}$	$S \subset \mathbb{F}$

$rank(A) \le r$	$\xrightarrow{\text{1 : } r}$	$r \overset{?}{<} \min\{m, n\}$
	$\xleftarrow{\text{2 : } U, V}$	$U \in \mathbb{B}_S^{m \times m}, V \in \mathbb{B}_S^{n \times n}$
		preconditioners of size $n^{1+o(1)}$
$w \in \mathbb{F}^{r+1} \neq 0$	$\xrightarrow{\text{3 : } w}$	$w \overset{?}{\neq} 0$
		$[I_{r+1}|0] \, UAV \begin{bmatrix} I_{r+1} \\ 0 \end{bmatrix} w \overset{?}{=} 0$

Protocol 4. Blackbox upper bound to the rank certificate [20]

Prover	Communication	Verifier
$H(\lambda) = f_u^{A,v}(\lambda),$		
$h(\lambda) = \rho_u^{A,v}(\lambda).$	$\xrightarrow{H, h}$	
$\phi, \psi \in \mathbb{F}[\lambda]$ with		
$\phi f_u^{A,v} + \psi \rho_u^{A,v} = 1,$	$\xrightarrow{\phi, \psi}$	$\deg(\phi) \overset{?}{\le} \deg(h) - 1,$
		$\deg(\psi) \overset{?}{\le} \deg(H) - 1.$
		Random $r_0 \in S \subseteq \mathbb{F}.$
		Checks $\text{GCD}(H(\lambda), h(\lambda)) = 1$ by $\phi(r_0)H(r_0) + \psi(r_0)h(r_0) \overset{?}{=} 1.$
Computes w such that	$\xleftarrow{r_1}$	Random $r_1 \in S \subseteq \mathbb{F}.$
$(r_1 I_n - A)w = v.$	$\xrightarrow{w}$	Checks $(r_1 I_n - A)w \overset{?}{=} v$ and $(u^T w)H(r_1) \overset{?}{=} h(r_1).$
		Returns $f_u^{A,v}(\lambda) = H(\lambda).$

Protocol 5. Certificate for $f_u^{A,v}$ [22]

	Prover	Communication	Verifier
1.	Form $B = DA$ with $D \in S^n \subseteq \mathbb{F}^{*n}$ and $u, v \in S^n$, s.t. $\deg(f_u^{B,v}) = n$.	$\xrightarrow{\quad D, u, v \quad}$	
		Protocol 5	
2.		$\xrightarrow{\quad H, h, \phi, \psi \quad}$	Checks:
3.		$\xleftarrow{\quad r_1 \quad}$	$\deg(H) \overset{?}{=} n$,
4.		$\xrightarrow{\quad w \quad}$	$H \overset{?}{=} f_u^{B,v}$, w.h.p.
5.			Returns $\dfrac{f_u^{B,v}(0)}{\det(D)}$.

Protocol 6. Determinant certificate for a non-singular blackbox [22]

	Prover	Communication	Verifier
Commitment	Result $r \in \mathbb{R}$	$\xrightarrow{\quad r \quad}$	
Challenge		$\xleftarrow{\quad \mathcal{I} \quad}$	$\mathcal{I} \overset{\$}{\leftarrow}$ maximal ideals
Response	Result $x \in \mathbb{R}/\mathcal{I}$ with field certificate $\mathcal{C}_{\mathbb{R}/I}$	$\xrightarrow{\quad x, \mathcal{C}_{\mathbb{R}/I} \quad}$	$x \overset{?}{\equiv} r \mod \mathcal{I}$ and $\mathcal{C}_{\mathbb{R}/I}(x) \overset{?}{=}$ valid

Protocol 7. Certification in a quotient field [20, Sects. 3.2 and 4.4].

– For **integral matrices**, if the true result v is bounded in magnitude, then only a finite number of prime numbers will divide the difference between the commitment r and the result. Therefore the result can be checked over a *small* prime field [20, Theorem 5].

– For **polynomial matrices**, if the true $v(X)$ result's degree is bounded, then only a finite number of evaluation points can be roots of the difference polynomial between the committed one $r(X)$ and the result. Therefore the result can be checked in the ground field at a *small* evaluation point [20, Theorem 2].

The latter results allows, for instance, to certify the global optimization problems of Sect. 1.2. This is illustrated in Fig. 4, where many of the reductions presented here are recalled.

3.5   Non-interactive Certificates

The certificates in Sects. 3.1 and 3.2 are non-interactive: all the communications can be recorded and publicly verified later.

On the contrary the certificates of Sects. 3.3 and 3.4 are interactive: the Verifier chooses some random bits during the computation of the certificate. Non-interactivity can be recovered via Fiat-Shamir scheme: any random bits are gen-

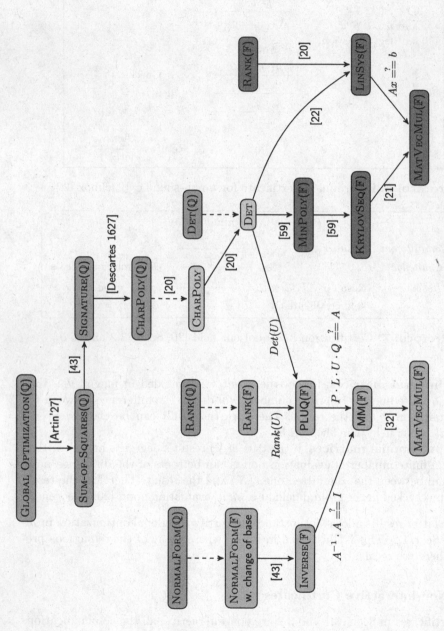

Fig. 4. Global optimization via problem-specific interactive certificates: dense (purple) or sparse (red) algebraic problems, as well as over the reals (green) or oblivious (yellow). (Color figure online)

erated by cryptographic hashes of the inputs and all the previous intermediate commitments. Soundness is then subject to standard cryptographic assumptions.

- For sparse or structured problems fewer results exist without this assumption, or with worse complexity bounds:

- For the minimal polynomial (scalar or matrix) or the determinant, non-interactive certificates exist, but with communications and computational cost $O(n\sqrt{\mu(A)})$ instead of $\mu(A) + n^{1+o(1)}$ [21].
- Non-interactive certificates can also verify polynomial minimal approximant bases in $O(mD + m^\omega)$, where D is the sum of the column degrees of the output [37].

4 Some Open Problems

We conclude this survey with some open problems in the area of problem specific linear algebra certificates:

- **Sparse Smith form:** for dense matrices, one can interactively certify any normal form via a Freivalds certificate on a randomly chosen modular factorization. With sparse matrices, even the modular projection of the change of base can be too large. In that setting extending protocols for the rank or the determinant to deal with the Smith form should be possible.
- **Non integral domains certificates:** more generally, how to efficiently certify some properties when there is no quotients or if those properties do not carry over (e.g., Smith form)?
- We have defined certificates resisting a malicious server with unbounded power. This is error detection with unbounded number of errors. Thus the question of the complexity of **problem specific unbounded error correction** also arises. This path again was first taken for matrix multiplication [35] and was recently extended to the matrix inverse [51].

Acknowledgment. I thank Brice Boyer, Pascal Lafourcade, Shafi Goldwasser, Erich Kaltofen, Julio López Fenner, David Lucas, Vincent Neiger, Jean-Baptiste Orfila, Clément Pernet, Maxime Puys, Jean-Louis Roch, Dan Roche, Guy Rothblum, Justin Thaler, Emmanuel Thomé, Gilles Villard, Lihong Zhi and an anonymous referee for their helpful comments.

References

1. Aaronson, S., Wigderson, A.: Algebrization: a new barrier in complexity theory. ACM Trans. Comput. Theory **1**(1), 2:1–2:54 (2009). https://doi.org/10.1145/1490270.1490272
2. Ábrahám, E., et al.: SC²: satisfiability checking meets symbolic computation. In: Kohlhase, M., Johansson, M., Miller, B., de de Moura, L., Tompa, F. (eds.) CICM 2016. LNCS (LNAI), vol. 9791, pp. 28–43. Springer, Cham (2016). https://doi.org/10.1007/978-3-319-42547-4_3. https://members.loria.fr/PFontaine/Abraham1.pdf

3. Arora, S., Safra, S.: Probabilistic checking of proofs; a new characterization of NP. In: 33rd Annual Symposium on Foundations of Computer Science, 24–27 October 1992, pp. 2–13. IEEE, Pittsburgh (1992)

4. Arreche, C. (ed.): ISSAC 2018, Proceedings of the 2018 ACM International Symposium on Symbolic and Algebraic Computation, New York, USA. ACM Press, New York, July 2018

5. Babai, L.: Trading group theory for randomness. In: Sedgewick [54], pp. 421–429. https://doi.org/10.1145/22145.22192

6. Babai, L., Fortnow, L., Lund, C.: Nondeterministic exponential time has two-prover interactive protocols. In: Proceedings of the 31st Annual Symposium on Foundations of Computer Science, vol. 1, pp. 16–25, October 1990. https://doi.org/10.1109/FSCS.1990.89520

7. Bangerter, E., Camenisch, J., Maurer, U.: Efficient proofs of knowledge of discrete logarithms and representations in groups with hidden order. In: Vaudenay, S. (ed.) PKC 2005. LNCS, vol. 3386, pp. 154–171. Springer, Heidelberg (2005). https://doi.org/10.1007/978-3-540-30580-4_11

8. Beame, P.W., Cook, S.A., Hoover, H.J.: Log depth circuits for division and related problems. SIAM J. Comput. **15**, 994–1003 (1986). https://doi.org/10.1137/0215070

9. Bellare, M., Rogaway, P.: Random oracles are practical: a paradigm for designing efficient protocols. In: Ashby, V. (ed.) Proceedings of the 1st ACM Conference on Computer and Communications Security, pp. 62–73. ACM Press, Fairfax, November 1993. http://www-cse.ucsd.edu/users/mihir/papers/ro.pdf

10. Ben-Sasson, E., et al.: Computational integrity with a public random string from quasi-linear PCPs. In: Coron, J.-S., Nielsen, J.B. (eds.) EUROCRYPT 2017, Part III. LNCS, vol. 10212, pp. 551–579. Springer, Cham (2017). https://doi.org/10.1007/978-3-319-56617-7_19

11. Ben-Sasson, E., Bentov, I., Horesh, Y., Riabzev, M.: Scalable, transparent, and post-quantum secure computational integrity. Cryptology ePrint Archive, Report 2018/046 (2018). https://eprint.iacr.org/2018/046

12. Blum, M., Kannan, S.: Designing programs that check their work. J. ACM **42**(1), 269–291 (1995). http://www.icsi.berkeley.edu/pubs/techreports/tr-88-009.pdf

13. Bootle, J., Cerulli, A., Chaidos, P., Groth, J., Petit, C.: Efficient zero-knowledge arguments for arithmetic circuits in the discrete log setting. In: Fischlin, M., Coron, J.-S. (eds.) EUROCRYPT 2016, Part II. LNCS, vol. 9666, pp. 327–357. Springer, Heidelberg (2016). https://doi.org/10.1007/978-3-662-49896-5_12

14. Bünz, B., Bootle, J., Boneh, D., Poelstra, A., Wuille, P., Maxwell, G.: Bulletproofs: short proofs for confidential transactions and more. In: 2018 IEEE Symposium on Security and Privacy (SP), pp. 319–338 (2018). https://doi.org/10.1109/SP.2018.00020

15. Calude, C.S., Thompson, D.: Incompleteness, undecidability and automated proofs. In: Gerdt, V.P., Koepf, W., Seiler, W.M., Vorozhtsov, E.V. (eds.) CASC 2016. LNCS, vol. 9890, pp. 134–155. Springer, Cham (2016). https://doi.org/10.1007/978-3-319-45641-6_10

16. Chyzak, F., Mahboubi, A., Sibut-Pinote, T., Tassi, E.: A computer-algebra-based formal proof of the irrationality of $\zeta(3)$. In: ITP - 5th International Conference on Interactive Theorem Proving, Vienna, Austria (2014). https://hal.inria.fr/hal-00984057

17. Cramer, R., Damgård, I., Nielsen, J.B.: Multiparty computation from threshold homomorphic encryption. In: Pfitzmann, B. (ed.) EUROCRYPT 2001. LNCS, vol. 2045, pp. 280–300. Springer, Heidelberg (2001). https://doi.org/10.1007/3-540-44987-6_18

18. DeMillo, R.A., Lipton, R.J.: A probabilistic remark on algebraic program testing. Inf. Proces. Lett. **7**(4), 193–195 (1978). https://doi.org/10.1016/0020-0190(78)90067-4

19. Dumas, J.G., Giorgi, P., Elbaz-Vincent, P., Urbańska, A.: Parallel computation of the rank of large sparse matrices from algebraic k-theory. In: Moreno-Maza, M., Watt, S. (eds.) PASCO 2007, Proceedings of the 3rd ACM International Workshop on Parallel Symbolic Computation, pp. 43–52. Waterloo University, Ontario, July 2007. http://hal.archives-ouvertes.fr/hal-00142141

20. Dumas, J.G., Kaltofen, E.: Essentially optimal interactive certificates in linear algebra. In: Nabeshima [46], pp. 146–153. https://doi.org/10.1145/2608628.2608644, http://hal.archives-ouvertes.fr/hal-00932846

21. Dumas, J.G., Kaltofen, E., Thomé, E.: Interactive certificate for the verification of Wiedemann's Krylov sequence: application to the certification of the determinant, the minimal and the characteristic polynomials of sparse matrices. Technical report, IMAG-hal-01171249 arXiv cs.SC/1507.01083, January 2016. http://hal.archives-ouvertes.fr/hal-01171249

22. Dumas, J.G., Kaltofen, E., Thomé, E., Villard, G.: Linear time interactive certificates for the minimal polynomial and the determinant of a sparse matrix. In: Gao [34], pp. 199–206. https://doi.org/10.1145/2930889.2930908, http://hal.archives-ouvertes.fr/hal-01266041

23. Dumas, J.G., Kaltofen, E., Villard, G., Zhi, L.: Polynomial time interactive proofs for linear algebra with exponential matrix dimensions and scalars given by polynomial time circuits. In: Safey El Din [52], pp. 125–132. https://doi.org/10.1145/3087604.3087640, http://ljk.imag.fr/membres/Jean-Guillaume.Dumas/Publications/DKVZ17.pdf

24. Dumas, J.G., Lucas, D., Pernet, C.: Certificates for triangular equivalence and rank profiles. In: Safey El Din [52], pp. 133–140. https://doi.org/10.1145/3087604.3087609, http://hal.archives-ouvertes.fr/hal-01466093

25. Dumas, J.-G., Zucca, V.: Prover efficient public verification of dense or sparse/structured matrix-vector multiplication. In: Pieprzyk, J., Suriadi, S. (eds.) ACISP 2017. LNCS, vol. 10343, pp. 115–134. Springer, Cham (2017). https://doi.org/10.1007/978-3-319-59870-3_7. http://hal.archives-ouvertes.fr/hal-01503870

26. Eberly, W.: A new interactive certificate for matrix rank. Technical report 2015-1078-11, University of Calgary, June 2015. http://prism.ucalgary.ca/bitstream/1880/50543/1/2015-1078-11.pdf

27. Eberly, W.: Selecting algorithms for black box matrices: checking for matrix properties that can simplify computations. In: Gao [34]

28. Elkhiyaoui, K., Önen, M., Azraoui, M., Molva, R.: Efficient techniques for publicly verifiable delegation of computation. In: Proceedings of the 11th ACM on Asia Conference on Computer and Communications Security, ASIA CCS 2016, pp. 119–128. ACM, New York (2016). https://doi.org/10.1145/2897845.2897910

29. Fiat, A., Shamir, A.: How To Prove Yourself: Practical Solutions to Identification and Signature Problems. In: Odlyzko, A.M. (ed.) CRYPTO 1986. LNCS, vol. 263, pp. 186–194. Springer, Heidelberg (1987). https://doi.org/10.1007/3-540-47721-7_12. http://www.cs.rit.edu/~jjk8346/FiatShamir.pdf

30. Fiore, D., Fournet, C., Ghosh, E., Kohlweiss, M., Ohrimenko, O., Parno, B.: Hash first, argue later: adaptive verifiable computations on outsourced data. In: Weippl, E.R., Katzenbeisser, S., Kruegel, C., Myers, A.C., Halevi, S. (eds.) Proceedings of the 2016 ACM SIGSAC Conference on Computer and Communications Security, Vienna, Austria, 24–28 October 2016, pp. 1304–1316. ACM (2016). http://doi.acm.org/10.1145/2976749.2978368

31. Fiore, D., Gennaro, R.: Publicly verifiable delegation of large polynomials and matrix computations, with applications. In: Proceedings of the 2012 ACM Conference on Computer and Communications Security, CCS 2012, pp. 501–512. ACM, New York (2012). https://doi.org/10.1145/2382196.2382250

32. Freivalds, R.: Fast probabilistic algorithms. In: Bečvář, J. (ed.) MFCS 1979. LNCS, vol. 74, pp. 57–69. Springer, Heidelberg (1979). https://doi.org/10.1007/3-540-09526-8_5

33. Furer, M., Goldreich, O., Mansour, Y., Sipser, M., Zachos, S.: On completeness and soundness in interactive proof systems. In: Micali, S. (ed.) Randomness and Computation. Advances in Computing Research, vol. 5, pp. 429–442. JAI Press, Greenwich (1989). http://www.wisdom.weizmann.ac.il/~oded/PS/fgmsz.ps

34. Gao, X.S. (ed.): ISSAC 2016, Proceedings of the 2016 ACM International Symposium on Symbolic and Algebraic Computation, Waterloo, Canada. ACM Press, New York, July 2016

35. Gąsieniec, L., Levcopoulos, C., Lingas, A., Pagh, R., Tokuyama, T.: Efficiently correcting matrix products. Algorithmica **79**, 1–16 (2016). https://doi.org/10.1007/s00453-016-0202-3

36. Gentry, C., Groth, J., Ishai, Y., Peikert, C., Sahai, A., Smith, A.: Using fully homomorphic hybrid encryption to minimize non-interactive zero-knowledge proofs. J. Cryptol. **28**, 1–24 (2014). https://doi.org/10.1007/s00145-014-9184-y

37. Giorgi, P., Neiger, V.: Certification of minimal approximant bases. In: Arreche [4]

38. Goldwasser, S., Kalai, Y.T., Rothblum, G.N.: Delegating computation: interactive proofs for muggles. In: Dwork, C. (ed.) STOC 2008, Proceedings of the 40th Annual ACM Symposium on Theory of Computing, Victoria, British Columbia, Canada, pp. 113–122. ACM Press, May 2008. https://doi.org/10.1145/1374376.1374396, http://research.microsoft.com/en-us/um/people/yael/publications/2008-delegatingcomputation.pdf

39. Goldwasser, S., Micali, S., Rackoff, C.: The knowledge complexity of interactive proof-systems. In: Sedgewick [54], pp. 291–304. https://doi.org/10.1145/22145.22178

40. Kaltofen, E., Trager, B.: Computing with polynomials given by black boxes for their evaluations: greatest common divisors, factorization, separation of numerators and denominators. J. Symb. Comput. **9**(3), 301–320 (1990). http://www.math.ncsu.edu/~kaltofen/bibliography/90/KaTr90.pdf

41. Kaltofen, E.: Analysis of Coppersmith's block Wiedemann algorithm for the parallel solution of sparse linear systems. Math. Comput. **64**(210), 777–806 (1995). https://doi.org/10.2307/2153451

42. Kaltofen, E., Pernet, C.: Sparse polynomial interpolation codes and their decoding beyond half the minimum distance. In: Nabeshima [46]. http://arxiv.org/abs/1403.3594

43. Kaltofen, E.L., Nehring, M., Saunders, B.D.: Quadratic-time certificates in linear algebra. In: Leykin, A. (ed.) ISSAC 2011, Proceedings of the 2011 ACM International Symposium on Symbolic and Algebraic Computation, San Jose, California, USA, pp. 171–176. ACM Press, New York, June 2011. http://www.math.ncsu.edu/~kaltofen/bibliography/11/KNS11.pdf

44. Kimbrel, T., Sinha, R.K.: A probabilistic algorithm for verifying matrix products using $O(n^2)$ time and $\log_2 n + O(1)$ random bits. Inf. Proces. Lett. **45**(2), 107–110 (1993). ftp://trout.cs.washington.edu/tr/1991/08/UW-CSE-91-08-06.pdf
45. Lund, C., Fortnow, L., Karloff, H., Nisan, N.: Algebraic methods for interactive proof systems. J. ACM **39**(4), 859–868 (1992). https://doi.org/10.1145/146585.146605
46. Nabeshima, K. (ed.): ISSAC 2014, Proceedings of the 2014 ACM International Symposium on Symbolic and Algebraic Computation, Kobe, Japan. ACM Press, New York, Jul 2014
47. Ng, E.W. (ed.): Symbolic and Algebraic Computation. LNCS, vol. 72. Springer, Heidelberg (1979). https://doi.org/10.1007/3-540-09519-5
48. Parno, B., Howell, J., Gentry, C., Raykova, M.: Pinocchio: nearly practical verifiable computation. In: Proceedings of the 2013 IEEE Symposium on Security and Privacy, SP 2013, pp. 238–252. IEEE Computer Society, Washington, DC (2013). https://doi.org/10.1109/SP.2013.47
49. Parno, B., Raykova, M., Vaikuntanathan, V.: How to delegate and verify in public: verifiable computation from attribute-based encryption. In: Cramer, R. (ed.) TCC 2012. LNCS, vol. 7194, pp. 422–439. Springer, Heidelberg (2012). https://doi.org/10.1007/978-3-642-28914-9_24
50. Reingold, O., Rothblum, G.N., Rothblum, R.D.: Constant-round interactive proofs for delegating computation. In: Wichs, D., Mansour, Y. (eds.) Proceedings of the 48th Annual ACM SIGACT Symposium on Theory of Computing, STOC 2016, Cambridge, MA, USA, 18–21 June 2016, pp. 49–62. ACM (2016). https://doi.org/10.1145/2897518.2897652, http://dl.acm.org/citation.cfm?id=2897518
51. Roche, D.: Error correction in fast matrix multiplication and inverse. In: Arreche [4]
52. Safey El Din, M. (ed.): ISSAC 2017, Proceedings of the 2017 ACM International Symposium on Symbolic and Algebraic Computation, Kaiserslautern, Deutschland. ACM Press, New York, July 2017
53. Schwartz, J.T.: Probabilistic algorithms for verification of polynomial identities. In: Ng [47], pp. 200–215. https://doi.org/10.1007/3-540-09519-5_72
54. Sedgewick, R. (ed.): STOC 1985, ACM Symposium on Theory of Computing, Providence, Rhode Island, USA. ACM Press, New York, May 1985
55. Shamir, A.: IP = PSPACE. J. ACM **39**(4), 869–877 (1992). https://doi.org/10.1145/146585.146609
56. Storjohann, A.: Integer matrix rank certification. In: May, J.P. (ed.) ISSAC 2009, Proceedings of the 2009 ACM International Symposium on Symbolic and Algebraic Computation, Seoul, Korea, pp. 333–340. ACM Press, New York, Jul 2009. https://cs.uwaterloo.ca/~astorjoh/issac09.pdf
57. Thaler, J.: Time-optimal interactive proofs for circuit evaluation. In: Canetti, R., Garay, J.A. (eds.) CRYPTO 2013. LNCS, vol. 8043, pp. 71–89. Springer, Heidelberg (2013). https://doi.org/10.1007/978-3-642-40084-1_5
58. Walfish, M., Blumberg, A.J.: Verifying computations without reexecuting them. Commun. ACM **58**(2), 74–84 (2015). https://doi.org/10.1145/2641562
59. Wiedemann, D.H.: Solving sparse linear equations over finite fields. IEEE Trans. Inf. Theory **32**(1), 54–62 (1986). https://doi.org/10.1109/TIT.1986.1057137
60. Zhang, Y., Blanton, M.: Efficient secure and verifiable outsourcing of matrix multiplications. In: Chow, S.S.M., Camenisch, J., Hui, L.C.K., Yiu, S.M. (eds.) ISC 2014. LNCS, vol. 8783, pp. 158–178. Springer, Cham (2014). https://doi.org/10.1007/978-3-319-13257-0_10
61. Zippel, R.: Probabilistic algorithms for sparse polynomials. In: Ng [47], pp. 216–226. https://doi.org/10.1007/3-540-09519-5_73

On Unimodular Matrices of Difference Operators

S. A. Abramov[✉] and D. E. Khmelnov

Dorodnicyn Computing Centre,
Federal Research Center "Computer Science and Control",
Russian Academy of Sciences, Vavilova, 40, Moscow 119333, Russia
{sergeyabramov,dennis_khmelnov}@mail.ru

Abstract. We consider matrices $L \in \mathrm{Mat}_n(K[\sigma, \sigma^{-1}])$ of scalar difference operators, where K is a difference field of characteristic 0 with an automorphism σ. We discuss approaches to compute the dimension of the space of those solutions of the system of equations $L(y) = 0$ that belong to an adequate extension of K. On the base of one of those approaches, we propose a new algorithm for computing $L^{-1} \in \mathrm{Mat}_n(K[\sigma, \sigma^{-1}])$ whenever it exists. We investigate the worst-case complexity of the new algorithm, counting both arithmetic operations in K and shifts of elements of K. This complexity turns out to be smaller than in the earlier proposed algorithms for inverting matrices of difference operators.

Some experiments with our implementation in Maple of the algorithm are reported.

1 Introduction

Matrix calculus has wide application in various branches of science. Testing whether a given matrix over a field or ring is invertible and computing the inverse matrix are classical mathematical problems. Below, we consider these problems for matrices whose entries belong to the ring (non-commutative) of scalar linear difference operators with coefficients from a difference field K of characteristic 0 with an automorphism (shift) σ. We discuss some new algorithms for solving these problems. These problems can be also solved by well-known algorithms proposed originally for more general problems. The new algorithms below have lower complexity.

In the case of matrices of operators, the term "*unimodular* matrix" is usually used instead of "invertible matrix". This term will be used throughout this paper.

In the differential case when the ground field K is a differential field of characteristic 0 with a derivation $\delta = {}'$ and when the matrix entries are scalar linear differential operators over K, algorithms for the unimodularity testing of a matrix and computing its inverse were considered in [2]. For a given matrix L, the algorithms discussed below rely on determining the dimension of the solution space V_L of the corresponding system of equations under the assumption that the components of solutions belong to the Picard–Vessiot extension of K associated

© Springer Nature Switzerland AG 2018
V. P. Gerdt et al. (Eds.): CASC 2018, LNCS 11077, pp. 18–31, 2018.
https://doi.org/10.1007/978-3-319-99639-4_2

with L (see [16]). A matrix L of operators, when L is of full rank (the rows of L are independent over the ring of scalar linear operators) is unimodular if and only if $\dim V_L = 0$, i.e., V_L is the zero space (see [4]).

There are two significant dissimilarities between the differential and difference cases. One of them gives an advantage to the differential case, the other to the difference case. The differential system $y' = Ay$ has the n-dimensional solution space in the universal differential extension, regardless of the form (singular or non-singular) of the $n \times n$-matrix A [17]. But in the difference case, the non-singularity of A is required. However, the difference case has the advantage that the automorphism σ has the inverse in $K[\sigma, \sigma^{-1}]$, while the differentiation δ is not invertible in $K[\delta]$.

It is worth noting that some algorithms for solving the "difference problems" formulated above have been proposed in [3]. The algorithms below have lower complexity (this is the novelty of the results) due to the usage of the EG-eliminations algorithm [1,6,7] as an auxiliary tool instead of the algorithm Row-Reduction [11]. The obstacle for such a replacement in the differential case, is the absence of the inverse element for δ in the ring $K[\delta]$.

The problems of unimodularity testing and inverse matrix construction can be solved by applying various other algorithms. For example, the Jacobson and Hermite forms of the given operator matrix can be constructed; their definitions can be found in [13,15]. The complexity of the algorithms is greater than the complexity of the algorithms in this paper and in [3]. Of course, the algorithms in [13,15] are intended for more general problems, and the algorithms in this paper and in [3] have advantages only for unimodularity recognition and the construction of an inverse operator matrix.

We use the following notation. The ring of $n \times n$-matrices (n is a positive integer) with elements from a ring or field R is denoted by $\mathrm{Mat}_n(R)$. If M is an $n \times n$-matrix, then $M_{i,*}$ with $1 \leqslant i \leqslant n$ is the $1 \times n$-matrix equal to the ith row of M. The diagonal $n \times n$-matrix with diagonal elements $r_1, \ldots, r_n$ is denoted by $\mathrm{diag}(r_1, \ldots, r_n)$, and I_n is the $n \times n$ identity matrix.

The proposed algorithms are presented in Sect. 3. Their implementation in Maple and some experiments are described in Sect. 5.

2 Preliminaries

2.1 Adequate Difference Extensions

As usual, a *difference ring* K is a commutative ring with identity and an automorphism σ (which will frequently be referred to as a *shift*). If K is additionally a field, then it is called a *difference field*. We will assume that the considered difference fields are of characteristic 0. The *ring of constants* of a difference ring K is $\mathrm{Const}(K) = \{c \in K \mid \sigma c = c\}$. If K is a difference field, then $\mathrm{Const}(K)$ is a subfield of K. Let K be a difference field with an automorphism σ, and let Λ be a difference ring extension of K (on K, the corresponding automorphism of Λ coincides with σ; for this automorphism of Λ, we use the same notation σ).

Definition 1. The ring Λ which is a difference ring extension of a field K is an *adequate* difference extension of K if Const (Λ) is a field and an arbitrary system

$$\sigma y = Ay, \ y = (y_1, \ldots, y_n)^T \tag{1}$$

with a nonsingular $A \in \mathrm{Mat}\,_n(K)$ has in Λ^n the linear solution space over Const (Λ) of dimension n.

The non singularity of A in this definition is essential: e.g., if the first row of A is zero, then the entry y_1 in any solution of the system (1) is zero as well.

Note that the q-difference case [10] is covered by the general difference case.

If Const (K) is algebraically closed, then there exists a unique (up to a difference isomorphism, i.e., an isomorphism commuting with σ) adequate extension Ω such that Const (Ω) = Const (K), which is called the *universal* difference (Picard-Vessiot) ring extension of K. The complete proof of its existence is not easy (see [16, Sect. 1.4]), while the existence of an adequate difference extension for an arbitrary difference field can be rather easily proved (see [5, Sect. 5.1]). However, it should be emphasized that, for an adequate extension, the equality Const (Λ) = Const (K) is not guaranteed; in the general case, Const (K) is a proper subfield of Const (Λ). The assertion that a universal difference extension exists for an arbitrary difference field of characteristic 0 is not true if the extension is understood as a field. Franke's well-known example [12] is the scalar equation over a field with an algebraically closed field of constants. This equation has no nontrivial solutions in any difference extension having an algebraically closed field of constants.

In the sequel, Λ denotes a fixed adequate difference extension of a difference field with an automorphism σ.

2.2   Orders of Difference Operators

A scalar difference operator is an element of the ring $K[\sigma, \sigma^{-1}]$. Given a nonzero scalar operator $f = \sum a_i \sigma^i$, its *leading* and *trailing* orders are defined as

$$\overline{\mathrm{ord}}\,f = \max\{i \mid a_i \neq 0\}, \ \underline{\mathrm{ord}}\,f = \min\{i \mid a_i \neq 0\},$$

and the *order* of f is defined as

$$\mathrm{ord}\,f = \overline{\mathrm{ord}}\,f - \underline{\mathrm{ord}}\,f.$$

Set $\overline{\mathrm{ord}}\,0 = -\infty$, $\underline{\mathrm{ord}}\,0 = \infty$, and $\mathrm{ord}\,0 = -\infty$.

For a finite set F of scalar operators (a vector, matrix, matrix row etc), $\overline{\mathrm{ord}}\,F$ is defined as the maximum of the leading orders of its elements; $\underline{\mathrm{ord}}\,F$ is defined as the minimum of the trailing orders of its elements; finally, $\mathrm{ord}\,F$ is defined as $\overline{\mathrm{ord}}\,F - \underline{\mathrm{ord}}\,F$. A matrix of difference operators is a matrix from $\mathrm{Mat}\,_n(K[\sigma, \sigma^{-1}])$. In the sequel, such a matrix of difference operators is associated with some matrices belonging to $\mathrm{Mat}_n(K)$. To avoid confusion of terminology, matrices of difference operators will be briefly referred to as *operators*. The case of scalar operators will be considered separately. An operator is of full rank (or is a full rank operator) if its rows are linearly independent over $K[\sigma, \sigma^{-1}]$. Same-length rows $u_1, \ldots, u_s$ with components belonging to $K[\sigma, \sigma^{-1}]$ are called *linearly dependent* (over $K[\sigma, \sigma^{-1}]$) if there exist $f_1, \ldots, f_s \in K[\sigma, \sigma^{-1}]$ not all zero such that $f_1 u_1 + \cdots + f_s u_s = 0$; otherwise, these rows are called *linearly independent* (over $K[\sigma, \sigma^{-1}]$). If

$$L \in \mathrm{Mat}_n(K[\sigma, \sigma^{-1}]), \ l = \overline{\mathrm{ord}}\, L, \ t = \underline{\mathrm{ord}}\, L,$$

and L is nonzero, then it can be represented in the expanded form as

$$L = A_l \sigma^l + A_{l-1} \sigma^{l-1} + \cdots + A_t \sigma^t, \tag{2}$$

where $A_l, A_{l-1}, \ldots, A_t \in \mathrm{Mat}_n(K)$, and A_l, A_t (the *leading* and *trailing* matrices of the original operator) are nonzero.

Let the leading and trailing row orders of an operator L be $\alpha_1, \ldots, \alpha_n$ and $\beta_1, \ldots, \beta_n$, respectively. The *frontal* matrix of L is the leading matrix of the operator PL, where

$$P = \mathrm{diag}(\sigma^{l-\alpha_1}, \ldots, \sigma^{l-\alpha_n}), \ l = \overline{\mathrm{ord}}\, L.$$

Accordingly, the *rear* matrix of L is the trailing matrix of the operator QL, where

$$Q = \mathrm{diag}(\sigma^{t-\beta_1}, \ldots, \sigma^{t-\beta_n}), \ t = \underline{\mathrm{ord}}\, L.$$

If $\alpha_i = -\infty$ (resp. $\beta_i = \infty$) then the i-th row of P (resp. Q) is zero.

We say that L is *strongly reduced* if its frontal and rear matrices are both nonsingular.

Definition 2. An operator $L \in \mathrm{Mat}_n(K[\sigma, \sigma^{-1}])$ is *unimodular* or *invertible* if there exists an inverse $L^{-1} \in \mathrm{Mat}_n(K[\sigma, \sigma^{-1}])$: $LL^{-1} = L^{-1}L = I_n$. The group of unimodular $n \times n$-operators is denoted by Υ_n. Two operators are said to be *equivalent* if $L_1 = UL_2$ for some $U \in \Upsilon_n$.

If L has a zero row (in such a case, its frontal and rear matrices have also zero rows) then L is not of full rank, and is not unimodular: suppose, e.g., that the first row of L is zero, then for any $M \in \mathrm{Mat}_n(K[\sigma, \sigma^{-1}])$, the first row of the product LM is also zero, thus, the equality $LM = I_n$ is impossible. Similarly, if $U \in \Upsilon_n$ and UL has a zero row then $L \notin \Upsilon_n$.

Let V_L denote the space of the solutions of the system $L(y) = 0$ that belong to Λ^n (see Sect. 2.1). For brevity, V_L is sometimes called the solution space of L.

For the difference case, Theorem 1 from [2] can be reformulated as follows.

Theorem 1. *Let $L \in Mat_n(K[\sigma, \sigma^{-1}])$ be of full rank. Then*

(i) *If L is strongly reduced, then* $\dim V_L = \sum_{i=1}^{n} \operatorname{ord} L_{i,*}$.
(ii) $L \in \Upsilon_n$ *iff* $V_L = 0$.

The proof is based on [4, 5].

2.3   Complexity

Besides the complexity as *the number of arithmetic operations* (the *arithmetic complexity*) one can consider *the number of shifts* in the worst case (the *shift complexity*).

Thus, we will consider two complexities. This is similar to the situation with sorting algorithms, when we consider separately the complexity as the number of comparisons and, resp. the number of swaps in the worst case.

We can also consider the *full algebraic complexity* as *the total number of all operations* in the worst case.

Supposing that $L \in \operatorname{Mat}_n(K[\sigma, \sigma^{-1}]), \operatorname{ord} L = d$, each of the mentioned complexities is a function of n and d.

In asymptotic complexity estimates, along with the O notation we use the Θ notation (see [14]): the relation $f(n, d) = \Theta(g(n, d))$ is equivalent to

$$f(n, d) = O(g(n, d)) \text{ and } g(n, d) = O(f(n, d)).$$

Note that the full complexity of an algorithm counting operations of two different types in the worst case is not, in general, equal to the sum of two complexities, counting operations of the first and, resp. second type. We can claim only that the full complexity does not exceed that sum. If for the first and second complexities we have asymptotic estimates $\Theta(f(n, d))$ and $\Theta(g(n, d))$ then for the full complexity we have the estimate $O(f(n, d) + g(n, d))$. To this we can add that if for the first and second complexities we have estimates $O(f(n, d))$ and $O(g(n, d))$ then we have the estimate $O(f(n, d) + g(n, d))$ for the full complexity.

2.4   EG-Eliminations (Family of EG-Algorithms)

Definition 3. Let the ith row of the frontal matrix of $L \in \operatorname{Mat}_n(K[\sigma, \sigma^{-1}])$ be non-zero and have the form

$$\underbrace{(0, \ldots, 0}_{k}, a, \ldots, b),$$

$0 \leqslant k \leqslant n$, $a \neq 0$. In this case, k is the *indent* of the ith row of L.

The algorithm EG_σ (the version published in [1]) is as follows:

Algorithm: EG_σ
Input: An operator $L \in \text{Mat}_n(K[\sigma, \sigma^{-1}])$ whose leading matrix has no zero row.
Output: An equivalent operator having an upper triangle leading matrix (that operator is also denoted by L) or the message "is not of full rank".
while L has rows with equal indents do

(Reduction) Let some rows r_1, r_2 of L have the same indent k. Then compute $v \in K$ such that the indent of the row

$$r = r_1 - vr_2 \tag{3}$$

is greater than k or $\underline{\text{ord}}\, r < \underline{\text{ord}}\, L$ (the computation of v uses one arithmetic operation); if r is zero row of L then STOP with the message "is not of full rank"fi; The row from r_1, r_2 which has the smaller trailing order, must be replaced by r (if $\text{ord}\, r_1 = \text{ord}\, r_2$ then any of r_1, r_2 can be taken for the replacement);

(Shift) If $\overline{\text{ord}}\, r < \overline{\text{ord}}\, L$ then apply $\sigma^{\overline{\text{ord}}\, L - \overline{\text{ord}}\, r}$ to r in L
od;
Return L.   □

Thus, each step of the algorithm EG_σ is a combination "reduction + shift". All the steps are unimodular since the operator σ^{-1} is the inverse for σ.

Example 1.

$$L = \begin{pmatrix} 1 & -\frac{1}{x}\sigma \\ \frac{x^2}{2} & -\frac{x}{2}\sigma + 1 \end{pmatrix} = \begin{pmatrix} 0 & -\frac{1}{x} \\ 0 & -\frac{x}{2} \end{pmatrix}\sigma + \begin{pmatrix} 1 & 0 \\ \frac{x^2}{2} & 1 \end{pmatrix}. \tag{4}$$

By applying the algorithm EG_σ, the operator L is transformed as follows:

$$\begin{pmatrix} 0 & -\frac{1}{x} \\ 0 & -\frac{x}{2} \end{pmatrix}\sigma + \begin{pmatrix} 1 & 0 \\ -\frac{x}{2} & 1 \end{pmatrix} \xrightarrow{1} \begin{pmatrix} 0 & -\frac{1}{x} \\ 0 & 0 \end{pmatrix}\sigma + \begin{pmatrix} 1 & 0 \\ 0 & 1 \end{pmatrix} \xrightarrow{2}$$

$$\begin{pmatrix} 0 & -\frac{1}{x} \\ 0 & 1 \end{pmatrix}\sigma + \begin{pmatrix} 1 & 0 \\ 0 & 0 \end{pmatrix} \xrightarrow{3} \begin{pmatrix} 0 & 0 \\ 0 & 1 \end{pmatrix}\sigma + \begin{pmatrix} 1 & 0 \\ 0 & 0 \end{pmatrix} \xrightarrow{4} \begin{pmatrix} 1 & 0 \\ 0 & 1 \end{pmatrix}\sigma + \begin{pmatrix} 0 & 0 \\ 0 & 0 \end{pmatrix}.$$

Here

$$1.\ L_{2,*} := \frac{-x^2}{2}L_{1,*} + L_{2,*},$$
$$2.\ L_{2,*} := \sigma L_{2,*},$$
$$3.\ L_{1,*} := L_{1,*} + \frac{1}{x}L_{2,*}, \tag{5}$$
$$4.\ L_{1,*} := \sigma L_{1,*}.$$

As the result of this transformation, we obtain the operator $\begin{pmatrix} 1 & 0 \\ 0 & 1 \end{pmatrix}\sigma$, i.e.,

$$\begin{pmatrix} \sigma & 0 \\ 0 & \sigma \end{pmatrix}. \tag{6}$$

By analogy with EG_σ we can propose an algorithm $\mathrm{EG}_{\sigma^{-1}}$ in which the trailing matrix of the operator is considered instead of its leading matrix.

Proposition 1. *The arithmetic complexity of the algorithms* EG_σ, $\mathrm{EG}_{\sigma^{-1}}$ *is*

$$\Theta(n^3 d^2), \tag{7}$$

the shift complexity is

$$\Theta(n^2 d^2). \tag{8}$$

Correspondingly, the full algebraic complexity is

$$O(n^3 d^2). \tag{9}$$

See [3, Sect. 5.4] for the proof.

3   Unimodularity Testing, Computing Inverse Operator

3.1   Unimodularity Testing

Proposition 2. *Let the rear matrix of an operator* $L \in Mat_n(K[\sigma, \sigma^{-1}])$ *be non-singular. Then applying* EG_σ *to* L *gives an operator having a non-singular rear matrix.*

Proof. Let us prove that one step of EG_σ does not change the determinant of the rear matrix of L. Indeed, let the reduction stage of this step change a row r_1 of L and before this step, we have $\underline{\mathrm{ord}}\, r_1 = \beta$. The row r_1 is replaced by a sum of r_1 and another row r_2, multiplied by $v \in K$: $r_1 := r_1 + vr_2$. The inequality $\underline{\mathrm{ord}}\, r_2 \geqslant \beta$ holds. If $\underline{\mathrm{ord}}\, r_2 > \beta$ then the rear matrix gets no change. If $\underline{\mathrm{ord}}\, r_2 = \beta$ then the determinant of the rear matrix gets no change since the shift stage does not change the rear matrix.

The following algorithm can be verified by means of Theorem 1 and Proposition 2:

Algorithm: Unimodularity testing (this algorithm has been described in [3])
Input: An operator $L \in \mathrm{Mat}_n(K)$.
Output: "is unimodular" or "is not unimodular" depending on whether L is unimodular or not.
if $\mathrm{EG}_{\sigma^{-1}}$ did not find that L is not of full rank and $\underline{\mathrm{ord}}\, r = \overline{\mathrm{ord}}\, r$ for each row r of $\mathrm{EG}_\sigma(\mathrm{EG}_{\sigma^{-1}}(\mathrm{L}))$ then Return "is unimodular" otherwise Return "is not unimodular" fi.   □

Example 2. Let L be again as in Example 1, i.e., of the form (4). The rear matrix coincides with the trailing one, and is nonsingular. By applying the algorithm EG_σ, the operator L is transformed to $\tilde{L}$ of the form (6). We have $\dim V_{\tilde{L}} = 0$. Thus, the original operator L is unimodular.

Proposition 3. *The arithmetic, shift and full algebraic complexities of the algorithm* Unimodularity testing *are, resp. (7), (8), and (9).*

Proof. This follows from Proposition 1 and the fact that the values of n, d are not increased after applying $\mathrm{EG}_{\sigma^{-1}}$.

3.2 Inverse Operator

Algorithm: ExtEG$_\sigma$
Input: Operators $J, L \in \text{Mat}_n(K)$.
Output: The operator $M = UJ$, where U is such that $\text{EG}_\sigma(L) = UL$.
Apply EG_σ to L, and repeat in parallel the application of all the operations to J. $\Box$

Note that in the case when we use I_n as J, we obtain M which is equal to U.

By analogy with ExtEG_σ we can propose an algorithm $\text{ExtEG}_{\sigma^{-1}}$ in which the trailing matrix of the operator is considered instead of its leading matrix.

Proposition 4. *We have* $\text{ord}\, U \leqslant nd$ *on each step of applying of* ExtEG_σ *to* $L \in \text{Mat}_n(K[\sigma, \sigma^{-1}])$, $\text{ord}\, L = d$.

Proof. If in a step of the algorithm the shift σ^k of a row r was performed, then the order of U will be increased by no more than $|k|$, while the order of the shifted row is decreased by $|k|$. This implies that $\text{ord}\, U$ after any step of ExtEG_σ does not exceed the sum of the orders of all rows of L. The order of each row does not exceed d and the sum of the orders of all rows of L does not exceed nd.

Proposition 5. *Both arithmetic and shift complexities of each of the algorithms* ExtEG_σ, $\text{ExtEG}_{\sigma^{-1}}$ *can be estimated by* $O(n^4 d^2)$. *The full complexity is* $O(n^4 d^2)$ *as well.*

Proof. When one applies EG_σ or $\text{EG}_{\sigma^{-1}}$ to $L \in \text{Mat}_n(K[\sigma, \sigma^{-1}])$, $\text{ord}\, L = d$, then the operation (3) is performed at most $n \cdot nd$ times. By Proposition 4, when we compute U, each operation (3) uses at most $O(n \cdot nd)$ arithmetic operations, i.e., $O(n^2 d)$ arithmetic operations. Totally, the number of arithmetic operations is $O(n^2 d \cdot n^2 d)$, i.e. $O(n^4 d^2)$.

The shift complexity of each of EG_σ, $\text{EG}_{\sigma^{-1}}$ is $O(n^2 d^2)$. When we substitute nd for d (by Proposition 4) we obtain $O(n^4 d^2)$.

The estimate $O(n^4 d^2)$ for the full complexity follows from the obtained estimates for the arithmetic and shift complexities.

Algorithm: Inverse operator
Input: An operator $L \in \text{Mat}_n(K)$.
Output: The inverse of L or the message "is not unimodular".
$U := I_n$; $(U, L) := \text{ExtEG}_{\sigma^{-1}}(U, L)$; $(U, L) := \text{ExtEG}_\sigma(U, L)$;
if $\underline{\text{ord}}\, r \neq \overline{\text{ord}}\, r$ for at least one row r of L then STOP with the message "is not unimodular"
fi;
Let $\beta_1, \ldots, \beta_n$ be the trailing orders of rows of L,
thus $L = MD$ with $M \in \text{Mat}_n(K)$, $D = \text{diag}(\sigma^{\beta_1}, \ldots, \sigma^{\beta_n})$;
$L^{-1} := \text{diag}(\sigma^{-\beta_1}, \ldots, \sigma^{-\beta_n}) M^{-1} U$. $\Box$

Example 3. Consider again operator (4). To find L^{-1} after getting the operator $\tilde{L}$ (as it was shown in Example 2), we first apply (5) to I_2. We get

$$
\begin{pmatrix} 1 & 0 \\ 0 & 1 \end{pmatrix} \xrightarrow{1} \begin{pmatrix} 1 & 0 \\ -\frac{x^2}{2} & 1 \end{pmatrix} \xrightarrow{2} \begin{pmatrix} 1 & 0 \\ -\frac{(x+1)^2}{2}\sigma & \sigma \end{pmatrix} \xrightarrow{3}
$$

$$
\begin{pmatrix} 1 - \frac{(x+1)^2}{2x}\sigma & \frac{1}{x}\sigma \\ -\frac{(x+1)^2}{2}\sigma & \sigma \end{pmatrix} \xrightarrow{4} \begin{pmatrix} \sigma - \frac{(x+2)^2}{2(x+1)}\sigma^2 & \frac{1}{x+1}\sigma^2 \\ -\frac{(x+1)^2}{2}\sigma & \sigma \end{pmatrix}.
$$

We get

$$
L^{-1} = \mathrm{diag}(\sigma^{-1}, \sigma^{-1})\, I_2^{-1} \begin{pmatrix} \sigma - \frac{(x+2)^2}{2(x+1)}\sigma^2 & \frac{1}{x+1}\sigma^2 \\ -\frac{(x+1)^2}{2}\sigma & \sigma \end{pmatrix} = \begin{pmatrix} 1 - \frac{(x+1)^2}{2x}\sigma & \frac{1}{x}\sigma \\ -\frac{x^2}{2} & 1 \end{pmatrix}.
$$

Proposition 6. *The estimate $O(n^4 d^2)$ holds for all of the arithmetic, shift, and full complexities of the algorithm* Inverse operator.

Proof. The statement follows from Proposition 5.

4   Other Versions of EG and Inverse Operator

The algorithm Inverse operator proposed in this paper is based on the version [1] of the EG-eliminations algorithm as an auxiliary tool. Another variant of the algorithm for constructing the inverse operator has been proposed in [3], it is based on a version (named RR in [3]) of the Row-Reduction algorithm [11] as an auxiliary tool. For a given operator, the algorithm RR_σ constructs an equivalent operator that has a nonsingular frontal matrix. Similarly, the algorithm $RR_{\sigma^{-1}}$ constructs an equivalent operator that has a nonsingular rear matrix. The arithmetic complexity of the algorithms presented in this paper and, resp. in [3], is the same, however, the shift complexity (and, hence, the full algebraic complexity) of the new algorithm is lower: $O(n^4 d^2)$ instead of $\Theta(n^4 d^3)$.

Some other versions of the algorithms belonging to the EG-eliminations family [6,7], whose full complexity does not differ much from the full complexity of the above considered version, can be to some extent more convenient for implementation. This question has been discussed in [8]. In our Maple-implementation of the Inverse operator algorithm represented below, we use elements of various variants of EG-eliminations. (It is well known that an algorithm that looks the best in terms of complexity theory is not necessarily the best in computational practice.)

5   Implementation and Experiments

The implementation[1] is performed in Maple [18]. The existing implementation of the algorithm EG described in [9] is taken as a starting point. The procedure is adjusted to the difference case and to provide extended versions, both ExtEG_σ and $\text{ExtEG}_{\sigma^{-1}}$. On top of the procedure for ExtEG_σ and $\text{ExtEG}_{\sigma^{-1}}$, the procedure IsUnimodular to test the unimodularity of an operator and to compute its inverse is implemented.

An operator $L = A_l \sigma^l + A_{l-1} \sigma^{l-1} + \cdots + A_t \sigma^t$ is specified at the input of the procedures as the list

$$[A, l, t], \tag{10}$$

where A is an *explicit matrix*

$$A = (A_l | A_{l-1} | \ldots | A_t) \tag{11}$$

of size $n \times n(l - t + 1)$. The explicit matrix A is defined by means of the standard Maple object Matrix. The entries of the explicit matrix are rational functions of one variable, which are also specified in a standard way accepted in Maple. If $t = 0$ then the input may be given alternatively just by the explicit matrix A.

The procedure IsUnimodular returns true or false as the result of checking the unimodularity of the given operator, its inverse operator is returned additionally being assigned to a given variable name (an optional input parameter of the procedure). The inverse operator is also represented by the list of its explicit matrix and its leading and trailing orders. If the optional variable name is not given, then the procedure uses the algorithm Unimodularity Testing from Sect. 3.1, otherwise the algorithm Inverse Operator from Sect. 3.2 is used.

Example 4. We apply the procedure IsUnimodular to the operator matrix (4) considered in Examples 1–3. The explicit matrix for the operator is

$$\begin{pmatrix} 0 & -\frac{1}{x} & 1 & 0 \\ 0 & -\frac{x}{2} & \frac{x^2}{2} & 1 \end{pmatrix},$$

with $l = 1$ and $t = 0$. The procedure is applied twice: first time just for checking the unimodularity, and the second time, for computing the inverse operator as well. One can see that the result of the application coincides with the result presented in Example 3 (the computation time is also presented):

---

[1] Available at http://www.ccas.ru/ca/egrrext.

```
> L := Matrix([[0, -1/x, 1, 0], [0, -x/2, x^2/2, 1]]);
```

$$L := \begin{bmatrix} 0 & -\frac{1}{x} & 1 & 0 \\ 0 & -\frac{x}{2} & \frac{x^2}{2} & 1 \end{bmatrix}$$

```
> st:=time(): IsUnimodular(L, x); time()-st;
```

$$true$$

$$0.032$$

```
> st:=time(): IsUnimodular(L, x, 'InvL'); time()-st;
```

$$true$$

$$0.063$$

```
> InvL;
```

$$\left[\left[\begin{matrix} -\frac{(x+1)^2}{2x} & \frac{1}{x} & 1 & 0 \\ 0 & 0 & -\frac{x^2}{2} & 1 \end{matrix} \right], 1, 0 \right]$$

Example 5. Consider the operator

$$\begin{pmatrix} \sigma^{-1} & -\frac{1}{x-1} \\ \frac{x^2}{2} & -\frac{x}{2}\sigma + 1 \end{pmatrix}.$$

The explicit matrix for the operator is

$$\begin{pmatrix} 0 & 0 & 0 & -\frac{1}{x-1} & 1 & 0 \\ 0 & -\frac{x}{2} & \frac{x^2}{2} & 1 & 0 & 0 \end{pmatrix}$$

with $l = 1$ and $t = -1$. The procedure IsUnimodular is applied twice again: first time just for checking the unimodularity, and the second time, for computing the inverse operator as well. The computation time is also presented.

```
> L:= Matrix([[0, 0, 0,-1/(x-1), 1, 0], [0, -x/2, x^2/2, 1, 0, 0]]);
```

$$L := \begin{bmatrix} 0 & 0 & 0 & -\frac{1}{x-1} & 1 & 0 \\ 0 & -\frac{x}{2} & \frac{x^2}{2} & 1 & 0 & 0 \end{bmatrix}$$

```
> st:=time(): IsUnimodular([L, 1, -1], x); time()-st;
```

$$\textit{true}$$

$$0.078$$

```
> st:=time(): IsUnimodular([L, 1, -1], x, 'InvL'); time()-st;
```

$$\textit{true}$$

$$0.109$$

```
> InvL;
```

$$\left[\left[\begin{matrix} -\frac{(x+1)^2}{2x} & 0 & 1 & \frac{1}{x} & 0 & 0 \\ 0 & 0 & -\frac{x^2}{2} & 0 & 0 & 1 \end{matrix}\right], 2, 0\right]$$

It means that

$$\begin{pmatrix} \sigma^{-1} & -\frac{1}{x-1} \\ \frac{x^2}{2} & -\frac{x}{2}\sigma + 1 \end{pmatrix}^{-1} = \begin{pmatrix} -\frac{(x+1)^2}{2x}\sigma^2 + \sigma & \frac{1}{x}\sigma \\ \frac{x^2}{2}\sigma & 1 \end{pmatrix}.$$

In addition, a series of experiments has been executed.

Example 6. Consider the following $n \times n$-operator with $n = 2k$:

$$M = \begin{pmatrix} I_k & A \\ 0_k & I_k \end{pmatrix}, \tag{12}$$

where 0_k is the zero $k \times k$-matrix, $A \in \text{Mat}_k(K[\sigma, \sigma^{-1}])$ is an arbitrary operator. The operator (12) is unimodular for any A, its inverse operator is

$$M^{-1} = \begin{pmatrix} I_k & -A \\ 0_k & I_k \end{pmatrix}. \tag{13}$$

For each experiment, we have generated an operator A whose entries are scalar difference operators having random rational function coefficients with the numerators and denominators of the degree up to 2. We compute the inverse for M of the form (12). The order of A, and hence, the order of M varies as $d = 1, 2, 4, 6, 8, 10$ and the number of rows of M varies as $n = 4, 6, 8, 10$ (hence, the number of rows of A varies as $k = 2, 3, 4, 5$). The inverse of M is calculated in each experiment by IsUnimodular. The results are presented in Table 1.

The table shows that the computation time in general corresponds to the complexity estimates (it should not be exact since the estimates are for the worst

Table 1. Results of the experiments, in seconds

	d = 1	d = 2	d = 4	d = 6	d = 8	d = 10
n = 4	0.125	0.188	0.500	0.969	2.078	2.906
n = 6	0.282	0.593	1.734	6.563	79.375	92.562
n = 8	0.516	1.500	37.938	94.813	427.375	1836.547
n = 10	0.703	5.562	910.218	1006.797	7576.063	13372.172

case and asymptotical). However, the computing time starts to increase faster than expected with the growth of n and d. It is again caused by the significant growth of the size of the elements of the matrix in the course of the computation. The size of the elements in M^{-1} is equal to the size of the elements in M in these experiments, so the coefficients of the elements are rational functions with the numerators and denominators of the degree up to 2. But in the course of the computation, the elements of the matrix have coefficients with the numerators and denominators of the degree up to several dozens for the smaller n and d, and up to several hundreds and even more than a thousand for the greater n and d.

6   Conclusion

In this paper, we have presented some new algorithms for solving problems for matrices whose entries belong to the non-commutative ring of scalar linear difference operators with coefficients from a difference field K of characteristic 0 with an automorphism σ. Some algorithms for solving the difference problems formulated in the paper had been proposed in [3]. The algorithms in the present paper have lower complexity due to the usage of the EG-eliminations algorithm as an auxiliary tool instead of Row-Reduction algorithm. The implementation of the algorithm in Maple was done and some experiments were reported. The experimental results show that the computation time corresponds in general to the complexity estimates from the proposed theory.

From our work, new questions arise (they were earlier formulated in [3]). For example, it is not clear, whether the problem of inverting can be reduced to the matrix multiplication problem (similarly to the "commutative" case)?

One more question: whether there exists an algorithm for such $n \times n$-matrices inverting with the full complexity $O(n^a d^b)$, with $a < 3$? It is possible to prove by the usual way that the matrix multiplication can be reduced to the problem of the matrix inverting (we have in mind the difference matrices). However, it is not so easy to prove that the problem of the matrix inverting can be reduced to the problem of the matrix multiplication.

We will continue to investigate this line of enquiry.

Acknowledgments. The authors are thankful to anonymous referees for useful comments. Supported in part by the Russian Foundation for Basic Research, project No. 16-01-00174.

References

1. Abramov, S.: EG-eliminations. J. Differ. Equ. Appl. **5**, 393–433 (1999)
2. Abramov, S.A.: On the differential and full algebraic complexities of operator matrices transformations. In: Gerdt, V.P., Koepf, W., Seiler, W.M., Vorozhtsov, E.V. (eds.) CASC 2016. LNCS, vol. 9890, pp. 1–14. Springer, Cham (2016). https://doi.org/10.1007/978-3-319-45641-6_1
3. Abramov, S.: Inverse linear difference operators. Comput. Math. Math. Phys. **57**, 1887–1898 (2017)
4. Abramov, S.A., Barkatou, M.A.: On the dimension of solution spaces of full rank linear differential systems. In: Gerdt, V.P., Koepf, W., Mayr, E.W., Vorozhtsov, E.V. (eds.) CASC 2013. LNCS, vol. 8136, pp. 1–9. Springer, Cham (2013). https://doi.org/10.1007/978-3-319-02297-0_1
5. Abramov, S., Barkatou, M.: On solution spaces of products of linear differential or difference operators. ACM Commun. Comput. Algebra **4**, 155–165 (2014)
6. Abramov, S., Bronstein, M.: On solutions of linear functional systems. In: ISSAC 2001 Proceedings, pp. 1–6 (2001)
7. Abramov, S., Bronstein, M.: Linear algebra for skew-polynomial matrices. Rapport de Recherche INRIA RR-4420, March 2002 (2002). http://www.inria.fr/RRRT/RR-4420.html
8. Abramov, S.A., Glotov, P.E., Khmelnov, D.E.: A scheme of eliminations in linear recurrent systems and its applications. Trans. Fr.-Russ. Lyapunov Inst. **3**, 78–89 (2001)
9. Abramov, S., Khmelnov, D., Ryabenko, A.: Procedures for searching local solutions of linear differential systems with infinite power series in the role of coefficients. Program. Comput. Softw. **42**(2), 55–64 (2016)
10. Andrews, G.E.: q-Series: Their Development and Application in Analysis, Number Theory, Combinatorics, Physics, and Computer Algebra. CBMS Regional Conference Series, vol. 66. AMS, Providence (1986)
11. Beckermann, B., Cheng, H., Labahn, G.: Fraction-free row reduction of matrices of Ore polynomials. J. Symb. Comput. **41**, 513–543 (2006)
12. Franke, C.H.: Picard-Vessiot theory of linear homogeneous difference equations. Trans. Am. Math. Soc. **108**, 491–515 (1986)
13. Giesbrecht, M., Sub Kim, M.: Computation of the Hermite form of a matrix of Ore polynomials. J. Algebra **376**, 341–362 (2013)
14. Knuth, D.E.: Big Omicron and big Omega and big Theta. ACM SIGACT News **8**(2), 18–23 (1976)
15. Middeke, J.: A polynomial-time algorithm for the Jacobson form for matrices of differential operators. Technical report No. 08–13 in RISC Report Series (2008)
16. van der Put, M., Singer, M.F.: Galois Theory of Difference Equations. LNM, vol. 1666. Springer, Heidelberg (1997). https://doi.org/10.1007/BFb0096118
17. van der Put, M., Singer, M.F.: Galois Theory of Linear Differential Equations. Grundlehren der mathematischen Wissenschaften, vol. 328. Springer, Heidelberg (2003). https://doi.org/10.1007/978-3-642-55750-7
18. Maple online help. https://www.maplesoft.com/support/help/

Sparse Polynomial Arithmetic
with the BPAS Library

Mohammadali Asadi, Alexander Brandt$^{(\boxtimes)}$, Robert H. C. Moir,
and Marc Moreno Maza

Department of Computer Science, The University of Western Ontario,
London, Canada
{masadi4,abrandt5,rmoir3}@uwo.ca, moreno@csd.uwo.ca

Abstract. We discuss algorithms for pseudo-division and division with
remainder of multivariate polynomials with sparse representation. This
work is motivated by the computations of normal forms and pseudo-
remainders with respect to regular chains. We report on the implementation of those algorithms with the BPAS library.

1 Introduction

General-purpose polynomial system solvers, like MAPLE's `solve` command, combine different algorithms using various polynomial data-types. Consider, as input
for such a solver, a polynomial system coming from a real life application, typically consisting of sparse multivariate polynomials with rational number coefficients. A pre-processing phase, using sparse polynomial data-types, attempts to
reduce the number of equations, variables or the total degree, say by exploiting
properties like symmetries. Then a *core engine*, say based on Gröbner bases, a
homotopy method, or triangular decompositions, determines a representation of
the real or complex solutions of the input system; this step generally requires
a change of polynomial representation (e.g. dense data-types) together with a
change of coefficient type (e.g. to finite fields when modular methods are used).
Finally, the representation computed by the core engine is converted to one
which is more "explicit" or convenient to an end-user; in fact, a return to the
original sparse polynomial data-type is likely to take place.

Core engines of polynomial systems solvers have driven a large body of work
in the computer algebra community. In particular, algorithms and implementation techniques supporting the polynomial and matrix data-types used by
those core engines have received great attention. In contrast, until a decade ago,
the implementation of sparse polynomial arithmetic, which is the default data-
type for general-purpose computer algebra systems, like MAPLE, MATHEMATICA,
SAGE, and SINGULAR, was often less optimized. Nevertheless, we should mention
pioneer works like the seminal article of Johnson [11] in 1974.

Research works conducted in the last decade on sparse polynomial arithmetic
operations[1] and data-types can essentially be categorized into two streams. The

---

[1] Polynomial arithmetic operations refers here to addition, multiplication, division and
pseudo-division.

© Springer Nature Switzerland AG 2018
V. P. Gerdt et al. (Eds.): CASC 2018, LNCS 11077, pp. 32–50, 2018.
https://doi.org/10.1007/978-3-319-99639-4_3

first one deals primarily with algebraic complexity, see the works of van der Hoeven and Lecerf [10] and those of Arnold and Roch [1]. The latter focuses on implementation techniques, see the works of Monagan and Pearce [15,19], and those of Gastineau and Laskar [5,6]. The present work subscribes to this second stream. We are motivated by obtaining efficient implementation of triangular decomposition algorithms based on the theory of regular chains [4]. To be precise, we aim at adapting the algorithms of the RegularChains library [13] to the *Basic Polynomial Algebra Subprograms* (BPAS). This latter library is written mainly in C language, for high performance, wrapped in a C++ interface to make use of object-oriented programming and for end-user usability. The Cilk extension [12] is used for multi-threading, targeting multi-core processors. BPAS is already equipped with parallel dense polynomial arithmetic over finite fields [20] and the integers [3]. BPAS is publicly available in source at www.bpaslib.org.

We report in this paper on the implementation with the BPAS library of elementary arithmetic operations for multivariate polynomials represented with sparse data-types. In Sect. 2, we start by discussing multiplication and division with remainder, following the papers [11,15,19]. Then, we propose an algorithm for pseudo-division using similar principles. Our presentation of both division with remainder and pseudo-division has two levels: one abstract level independent of the supporting data-structures (see Algorithms 1 and 3) and one level taking advantage of heap data-structures (see Algorithms 2 and 4). This presentation allows us to formally prove those algorithms.

In Sect. 3, we discuss the implementation of the algorithms presented in Sect. 2 within the BPAS library; we highlight the differences between our implementation and that realized in MAPLE by Monagan and Pearce. Note that, currently, all the BPAS code for sparse polynomial arithmetic is entirely serial C code, that is, multi-threading is not used yet. We stress the fact that, while algorithms for division with remainder (Algorithms 1 and 2) may look similar to their counterparts for pseudo-division (Algorithms 3 and 4), implementation of the latter is by far more challenging than that of the former. Indeed, pseudo-division is essentially a univariate operation. Thus, when used in the context of multivariate polynomials, careful data-structure manipulations are needed to optimize both memory usage and access time to terms of polynomials, see Sect. 3.5. Section 4 gathers our experimental results. For multivariate polynomials over the integers (for which both BPAS and MAPLE have optimized implementation), BPAS is usually faster with a speedup factor typically between 2 to 3, see Figs. 5, 6 and 8. For multivariate polynomials over the rational numbers (for which only BPAS has an optimized implementation), BPAS is faster than MAPLE by 2 to 3 orders of magnitude, see Figs. 3, 4 and 7. This is particularly true for the computation of normal forms, see Fig. 9.

2   Sparse Polynomial Arithmetic

For the treatment of sparse polynomial arithmetic we require both a distributed and recursive view of polynomials, depending on the operation. For a distributed polynomial $a \in \mathbb{D}[x_1, \ldots, x_m]$, for an integral domain $\mathbb{D}$ and variable ordering $x_1 < x_2 < \cdots < x_m$, we use the notation $a = \sum_{i=1}^{n_a} A_i = \sum_{i=1}^{n_a} a_i X^{\alpha_i}$, where n_a is the number of (non-zero) terms, $0 \neq a_i \in \mathbb{D}$, α_i is an exponent vector for the variables $X = (x_1, \ldots, x_m)$. A term of a is represented by $A_i = a_i X^{\alpha_i}$. We assume that the terms are ordered (decreasing) lexicographically, so that $\mathrm{lc}(a) = a_1$ is the *leading coefficient* of a, $\mathrm{lm}(a) = X^{\alpha_1}$ is the *leading monomial* of a, and $\mathrm{lt}(a) = a_1 X^{\alpha_1}$ is the *leading term* of a. If a is not constant, the greatest variable appearing in a is the *main variable* of a (denoted $\mathrm{mvar}(a)$). Given a term A_i of a, $\mathrm{coef}(A_i) = a_i$ is the coefficient, $\mathrm{expn}(A_i) = \alpha_i$ is the exponent vector, and $\deg(A_i, x_j)$ is the component of α_i corresponding to x_j. Then, $\deg(a, x_j)$ is the maximum value of $\deg(A_i, x_j)$ among all terms A_i of a.

For a recursive view of a non-constant polynomial $a \in \mathbb{D}[x_1, \ldots, x_m]$, again with $x_1 < x_2 < \cdots < x_m$, we view a as a univariate polynomial in $R[x_j]$, with $x_j = \mathrm{mvar}(a)$ is the largest variable occurring in a, and where $R = \mathbb{D}[x_1, \ldots, x_{j-1}]$. Viewed in $R[x_j]$, the leading coefficient of a is the *initial* of a (denoted $\mathrm{init}(a)$). Given a term A_i of $a \in R[x_j]$, $\mathrm{coef}(A_i) \in \mathbb{D}[x_1, \ldots, x_{j-1}]$ and $\mathrm{expn}(A_i) = \deg(A_i, x_j)$.

Addition (or subtraction) of two polynomials requires joining the terms of the two summands, combining terms with identical exponents (with possible cancellation) and then sorting the terms of the sum. A naïve approach is to compute the sum $a + b$ term-by-term, adding a term of the addend (b) to the augend (a), sorting at each step, in a manner similar to *insertion sort*. An efficient algorithm instead uses *merge sort*, taking advantage of the fact that the terms of a and b are already ordered. For details of the algorithm see [11, p. 65].

Multiplication of two polynomials requires generating the terms of the product, combining terms with equal exponents and sorting the product terms. A naïve approach is to compute the product $a \cdot b$ (where a has n_a terms and b has n_b terms) by distributing each term of the multiplier (a) over the multiplicand (b) and combining like terms: $c = a \cdot b = (a_1 X^{\alpha_1} \cdot b) + (a_2 X^{\alpha_2} \cdot b) + \cdots$. This is inefficient because all $n_a n_b$ terms are generated, whether or not they have equal exponents, and the $n_a n_b$ terms must be sorted. Again, following Johnson [11], we can obtain more efficient algorithms by generating terms in sorted order.

We make good use of the sparse data structure for $a = \sum_{i=1}^{n_a} a_i X^{\alpha_i}$, and $b = \sum_{j=1}^{n_b} b_j X^{\beta_j}$, by observing that for given α_i and β_j, we always have that $X^{\alpha_{i+1} + \beta_j}$ and $X^{\alpha_i + \beta_{j+1}}$ are less than $X^{\alpha_i + \beta_j}$ in the term order. Given that $X^{\alpha_i + \beta_j} > X^{\alpha_i + \beta_{j+1}}$ we can generate terms of the product in order by merging n_a "streams" of terms obtained by multiplying a single term of a distributed over b,

$$a \cdot b = \begin{cases} (a_1 \cdot b_1) X^{\alpha_1 + \beta_1} + (a_1 \cdot b_2) X^{\alpha_1 + \beta_2} + (a_1 \cdot b_3) X^{\alpha_1 + \beta_3} + \ldots \\ (a_2 \cdot b_1) X^{\alpha_2 + \beta_1} + (a_2 \cdot b_2) X^{\alpha_2 + \beta_2} + (a_2 \cdot b_3) X^{\alpha_2 + \beta_3} + \ldots \\ \vdots \\ (a_{n_a} \cdot b_1) X^{\alpha_{n_a} + \beta_1} + (a_{n_a} \cdot b_2) X^{\alpha_{n_a} + \beta_2} + (a_{n_a} \cdot b_3) X^{\alpha_{n_a} + \beta_3} + \ldots \end{cases}$$

and then selecting the maximum term from the heads of the streams. The new head of the stream where a term is removed is then the term to its right in that stream. We can efficiently handle this sub-problem of selecting the maximum term by storing the heads of the streams in a priority queue, which we implement using a binary max-heap. We minimize the size of the heap by choosing the order of multiplicative factors such that $n_a \leq n_b$, which we are free to do since multiplication is commutative. Because the heap multiplication algorithm was specified completely by Johnson, we refer the reader to [11], which discusses the algorithm and provides pseudo-code.

2.1 Division

We now consider the problem of multivariate division, where the input polynomials $a, b \in \mathbb{D}[x_1, \ldots, x_m]$, with $b \notin \mathbb{D}$ being the divisor and a the dividend. We assume that $\mathbb{D}$ is a field. Hence $\{b\}$ is a Gröbner basis of the ideal it generates. Thus, we can specify the output as $q, r \in \mathbb{D}[x_1, \ldots, x_m]$ satisfying $a = qb + r$, such that r is reduced with respect to b treated as a Gröbner basis.

Division presents a more tricky problem in terms of heap-optimization. We must compute terms of the quotient and remainder in order, and produce terms of the product qb in order, as terms of q are generated in the execution of the algorithm. To see how this can be done without a heap, consider Algorithm 1, which computes q and r term by term by computing $\tilde{r} = \mathrm{lt}(a - qb - r)$ at each step. This works for multivariate division because introducing a new quotient term whenever $\mathrm{lt}(b) \mid \tilde{r}$ ensures that any subsequent terms of $a - qb - r$ that do not satisfy this condition will be remainder terms. This allows terms of both q and r to be computed in order.

Proposition 1. *Algorithm 1 terminates and is correct.*

Proof. It is enough to show that for each iteration of the loop, the term $\tilde{r}$ decreases strictly. It follows from the axioms of a term order that $\tilde{r}$ becomes zero after finitely many iterations. We denote the values of the variables of Algorithm 1 on the i-th iteration by superscripts. For each i, depending on whether or not $\mathrm{lt}(b) \mid \tilde{r}^{(i)}$ holds, we have two possibilities:

- $Q_\ell = \tilde{r}^{(i)}/B_1$, where Q_ℓ is a new quotient term;
- or $R_k = \tilde{r}^{(i)}$, where R_k is a new remainder term.

We provide the proof for the first case. The second case is similar but essentially trivial. Since $\tilde{r}^{(i)} = Q_\ell B_1$ holds by assumption, we have

$$
\begin{aligned}
\tilde{r}^{(i+1)} = \mathrm{lt}(a - q^{(i+1)}b - r^{(i+1)}) &= \mathrm{lt}(a - ([q^{(i)} + Q_\ell]b + r^{(i)})) \\
&= \mathrm{lt}(a - (q^{(i)}b + r^{(i)} + (\tilde{r}^{(i)} - \tilde{r}^{(i)}) + Q_\ell b)) \\
&= \mathrm{lt}([(a - q^{(i)}b - r^{(i)}) - \tilde{r}^{(i)}] - [Q_\ell(b - B_1)]) \\
&< \mathrm{lt}(\tilde{r}^{(i)}) = \tilde{r}^{(i)}.
\end{aligned}
$$

The remainder r is reduced with respect to $\{b\}$ because all terms R_k of r satisfy $\mathrm{lt}(b) \nmid R_k$ by construction. $\qquad\square$

Heap-optimization can then be applied to Algorithm 1 by using a heap to keep track of the computation of the product qb. This is a special case of heap multiplication. The major difference from multiplication, where all terms of both factors are known at the outset, is that q is computed as the algorithm proceeds, which forces q to be the multiplier and b the multiplicand. Thus, each stream consists of a term Q_ℓ of q distributed over b. Another difference from multiplication is that each stream is initiated with the term $Q_\ell B_2$, because $Q_\ell B_1$ need not be computed because it is canceled out by construction.

The management of the heap to compute a product ab requires a number of specialized functions. We provide here a simplified interface consisting of three functions. **heapInsert**(A_i, B_j) adds the product of A_i and B_j to the heap[2]. **heapPeek**() gets the exponent vector ε of the top element in the heap. **heapExtract**() removes the top element of the heap *and* inserts the next element of the stream from which the top element came from. That is, if there are any elements remaining in that stream. The key modification of Algorithm 1 to reach Algorithm 2 is to use terms of qb from the heap to compute $\tilde{r} = \mathrm{lt}(a - qb - r)$. This requires tracking three cases: (1) $\tilde{r}$ is an uncanceled term of a; (2) $\tilde{r}$ is a term of the difference $(a - r) - (qb)$; and (3) $\tilde{r}$ is a term of $-qb$ such that all remaining terms of $a - r$ are smaller in the term order.

Let ε be the exponent vector of the top term of the heap computation of qb. If the heap is empty, we let $\varepsilon = (-1, 0, \ldots, 0)$, which will be less than any exponent of any polynomial term on account of the first element being -1. We therefore abuse notation and write $\varepsilon = -1$ for an empty heap. Let A_k be the greatest uncanceled term of a. Then, the three cases correspond to conditions on the ordering of ε and $\mathrm{expn}(A_k)$. The term $\tilde{r}$ is an uncanceled term of a (case 1) either if the heap is empty (indicating that no terms of q have yet been computed or all terms of qb have been extracted) or if $\varepsilon > -1$ but $\varepsilon < \mathrm{expn}(A_k)$. In either of these two situations $\varepsilon < \mathrm{expn}(A_k)$ holds. The term $\tilde{r}$ is a term of the difference $(a - r) - (qb)$ (case 2) if both A_k and ε have the same exponent ($\varepsilon = \mathrm{expn}(A_k)$). And $\tilde{r}$ is a term of $-qb$ (case 3) whenever $\varepsilon > \mathrm{expn}(A_k)$ holds.

Algorithm 2 uses this observation to compute $\tilde{r}$ by adding a conditional to compare the ranks of ε and $\mathrm{expn}(A_k)$. Terms are only extracted from the heap when $\varepsilon \geq \deg(A_k)$ holds; and when a term is extracted the next term from the given stream, if there is one, is added to the heap (defined behaviour of **heapExtract**()). The adding of new terms to q and r is almost identical to Algorithm 1, except that for quotient terms we initiate a new stream starting with $Q_\ell B_2$ (because $Q_\ell B_1$ is canceled by construction). Together with Proposition 1, then, we have established the following proposition.

Proposition 2. *Algorithm 2 terminates and is correct.*   □

---

[2] Note that the heap need not actually store product terms but can simply store the indices of the two factors, with the product only computed when elements of the heap are removed. This strategy is needed for pseudodivision, discussed below, where the quotient terms are updated in the course of the algorithm.

Algorithm 1. DIVIDE(a,b)

$a, b \in \mathbb{D}[x_1, \ldots, x_m]$, mdeg($b$) > 0, return $q, r \in \mathbb{D}[x_1, \ldots, x_m]$ such that $a = qb + r$ where r is reduced with respect to the Gröbner basis $\{b\}$.

```
1: q := 0; r := 0
2: while r̃ := lt(a − qb − r) ≠ 0 do
3:     if lt(b) | r̃ then
4:         q := q + r̃/lt(b)
5:     else
6:         r := r + r̃
7:     end if
8: end while
9: return (q, r)
```

Algorithm 3. PSEUDODIVIDE(a,b,x)

$a, b \in \mathbb{D}[x]$, deg(b, x) > 0, returns $q, r \in \mathbb{D}[x]$ and $\ell \in \mathbb{N}$ such that $h^\ell a = qb + r$, with deg(r, x) $<$ deg(b, x).

```
1: q := 0; r := 0; h := lc(b); ℓ := 0; γ = deg(b.x)
2: while r̃ := lt(h^ℓ a − qb − r) ≠ 0 do
3:     if x^γ | r̃ then
4:         q := hq + r̃/x^γ; ℓ := ℓ + 1
5:     else
6:         r := r + r̃
7:     end if
8: end while
9: return (q, r, ℓ)
```

Algorithm 2. DIVIDE(a,b)

$a, b \in \mathbb{D}[x_1, \ldots, x_m]$, mdeg($b$) > 0, return $q, r \in \mathbb{D}[x_1, \ldots, x_m]$ such that $a = qb + r$ where r is reduced with respect to the Gröbner basis $\{b\}$.

```
 1: q := 0; r := 0
 2: k := 1; ℓ := 0
 3: while ε := heapPeek() > −1 or k ≤ n_a do
 4:     if ε < expn(A_k) then
 5:         r̃ := A_k
 6:         η := expn(A_k); k := k + 1
 7:     else if ε = expn(A_k) then
 8:         r̃ := A_k − heapExtract()
 9:         η := ε; k := k + 1
10:     else
11:         r̃ := −heapExtract()
12:         η := ε
13:     end if
14:     if expn(B_1) | η then
15:         ℓ := ℓ + 1; Q_ℓ := r̃/B_1; q := q + Q_ℓ
16:         heapInsert(Q_ℓ, B_2)
17:     else
18:         r := r + r̃
19:     end if
20: end while
21: return (q, r)
```

Algorithm 4. PSEUDODIVIDE(a,b,x)

$a, b \in \mathbb{D}[x]$, deg(b) > 0, returns $q, r \in \mathbb{D}[x]$ and $\ell \in \mathbb{N}$ such that $h^\ell a = qb + r$, with deg(r, x) $<$ deg(b, x).

```
 1: q := 0; r := 0; h := lc(b)
 2: ε := −1; s := 0
 3: k := 1; ℓ := 0; γ := deg(b, x)
 4: while ε := heapPeek() > −1 or k ≤ n_a do
 5:     if ε < deg(A_k, x) then
 6:         r̃ := h^ℓ A_k
 7:         η := deg(A_k, x); k := k + 1
 8:     else if ε = deg(A_k, x) then
 9:         r̃ := h^ℓ A_k − heapExtract()
10:         η := ε; k := k + 1
11:     else
12:         r̃ := −heapExtract()
13:         η := ε
14:     end if
15:     if deg(b, x) ≤ η then
16:         q := hq; ℓ := ℓ + 1; Q_ℓ := r̃/x^γ
17:         heapInsert(Q_ℓ, B_2); q := q + Q_ℓ
18:     else
19:         r := r + r̃
20:     end if
21: end while
22: return (q, r, ℓ)
```

2.2 Pseudo-Division

The pseudo-division algorithm is essentially univariate, and terms here are elements of $\mathbb{D}[x]$ for an arbitrary integral domain $\mathbb{D}$. Pseudo-division is essentially a fraction-free division: rather than dividing a by $h = \mathrm{lc}(b)$ for each term of the quotient q, it multiplies a by h. If the quotient ends up with ℓ terms, then the result must satisfy $h^\ell a = qb + r$.

An important consequence of pseudo-division being univariate is that all of the quotient terms are computed first and then all of the remainder terms are computed. This is because we can always carry out a pseudo-division step provided that $\deg(b, x) \leq \deg(\mathrm{lt}(h^\ell a - qb), x)$, where $\mathrm{lt}(h^\ell a - qb)$ is the equivalent of $\tilde{r}$ from Algorithm 1 when $r = 0$. Thus, we adopt the same symbol for it in Algorithm 3, which is the extension of Algorithm 1 to pseudo-division. The only difference in these algorithms is that each time we compute a new pseudo-quotient term we do so as $\tilde{r}/x^\gamma$, where $\gamma = \deg(b, x)$ (fraction free division), rather than $\tilde{r}/B_1 = \tilde{r}/(hx^\gamma)$ as before, and because we add a factor of h to a, we must also multiply the previous value of the quotient by h.

Proposition 3. *Algorithm 3 terminates and is correct.*

Proof. Similar to Proposition 1. The two cases here are $Q_\ell = \tilde{r}^{(i)}/x^\gamma$ and $R_k = \tilde{r}^{(i)}$. We consider the first case (the second case is similar and essentially trivial). In the first case $r^{(i)} = 0$, since quotient terms are still being computed, so that $\tilde{r}^{(i)} = \mathrm{lt}(h^\ell a - q^{(i)}b)$. Since $\tilde{r}^{(i)} = Q_\ell x^\gamma$ by assumption, $h\tilde{r}^{(i)} = Q_\ell B_1$, and we have

$$\begin{aligned}
\tilde{r}^{(i+1)} = \mathrm{lt}(h^{\ell+1}a - q^{(i+1)}b - r^{(i+1)}) &= \mathrm{lt}(h^{\ell+1}a - ([hq^{(i)} + Q_\ell]b)) \\
&= \mathrm{lt}(h^{\ell+1}a - (hq^{(i)}b + (h\tilde{r}^{(i)} - h\tilde{r}^{(i)}) + Q_\ell b)) \\
&= \mathrm{lt}(h[(h^\ell a - q^{(i)}b) - \tilde{r}^{(i)}] - [Q_\ell(b - B_1)]) \\
&< \mathrm{lt}(\tilde{r}^{(i)}) = \tilde{r}^{(i)}.
\end{aligned}$$

The condition $\deg(r, x) < \deg(b, x)$ is ensured because quotient terms are computed until $x^\gamma \nmid \tilde{r}$ holds, that is, until $\deg(h^\ell a - qb, x) < \deg(b, x)$ holds. □

Heap-optimization of Algorithm 3 proceeds in much the same way as for division. The only additional consideration required for Algorithm 4 is the accounting for factors of h in the computation of $\mathrm{lt}(h^\ell a - qb - r)$. This only requires adding as many factors of h to A_k that have been added to the quotient up to the current iteration. Since ℓ terms have been added to q, we multiply A_k by h^ℓ each time we use one of the terms. Additional factors of h are added when the previous quotient is multiplied by h prior to the computation of the next quotient term. Other than this, the shift from Algorithm 3 to Algorithm 4 follows the analogous shift between Algorithms 1 and 2 exactly. We therefore have the following.

Proposition 4. *Algorithm 4 terminates and is correct.*

Proof. The proof is a straightforward adaptation of the preceding observations and the proofs for Propositions 2 and 3. The key observation is the first main conditional statement in the while loop computes $\tilde{r} = \mathrm{lt}(h^\ell a - qb - r)$, where $r = 0$ until q has been computed, and the second main conditional computes a term of q or r from $\tilde{r}$ accordingly, following the structure of Algorithm 3. □

2.3 Multi-Divisor (Pseudo-)Division

One natural application of division with remainder of multivariate polynomials is the computation of normal forms with respect to Gröbner bases. Moreover, the computation of pseudo-quotient and pseudo-remainder of a polynomial with respect to multiple polynomials (or a triangular set) is also natural. Normal forms can be computed by Algorithms 5 and 7 in Appendix A while pseudo-division by a triangular set can be computed by Algorithms 6 and 8. Section 4 includes benchmarks of those four algorithms implemented with the BPAS library.

3   Implementation and Optimizations

With the ever-increasing gap between processor speeds and memory-access time, our implementation techniques focus on memory usage and management. Our implementations effectively traverse memory while making use of memory-efficient data structures with good data locality. In this section we consider polynomial representations and corresponding data structures (Sect. 3.1), addition and multiplication (Sect. 3.2), heap-optimizations (Sect. 3.3), division (Sect. 3.4), and lastly, pseudo-division (Sect. 3.5).

3.1   Polynomial Representations

The simplest scheme to represent a polynomial sparsely would be a linked list where each node in the list is a single term of that polynomial. This representation makes handling and manipulating terms very easy with simple pointer manipulation. However, the indirection created by pointers and (possibly) poor locality of successive nodes in the list makes this scheme inefficient for memory usage. Rather, packing the polynomial terms into an array removes the overhead of linked list pointers and improves locality. We call this array-based representation of a polynomial an *alternating array* following the terminology introduced in 1997 in the BasicMath library, part of the European Project FRISCO https:// cordis.europa.eu/project/rcn/31471_en.html; see also [2].

The alternating array representation packs terms side-by-side in an array, effectively alternating between coefficients and monomials. A coefficient and its corresponding monomial are thus optimally local in memory with respect to each other. Similar schemes have been used in MAPLE [18,19]. In the case of MAPLE, their scheme uses pointers into a parallel array to store the multi-precision integer coefficient, whereas we store the multi-precision coefficients directly in the array. Moreover, for this efficient data structure MAPLE is limited to integer polynomials while all other polynomials use an old sum-of-products form [18]. In contrast, our alternating array representation in the BPAS library supports both integer and rational number coefficients.

Coefficients are represented easily using GMP multi-precision numbers [7]. As for monomials, we use *exponent packing*. Using bit-masks and shifts, multiple integers, each of small absolute value, can effectively be stored in a single 64-bit machine word. The idea of exponent packing has been employed at least since ALTRAN in the late 60s [8] and more recently in [10,16]. Some systems, such as MAPLE, also encode the total degree of the monomial in the single 64-bit word. This scheme wastes bits which could be used for additional variables or higher degrees. In particular, monomials are limited to 21 variables each with a maximum degree of 3 [18]. Our representation does not encode total degree, therefore we can encode up to 32 variables, each of maximum degree 3. Moreover, in polynomial system solving, degrees of lower ordered variables often increase much quicker than those of high ordered variables. Thus, in our implementation, we pack exponents disproportionately within the machine word, giving more bits to lower ordered variables, ensuring all 64 bits are made useful.

It is worth noting that our sparse representations are used for all of our algorithms, including division and pseudo-division, where (pseudo-)quotients and (pseudo-)remainders are often much more dense than the divisor and dividend. However, since we are working with multivariate polynomials, a dense representation would grow exponentially with the number of variables and, therefore, our sparse representation is still worthwhile and efficient.

3.2   Addition and Multiplication

For these two simple operations, we just point out a few implementation tricks. An "in-place" addition (subtraction) can be implemented with our alternating array representation. This is not strictly in-place, as that would involve far too much memory movement and swapping of elements, resulting in poor locality and poor performance. Instead, we can pre-allocate a destination array as with an "out-of-place" addition algorithm, but, rather than copying coefficients, we reuse the underlying GMP data. With modestly-sized coefficients, less than 192 bits each, the savings can reach 20% compared to the out of place implementation.

As for multiplication, we pre-allocate the maximum possible space for the product $(n_a \cdot n_b)$. Assuming that a has fewer terms than b, we pre-allocate space in the heap for exactly n_a elements as that will be the exact number of streams to consider. This minimizes memory movement and reallocation required throughout the computation of appending product terms to the product polynomial. If the product terms were to out-grow some initial conservative pre-allocation the reallocation and memory movement could result in a large overhead.

3.3   Heap-Optimizations

The performance of our code is very dependent on the implementation of its data-structures, and in particular, heaps. Aside from coefficient arithmetic, all of the work for multiplying terms comes from obtaining the ordering of product terms. Hence, the heap, whose purpose is to produce terms in the required order, takes the majority of the effort of our algorithm. Our implementation of heaps includes all the techniques reported in [16], including the technique of *chaining*. We mention an additional trick used in our code. With chaining, the coefficients of the product terms are already not stored directly in the heap, but they still play a role in overall auxiliary memory needed for the algorithm. With our alternating array representation of polynomials it is very easy to directly index the operand polynomials to access the appropriate coefficient. Thus, our heap only stores the *indices* of the operand coefficients which together form the coefficient of the particular product term (Fig. 1). This reduces the memory required for each coefficient from 32 bytes, in the case of rational number coefficients, down to 8 bytes. Similar schemes using pointers to coefficients have been examined in [16,19] but indices are even more succinct than pointers.

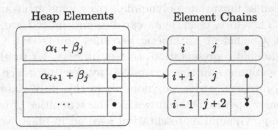

Fig. 1. A heap of product terms, showing element chaining and index-based storing of coefficients. In this case, terms $A_{i+1} \cdot B_j$ and $A_{i-1} \cdot B_{j+2}$ have equal monomials and are chained together.

3.4 Division

Division is essentially a direct application of multiplication. We again use heaps, with all of its optimizations, using the production of product terms in-order to produce the terms of the quotient and remainder in-order. Division varies from multiplication as instead of producing the product terms of the two input operands, we must produce product terms between the divisor and the continually updating quotient. This poses problems for memory management as we do not know ahead of time the sizes of the quotient or remainder. In multiplication we are able to pre-allocate $n_a \cdot n_b$ space for the product as that is the known maximum number of product terms. The indeterminate number of quotient and remainder terms does not allow for such one-time allocation and we must continually check for producing more terms than the number for which we have allocated space. We begin by allocating n_a space for the quotient and remainder, as generally the dividend is larger than the divisor. Then, if more terms are produced than we have currently allocated for, we double the current allocation.

Whenever we reallocate space for the quotient we also reallocate space for the same number of terms in the heap. Recall the maximum number of terms in the heap is equal to the number of quotient terms (as we distribute terms of the quotient over the divisor in the multiplication). So, we are safe in doing this memory allocation for the heap even if it does not make use of it all. This has benefits for performance as we do not need to check for overflow on each insert into the heap; it is guaranteed to have enough space.

3.5 Pseudo-Division

As seen in Sect. 2 the algorithm for division can easily be adapted for pseudo-division. With only the modification of multiplying the dividend and quotient by the divisor's initial, we obtained an algorithm for pseudo-division that efficiently produces terms in order. However, the implementation between these two algorithms is very different. In essence, pseudo-division is a univariate operation, viewing the input multivariate polynomials recursively. That is, the dividend and divisor are seen as univariate polynomials over some arbitrary (polynomial) integral domain. Therefore, coefficients can be, and indeed are, entire polynomials themselves. Coefficient arithmetic becomes non-trivial. Our distributed multivariate polynomial representation, as seen in Sect. 3.1 would be inefficient to traverse and manipulate in this recursive way. We introduce a new polynomial representation to easily view polynomials in this univariate, recursive way in order to efficiently operate on them within the semantics of pseudo-division.

This recursive polynomial representation uses an in-place, very fast conversion between the normal distributed representation and the recursive one. This amounts to minimal overhead and allows the same polynomials to be easily used as operands to pseudo-division or any other arithmetic operation. Of course, an in-place conversion is beneficial to avoid memory movement and reduce the working memory required for the algorithm.

To view the polynomial recursively, we begin by blocking the alternating array representation of the distributed polynomial based on degrees of the main variable. Each block groups together terms which have equal degree with respect to the main variable. As our polynomials are ordered lexicographically, then all terms are already in order with respect to the degree of the main variable, and, moreover, within a block, all terms are also sorted lexicographically with respect to all of the remaining variables. Because of this, we can create these blocks in-place, without any memory movement, simply by maintaining the offset into the array for the beginning of each block.

Next, we create a secondary alternating array to store these offsets. This array alternates between an exponent of the main variable and a pointer to the original array which is offset to point to the beginning of the block that corresponds to the preceding main variable exponent. Note that we also store the size of each block. This is convenient when we need to do coefficient arithmetic as those coefficients are themselves polynomials that must know their size to perform arithmetic. In addition, as we traverse the array to determine the blocks, we zero out the degree of the main variable for every monomial. This ensures that the degree of the main variable does not pollute the polynomial coefficient arithmetic. Figure 2 shows this secondary array structure along with the original array, highlighting the conversion process.

These two alternating arrays together exactly and efficiently represent the recursive view of a polynomial, having coefficients from an arbitrary polynomial ring and univariate monomials. The secondary alternating array requires little additional memory. It will have size equal to the number of unique values of degree of the main variable in the distributed polynomial. In practice, with sparse polynomials, this number is quite small. In the absolute worst case, for integer polynomials that are fully dense with respect to the main variable, this secondary array requires $O(\frac{2}{3}n)$ additional space. With multi-precision coefficients and/or rational number coefficients, this fraction becomes much smaller. This additional space becomes increasingly insignificant as the integers (rational numbers) grow in size, as they always do in pseudo-division calculations.

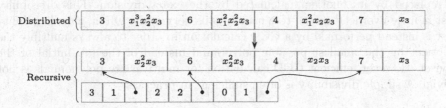

Fig. 2. A distributed polynomial representation converted to the recursive polynomial representation, showing the additional secondary array. The secondary array alternates between: (1) exponent of the main variable, (2) size of the coefficient polynomial, and (3) a pointer to the coefficient polynomial which is simply an offset into the original distributed polynomial.

With the recursive view of a polynomial efficiently implemented, it is then important to consider efficiency of coefficient arithmetic. As coefficients are now full polynomials there is more overhead in manipulating them and performing arithmetic. One important implementation detail is to perform the addition (and subtraction) of like-terms in-place. Such combinations occur when computing the leading term of $h^\ell a - qb$ and when combining like-terms in the quotient-divisor product. In-place addition, as described in the previous sub-section, allows for the re-use of underlying GMP data. Therefore, performance of in-place addition compared to out-of-place becomes increasingly better as coefficients grow throughout the pseudo-division algorithm.

Similarly, the update of the quotient by multiplying by the initial of the divisor, requires a multiplication of full polynomials. If we wish to save on memory movement we should perform this multiplication in place. However, notice that, in our recursive representation, coefficient polynomials are tightly packed in a continuous array. To modify them in place would require shifting all following coefficients down the array to make room for the strictly large product polynomial. To avoid this unnecessary memory movement, we modify the recursive data structure exclusively for the quotient polynomial. We break the continuous array of coefficients into many arrays, one for each coefficient. This allows them to grow without displacing the following coefficients. At the end of the algorithm, once the quotient has finished being produced, we collect and compact all of these disjoint polynomials into a single, packed array. In contrast, the remainder is never updated once its terms are produced. Moreover, we do not require any recursively viewed operations on the remainder. Hence, as terms of the remainder are produced, we store them directly in the normal, distributed representation, avoiding conversion out of the recursive representation and any memory overhead of the additional recursive array.

Lastly, our final optimization is common among other sparse pseudo-division algorithms. We perform a divisibility test between a newly produced quotient term and the initial of the divisor. If division is exact, we avoid one multiplication of the quotient with the divisor's initial, and the newly produced quotient term is replaced by its quotient calculated by the exact division. This divisibility test is little overhead as the test usually fails very early. Often, this divisibility test is instead performed by a GCD calculation in order to always multiply the quotient by the smallest possible polynomial instead of the full initial of the divisor. However, efficient GCD calculation for multivariate polynomials is not trivial. A simple divisibility is often sufficient in practice.

4   Experimentation

For univariate polynomials sparsity is easily defined as the maximum degree difference between successive polynomial terms. Though sparsity is not so easily defined for multivariate polynomials, we propose the following adaptation of the univariate case to the multivariate one, inspired by Kronecker substitution. Let $f \in \mathbb{D}[x_1, \ldots, x_m]$ be non-zero and define $r = \max(\deg(f, x_i), 1 \leq i \leq m) + 1$. Then, every exponent vector $e = (e_1, \ldots, e_m)$ of a term of f can be identified with the radix r representation of the integer $z(e) = e_1 + e_2 r + \cdots + e_m r^{m-1}$. We call *sparsity* of f the smallest integer s which is greater or equal to $z(e) - z(e')$, where e, e' are any two consecutive exponent vectors of f. If f has n terms then we have $r^m \leq n s$. For our experiments, sparse polynomials were randomly generated using the following parameters: number of variables m, number of terms n, sparsity s, and maximum number of bits in any coefficient. Then, exponent vectors are generated as radix r representations with m digits and r computed as $\lfloor \sqrt[m]{s \cdot m} \rfloor$.

We compare our implementation against MAPLE for both integer polynomials and rational number polynomials. Over the past 10 years or so, MAPLE has become the leader in integer polynomial arithmetic thanks to the extensive work of Monagan and Pearce [15–17,19]. Benchmarks there provide clear indication that their implementation outperforms many other computer algebra systems including: TRIP, MAGMA, SINGULAR, and PARI. Moreover, other common systems like FLINT [9] and NTL [21] provide only univariate polynomial implementations, meaning the comparison against our multivariate implementation would be unfair. Therefore, we compare our implementations against the leading high-performance implementation that is provided by MAPLE in particular, MAPLE 2017.

We consider multiplication and division over $\mathbb{Q}$ (Figs. 3 and 4), multiplication and division over $\mathbb{Z}$ (Figs. 5 and 6), pseudo-division over $\mathbb{Q}$ and $\mathbb{Z}$ (Figs. 7 and 8), and multi-divisor normal form and pseudo-division computation over $\mathbb{Q}$ (Fig. 9 and 10). In all cases (except dense integer multiplication) BPAS performs favourably over MAPLE. We note that random instances of division do not provide smooth results due to varying sizes of resulting quotients and remainders. Our benchmarks were collected using an Intel Xeon X560 processor at 2.67 GHz, 32 KB L1 data cache, 256 KB L2 cache, 12288 KB L3 cache, and 48 GB of RAM.

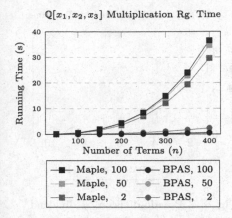

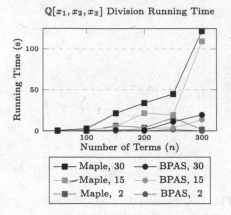

Fig. 3. $\mathbb{Q}$ multiplication. Sparsity varies as noted in the legend, # coefficient bits is 128.

Fig. 4. $\mathbb{Q}$ division. Sparsity varies as noted in the legend, # of divisor terms is $n/2$, # coefficient bits is 128.

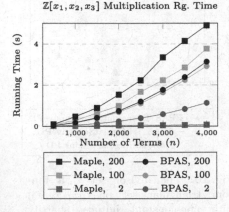

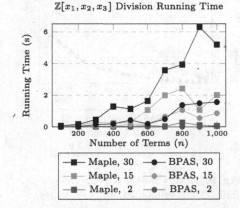

Fig. 5. $\mathbb{Z}$ multiplication. Sparsity varies as noted in the legend, # coefficient bits is 128.

Fig. 6. $\mathbb{Z}$ division. Sparsity varies as noted in the legend, # divisor terms is $n/2$, # coefficient bits is 128.

It is clear from these benchmarks that having optimized data structures and fundamental algorithms is important. For polynomials over the rational numbers, where MAPLE lacks an optimized implementation, our code performs orders of magnitude better. Even for MAPLE's optimized implementation of polynomials over the integers, our code still performs at a fraction of the time. This performance savings is substantial and is very apparent when comparing normal forms (see Fig. 9). With the repeated division required for normal forms, an optimized division algorithm results in extensive performance gains.

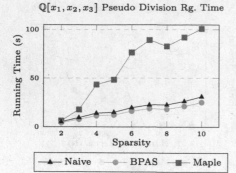

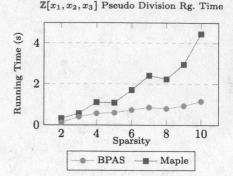

Fig. 7. $\mathbb{Q}$ Pseudo-division. # dividend terms is 175, # divisor terms is 50.

Fig. 8. $\mathbb{Z}$ Pseudo-division. # dividend terms is 175, # divisor terms is 50.

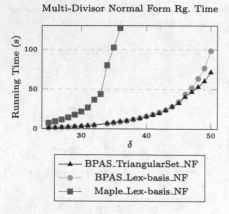

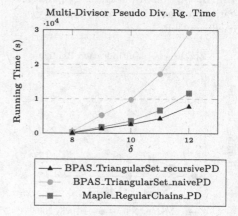

Fig. 9. The divisor set is a random normalized triangular set of $\mathbb{Q}[x_1, x_2, x_3]$ and $deg(a, x_1) - deg(t_3, x_1) = \delta^3$, $deg(a, x_2) - deg(t_2, x_2) = lg(\delta)^3$, $deg(a, x_3) - deg(t_1, x_3) = lg(\delta)^3$ and sparsity 2. BPAS implements Algorithms 5 and 7, see Appendix A.

Fig. 10. The divisor set is a random triangular set of $\mathbb{Q}[x_1, x_2, x_3]$ with non-constant initials, sparsity 2 and $deg(a, x_1) - deg(t_3, x_1) = \delta^3$, $deg(a, x_2) - deg(t_2, x_2) = lg(\delta)^3$, $deg(a, x_3) - deg(t_1, x_3) = lg(\delta)^3$. BPAS uses Algorithms 6 and 8.

5   Conclusion

The open-source library Basic Polynomial Algebra Subprograms (BPAS) provides high performance implementations of sparse multivariate polynomial arithmetic, over $\mathbb{Z}$ and $\mathbb{Q}$, including addition, multiplication, division, and pseudo-division, using highly efficient data structures and algorithms. These fundamental operations were extended to the mid-level algorithms of multi-divisor division (normal form) and multi-divisor pseudo-division. Their performance against the leader in polynomial arithmetic, MAPLE, was shown to be a 2–3 times (or order of magnitude for $\mathbb{Q}$) better. The optimization of these fundamental operations will become the basis for efficient computations with regular chains.

Acknowledgments. The authors would like to thank IBM Canada Ltd (CAS project 880) and NSERC of Canada (CRD grant CRDPJ500717-16).

A   Appendix

Let $\mathbb{K}$ be a field. If B is a Gröbner basis of $\mathbb{K}[x_1, \ldots, x_m]$ Algorithm 5 computes the normal form of a polynomial $a \in \mathbb{K}[x_1, \ldots, x_m]$ (together with the quotients) w.r.t. B; the principle is direct (or naïve). Alternatively, when B is a zero-dimensional normalized (thus so-called Lazard) triangular set, one can use Algorithm 7, the recursive principle of which is taken from [14]. Some details are given here. For computing the normal form of polynomial $a \in \mathbb{K}[x_1, \ldots, x_m]$ with respect to a Lazard triangular set $T = \{t_1, \ldots, t_m\} \subset \mathbb{K}[x_1, \ldots, x_m]$, Algorithm 7 uses the recursive representation of polynomials. If $m = 1$, the result is obtained by applying Algorithm 2. Otherwise, the coefficients of a with respect to x_m are reduced w.r.t. to $\{t_1, \ldots, t_{m-1}\}$ by means of a recursive call (Lines 4–11 of the pseudo-code), yielding a polynomial r. Then, r is divided by t_m by applying Algorithm 2, see Line 12, yielding a new polynomial r. Finally, the coefficients w.r.t. x_m of this new polynomial r are reduced w.r.t. to $\{t_1, \ldots, t_{m-1}\}$, by means of a second recursive call, see Lines 13–16.

Algorithm 5. NORMALFORM (a,B)

Given $a, b_1, \ldots, b_N \in \mathbb{K}[x_1, \ldots, x_m]$, $B = \{b_1, \ldots, b_N\}$ a Gröbner basis, returns $q_1, \ldots, q_N, r \in \mathbb{K}[x_1, \ldots, x_m]$ such that $a = q_1 b_1 + \cdots + q_N b_N + r$ where r is reduced with respect to B.

```
 1: h := a; r := 0
 2: while h ≠ 0 do
 3:     i = 1;
 4:     while i ≤ N do
 5:         if lm(bᵢ) | lm(h) then
 6:             qᵢ := qᵢ + lt(h)/lt(bᵢ)
 7:             h := h − lt(h)/lt(bᵢ) bᵢ
 8:             i := 1
 9:         else
10:             i := i + 1
11:         end if
12:     end while
13:     r := r + lt(h)
14:     h := h − lt(h)
15: end while
16: return (q₁, ..., q_N, r)
```

Algorithm 6. NAÏVETSPD (a,T)

Given $a, t_1, \ldots, t_N \in \mathbb{K}[x_1, \ldots, x_m]$, $T = \{t_1, \ldots, t_N\}$, with $\mathrm{mvar}(t_1) < \cdots < \mathrm{mvar}(t_N)$, returns $q_1, \ldots, q_N, r, h \in \mathbb{K}[x_1, \ldots, x_m]$ such that $ha = q_1 t_1 + \cdots + q_N t_N + r$ where r is reduced with respect to the triangular set T (in the sense that $r = 0$ or $\deg(r, \mathrm{mvar}(t_j)) < \deg(t_j, \mathrm{mvar}(t_j))$, $1 \le j \le N$) and h is a product of powers of the initials of the polynomials of T.

```
 1: r := a; h := 1
 2: for i = 1, ..., N do
 3:     v := mvar(T_{N−i+1})
 4:     (Q, r, e) := PSEUDODIVIDE(r, T_{N−i+1}, v)
 5:     H := init(T_{N−i+1})ᵉ
 6:     h := H h
 7:     for j = 1, ..., N do
 8:         qⱼ := qⱼ H
 9:     end for
10:     qᵢ := qᵢ + Q;
11: end for
12: return (q₁, ..., q_N, r, h)
```

48   M. Asadi et al.

Algorithm 7. TSNF (a, T)

Given $a \in \mathbb{K}[x_1,\ldots,x_m]$, $T = \{t_1,\ldots,t_m\} \subset \mathbb{K}[x_1,\ldots,x_m]$, with $\mathrm{mvar}(t_1) = x_1 < \cdots < \mathrm{mvar}(t_m) = x_m$ and $\mathrm{init}(t_1),\ldots,\mathrm{init}(t_m) \in \mathbb{K}$, returns $q_1,\ldots,q_m, r \in \mathbb{K}[x_1,\ldots,x_m]$ such that $a = q_1 t_1 + \cdots + q_m t_m + r$ where r is reduced (in the sense of Gröbner bases) with respect to the Lazard triangular set T.

1: **if** $m = 1$ **then**
2: $\quad (q_1, r) := \mathrm{DIVIDE}(a, t_1)$
3: **else**
4: $\quad$ **for** $i = 0, \ldots, \deg(a, x_m)$ **do**
5: $\quad\quad (\{Q_j[i]\}_{j=1}^{m-1}, R[i]) := \mathrm{TSNF}(\mathrm{coef}(a, x_m, i), \{t_j\}_{j=1}^{m-1})$
6: $\quad$ **end for**
7: $\quad q_1 := 0; \ldots; q_m := 0$
8: $\quad r := \sum_i R[i] x_m^{\,i}$
9: $\quad$ **for** $j = 1, \ldots, m-1$ **do**
10: $\quad\quad q_j := q_j + \sum_i Q_j[i](x_m)^i$
11: $\quad$ **end for**
12: $\quad (\tilde{q}, r) := \mathrm{DIVIDE}(r, t_m); q_m := q_m + \tilde{q}$
13: $\quad$ **for** $i = 0, \ldots, \deg(r, x_m)$ **do**
14: $\quad\quad (\{Q_j[i]\}_{j=1}^{m-1}, R[i]) := \mathrm{TSNF}(\mathrm{coef}(r, x_m, i), \{t_j\}_{j=1}^{m-1})$
15: $\quad$ **end for**
16: $\quad$ execute Lines 8-11
17: **end if**
18: **return** $(q_1, \ldots, q_m, r)$

Algorithm 8. RECTSPD (a, T)

Same input and output specifications as Algorithm 6.

1: **if** $N = 1$ **then**
2: $\quad (q_1, r, e) := \mathrm{PSEUDODIVIDE}(a, t_1, \mathrm{mvar}(t_1)); h = \mathrm{init}(t_1)^e$
3: **else**
4: $\quad v := \mathrm{mvar}(t_N)$
5: $\quad$ **for** $i = 0, \ldots, \deg(a, v)$ **do**
6: $\quad\quad (\{Q_j[i]\}_{j=1}^{N-1}, R[i], H[i]) := \mathrm{RECTSPD}(\mathrm{coef}(a, v, i), \{t_j\}_{j=1}^{N-1})$
7: $\quad$ **end for**
8: $\quad q_1 := 0; \ldots; q_N := 0$
9: $\quad H_1 := \mathrm{lcm}(H[i], 0 \leq i \leq \deg(a, v))$
10: $\quad r := \sum_i \frac{H_1}{H[i]} R[i] v^i$
11: $\quad$ **for** $j = 1, \ldots, N-1$ **do**
12: $\quad\quad q_j := q_j + \sum_i \frac{H_1}{H[i]} Q_j[i] v^i$
13: $\quad$ **end for**
14: $\quad (\tilde{q}, r, \tilde{e}) := \mathrm{PSEUDODIVIDE}(r, t_N, v); \tilde{h} = \mathrm{init}(t_N)^{\tilde{e}}$
15: $\quad$ **for** $j = 1, \ldots, N-1$ **do**
16: $\quad\quad q_j := q_j \tilde{h}$
17: $\quad$ **end for**
18: $\quad q_N := q_N + \tilde{q}$
19: $\quad$ **for** $i = 0, \ldots, \deg(r, v)$ **do**
20: $\quad\quad (\{Q_j[i]\}_{j=1}^{N-1}, R[i], H[i]) := \mathrm{RECTSPD}(\mathrm{coef}(r, v, i), \{t_j\}_{j=1}^{N-1})$
21: $\quad$ **end for**
22: $\quad H_2 := \mathrm{lcm}(H[i], 0 \leq i \leq \deg(r, v))$
23: $\quad$ **for** $j = 1, \ldots, N$ **do**
24: $\quad\quad q_j := q_j H_2$
25: $\quad$ **end for**
26: $\quad$ execute Lines 10-13 with H_2 replacing H_1
27: $\quad h := H_1 \tilde{h} H_2$
28: **end if**
29: **return** $(q_1, \ldots, q_N, r, h)$

Algorithm 6 is a direct (or naïve) procedure for computing the pseudo-remainder and the pseudo-quotients of a polynomial $a \in \mathbb{K}[x_1,\ldots,x_m]$ by a triangular set $T = \{t_1, \ldots, t_N\}$. Note that T may not be zero-dimensional, that is, $N < m$ may hold. Moreover, T may not be normalized; in particular its initials may not be constant. Algorithm 8 is a recursive version of Algorithm 6 following the same principles as Algorithm 7 and calling Algorithm 4 at Line 14.

References

1. Arnold, A., Roche, D.S.: Output-sensitive algorithms for sumset and sparse polynomial multiplication. In: Proceedings of ISSAC 2015, pp. 29–36 (2015). http://doi.acm.org/10.1145/2755996.2756653
2. Bronstein, M., Moreno Maza, M., Watt, S.: Generic programming techniques in ALDOR. In: Proceedings of AWFS 2007, pp. 72–77 (2007)
3. Chen, C., Covanov, S., Mansouri, F., Maza, M.M., Xie, N., Xie, Y.: Parallel integer polynomial multiplication. CoRR abs/1612.05778 (2016). http://arxiv.org/abs/1612.05778
4. Chen, C., Moreno Maza, M.: Algorithms for computing triangular decomposition of polynomial systems. J. Symb. Comput. **47**(6), 610–642 (2012). https://doi.org/10.1016/j.jsc.2011.12.023
5. Gastineau, M., Laskar, J.: Highly scalable multiplication for distributed sparse multivariate polynomials on many-core systems. In: Gerdt, V.P., Koepf, W., Mayr, E.W., Vorozhtsov, E.V. (eds.) CASC 2013. LNCS, vol. 8136, pp. 100–115. Springer, Cham (2013). https://doi.org/10.1007/978-3-319-02297-0_8
6. Gastineau, M., Laskar, J.: Parallel sparse multivariate polynomial division. In: Proceedings of PASCO 2015, pp. 25–33 (2015). http://doi.acm.org/10.1145/2790282.2790285
7. Granlund, T., et al.: GNU MP 6.0 Multiple Precision Arithmetic Library. Samurai Media Limited (2015)
8. Hall Jr., A.D.: The ALTRAN system for rational function manipulation-a survey. In: Proceedings of the Second ACM Symposium on Symbolic and Algebraic Manipulation, pp. 153–157. ACM (1971)
9. Hart, W., Johansson, F., Pancratz, S.: FLINT: Fast Library for Number Theory, v. 2.4.3. http://flintlib.org
10. van der Hoeven, J., Lecerf, G.: On the bit-complexity of sparse polynomial and series multiplication. J. Symb. Comput. **50**, 227–254 (2013). https://doi.org/10.1016/j.jsc.2012.06.004
11. Johnson, S.C.: Sparse polynomial arithmetic. ACM SIGSAM Bull. **8**(3), 63–71 (1974)
12. Leiserson, C.E.: Cilk. In: Padua, D. (ed.) Encyclopedia of Parallel Computing, pp. 273–288. Springer, Boston (2011). https://doi.org/10.1007/978-0-387-09766-4_289
13. Lemaire, F., Maza, M.M., Xie, Y.: The regularchains library in MAPLE. ACM SIGSAM Bull. **39**(3), 96–97 (2005). https://doi.org/10.1145/1113439.1113456
14. Li, X., Maza, M.M., Schost, É.: Fast arithmetic for triangular sets: from theory to practice. J. Symb. Comput. **44**(7), 891–907 (2009). https://doi.org/10.1016/j.jsc.2008.04.019
15. Monagan, M.B., Pearce, R.: Parallel sparse polynomial multiplication using heaps. In: ISSAC, pp. 263–270 (2009)
16. Monagan, M., Pearce, R.: Polynomial division using dynamic arrays, heaps, and packed exponent vectors. In: Ganzha, V.G., Mayr, E.W., Vorozhtsov, E.V. (eds.) CASC 2007. LNCS, vol. 4770, pp. 295–315. Springer, Heidelberg (2007). https://doi.org/10.1007/978-3-540-75187-8_23

17. Monagan, M., Pearce, R.: Parallel sparse polynomial division using heaps. In: Proceedings of PASCO 2010, pp. 105–111. ACM (2010)
18. Monagan, M., Pearce, R.: The design of Maple's sum-of-products and POLY data structures for representing mathematical objects. ACM Commun. Comput. Algebra **48**(3/4), 166–186 (2015)
19. Monagan, M.B., Pearce, R.: Sparse polynomial division using a heap. J. Symb. Comput. **46**(7), 807–822 (2011). https://doi.org/10.1016/j.jsc.2010.08.014
20. Moreno Maza, M., Xie, Y.: Balanced dense polynomial multiplication on multicores. Int. J. Found. Comput. Sci. **22**(5), 1035–1055 (2011)
21. Shoup, V., et al.: NTL: A library for doing number theory. www.shoup.net/ntl/

Computation of Pommaret Bases
Using Syzygies

Bentolhoda Binaei[1], Amir Hashemi[1,2]([✉]), and Werner M. Seiler[3] (iD)

[1] Department of Mathematical Sciences, Isfahan University of Technology,
84156-83111 Isfahan, Iran
h.binaei@math.iut.ac.ir, Amir.Hashemi@cc.iut.ac.ir
[2] School of Mathematics, Institute for Research in Fundamental Sciences (IPM),
19395-5746 Tehran, Iran
[3] Institut für Mathematik, Universität Kassel, Heinrich-Plett-Straße 40,
34132 Kassel, Germany
seiler@mathematik.uni-kassel.de

Abstract. We investigate the application of syzygies for efficiently computing (finite) Pommaret bases. For this purpose, we first describe a nontrivial variant of Gerdt's algorithm [10] to construct an involutive basis for the input ideal as well as an involutive basis for the syzygy module of the output basis. Then we apply this new algorithm in the context of Seiler's method to transform a given ideal into quasi stable position to ensure the existence of a finite Pommaret basis [19]. This new approach allows us to avoid superfluous reductions in the iterative computation of Janet bases required by this method. We conclude the paper by proposing an involutive variant of the signature based algorithm of Gao et al. [8] to compute simultaneously a Gröbner basis for a given ideal and for the syzygy module of the input basis. All the presented algorithms have been implemented in MAPLE and their performance is evaluated via a set of benchmark ideals.

1 Introduction

Gröbner bases provide a powerful computational tool for a wide variety of problems connected to multivariate polynomial ideals. Together with the first algorithm to compute them, they were introduced by Buchberger in his PhD thesis [3]. Later on, he discovered two criteria to improve his algorithm [2] by omitting superfluous reductions. In 1983, Lazard [15] developed a new approach by using linear algebra techniques to compute Gröbner bases. In 1988, Gebauer and Möller [9], by interpreting Buchberger's criteria in terms of syzygies, presented an efficient way to improve Buchberger's algorithm. Furthermore, Möller et al. [16] extended this idea and described the first *signature-based* algorithm to compute Gröbner bases. In 1999, Faugère [6], by applying fast linear algebra on sparse matrices, found his F_4 algorithm to compute Gröbner bases. Then, he introduced the well-known F_5 algorithm [7] that uses two new criteria (F_5 and

© Springer Nature Switzerland AG 2018
V. P. Gerdt et al. (Eds.): CASC 2018, LNCS 11077, pp. 51–66, 2018.
https://doi.org/10.1007/978-3-319-99639-4_4

IsRewritten) based on the idea of signatures and that performs no useless reduction as long as the input polynomials define a (semi-)regular sequence. Finally, Gao et al. [8] presented a new approach to compute simultaneously Gröbner bases for an ideal and its syzygy module.

Involutive bases may be considered as a special kind of non-reduced Gröbner bases with additional combinatorial properties. They originate from the works of Janet [14] on the analysis of partial differential equations. By evolving related methods used by Pommaret [17], the notion of *involutive polynomial bases* was introduced by Zharkov and Blinkov [22]. Later, Gerdt and Blinkov [11] generalised these ideas to the concepts of *involutive divisions* and *involutive bases* for polynomial ideals to produce an effective alternative approach to Buchberger's algorithm (for the efficiency analysis of an implementation of Gerdt's algorithm [10], we refer to the web pages http://invo.jinr.ru). Recently, Gerdt et al. [12] proposed a signature-based approach to compute involutive bases.

In this article we discuss effective approaches to compute involutive bases and in particular *Pommaret bases*. These bases are a special kind of involutive bases introduced by Zharkov and Blinkov [22]. While finite Pommaret bases do not always exist, every ideal in a sufficiently generic position has one (see [13] for an extensive discussion of this topic). A finite Pommaret basis reflects many (homological) properties of the ideal it generates. For example, many invariants like dimension, depth and Castelnuovo-Mumford regularity can be easily read off from it. We note that all these invariants remain unchanged under coordinate transformations. We refer to [20] for a comprehensive overview of the theory and applications of Pommaret bases.

We first propose a variant of Gerdt's algorithm to compute an involutive basis which simultaneously determines an involutive basis for the syzygy module of the output basis. Based on it, we improve Seiler's method [19] to compute a linear change of coordinates which brings the input ideal into a generic position so that the new ideal has a finite Pommaret basis. Then, as a related work, we describe an involutive version of the approach by Gao et al. [8] to compute simultaneously Gröbner bases of a given ideal and of the syzygy module of the input basis. All the algorithms described in this paper have been implemented in MAPLE and their efficiency is illustrated via a set of benchmark ideals.

This paper is organized as follows. In Sect. 2, we review basic definitions and notations related to involutive bases. Section 3 is devoted to a variant of Gerdt's algorithm which also computes an involutive basis for the syzygy module of the output basis. In Sect. 4, we show how to apply it in the computation of Pommaret bases. Finally in Sect. 5, we conclude by presenting an involutive variant of the algorithm of Gao et al. by combining it with Gerdt's algorithm.

2 Preliminaries

In this section, we review basic notations and preliminaries needed in the subsequent sections. Throughout this paper, we assume that $\mathcal{P} = \Bbbk[x_1, \ldots, x_n]$ is the polynomial ring over an infinite field $\Bbbk$. We consider polynomials $f_1, \ldots, f_k \in \mathcal{P}$

and the ideal $\mathcal{I} = \langle f_1, \ldots, f_k \rangle$ generated by them. The total degree and the degree w.r.t. a variable x_i of a polynomial in $f \in \mathcal{P}$ are denoted by $\deg(f)$ and $\deg_i(f)$, respectively. In addition, $\mathcal{M} = \{x_1^{\alpha_1} \cdots x_n^{\alpha_n} \mid \alpha_i \geq 0, 1 \leq i \leq n\}$ stands for the monoid of all monomials in $\mathcal{P}$. We use throughout the reverse degree lexicographic ordering with $x_n \prec \cdots \prec x_1$. The leading monomial of a given polynomial $f \in \mathcal{P}$ w.r.t. $\prec$ is denoted by $\mathrm{LM}(f)$. If $F \subset \mathcal{P}$ is a finite set of polynomials, $\mathrm{LM}(F)$ denotes the set $\{\mathrm{LM}(f) \mid f \in F\}$. The leading coefficient of f, denoted by $\mathrm{LC}(f)$, is the coefficient of $\mathrm{LM}(f)$. The leading term of f is defined to be $\mathrm{LT}(f) = \mathrm{LM}(f)\,\mathrm{LC}(f)$. A finite set $G = \{g_1, \ldots, g_t\} \subset \mathcal{P}$ is called a *Gröbner basis* of $\mathcal{I}$ w.r.t $\prec$ if $\mathrm{LM}(\mathcal{I}) = \langle \mathrm{LM}(g_1), \ldots, \mathrm{LM}(g_t) \rangle$ where $\mathrm{LM}(\mathcal{I}) = \langle \mathrm{LM}(f) \mid f \in \mathcal{I} \rangle$. We refer e.g. to the book of Cox et al. [4] for further details on Gröbner bases.

An analogous notion of Gröbner bases may be defined for sub-modules of $\mathcal{P}^t$ for some t, see [5]. In this direction, let us recall some basic notations and results. Let $\{\mathbf{e}_1, \ldots, \mathbf{e}_t\}$ be the standard basis of $\mathcal{P}^t$. A module monomial in $\mathcal{P}^t$ is an element of the form $x^\alpha \mathbf{e}_i$ for some i, where x^α is a monomial in $\mathcal{P}$. So, each $f \in \mathcal{P}^t$ can be written as a $\Bbbk$-linear combination of module monomials in $\mathcal{P}^t$. A total ordering $<$ on the set of monomials of $\mathcal{P}^t$ is called a *module monomial ordering* if the following conditions are satisfied:

- if $\mathbf{m}$ and $\mathbf{n}$ are two module monomials such that $\mathbf{n} < \mathbf{m}$ and $x^\alpha \in \mathcal{P}$ is a monomial then $x^\alpha \mathbf{n} < x^\beta \mathbf{m}$,
- $<$ is a well-ordering.

In addition, we say that $x^\alpha \mathbf{e}_i$ divides $x^\beta \mathbf{e}_j$ if $i = j$ and x^α divides x^β. Based on these definitions, one is able to extend the theory of Gröbner bases to sub-modules of the $\mathcal{P}$-modules of finite rank. Some well-known examples of module monomial orderings are term over position (TOP), position over term (POT) and the Schreyer ordering.

Definition 1. *Let $\{g_1, \ldots, g_t\} \subset \mathcal{P}$ and $\prec$ a monomial ordering on $\mathcal{P}$. We define the* Schreyer module ordering *on $\mathcal{P}^t$ as follows: We write $x^\alpha \mathbf{e}_i \prec_s x^\beta \mathbf{e}_j$ if either $\mathrm{LM}(x^\alpha g_i) \prec \mathrm{LM}(x^\beta g_j)$, or $\mathrm{LM}(x^\alpha g_i) = \mathrm{LM}(x^\beta g_j)$ and $j < i$.*

Schreyer proposed in his master thesis [18] a slight modification of Buchberger's algorithm to compute a Gröbner basis for the syzygy module of a Gröbner basis.

Definition 2. *Let us consider $G = (g_1, \ldots, g_t) \in \mathcal{P}^t$. The (first) syzygy module of G is defined to be $\mathrm{Syz}(G) = \{(h_1, \ldots, h_t) \mid h_i \in \mathcal{P}, \sum_{i=1}^t h_i g_i = 0\}$.*

Let $G = \{g_1, \ldots, g_t\}$ be a Gröbner basis. By Buchberger's criterion, each S-polynomial has a standard representation: $\mathrm{SPoly}(g_i, g_j) = a_{ji} m_{ji} g_i - a_{ij} m_{ij} g_j = h_{ij1} g_1 + \cdots + h_{ijt} g_t$ where $a_{ji}, a_{ij} \in \Bbbk$, $h_{ijl} \in \mathcal{P}$ and m_{ji}, m_{ij} are monomials. Let $\mathbf{S}_{ij} = a_{ji} m_{ji} \mathbf{e}_i - a_{ij} m_{ij} \mathbf{e}_j - h_{ij1} \mathbf{e}_1 - \cdots - h_{ijt} \mathbf{e}_t$ be the corresponding syzygy.

Theorem 1 (Schreyer's Theorem). *With the above introduced notations, the set $\{\mathbf{S}_{ij} \mid 1 \leq i < j \leq t\}$ is a Gröbner basis for $\mathrm{Syz}(g_1, \ldots, g_t)$ w.r.t. $\prec_s$.*

Example 1. Let $F = \{xy - x, x^2 - y\} \subset \Bbbk[x, y]$. The Gröbner basis of F w.r.t. $x \prec_{dlex} y$ is $G = \{g_1 = xy - x, g_2 = x^2 - y, g_3 = y^2 - y\}$ and the Gröbner basis of $\mathrm{Syz}(g_1, g_2, g_3)$ is $\{(x, -y + 1, -1), (-x, y^2 - 1, -x^2 + y + 1), (y, 0, -x)\}$.

If $F = \{f_1, \ldots, f_k\}$ is *not* a Gröbner basis, Wall [21] proposed an effective method to compute $\mathrm{Syz}(F)$. If the extended set $G = f_1, \ldots, f_k, f_{k+1}, \ldots, f_t$ is a Gröbner basis of $\langle F \rangle$, then $\mathrm{Syz}(F) = \{A\mathbf{s} \mid \mathbf{s} \in \mathrm{Syz}(G)\}$ where A is a matrix such that $G = FA$.

We conclude this section by recalling some definitions and results from the theory of involutive bases (see [10, 20] for more details). Given a set of polynomials, an involutive division partitions the variables into two disjoint subsets of *multiplicative* and *non-multiplicative* variables.

Definition 3. *An* involutive division $\mathcal{L}$ *is given on* $\mathcal{M}$ *if for any finite set* $U \subset \mathcal{M}$ *and any* $u \in U$, *the set of variables is partitioned into the subsets of multiplicative variables* $M_{\mathcal{L}}(u, U)$ *and non-multiplicative variables* $NM_{\mathcal{L}}(u, U)$ *such that the following conditions hold where* $\mathcal{L}(u, U)$ *denotes the monoid generated by* $M_{\mathcal{L}}(u, U)$:

1. $v, u \in U$, $u\mathcal{L}(u, U) \cap v\mathcal{L}(v, U) \neq \emptyset \Rightarrow u \in v\mathcal{L}(v, U)$ *or* $v \in u\mathcal{L}(u, U)$,
2. $v \in U$, $v \in u\mathcal{L}(u, U) \Rightarrow \mathcal{L}(v, U) \subset \mathcal{L}(u, U)$,
3. $V \subset U$ *and* $u \in V \Rightarrow \mathcal{L}(u, U) \subset \mathcal{L}(u, V)$.

We shall write $u \mid_{\mathcal{L}} w$ *if* $w \in u\mathcal{L}(u, U)$. *In this case,* u *is called an* $\mathcal{L}$-involutive *divisor of* w *and* w *an* $\mathcal{L}$-involutive *multiple of* u.

We recall the definitions of the Janet and Pommaret division, respectively.

Example 2. Let $U \subset \mathcal{P}$ be a finite set of monomials. For each sequence $d_1, \ldots, d_n$ of non-negative integers and for each $1 \leq i \leq n$ we define

$$[d_1, \ldots, d_i] = \{u \in U \mid d_j = \deg_j(u), \ 1 \leq j \leq i\}.$$

The variable x_1 is Janet multiplicative (denoted by $\mathcal{J}$-multiplicative) for $u \in U$ if $\deg_1(u) = \max\{\deg_1(v) \mid v \in U\}$. For $i > 1$ the variable x_i is Janet multiplicative for $u \in [d_1, \ldots, d_{i-1}]$ if $\deg_i(u) = \max\{\deg_i(v) \mid v \in [d_1, \ldots, d_{i-1}]\}$.

Example 3. For $u = x_1^{d_1} \cdots x_k^{d_k}$ with $d_k > 0$ the variables $\{x_k, \ldots, x_n\}$ are considered as Pommaret multiplicative (denoted by $\mathcal{P}$-multiplicative) and the other variables as Pommaret non-multiplicative. For $u = 1$ all the variables are multiplicative. The integer k is called the *class* of u and is denoted by $\mathrm{cls}(u)$.

Definition 4. *The set* $F \subset \mathcal{P}$ *is called* involutively head autoreduced *if for each* $f \in F$ *there is no* $h \in F \setminus \{f\}$ *with* $\mathrm{LM}(h) \mid_{\mathcal{L}} \mathrm{LM}(f)$.

Definition 5. *Let* $I \subset \mathcal{P}$ *be an ideal and* $\mathcal{L}$ *an involutive division. An involutively head autoreduced subset* $H \subset I$ *is an* involutive basis *for* I *if for all* $f \in I$ *there exists* $h \in H$ *so that* $\mathrm{LM}(h) \mid_{\mathcal{L}} \mathrm{LM}(f)$.

Example 4. For the ideal $\mathcal{I} = \langle xy, y^2, z \rangle \subset \mathbb{k}[x, y, z]$ the set $\{xy, y^2, z, xz, yz\}$ is a Janet basis, but there exists only an infinite Pommaret basis of the form $\{xy, y^2, z, xz, yz, x^2y, x^2z, \ldots, x^ky, x^kz, \ldots\}$. One can show that every ideal has a finite Janet basis, i. e. the Janet division is Noetherian.

Gerdt [10] proposed an efficient algorithm to construct involutive bases using a completion process where prolongations of given elements by non-multiplicative variables are reduced. This process terminates in finitely many steps for any Noetherian division. In addition, Seiler [19] characterized the ideals having finite Pommaret bases by relating them to the notion of quasi stability. More precisely, a given ideal has a finite Pommaret basis iff it is in *quasi stable position* (or equivalently if the coordinates are δ-regular) see [19, proposition 4.4].

Definition 6. *A monomial ideal* $\mathcal{I}$ *is called* quasi stable *if for any monomial* $m \in \mathcal{I}$ *and all integers* i, j, s *with* $1 \leq j < i \leq n$ *and* $s > 0$, *if* $x_i^s \mid m$ *there exists an integer* $t \geq 0$ *such that* $x_j^t m / x_i^s \in \mathcal{I}$. *A homogeneous ideal* $\mathcal{I}$ *is in* quasi stable position *if* $\mathrm{LM}(\mathcal{I})$ *is quasi stable.*

3   Computation of Involutive Basis for Syzygy Module

We present now an effective approach to compute, for a given ideal, simultaneously involutive bases of the ideal and of its syzygy module. We first recall some related concepts and facts from [19]. In loc. cit., an involutive version of Schreyer's theorem is stated where S-polynomials are replaced by non-multiplicative prolongations and an involutive normal form algorithm is used.

More precisely, let $H \subset \mathcal{P}^t$ be a finite set for some $t \in \mathbb{N}$, $\prec_s$ the corresponding Schreyer ordering and $\mathcal{L}$ an involutive division. We divide H into t disjoint subsets $H_i = \{\mathbf{h} \in H \mid \mathrm{LM}(\mathbf{h}) = x^\alpha \mathbf{e}_i, x^\alpha \in \mathcal{M}\}$. In addition, for each i, let $B_i = \{x^\alpha \in \mathcal{M} \mid x^\alpha \mathbf{e}_i \in \mathrm{LM}(H_i)\}$. We assign to each $\mathbf{h} \in H_i$ the multiplicative variables $M_{\mathcal{L}, H, \prec}(\mathbf{h}) = \{x_i \mid x_i \in M_{\mathcal{L}, B_i}(x^\alpha) \text{ with } \mathrm{LM}(\mathbf{h}) = x^\alpha \mathbf{e}_i\}$. Then, the definition of involutive bases for sub-modules proceeds as for ideals.

Let $H = \{h_1, \ldots, h_t\} \subset \mathcal{P}$ be an involutive basis. Let $h_i \in H$ be an arbitrary element and x_k a non-multiplicative variable of it. From the definition of involutive bases, there exists a unique j such that $\mathrm{LM}(h_j) \mid x_k \mathrm{LM}(h_i)$. We order the elements of H in such a way that $i < j$ (which is always possible for a continuous division [19, Lemma 5.5]). Then we find a unique involutive standard representation $x_k h_i = \sum_{j=1}^t p_j^{(i,k)} h_j$ where $p_j^{(i,k)} \in \mathbb{k}[M_{\mathcal{L}, H, \prec}(h_j)]$ and the corresponding syzygy $\mathbf{S}_{i,k} = x_k \mathbf{e}_i - \sum_{j=1}^t p_j^{(i,k)} \mathbf{e}_j \in \mathcal{P}^t$. We denote the set of all thus obtained syzygies by $H_{\mathrm{Syz}} = \{\mathbf{S}_{i,k} \mid 1 \leq i \leq t; x_k \in \mathrm{NM}_{\mathcal{L}, H, \prec}(h_i)\}$. An involutive division $\mathcal{L}$ is of *Schreyer type* if all sets $NM_{\mathcal{L}, H, \prec}(h)$ with $h \in H$ are again involutive bases for the ideals defined by them. Both the Janet and the Pommaret divisions are of Schreyer type.

Theorem 2. (*[19, Theorem 5.10]*) *With the above notations, let* $\mathcal{L}$ *be a continuous involutive division of Schreyer type w.r.t.* $\prec$ *and* H *an involutive basis. Then* H_{Syz} *is an* $\mathcal{L}$-*involutive basis for* $\mathrm{Syz}(H)$ *w.r.t.* $\prec_s$.

We now present a non-trivial variant of Gerdt's algorithm [10] computing simultaneously a minimal involutive basis for the input ideal and an involutive basis for the syzygy module of this basis. It uses an analogous idea as the algorithm given in [1]. However, since we aim at determining also a syzygy module, we must save the traces of all reductions and for this reason we cannot use the syzygies to remove useless reductions.

Algorithm 1. INVBASIS

Input: A finite set $F \subset \mathcal{P}$; an involutive division $\mathcal{L}$; a monomial ordering $\prec$
Output: A minimal $\mathcal{L}$-basis for $\langle F \rangle$ and an $\mathcal{L}$-basis for syzygy module of this basis.
 1: $F :=$sort$(F, \prec)$
 2: $T := \{(F[1], F[1], \emptyset, \mathbf{e}_1, false)\}$
 3: $Q := \{(F[i], F[i], \emptyset, \mathbf{e}_i, false) \mid i = 2, \ldots, |F|\}$
 4: $S := \{\}$ and $j := |F|$
 5: **while** $Q \neq \emptyset$ **do**
 6:     $Q :=$sort$(Q, \prec_s)$
 7:     select and remove $p := Q[1]$ from Q
 8:     $h :=$ INVNORMALFORM$(p, T, \mathcal{L}, \prec)$
 9:     **if** $h[1] = 0$ **then**
10:         $S := S \cup \{h[2]\}$
11:     **end if**
12:     **if** $h[1] = 0$ and LM(Poly(p)) = LM(Anc(p)) **then**
13:         $Q := \{q \in Q \mid$ Anc$(q) \neq$ Poly(p) or $q[5] = true\}$
14:     **end if**
15:     **if** $p[5] = true$ **then**
16:         $q :=$UPDATE(q, p) for each $q \in T$
17:     **end if**
18:     **if** $h[1] \neq 0$ and LM(Poly(p)) $\neq$ LM(h) **then**
19:         **for** $q \in T$ with proper conventional division LM$(h[1])$ | LM(Poly(q)) **do**
20:             $Q := Q \cup \{[q[1], q[2], q[3], q[4], true]\}$
21:             $T := T \setminus \{q\}$
22:         **end for**
23:         $j := j + 1$ and $T := T \cup \{(h[1], h[1], \emptyset, \mathbf{e}_j, false)\}$
24:     **else**
25:         $T := T \cup \{(h[1], \text{Anc}(p), \text{NM}(p), h[2], false)\}$
26:     **end if**
27:     **for** $q \in T$ and $x \in NM_{\mathcal{L}}($LM(Poly(q)), LM(Poly(T))$) \setminus$ NM(q) **do**
28:         $Q := Q \cup \{(x. \text{Poly}(q), \text{Anc}(q), \emptyset, x. \text{Rep}(q), false)\}$
29:         NM$(q) :=$ NM$(q) \cup NM_{\mathcal{L}}($LM(Poly$(q)$), LM(Poly$(T)$)$) \cup \{x\}$
30:     **end for**
31: **end while**
32: **return** $(\text{Poly}(T), \{\text{Rep}(p) - \mathbf{e}_{\text{index}(p)} \mid p \in T\} \cup S)$

The algorithm INVBAS relies on the following data structure for polynomials. To each polynomial f, we associate a quintuple $p = (f, g, V, \mathbf{q}, flag)$. The first entry $f = $ Poly(p) is the polynomial itself, $g = $ Anc(p) is the ancestor of f

(realised as a pointer to the quintuple associated with the ancestor) and $V = \mathrm{NM}(p)$ is its list of already processed non-multiplicative variables. The fourth entry $\mathbf{q} = \mathrm{Rep}(p)$ denotes the representation of f in our current basis, i.e. if $\mathbf{q} = \sum_{r \in T \cup Q} h_r \mathbf{e}_{\mathrm{index}(r)}$ then $f = \sum_{r \in T \cup Q} h_r \, \mathrm{Poly}(r)$ where $h_r \in \mathcal{P}$ and $\mathrm{index}(r)$ gives the position of r in the current list $T \cup Q$. The final entry is a boolean flag. If $flag = true$ then at some stage of the algorithm p has been moved from T to Q, otherwise $flag = false$. We denote by $\mathrm{Sig}(p) = \mathrm{LM}_{\prec_s}(\mathrm{Rep}(p))$ the signature of p. By an abuse of notation, $\mathrm{Sig}(f)$ also denotes $\mathrm{Sig}(p)$. The same holds for the Rep function. If P is a set of quintuples, we denote by $\mathrm{Poly}(P)$ the set $\{\mathrm{Poly}(p) \mid p \in P\}$. In addition, the functions $\mathrm{sort}(X, \prec)$ and $\mathrm{sort}(X, \prec_s)$ sort X in increasing order according to $\mathrm{LM}(X)$ w.r.t. $\prec$ and $\{\mathrm{Sig}(p) \mid p \in X\}$ w.r.t. $\prec_s$, respectively. We remark that in the original form of Gerdt's algorithm [10] the function $\mathrm{sort}(Q, \prec)$ was applied to sort the set of all non-multiplicative prolongations, however, in our experiments we observed that using $\mathrm{sort}(Q, \prec_s)$ increased the performance of the algorithm.

Obviously, the representation of each polynomial must be updated whenever the set $T \cup Q$ changes in a non-trivial way. We remark that elements of Q can appear non-trivially in the representations of polynomials only if they have been elements of T at an earlier stage of the algorithm (recall that such a move is noted in the flag of each quintuple), as all reductions are performed w.r.t. T only. If updates are necessary, then they are performed by the function UPDATE. Involutive normal forms are computed with the help of the following subalgorithm taking care of the representations.

Algorithm 2. INVNORMALFORM

Input: A quintuple p; a set of quintuples T; a division $\mathcal{L}$; a monomial ordering $\prec$
Output: A normal form of p w.r.t. T and its new representation.
  $h := \mathrm{Poly}(p)$ and $G := \mathrm{Poly}(T)$ and $\mathbf{q} := \mathrm{Rep}(p)$
  while h contains a monomial m which is $\mathcal{L}$-divisible by $g \in G$ **do**
    if $m = \mathrm{LM}(\mathrm{Poly}(p))$ and $\mathrm{C1}(h, g)$ **then**
      return $([0, \mathrm{Anc}(p)\,\mathrm{Rep}(\mathrm{Anc}(g)) - \mathrm{Anc}(g)\,\mathrm{Rep}(\mathrm{Anc}(p))])$
    end if
    $h := h - (cm/\mathrm{LT}(g)).g$ where c is the coefficient of m in h
    $\mathbf{q} := \mathbf{q} - (cm/\mathrm{LT}(g))\,\mathrm{Rep}(g)$
  end while
  return $([h, \mathbf{q}])$

Here we apply the involutive form of Buchberger's first criterion [10]. We say that $\mathrm{C1}(p, g)$ is true if $\mathrm{LM}(\mathrm{Anc}(p))\,\mathrm{LM}(\mathrm{Anc}(g)) = \mathrm{LM}(\mathrm{Poly}(p))$.

Theorem 3. *If $\mathcal{L}$ is a Noetherian continuous involutive division of Schreyer type then* INVBASIS *terminates in finitely many steps and returns a minimal involutive basis for its input ideal and also an involutive basis for the syzygy module of the constructed basis.*

Proof. The termination of the algorithm is ensured by the termination of Gerdt's algorithm, see [10]. Let us now deal with its correctness. We first note that if an element p is removed by Buchberger's criteria, then it is superfluous and by [10, Theorem 2] the set $\mathrm{Poly}(T)$ forms a minimal involutive basis for $\langle F \rangle$. Thus, it remains to show that $R = \{\mathrm{Rep}(p) - \mathbf{e}_{\mathrm{index}(p)} \mid p \in T\} \cup S$ is an involutive basis for $\mathrm{Poly}(T) = \{h_1, \ldots, h_t\}$ w.r.t. $\prec_s$. Using Theorem 2, we must show that the representation of each non-multiplicative prolongation of the elements of $\mathrm{Poly}(T)$ appears in R. Let us consider $h_i \in \mathrm{Poly}(T)$ and a non-multiplicative variable x_k for it. Then, due to the structure of the algorithm, $x_k h_i$ is created and studied in the course of the algorithm.

Now, four cases can occur. If $x_k h_i$ reduces to zero then we can write $x_k h_i = \sum_{j=1}^{t} p_j^{(i,k)} h_j$ where $p_j^{(i,k)} \in \Bbbk[M_{\mathcal{L},H,\prec}(h_j)]$. Therefore the representation $x_k \mathbf{e}_i - \sum_{j=1}^{t} p_j^{(i,k)} \mathbf{e}_j \in \mathcal{P}^t$ is added to S and consequently it appears in R. If the involutive normal form of $x_k h_i$ is non-zero then we can write $x_k h_i = \sum_{j=1}^{t} p_j^{(i,k)} h_j + h_\ell$ where $p_j^{(i,k)} \in \Bbbk[M_{\mathcal{L},H,\prec}(h_j)]$. In this case, we add h_ℓ into T and the representation component of $x_k h_i$ is updated to $x_k \mathbf{e}_i - \sum_{j=1}^{t} p_j^{(i,k)} \mathbf{e}_j$. Then, as we can see in the output of the algorithm, $x_k \mathbf{e}_i - \sum_{j=1}^{t} p_j^{(i,k)} \mathbf{e}_j - \mathbf{e}_\ell$ appears in R as the syzygy corresponding to $x_k h_i$.

The third case that may occur is that $x_k h_i$ is removed by Buchberger's first criterion. Assume that p is the quintuple associated to $x_k h_i$ and g is another quintuple so that $\mathrm{C1}(p,g)$ is true. It follows that $\mathrm{LM}(\mathrm{Anc}(p))\, \mathrm{LM}(\mathrm{Anc}(g)) = \mathrm{LM}(\mathrm{Poly}(p))$ holds. We may let $x_k h_i = u\, \mathrm{Anc}(p)$, $\mathrm{Poly}(g) = v\, \mathrm{Anc}(g)$ and $\mathrm{LM}(x_k h_i) = m\, \mathrm{LM}(g)$ for some monomials u and v and term m (assume that the polynomials are monic). Thus,

$$x_k h_i - m\, \mathrm{Poly}(g) = u\, \mathrm{Anc}(p) - mv\, \mathrm{Anc}(g).$$

As $\mathrm{LM}(\mathrm{Anc}(p))\, \mathrm{LM}(\mathrm{Anc}(g)) = \mathrm{LCM}(\mathrm{LM}(\mathrm{Anc}(p)), \mathrm{LM}(\mathrm{Anc}(g)))$, Buchberger's first criterion applied to $\mathrm{Anc}(p)$ and $\mathrm{Anc}(g)$ yields that $\mathrm{Anc}(p)\, \mathrm{Rep}(\mathrm{Anc}(g)) - \mathrm{Anc}(g)\, \mathrm{Rep}(\mathrm{Anc}(p))$ is the corresponding syzygy which is added to S.

The last case to be considered is that $x_k h_i$ is removed by the second **if**-loop in the main algorithm. In this case, we conclude that $\mathrm{Anc}(p)$ is reduced to zero and in consequence h_i is reduced to zero. So, h_i is a useless polynomial and we do not need to keep $x_k h_i$ which ends the proof.   □

Remark 1. There also exists an involutive version of Buchberger's second criterion [10]: $\mathrm{C2}(p,g)$ is true if $\mathrm{LCM}(\mathrm{LM}(\mathrm{Anc}(p)), \mathrm{LM}(\mathrm{Anc}(g)))$ properly divides $\mathrm{LM}(\mathrm{Poly}(p))$. We cannot use this criterion in the INVNORMALFORM algorithm. A non-multiplicative prolongation $x_k h_i$ removed by it is surely useless in the sense that it is not needed for determining the involutive basis of $\mathcal{I}$, but it can nevertheless be necessary for the construction of its syzygy module.

Example 5. Let us consider the ideal $\mathcal{I}$ generated by $F = \{f_1 = z^2, f_2 = zy, f_3 = xz - y, f_4 = y^2, f_5 = xy - y, f_6 = x^2 - x + z\} \subset \Bbbk[x, y, z]$ from [19, Example 5.6]. Then, F is a Janet basis w.r.t. $z \prec y \prec x$. Since x, y are non-multiplicative variables for f_1, f_2, f_3 and x is non-multiplicative variable for f_4, f_5 then the following set is a Janet basis for the syzygy module of F: $\{y\mathbf{e}_1 - z\mathbf{e}_2, x\mathbf{e}_1 - z\mathbf{e}_3 - \mathbf{e}_2, y\mathbf{e}_2 - z\mathbf{e}_4, x\mathbf{e}_2 - z\mathbf{e}_5 - \mathbf{e}_2, y\mathbf{e}_3 - z\mathbf{e}_5 + \mathbf{e}_4 - \mathbf{e}_2, x\mathbf{e}_3 - z\mathbf{e}_6 + \mathbf{e}_5 - \mathbf{e}_3 + \mathbf{e}_1, x\mathbf{e}_4 - y\mathbf{e}_5 - \mathbf{e}_4, x\mathbf{e}_5 - y\mathbf{e}_6 + \mathbf{e}_2\}$.

4   Application to Pommaret Basis Computation

In this section we show how to apply the approach presented in the preceding section in the computation of Pommaret bases. The Pommaret division is not Noetherian and thus a given ideal may not have a finite Pommaret basis. However, a generic linear change of variables transforms the ideal into quasi stable position where a finite Pommaret basis exists. Seiler [19] proposed a deterministic algorithm to compute such a linear change by performing repeatedly an *elementary* linear change and then a test on the Janet basis of the transformed ideal. Now, to apply the method presented in this paper, we use the INVBASIS algorithm to compute a minimal Janet basis H for the input ideal and at the same time a Janet basis for $\mathrm{Syz}(H)$. Then, for each $h \in H$ we check whether there exists a variable which is Janet but not Pommaret multiplicative. If not, H is a Pommaret basis and we are done. Otherwise, we make an elementary linear change of variables, say ϕ. Then, we apply the following algorithm, NEXTINVBASIS, to compute a minimal Janet basis for the ideal generated by $\phi(H)$ by applying $\phi(\mathrm{Syz}(H))$ to remove superfluous reductions. We describe first the main procedure.

Algorithm 3. QUASISTABLE

Input: A finite set $F \subset \mathcal{P}$ of homogeneous polynomials and a monomial ordering $\prec$
Output: A linear change Φ so that $\langle \Phi(F) \rangle$ has a finite Pommaret basis
  $\Phi :=$ the identity map
  $J, S :=$ INVBASIS$(F, \mathcal{J}, \prec)$ and $A :=$ TEST$(\mathrm{LM}(J))$
  while $A \neq true$ **do**
    $\phi := A[3] \mapsto A[3] + cA[2]$ for a random choice of $c \in \Bbbk$
    $Temp :=$ NEXTINVBASIS$(\Phi \circ \phi(J), \Phi \circ \phi(S), \mathcal{J}, \prec)$
    $B :=$ TEST$(\mathrm{LM}(Temp))$
    if $B \neq A$ **then**
      $\Phi := \Phi \circ \phi$ and $A := B$
    end if
  end while
  return (Φ)

The function TEST receives a set of monomials forming a minimal Janet basis and returns true if it is a Pommaret basis, too. Otherwise, by [19, Proposition 2.10], there exists a monomial m in the set for which a Janet multiplicative variable (say x_ℓ) is not Pommaret multiplicative. In this case, the function returns $(false, x_\ell, \mathrm{cls}(m))$. Using these variables, we construct an elementary linear change of variables.

The NEXTINVBASIS algorithm is similar to the INVBASIS algorithm given above. However, the new algorithm computes only the involutive basis of the input ideal generated by a set H. In addition, in the new algorithm, we use $\mathrm{Syz}(H)$ to remove useless reductions. Below, only the differences between the two algorithms are exhibited.

Algorithm 4. NEXTINVBASIS

Input: A finite set $F \subset \mathcal{P}$; a generating set S for $\mathrm{Syz}(F)$; an involutive division $\mathcal{L}$; a monomial ordering $\prec$
Output: A minimal involutive basis for $\langle F \rangle$

   ⋮   {Lines 1–6 of INVBASIS}
   select and remove $p := Q[1]$ from Q
   if $\nexists \mathbf{s} \in S$ s.t $\mathrm{LM}_{\prec_s}(\mathbf{s}) \mid \mathrm{Sig}(p)$ then

      ⋮   {Lines 8–30 of INVBASIS}
   end if
   ⋮   {Lines 31/32 of INVBASIS}

Lemma 1. *Let $H \subset \mathcal{P}$ and S be a generating set for $\mathrm{Syz}(H)$. For any invertible linear change of variables ϕ, $\phi(S)$ generates $\mathrm{Syz}(\phi(H))$.*

Proof. Suppose that $H = \{h_1, \ldots, h_t\}$ and $S = \{\mathbf{s_1}, \ldots, \mathbf{s_\ell}\} \subset \mathcal{P}^t$. Let $\mathbf{s_i} = (p_{i1}, \ldots, p_{it})$. Since $p_{i1}h_1 + \cdots + p_{it}h_t = 0$ and ϕ is a ring homomorphism then $\phi(p_{i1})\phi(h_1) + \cdots + \phi(p_{it})\phi(h_t) = 0$ and therefore $\phi(\mathbf{s_i}) \in \mathrm{Syz}(\phi(H))$. Conversely, assume that $\mathbf{s} = (p_1, \ldots, p_t) \in \mathrm{Syz}(\phi(H))$. This shows that $p_1\phi(h_1) + \cdots + p_t\phi(h_t) = 0$. By invertibility of ϕ we have $(\phi^{-1}(p_1), \ldots, \phi^{-1}(p_t)) \in \mathrm{Syz}(H)$. From assumptions, we conclude that $(\phi^{-1}(p_1), \ldots, \phi^{-1}(p_t)) = g_1\mathbf{s_1} + \cdots + g_\ell\mathbf{s_\ell}$ for some $g_i \in \mathcal{P}$. By applying ϕ on both sides of this equality, we can deduce that $\mathbf{s}$ is generated by $\phi(S)$ and the proof is complete. ⊔⊓

Theorem 4. *The algorithm QUASISTABLE terminates in finitely many steps and returns for a given homogeneous ideal a linear change of variables s.t. the transformed ideal possesses a finite Pommaret basis.*

Proof. Seiler [19, Proposition 2.9] proved that for a generic linear change of variables ϕ, the ideal $\langle \phi(F) \rangle$ has a finite Pommaret basis. He also showed that the process of finding such a linear change, by applying elementary linear changes, terminates in finitely many steps, see [19, Remark 9.11] (or [13]). These arguments establish the finite termination of the algorithm. To prove the correctness, using Theorem 3, we must only show that if $p \in Q$ is removed by $\mathbf{s} \in S$ then it is superfluous. To this end, assume that $F = \{f_1, \ldots, f_k\}$ and $\mathbf{s} = (p_1, \ldots, p_k)$. Thus, we have $p_1 f_1 + \cdots + p_k f_k = 0$. On the other hand, we know that $\mathrm{LM}_{\prec_s}(\mathbf{s}) \mid \mathrm{Sig}(p)$. W.l.o.g., we may assume that $\mathrm{LM}_{\prec_s}(\mathbf{s}) = \mathrm{LM}(p_1)\mathbf{e_1}$. Therefore, $\mathrm{Poly}(p)$ can be written as a combination $g_1 f_1 + \cdots + g_k f_k$ such that $\mathrm{LM}(g_1)$ divides $\mathrm{LM}(p_1)$. Let $t = \mathrm{LM}(p_1)/\mathrm{LM}(g_1)$. We can write $\mathrm{LM}(g_1)f_1$ as a linear combination of some multiplications $m f_i$ where m is a monomial such that $m \mathbf{e_i}$ is strictly smaller than $\mathrm{LM}(g_1)\mathbf{e_1}$. It follows that p has an involutive representation provided that we study $tm f_i$ for each m and i. Since the signature of $tm f_i$ is strictly smaller than $t\,\mathrm{LM}(g_1)\mathbf{e_1} = \mathrm{Sig}(p)$, we are sure that no loop is performed and therefore p can be omitted. □

We have implemented the algorithm QUASISTABLE in MAPLE 17[1] and compared its performance with our implementation of the HDQUASISTABLE algorithm presented in [1] (it is a similar procedure applying a Hilbert driven technique). For this, we used some well-known examples from computer algebra literature. All computations were done over $\mathbb{Q}$ using the degree reverse lexicographical monomial ordering. The results are represented in the following tables where the time and memory columns indicate the consumed CPU time in seconds and amount of megabytes of used memory, respectively. The dim column refers to the dimension of the corresponding ideal. The columns corresponding to C_1 and C_2 show, respectively, the number of polynomials removed by the C_1 and C_2 criteria. The seventh column denotes the number of polynomials eliminated by the criterion related to signature applied in the NEXTINVBASIS algorithm (see [1] for more details). The eighth column shows the number of polynomials eliminated by the Hilbert driven technique which may be applied in the NEXTINVBASIS algorithm to remove useless reductions, (see [1] for more details). The ninth column shows the number of polynomials eliminated by the syzygy criterion described in the NEXTINVBASIS algorithm. The last three columns represent, respectively, the number of reductions to zero, the number of performed elementary linear changes and the maximum degree attained in the computations. The computations in this paper are performed on a personal computer with 2.60 GHz Pentium(R) Core(TM) Dual-Core CPU, 2 GB of RAM, 32 bits under the Windows 7 operating system.

[1] The MAPLE code of the implementations of our algorithms and examples are available at http://amirhashemi.iut.ac.ir/softwares.

Weispfenning94	time	memory	dim	C_1	C_2	SC	HD	Syz	redz	lin	deg
QUASISTABLE	4.5	255.5	2	0	0	0	34	10	41	1	14
HDQUASISTABLE	5.3	261.4	2	0	1	9	46	-	29	1	14

Liu	time	memory	dim	C_1	C_2	SC	HD	Syz	redz	lin	deg
QUASISTABLE	6.1	246.7	2	8	0	10	71	47	44	4	6
HDQUASISTABLE	8.9	346.0	2	6	3	25	125	-	60	4	6

Noon	time	memory	dim	C_1	C_2	SC	HD	Syz	redz	lin	deg
QUASISTABLE	74.1	3653.2.2	1	6	7	10	213	83	215	4	10
HDQUASISTABLE	72.3	3216.9.7	1	4	24	10	351	-	105	4	10

Katsura5	time	memory	dim	C_1	C_2	SC	HD	Syz	redz	lin	deg
QUASISTABLE	95.7	4719.2	5	49	0	0	257	56	115	3	8
HDQUASISTABLE	120.8	5527.7	5	44	4	6	420	-	122	3	8

Vermeer	time	memory	dim	C_1	C_2	SC	HD	Syz	redz	lin	deg
QUASISTABLE	175.5	8227.9	3	5	3	101	158	139	343	3	13
HDQUASISTABLE	192.5	8243.7	3	3	28	157	343	-	190	3	13

Butcher	time	memory	dim	C_1	C_2	SC	HD	Syz	redz	lin	deg
QUASISTABLE	290.6	12957.8	3	135	89	73	183	86	534	3	8
HDQUASISTABLE	433.1	17005.5	3	178	178	219	355	-	386	3	8

As one sees for some examples, some columns are different. It is worth noting that this difference may be due to the fact that the coefficients in the linear changes are chosen randomly and this may affect the behavior of the algorithm.

5   Involutive Variant of the GVW Algorithm

Gao et al. [8] described recently a new algorithm, the GVW algorithm, to compute simultaneously Gröbner bases for a given ideal and for the syzygy module of the given ideal basis. In this section, we present an involutive variant of this approach and compare its efficiency with the existing algorithms to compute involutive bases. For a review of the general setting of the signature based structure that we use in this paper, we refer to [8]. Let $\{f_1, \ldots, f_k\} \subset \mathcal{P}$ be a finite set of non-zero polynomials and $\{e_1, \ldots, e_k\}$ the standard basis for $\mathcal{P}^k$. Let us fix an involutive division $\mathcal{L}$ and a monomial ordering $\prec$. Our goal is to compute an involutive basis for $\mathcal{I} = \langle f_1, \ldots, f_k \rangle$ and a Gröbner basis for $\mathrm{Syz}(f_1, \ldots, f_k)$ w.r.t. $\prec_s$. Let us consider

$$\mathcal{V} = \{(\mathbf{u}, v) \in \mathcal{P}^k \times \mathcal{P} \mid u_1 f_1 + \cdots + u_k f_k = v \text{ with } \mathbf{u} = (u_1, \ldots, u_k)\}$$

as an $\mathcal{P}$-submodule of $\mathcal{P}^{k+1}$. For any pair $p = (\mathbf{u}, v) \in \mathcal{P}^k \times \mathcal{P}$, $\mathrm{LM}_{\prec_s}(\mathbf{u})$ is called the *signature* of p and is denoted by $\mathrm{Sig}(p)$. We define the involutive version of top-reduction defined in [8]. Let $p_1 = (\mathbf{u}_1, v_1)$, $p_2 = (\mathbf{u}_2, v_2) \in \mathcal{P}^k \times \mathcal{P}$. When v_2 is non-zero, we say p_1 is *involutively top-reducible* by p_2 if:

- v_1 is non-zero and $\mathrm{LM}(v_2)$ $\mathcal{L}$-divides $\mathrm{LM}(v_1)$ and
- $\mathrm{LM}(t\mathbf{u}_2) \preceq_s \mathrm{LM}(\mathbf{u}_1)$ where $t = \mathrm{LM}(v_1)/\mathrm{LM}(v_2)$.

The corresponding top-reduction is $p_1 - ctp_2 = (\mathbf{u}_1 - ct\mathbf{u}_2, v_1 - ctv_2)$ where $c = \mathrm{LC}(v_1)/\mathrm{LC}(v_2)$. Such a top-reduction is called *regular*, if $\mathrm{LM}(\mathbf{u}_1 - ct\mathbf{u}_2) = \mathrm{LM}(\mathbf{u}_1)$, and *super* otherwise.

Definition 7. *A finite subset $G \subset \mathcal{V}$ is called a* strong involutive basis *for $\mathcal{I}$ if every pair in $\mathcal{V}$ is involutively top-reducible by some pair in G. A strong involutive basis G is* minimal *if any other strong involutive basis G' of $\mathcal{I}$ satisfies $\mathrm{LM}(G) \subseteq \mathrm{LM}(G')$.*

Proposition 1. *Suppose that $G = \{(\mathbf{u}_1, v_1), \ldots, (\mathbf{u}_m, v_m)\}$ is a strong involutive basis for $\mathcal{I}$. Then $G_0 = \{\mathbf{u}_i \mid v_i = 0 , 1 \leq i \leq m\}$ is a Gröbner basis for $\mathrm{Syz}(f_1, \ldots, f_k)$, and $G_1 = \{v_1, \ldots, v_m\}$ is an involutive basis for $\mathcal{I}$.*

Proof. The proof is an easy consequence of the proof of [8, Proposition 2.2]. □

Let $p_1 = (\mathbf{u}_1, v_1)$ and $p_2 = (\mathbf{u}_2, v_2)$ be two pairs in $\mathcal{V}$. We say that p_1 is *covered* by p_2 if $\mathrm{LM}(\mathbf{u}_2)$ divides $\mathrm{LM}(\mathbf{u}_1)$ and $t\,\mathrm{LM}(v_2) \prec \mathrm{LM}(v_1)$ (strictly smaller) where $t = \mathrm{LM}(\mathbf{u}_1)/\mathrm{LM}(\mathbf{u}_2)$. Also, p is covered by G if it is covered by some pair in G. A pair $p \in \mathcal{V}$ is *eventually super reducible* by G if there is a sequence of regular top-reductions of p by G leading to $(\mathbf{u}', v')$ which is no longer regularly reducible by G but super reducible by G.

Theorem 5. *Let $G \subset \mathcal{V}$ be a finite set such that, for any module monomial $\mathbf{m} \in \mathcal{P}^k$, there is a pair $(\mathbf{u}, v) \in G$ such that $\mathrm{LM}(\mathbf{u}) \mid \mathbf{m}$. Then the following conditions are equivalent:*

1. *G is a strong involutive basis for $\mathcal{I}$,*
2. *any non-multiplicative prolongation of any element of G is eventually super top-reducible by G,*
3. *any non-multiplicative prolongation of any element in G is covered by G.*

Proof. The proof of all implications are similar to the proofs of the corresponding statements in [8, Theorem 2.4] except that we need some slight changes in the proof of $(3 \Rightarrow 1)$. We proceed by reductio ad absurdum. Assume that there is a pair $p = (\mathbf{u}, v) \in \mathcal{V}$ which is not involutively top-reducible by G and has minimal signature. Then, by assumption, there exists $p_1 = (\mathbf{u}_1, v_1) \in G$ such that $\mathrm{LM}(\mathbf{u}) = t\,\mathrm{LM}(\mathbf{u}_1)$ for some t. Select p_1 such that $t\,\mathrm{LM}(v_1)$ is minimal. Let us now consider tp_1. Two cases may happen: If all variables in t are multiplicative for p_1, then $p - tp_1$ has a signature smaller than p and by assumption it has a standard representation leading to a standard representation for p which is a contradiction. Otherwise, t has a non-multiplicative variable. Then, tp_1 is covered by a pair $p_3 = (\mathbf{u}_3, v_3) \in G$. This shows that $t_3\,\mathrm{LM}(v_3) \prec t\,\mathrm{LM}(v_1)$ with $t_3 = t\,\mathrm{LM}(\mathbf{u}_1)/\mathrm{LM}(\mathbf{u}_3)$. Therefore, the polynomial part of $t_3 p_3$ is smaller than tv_1 which contradicts the choice of p_1, and this ends the proof. □

Based on this theorem and similar to the structure of the GVW algorithm, we describe a variant of Gerdt's algorithm for computing strong involutive bases. The structure of the new algorithm is similar to the INVBASIS algorithm and therefore we omit the identical parts.

Algorithm 5. STINVBASIS

Input: A finite set $F \subset \mathcal{P}$; an involutive division $\mathcal{L}$; a monomial ordering $\prec$
Output: A minimal strong involutive basis for $\langle F \rangle$

$F := \text{sort}(F, \prec)$ and $T := \{(F[1], F[1], \emptyset, \mathbf{e}_1)\}$
$Q := \{(F[i], F[i], \emptyset, \mathbf{e}_i) \mid i = 2, \ldots, |F|\}$ and $H := \{\}$
while $Q \neq \emptyset$ **do**
    $Q := \text{sort}(Q, \prec_s)$ and select/remove the first element p from Q
    if p is not covered by G, T or H **then**
        $h := \text{INVTOPREDUCE}(p, T, \mathcal{L}, \prec)$
        if $\text{Poly}(h) = 0$ **then**
            $H := H \cup \{\text{Sig}(p)\}$
        end if
        if $\text{Poly}(h) = 0$ and $\text{LM}(\text{Poly}(p)) = \text{LM}(\text{Anc}(p))$ **then**
            $Q := \{q \in Q \mid \text{Anc}(q) \neq \text{Poly}(p)\}$
        end if
        if $\text{Poly}(h) \neq 0$ and $\text{LM}(\text{Poly}(p)) \neq \text{LM}(\text{Poly}(h))$ **then**
            $\vdots$ {Lines 19–25 of INVBAS}
        end if
        $\vdots$ {Lines 27–30 of INVBAS}
    end if
end while
return $(\text{Poly}(T), H)$

Algorithm 6. INVTOPREDUCE

Input: A quadruple p; a set of quadruples T; a division $\mathcal{L}$; a monomial ordering $\prec$
Output: A top-reduced form of p modulo T

$h := p$
while $\text{Poly}(h)$ has a term am with $a \in \Bbbk$ and $\text{LM}(\text{Poly}(q)) \mid_{\mathcal{L}} m$ with $q \in T$ **do**
    if $m/\text{LM}(\text{Poly}(q)) \text{Sig}(q) \prec_s \text{Sig}(p)$ **then**
        $\text{Poly}(h) := \text{Poly}(h) - am/\text{LT}(\text{Poly}(q)).\text{Poly}(q)$
        $\text{Rep}(h) := \text{Rep}(h) - am/\text{LT}(\text{Poly}(q)).\text{Rep}(q)$
    end if
end while
return (h)

The proof of the next theorem is a consequence of Theorem 5 and the termination and correctness of Gerdt's algorithm.

Theorem 6. *If $\mathcal{L}$ is Noetherian, then* STINVBASIS *terminates in finitely many steps returning a minimal strong involutive basis for its input ideal.*

We have implemented the STINVBASIS algorithm in MAPLE 17 and compared its performance with our implementation of INVOLUTIVEBASIS algorithm (see [1]) and VARGERDT algorithm (a variant of Gerdt's algorithm, see [12]).

Liu	time	memory	C_1	C_2	SC	cover	redz	deg
STINVBASIS	.390	14.806	-	-	-	17	20	6
INVOLUTIVEBASIS	.748	23.830	4	3	2	-	18	6
VARGERDT	1.653	64.877	6	3	-	-	18	19

Noon	time	memory	C_1	C_2	SC	cover	redz	deg
STINVBASIS	1.870	75.213	-	-	-	54	42	10
INVOLUTIVEBASIS	2.620	105.641	4	15	6	-	50	10
VARGERDT	12.32	454.573	6	9	-	-	56	10

Haas3	time	memory	C_1	C_2	SC	cover	redz	deg
STINVBASIS	157.623	6354.493	-	-	-	490	8	33
INVOLUTIVEBASIS	22.345	833.0	0	0	83	-	152	33
VARGERDT	137.733	5032.295	0	98	-	-	255	33

Sturmfels-Eisenbud	time	memory	C_1	C_2	SC	cover	redz	deg
STINVBASIS	2442.414	120887.953	-	-	-	634	29	8
INVOLUTIVEBASIS	24.70	951.070	28	103	95	-	81	6
VARGERDT	59.32	2389.329	43	212	-	-	91	6

Weispfenning94	time	memory	C_1	C_2	SC	cover	redz	deg
STINVBASIS	183.129	8287.044	-	-	-	588	28	18
INVOLUTIVEBASIS	1.09	45.980	0	1	9	-	28	10
VARGERDT	4.305	168.589	0	9	-	-	38	15

As we observe, the performance of the new algorithm is not in general better than that of the others. This is due to the signature-based structure of the new algorithm which does not allow to perform full normal forms.

Acknowledgments. The research of the second author was in part supported by a grant from IPM (No. 95550420). The work of the third author was partially performed as part of the H2020-FETOPEN-2016-2017-CSA project SC^2 (712689).

References

1. Binaei, B., Hashemi, A., Seiler, W.M.: Improved computation of involutive bases. In: Gerdt, V., Koepf, W., Seiler, W., Vorozhtsov, E. (eds.) CASC 2016. LNCS, vol. 9890, pp. 58–72. Springer, Cham (2016). https://doi.org/10.1007/978-3-319-45641-6_5
2. Buchberger, B.: A criterion for detecting unnecessary reductions in the construction of Gröbner-bases. In: Ng, E.W. (ed.) EUROSM 1979. LNCS, vol. 72, pp. 3–21. Springer, Heidelberg (1979). https://doi.org/10.1007/3-540-09519-5_52
3. Buchberger, B.: Ein Algorithmus zum Auffinden der Basiselemente des Restklassenringes nach einem nulldimensionalen Polynomideal. University of Innsbruck, Mathematisches Institut (Diss.), Innsbruck (1965)
4. Cox, D., Little, J., O'Shea, D.: Ideals, Varieties, and Algorithms, 3rd edn. Springer, New York (2007). https://doi.org/10.1007/978-0-387-35651-8
5. Cox, D.A., Little, J., O'Shea, D.: Using Algebraic Geometry. Graduate Texts in Mathematics, vol. 185, 2nd edn. Springer, New York (2005). https://doi.org/10.1007/b138611
6. Faugère, J.C.: A new efficient algorithm for computing Gröbner bases (F_4). J. Pure Appl. Algebra **139**(1–3), 61–88 (1999). https://doi.org/10.1016/S0022-4049(99)00005-5
7. Faugère, J.C.: A new efficient algorithm for computing Gröbner bases without reduction to zero (F_5). In: Proceedings of ISSAC 2002, pp. 75–83 (2002)

8. Gao, S., Volny, F.I., Wang, M.: A new framework for computing Gröbner bases. Math. Comput. **85**(297), 449–465 (2016). https://doi.org/10.1090/mcom/2969
9. Gebauer, R., Möller, H.: On an installation of Buchberger's algorithm. J. Symb. Comput. **6**(2–3), 275–286 (1988). https://doi.org/10.1016/S0747-7171(88)80048-8
10. Gerdt, V.P.: Involutive algorithms for computing Gröbner bases. In: Computational Commutative and Non-commutative Algebraic Geometry. Proceedings of the NATO Advanced Research Workshop, pp. 199–225. IOS Press, Amsterdam (2005)
11. Gerdt, V.P., Blinkov, Y.A.: Involutive bases of polynomial ideals. Math. Comput. Simul. **45**(5–6), 519–541 (1998). https://doi.org/10.1016/S0378-4754(97)00127-4
12. Gerdt, V.P., Hashemi, A., Alizadeh, B.M.: Involutive bases algorithm incorporating F_5 criterion. J. Symb. Comput. **59**, 1–20 (2013). https://doi.org/10.1016/j.jsc.2013.08.002
13. Hashemi, A., Schweinfurter, M., Seiler, W.: Deterministic genericity for polynomial ideals. J. Symb. Comput. **86**, 20–50 (2018)
14. Janet, M.: Sur les systèmes d'équations aux dérivées partielles. C. R. Acad. Sci. Paris **170**, 1101–1103 (1920)
15. Lazard, D.: Gröbner bases, Gaussian elimination and resolution of systems of algebraic equations. In: van Hulzen, J.A. (ed.) EUROCAL 1983. LNCS, vol. 162, pp. 146–156. Springer, Heidelberg (1983). https://doi.org/10.1007/3-540-12868-9_99
16. Möller, H., Mora, T., Traverso, C.: Gröbner bases computation using syzygies. In: Proceedings of ISSAC 1992, pp. 320–328 (1992)
17. Pommaret, J.: Systems of Partial Differential Equations and Lie Pseudogroups. Gordon and Breach Science Publishers, Philadelphia (1978)
18. Schreyer, F.O.: Die Berechnung von Syzygien mit dem verallgemeinerten Weierstrass'schen Divisionssatz. Master's thesis, University of Hamburg, Germany (1980)
19. Seiler, W.M.: A combinatorial approach to involution and δ-regularity. II: structure analysis of polynomial modules with Pommaret bases. Appl. Algebra Eng. Commun. Comput. **20**(3–4), 261–338 (2009). https://doi.org/10.1007/s00200-009-0101-9
20. Seiler, W.M.: Involution. The Formal Theory of Differential Equations and Its Applications in Computer Algebra. Springer, Berlin (2001). https://doi.org/10.1007/978-3-642-01287-7
21. Wall, B.: On the computation of syzygies. SIGSAM Bull. **23**(4), 5–14 (1989)
22. Zharkov, A., Blinkov, Y.: Involution approach to investigating polynomial systems. Math. Comput. Simul. **42**(4), 323–332 (1996). https://doi.org/10.1016/S0747-7171(88)80048-8

A Strongly Consistent Finite Difference Scheme for Steady Stokes Flow and its Modified Equations

Yury A. Blinkov[1], Vladimir P. Gerdt[2,3]([✉]), Dmitry A. Lyakhov[4], and Dominik L. Michels[4]

[1] Saratov State University, Saratov 413100, Russian Federation
BlinkovUA@info.sgu.ru
[2] Joint Institute for Nuclear Research, Dubna 141980, Russian Federation
Gerdt@jinr.ru
[3] Peoples' Friendship University of Russia, Moscow 117198, Russian Federation
[4] King Abdullah University of Science and Technology,
Thuwal 23955-6900, Kingdom of Saudi Arabia
{Dmitry.Lyakhov,Dominik.Michels}@kaust.edu.sa

Abstract. We construct and analyze a strongly consistent second-order finite difference scheme for the steady two-dimensional Stokes flow. The pressure Poisson equation is explicitly incorporated into the scheme. Our approach suggested by the first two authors is based on a combination of the finite volume method, difference elimination, and numerical integration. We make use of the techniques of the differential and difference Janet/Gröbner bases. In order to prove strong consistency of the generated scheme we correlate the differential ideal generated by the polynomials in the Stokes equations with the difference ideal generated by the polynomials in the constructed difference scheme. Additionally, we compute the modified differential system of the obtained scheme and analyze the scheme's accuracy and strong consistency by considering this system. An evaluation of our scheme against the established marker-and-cell method is carried out.

Keywords: Computer algebra · Difference elimination
Finite difference approximation · Janet basis · Modified equations
Stokes flow · Strong consistency

1 Introduction

In this paper, we consider the two-dimensional flow of an incompressible fluid described by the following system of partial differential equations (PDEs):

$$\begin{cases} F^{(1)} := u_x + v_y = 0, \\ F^{(2)} := p_x - \frac{1}{\text{Re}} \Delta u - f^{(1)} = 0, \\ F^{(3)} := p_y - \frac{1}{\text{Re}} \Delta v - f^{(2)} = 0. \end{cases} \tag{1}$$

© Springer Nature Switzerland AG 2018
V. P. Gerdt et al. (Eds.): CASC 2018, LNCS 11077, pp. 67–81, 2018.
https://doi.org/10.1007/978-3-319-99639-4_5

Here the velocities u and v, the pressure p, and the external forces $f^{(1)}$ and $f^{(2)}$ are functions in x and y; Re is the Reynolds number and $\Delta := \partial_{xx} + \partial_{yy}$ is the Laplace operator.

A flow that is governed by these equations is denoted in the literature as a Stokes flow or a creeping flow. Correspondingly, the PDE system (1) is called a Stokes system. It approximates the Navier–Stokes system for a two-dimensional incompressible steady flow when Re $\ll$ 1. The last condition makes the nonlinear inertia terms in the Navier–Stokes system much smaller then the viscous forces (cf. [16], Sect. 22·11), and neglecting of the nonlinear terms results in Eqs. (1). The fundamental mathematical theory of the Stokes flow is, e.g., presented in [14].

Our first aim is to construct, for a uniform and orthogonal grid, a finite difference scheme for the governing system (1) which contains a discrete version of the pressure Poisson equation and whose algebraic properties are strongly consistent (or s-consistent, for brevity) [9,12] with those of Eq. (1). For this purpose, we use the approach proposed in [7] based on a combination of the finite volume method, numerical integration, and difference elimination. For the generated scheme we apply the algorithmic criterion to verify its s-consistency. The last criterion was designed in [12] for linear PDE systems and then generalized in [9] to polynomially nonlinear systems. The computational experiments done in papers [2,3] with the Navier–Stokes equations demonstrated a substantial superiority in numerical behavior of s-consistent schemes over s-inconsistent ones.

The linearity of Eq. (1) not only makes the construction and analysis of its numerical solutions much easier than in the case of the Navier–Stokes equations, but also admits a fully algorithmic generation of difference schemes for Eq. (1) and their s-consistency verification. To perform related computations we use two Maple packages implementing the involutive algorithm (cf. [10]) for the computation of Janet and Gröbner bases: the package JANET [4] for linear differential systems and the package LDA [11] (Linear Difference Algebra) for linear difference systems.

Our second aim is to compute a modified differential system of the constructed difference scheme, i.e., modified Stokes flow, and to analyse the accuracy and consistency of the scheme via this differential system. Nowadays the method of modified equations suggested in [20] is widely used (see [6], Chap. 8 and [17], Sect. 5.5) in studying difference schemes. The method provides a natural and unified platform to study such basic properties of the scheme as order of approximation, consistency, stability, convergence, dissipativity, dispersion, and invariance. However, as far as we know, the methods for the computation of modified equations have not been extended yet to non-evolutionary PDE systems. We show how the extension can be done for our scheme by applying the technique of differential Janet/Gröbner bases.

The present paper is organized as follows. In Sect. 2, we generate for Eq. (1) a difference scheme by applying the approach of paper [7]. In Sect. 3, we show that our scheme is s-consistent and demonstrate s-inconsistency of another scheme obtained by a tempting compactification of our scheme. The computation of a modified Stokes system for our s-consistent scheme is described in Sect. 4. Here, we also show by the example of the s-inconsistent scheme of Sect. 3 how the modified Stokes system detects the s-inconsistency. Finally, a numerical benchmark against the marker-and-cell method is presented in Sect. 5 and some concluding remarks are given in Sect. 6.

2 Difference Scheme Generation for Stokes Flow

We consider the orthogonal and uniform solution grid with the grid spacing h and apply the approach of paper [7] to generate a difference scheme for Eq. (1).

Step 1. Completion to Involution (we refer to [19] and to the references therein for the theory of involution). We select the lexicographic POT (Position Over Term) [1] ranking with

$$x \succ y, \quad u \succ v \succ p \succ f^{(1)} \succ f^{(2)}. \tag{2}$$

Then the package JANET [4] outputs the following *Janet involutive form* of Eq. (1) which is the *minimal reduced differential Gröbner basis form*:

$$\begin{cases} F^{(1)} := \underline{u_x} + v_y = 0, \\ F^{(2)} := \underline{p_x} - \frac{1}{\text{Re}}\left(u_{yy} - v_{xy}\right) - f^{(1)} = 0, \\ F^{(3)} := \underline{p_y} - \frac{1}{\text{Re}}\left(v_{xx} + v_{yy}\right) - f^{(2)} = 0, \\ F^{(4)} := \underline{p_{xx}} + p_{yy} - f_x^{(1)} - f_y^{(2)} = 0. \end{cases} \tag{3}$$

We underlined the *leaders*, i.e., the highest ranking partial derivatives occurring in Eqs. (3). F^4 is the *pressure Poisson equation* which, being the integrability condition for system (1), is expressed in terms of its left-hand sides as

$$F^{(4)} := F_x^{(2)} + F_y^{(3)} + \frac{1}{\text{Re}}\left(F_{xx}^{(1)} + F_{yy}^{(1)}\right) = p_{xx} + p_{yy} - f_x^{(1)} - f_y^{(2)}. \tag{4}$$

Remark 1. The differential polynomial $F^{(2)}$ in Eq. (3) is $F^{(2)}$ in Eq. (1) reduced modulo the continuity equation $F^{(1)}$.

Step 2. Conversion into the Integral Form. We choose the following integration contour Γ as a "control volume" and rewrite equations $F^{(1)}, F^{(2)}$, and $F^{(3)}$ into the equivalent *integral form*

$$\begin{cases} \oint_\Gamma -v\,dx + u\,dy = 0, \\ \oint_\Gamma \frac{1}{\text{Re}}u_y\,dx + \left(p - \frac{1}{\text{Re}}u_x\right)dy - \iint_\Omega f^{(1)}dx\,dy = 0, \\ \oint_\Gamma -\left(p - \frac{1}{\text{Re}}v_y\right)dx - \frac{1}{\text{Re}}v_x\,dy - \iint_\Omega f^{(2)}dx\,dy = 0, \end{cases} \tag{5}$$

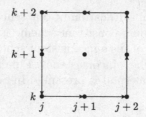

Fig. 1. Integration contour Γ (stencil 3×3).

where Ω is the internal area of the contour Γ.

It should be noted that we use in Eq. (5) the original form of $F^{(2)}$ given in Eq. (1) (see Remark 1) since we want to preserve at the discrete level the symmetry of system (1) under the swap transformation

$$\{x, u, f^{(1)}\} \longleftrightarrow \{y, v, f^{(2)}\}. \tag{6}$$

Step 3. Addition of Integral Relations for Derivatives. We add to system (5) the *exact integral relations* between the partial derivatives of velocities and the velocities themselves:

$$\begin{cases} \int\limits_{x_j}^{x_{j+1}} u_x dx = u(x_{j+1}, y) - u(x_j, y), & \int\limits_{y_k}^{y_{k+1}} u_y dy = u(x, y_{k+1}) - u(x, y_k), \\ \int\limits_{x_j}^{x_{j+1}} v_x dx = v(x_{j+1}, y) - v(x_j, y), & \int\limits_{y_k}^{y_{k+1}} v_y dy = v(x, y_{k+1}) - v(x, y_k). \end{cases} \tag{7}$$

Step 4. Numerical Evaluation of Integrals. We apply the midpoint rule for the contour integration in Eq. (5), the trapezoidal rule for the integrals (7) and approximate the double integrals as

$$f^{1(2)}_{i+1,k+1} 4h^2,$$

where h is the step of a square grid in the (x, y) plane.

As a result, we obtain the difference equations for the grid functions

$$u_{j,k} \approx u(jh, kh), \ v_{j,k} \approx v(jh, kh), \ p_{j,k} \approx p(jh, kh), \ f^{(1,2)}_{j,k} \approx f^{(1,2)}(jh, kh)$$

approximating functions $u(x, y)$, $v(x, y)$, $p(x, y)$, $f^{(1)}(x, y)$, $f^{(2)}(x, y)$, and the grid functions approximating partial derivatives

$$\begin{cases} u_{xj,k} \approx u_x(jh, kh), & u_{yj,k} \approx u_y(jh, kh), \\ v_{xj,k} \approx v_x(jh, kh), & v_{yj,k} \approx v_y(jh, kh), \end{cases}$$

where $j, k \in \mathbb{Z}$:

$$
\begin{cases}
(u_{j+2,\,k+1} - u_{j,\,k+1})\,2h + (v_{j+1,\,k+2} - v_{j+1,\,k})\,2h = 0, \\[6pt]
\dfrac{1}{\mathrm{Re}}\left(u_{y\,j+1,\,k} - u_{y\,j+1,\,k+2}\right)2h + \left(p_{j+2,\,k+1} - \dfrac{1}{\mathrm{Re}}u_{x\,j+2,\,k+1}\right)2h \\[6pt]
\quad -\left(p_{j,\,k+1} - \dfrac{1}{\mathrm{Re}}u_{x\,j,\,k+1}\right)2h - 4f^{(1)}_{j+1,\,k+1}h^2 = 0, \\[6pt]
\quad -\left(\left(p^n_{j+1,\,k} - \dfrac{1}{\mathrm{Re}}v^n_{y\,j+1,\,k}\right) - \left(p^n_{j+1,\,k+2} - \dfrac{1}{\mathrm{Re}}v^n_{y\,j+1,\,k+2}\right)\right)2h \\[6pt]
\quad +\left(-\dfrac{1}{\mathrm{Re}}v_{x\,j+2,\,k+1} + \dfrac{1}{\mathrm{Re}}v_{x\,j,\,k+1}\right)2h - 4f^{(2)}_{j+1,\,k+1}h^2 = 0, \\[6pt]
\dfrac{u_{x\,j+1,\,k} + u_{x\,j,\,k}}{2}h - u_{j+1,\,k} + u_{j,\,k} = 0, \\[6pt]
\dfrac{v_{x\,j+1,\,k} + v_{x\,j,\,k}}{2}h - v_{j+1,\,k} + v_{j,\,k} = 0, \\[6pt]
\dfrac{u_{y\,j,\,k+1} + u_{y\,j,\,k}}{2}h - u_{j,\,k+1} + u_{j,\,k} = 0, \\[6pt]
\dfrac{v_{y\,j,\,k+1} + v_{y\,j,\,k}}{2}h - v_{j,\,k+1} + v_{j,\,k} = 0.
\end{cases} \tag{8}
$$

Step 5. Difference Elimination of Derivatives. To eliminate the grid functions u_x, u_y, v_x, v_y for the partial derivatives of the velocities, we construct a difference Janet/Gröbner basis form of the set of linear difference polynomials in left-hand sides of Eq. (8) with the Maple package LDA [12] for the POT lexicographic ranking which is the difference analogue of the differential ranking used on Step 1:

$$
j \succ k, \quad u \succ v \succ p \succ f^{(1)} \succ f^{(2)}. \tag{9}
$$

The output of the LDA includes four difference polynomials not containing the grid functions u_x, u_y, v_x, v_y. These polynomials comprise a difference scheme. Being interreduced, this scheme does not reveal a desirable discrete analogue of symmetry under the transformation (6). Because of this reason, we prefer the following redundant but symmetric form of the scheme:

$$
\begin{cases}
\tilde{F}^{(1)} := \dfrac{u_{j+2,\,k+1} - u_{j,\,k+1}}{2h} + \dfrac{v_{j+1,\,k+2} - v_{j+1,\,k}}{2h} = 0, \\[8pt]
\tilde{F}^{(2)} := \dfrac{p_{j+2,\,k+1} - p_{j,\,k+1}}{2h} - \dfrac{1}{\mathrm{Re}}\Delta_1\left(u_{j,k}\right) - f^{(1)}_{j+1,\,k+1} = 0, \\[8pt]
\tilde{F}^{(3)} := \dfrac{p_{j+1,\,k+2} - p_{j+1,\,k}}{2h} - \dfrac{1}{\mathrm{Re}}\Delta_1\left(v_{j,k}\right) - f^{(2)}_{j+1,\,k+1} = 0, \\[8pt]
\tilde{F}^{(4)} := \Delta_2\left(p_{j,k}\right) - \dfrac{f^{(1)}_{j+3,\,k+2} - f^{(1)}_{j+1,\,k+2}}{2h} - \dfrac{f^{(2)}_{j+2,\,k+3} - f^{(2)}_{j+2,\,k+1}}{2h} = 0,
\end{cases} \tag{10}
$$

where Δ_1 and Δ_2 are discrete versions of the Laplace operator acting on a grid function $g_{j,\,k}$ as

$$\Delta_1\left(g_{j,\,k}\right) := \frac{g_{j+2,\,k+1} + g_{j+1,\,k+2} - 4g_{j+1,\,k+1} + g_{j+1,\,k} + g_{j,\,k+1}}{h^2}, \qquad (11)$$

$$\Delta_2\left(g_{j,\,k}\right) := \frac{g_{j+4,\,k+2} + g_{j+2,\,k+4} - 4g_{j+2,\,k+2} + g_{j+2,\,k} + g_{j,\,k+2}}{4h^2}. \qquad (12)$$

Remark 2. The difference equation $\tilde{F}^{(4)}$ of the system (10) can also be obtained (cf. [8]) from the integral form of F^4 in Eqs. (3)–(4) with the contour illustrated in Fig. 1 by using the midpoint rule for the contour integration of the p_x and p_y as well as for evaluation of the additional integrals

$$\int\limits_{x_j}^{x_{j+2}} p_x dx = p(x_{j+2}, y) - p(x_j, y), \qquad \int\limits_{y_k}^{y_{k+2}} p_y dy = p(x, y_{k+2}) - p(x, y_k), \qquad (13)$$

and the trapezoidal rule for the contour integration of $f^{(1)}$ and $f^{(2)}$.

The difference polynomials (10) approximate those in Eq. (3), and such correspondence between differential and difference Janet/Gröbner bases is a consequence of our choice of the differential (2) and difference (9) rankings.

3   Consistency Analysis

Let $\mathcal{R} = \mathbb{Q}(\mathrm{Re}, h)[u, v, p, f^{(1)}, f^{(2)}]$ be the *ring of differential polynomials* over the field of rational functions in Re and h. We consider the functions describing the Stokes flow (1) as *differential indeterminates* and their grid approximations as *difference indeterminates*. Respectively, we denote by $\tilde{\mathcal{R}}$ the *difference polynomial ring* whose elements are polynomials in the grid functions with the right-shift operators σ_1 and σ_2 acting as translations, for example,

$$\sigma_1 \circ u_{j,\,k} = u_{j+1,\,k}, \qquad \sigma_2 \circ u_{j,\,k} = u_{j,\,k+1}. \qquad (14)$$

We denote by $\mathcal{I} := \langle F^{(1)}, F^{(2)}, F^{(3)} \rangle \subset \mathcal{R}$ the *differential ideal* generated by the set of left-hand sides in (1) and by $\tilde{\mathcal{I}} := \langle \tilde{F}^{(1)}, \tilde{F}^{(2)}, \tilde{F}^{(3)}, \tilde{F}^{(4)} \rangle \subset \tilde{\mathcal{R}}$ the *difference ideal* generated by the left-hand sides of Eq. (10).

The elements in $\mathcal{I}$ vanish on solutions of the Stokes flow (1) and those in $\tilde{\mathcal{I}}$ vanish on solutions of (10). We refer to an element in $\mathcal{I}$ (respectively, in $\tilde{\mathcal{I}}$) as to a *consequence* of Eq. (1) (respectively, of Eq. (10)).

Definition 1. [12] *We shall say that a difference equation $\tilde{F} = 0$ implies the differential equation $F = 0$ and write $\tilde{F} \triangleright F$ when the Taylor expansion about a grid point yields*

$$\tilde{F} \xrightarrow[h \to 0]{} F \cdot h^k + O(h^{k+1}), \quad k \in \mathbb{Z}_{\geq 0}. \qquad (15)$$

It is clear that to approximate Eq. (3), the scheme (10) must be pairwise consistent with the involutive differential form (3). We call this sort of consistency *weak consistency*.

Definition 2. [12] *A difference polynomial set* $\{\tilde{F}^{(1)}, \tilde{F}^{(2)}, \tilde{F}^{(3)}, \tilde{F}^{(4)}\}$ *is weakly consistent or w-consistent with differential system* (3) *if*

$$(\forall\, 1 \le i \le 4)\, [\tilde{F}^{(i)} \triangleright F^{(i)}]. \tag{16}$$

The following definition establishes the consistency interrelation between the differential and difference ideals generated by Eqs. (1) and (10), respectively. If such a consistency holds, then it provides a certain inheritance of algebraic properties of Stokes flow by the difference scheme.

Definition 3. [9] *A finite difference approximation* $\tilde{F} := \{\tilde{F}^{(1)}, \ldots, \tilde{F}^{(m)}\}$ *to* (1) *is strongly consistent or s-consistent with Stokes flow* (1) *if*

$$(\forall \tilde{F} \in [\![\tilde{F}]\!])\, (\exists F \in \mathcal{I})\, [\tilde{F} \triangleright F], \tag{17}$$

where $[\![\tilde{F}]\!]$ *is a* perfect difference ideal [15] *generated by the elements in the difference approximation.*

Theorem 1. [9] *The s-consistency condition* (17) *holds if and only if a Gröbner basis* $\tilde{G}$ *of* $\tilde{\mathcal{I}}$ *satisfies*

$$(\forall \tilde{g} \in \tilde{G})\, (\exists g \in \langle F \rangle)\, [\tilde{g} \triangleright g]. \tag{18}$$

Corollary 1. *The difference scheme* (10) *is s-consistent with the Stokes system* (1).

Proof. By its construction, the set of difference polynomials in Eq. (10) is a Janet/Gröbner basis of the elimination ideal $\tilde{\mathcal{I}}_0 \cap \mathcal{R}$ where $\tilde{\mathcal{I}}_0$ is the difference ideal generated by the polynomials in Eq. (8) (cf. [1], Theorem 2.3.4). The same set is also a Janet/Gröbner basis for the ideal $\langle \tilde{F}^{(1)}, \tilde{F}^{(2)}, \tilde{F}^{(3)} \rangle$ and for the same POT ranking with $j \succ k$ and $u \succ v \succ p \succ f^{(1)} \succ f^{(2)}$. It is readily verified with the LDA package. Furthermore, it is easy to see that

$$\tilde{F}^{(i)} \triangleright F^{(i)}, \quad (i = 1 \div 4) \tag{19}$$

where $F^{(i)}$ are differential polynomials in Eq. (3). $\qquad\square$

Remark 3. For the computation of the image in mapping (19) one can use the command *ContinuousLimit* of the package LDA.

It is clear that s-consistency implies w-consistency. But the converse is not true. For the numerical simulation of the Stokes flow it is tempting to replace $\tilde{F}^{(4)}$ in Eq. (10) with a more compact discretization

$$\tilde{F}_1^{(4)} := \Delta_1\,(p_{j,k}) - \frac{f_{j+2,\,k+1}^{(1)} - f_{j,\,k+1}^{(1)}}{2h} - \frac{f_{j+1,\,k+2}^{(2)} - f_{j+1,\,k}^{(2)}}{2h} = 0. \tag{20}$$

Although this substitution preserves w-consistency since

$$\tilde{F}_1^{(4)} \triangleright F^{(4)}, \tag{21}$$

the scheme $\{\tilde{F}^{(1)}, \tilde{F}^{(2)}, \tilde{F}^{(3)}, \tilde{F}_1^{(4)}\}$ is not s-consistent.

Proposition 1. *The difference scheme* $\{\tilde{F}^{(1)}, \tilde{F}^{(2)}, \tilde{F}^{(3)}, \tilde{F}_1^{(4)}\}$ *is s-inconsistent.*

Proof. The difference polynomial (20) does not belong to the difference ideal $\tilde{\mathcal{I}}$ generated by the polynomial set in Eq. (10) since $\tilde{F}_1^{(4)}$ is irreducible modulo the ideal $\tilde{\mathcal{I}}$. This can be shown by the direct computation of the normal form of $\tilde{F}_1^{(4)}$ modulo the Janet basis (10) with the routine *InvReduce* of the Maple package LDA. □

Now let us analyse the s-consistency of $\{\tilde{F}^{(1)}, \tilde{F}^{(2)}, \tilde{F}^{(3)}, \tilde{F}_1^{(4)}\}$. The Janet/Gröbner basis of the difference ideal $\tilde{\mathcal{I}} := \langle \tilde{F}^{(1)}, \tilde{F}^{(2)}, \tilde{F}^{(3)}, \tilde{F}_1^{(4)} \rangle$ computed with LDA consists of seven elements. Four of them imply system (3) and the three remaining elements denoted by $\tilde{F}^{(5)}$, $\tilde{F}^{(6)}$, and $\tilde{F}^{(7)}$ are rather cumbersome difference equations which imply, respectively, the following differential ones

$$\begin{cases} F^{(5)} := f^{(1)}_{xxxxx} + f^{(1)}_{xyyyy} + f^{(2)}_{xxxxy} + f^{(2)}_{yyyyy} = 0, \\ F^{(6)} := f^{(1)}_{xxx} - f^{(1)}_{xyy} + f^{(2)}_{xxy} - f^{(2)}_{yyy} + 2\, p_{yyyy} = 0, \end{cases} \tag{22}$$

and $\tilde{F}^{(7)} \triangleright F^{(6)}$.

Equations (22) are not consequences of the Stokes equations since the differential polynomials $F^{(5)}$ and $F^{(6)}$ are irreducible modulo the differential ideal generated by the differential polynomials in Eq. (1). It follows that there are solutions to the Stokes equations which do not satisfy Eq. (22).

Remark 4. Equations (22) impose the limitations on the external forces which do not follow from the governing differential equations (1). This is a result of s-inconsistency.

4 Modified Stokes Flow

In the framework of the method of *modified equation* (cf. [17], Sect. 5.5), a numerical solution of the governing differential system (1), for given external forces $f^{(1)}$ and $f^{(2)}$, should be considered as a set of continuous differentiable functions $\{u, v, p\}$ whose values at the grid points satisfy the difference scheme (10). Since the difference Eq. (10) describe the differential ones (3) only approximately, we cannot expect that a continuous solution interpolating the grid values exactly satisfies Eq. (3). In reality, it satisfies another set of differential equations which we shall call the *modified steady Stokes flow* or *modified flow* for short.

Generally, the method of modified differential equation uses the representation of difference equations comprising the scheme as infinite order differential equations obtained by replacing the various shift operators in the difference equations by the Taylor series about a grid point. For equations of evolutionary type, the next step is to eliminate all derivatives with respect to the evolutionary variable of order greater than one. This step is done to obtain a kind of canonical form of the modified equation. Then, truncation of the order of the differential representations in the grid steps gives various modified equations ("differential approximations") of the difference scheme.

As we show, the fact that both equation systems are Gröbner bases of the ideals they generate and satisfy the condition (19) of s-consistency allows to develop a constructive procedure for the computation of the modified flow. Since the finite differences in the scheme (10) approximate the partial derivatives occurring in Eq. (3) with accuracy $\mathcal{O}(h^2)$, it would appear reasonable that the scheme would have the second order of accuracy. For this reason, we restrict ourselves to the computation of the second order modified flow.

The Taylor expansions of the difference polynomials in Eq. (10) at the grid point $(-h, -h)$ for $\tilde{F}^{(1)}, \tilde{F}^{(2)}, \tilde{F}^{(3)}, \tilde{F}_1^{(4)}$, and at the point $(-2h, -2h)$ for $\tilde{F}^{(4)}$ read

$$
\begin{cases}
\tilde{F}^{(1)} := u_x + v_y + \frac{h^2 u_{xxx}}{6} + \frac{h^2 v_{yyy}}{6} + \mathcal{O}(h^4) = 0, \\
\tilde{F}^{(2)} := p_x - \frac{1}{\mathrm{Re}} u_{xx} - \frac{1}{\mathrm{Re}} u_{yy} - f^{(1)} + \frac{h^2 p_{xxx}}{6} - \frac{h^2 u_{xxxx}}{12\,\mathrm{Re}} \\
\qquad - \frac{h^2 u_{yyyy}}{12\,\mathrm{Re}} + \mathcal{O}(h^4) = 0, \\
\tilde{F}^{(3)} := p_y - \frac{1}{\mathrm{Re}} v_{xx} - \frac{1}{\mathrm{Re}} v_{yy} - f^{(2)} + \frac{h^2 p_{yyy}}{6} - \frac{h^2 v_{xxxx}}{12\,\mathrm{Re}} \\
\qquad - \frac{h^2 v_{yyyy}}{12\,\mathrm{Re}} + \mathcal{O}(h^4) = 0, \\
\tilde{F}^{(4)} := p_{xx} + p_{yy} - f_x^{(1)} - f_y^{(2)} - \frac{h^2 f_{xxx}^{(1)}}{6} - \frac{h^2 f_{yyy}^{(2)}}{6} \\
\qquad + \frac{h^2 p_{xxxx}}{3} + \frac{h^2 p_{yyyy}}{3} + \mathcal{O}(h^4) = 0,
\end{cases}
\tag{23}
$$

where the terms of order h^2 are written explicitly. The calculation of the right-hand sides in Eq. (23) as well as the computation of the expressions given below was done with the use of freely available Python library SYMPY (http://www.sympy.org/) for symbolic mathematics.

Remark 5. The Taylor expansions of the s-consistent difference scheme (10) and of the s-inconsistent scheme $\{\tilde{F}^{(1)}, \tilde{F}^{(2)}, \tilde{F}^{(3)}, \tilde{F}_1^{(4)}\}$ over the chosen grid points contain only the even powers of h. It follows immediately from the fact that all the finite differences occurring in the equations of both schemes are the central difference approximations of the partial derivatives occurring in (3).

Furthermore, we reduce the terms of order h^2 in the right-hand sides of (23) modulo the differential Janet/Gröbner basis (10). This reduction will give us a *canonical form* of the second order modified flow, since given a Gröbner basis, the *normal form* of a polynomial modulo this basis is uniquely defined (cf. [1], Sect. 2.1). The normal form can be computed with the command *InvReduce* using the Maple package JANET.

Thus, the Taylor expansion of the difference polynomials yields the *second order modified Stokes flow* as follows:

$$
\begin{cases}
\tilde{F}^{(1)} := u_x + v_y + \dfrac{h^2\,\mathrm{Re}f_y^{(2)}}{6} - \dfrac{h^2\,\mathrm{Re}\,p_{yy}}{6} + \dfrac{h^2 v_{yyy}}{3} + \mathcal{O}(h^4) = 0, \\[2mm]
\tilde{F}^{(2)} := p_x + \dfrac{1}{\mathrm{Re}}v_{xy} - \dfrac{1}{\mathrm{Re}}u_{yy} - f^{(1)} + \dfrac{h^2 f_{xx}^{(1)}}{6} + \dfrac{h^2 f_{yy}^{(1)}}{4} + \dfrac{h^2 f_{xy}^{(2)}}{4} \\[2mm]
\qquad - \dfrac{h^2 p_{xyy}}{2} + \dfrac{h^2 u_{yyyy}}{6\,\mathrm{Re}} + \mathcal{O}(h^4) = 0, \\[2mm]
\tilde{F}^{(3)} := p_y - \dfrac{1}{\mathrm{Re}}v_{xx} - \dfrac{1}{\mathrm{Re}}v_{yy} - f^{(2)} - \dfrac{h^2 f_{xy}^{(1)}}{12} + \dfrac{h^2 f_{xx}^{(2)}}{12} - \dfrac{h^2 f_{yy}^{(2)}}{6} \\[2mm]
\qquad + \dfrac{h^2 p_{yyy}}{3} - \dfrac{h^2 v_{yyyy}}{6\,\mathrm{Re}} + \mathcal{O}(h^4) = 0, \\[2mm]
\tilde{F}^{(4)} := p_{xx} + p_{yy} - f_x^{(1)} - f_y^{(2)} + \dfrac{h^2 f_{xxx}^{(1)}}{6} - \dfrac{h^2 f_{xyy}^{(1)}}{3} \\[2mm]
\qquad + \dfrac{h^2 f_{xxy}^{(2)}}{3} - \dfrac{h^2 f_{yyy}^{(2)}}{2} + \dfrac{2h^2 p_{yyyy}}{3} + \mathcal{O}(h^4) = 0.
\end{cases}
\tag{24}
$$

Remark 6. Note that the symmetry under the swap transformation (6) that holds in Eq. (23) does not hold in Eq. (24). This symmetry breaking is a typical effect of the application of the Gröbner reduction to symmetric systems and caused by the non-symmetry of the term ordering.

As we know, Stokes flow (1) satisfies the integrability condition (4) which we rewrite as

$$
F_x^{(2)} + F_y^{(3)} + \frac{1}{\mathrm{Re}}\left(F_{xx}^{(1)} + F_{yy}^{(1)}\right) - F^{(4)} = 0.
\tag{25}
$$

Substitution of the Taylor expansions (24) into the equality (25) shows that the sum of the second-order terms explicitly written in formulae (24) is equal to zero. The following proposition shows that this is a consequence of the s-consistency of the scheme.

Proposition 2. *Given a uniform and orthogonal solution grid with a spacing h, a w-consistent difference scheme for Eq. (3) is s-consistent only if its Taylor expansion based on the central-difference formulas for derivatives and reduced modulo system (3), after its substitution into the left-hand side of the equality (25) vanishes for every order in h^2.*

Proof. Let $\tilde{G} := \{\tilde{G}^{(1)}, \tilde{G}^{(2)}, \tilde{G}^{(3)}, \tilde{G}^{(4)}\}$ be a set of s-consistent difference approximations to the differential polynomials $F^{(1)}, F^{(2)}, F^{(3)}, F^{(4)}$ in the Janet/Gröbner basis (3). The w-consistency of G implies the central difference Taylor expansion

$$
\tilde{G}^{(i)} = F^{(i)} + \sum_{m=1}^{\infty} h^{2m} r_m^{(i)}, \quad r_m^{(i)} \in \mathbb{Q}(\mathrm{Re})[u, v, p, f^{(1)}, f^{(2)}] \quad (i = 1 \div 4).
\tag{26}
$$

We consider the family of difference polynomials ($m \in \mathbb{N}_{\geq 1}$)

$$
\tilde{G}_0^{(m)} := D_1^{(m)}\tilde{G}^{(2)} + D_2^{(m)}\tilde{G}^{(3)} + \frac{1}{\mathrm{Re}}\left(D_{1,1}^{(m)}\tilde{G}^{(1)} + D_{2,2}^{(m)}\tilde{G}^{(1)}\right) - \tilde{G}^{(4)}
\tag{27}
$$

with the central-difference operators $D_1^{(m)}, D_2^{(m)}, D_{1,1}^{(m)}, D_{2,2}^{(m)}$ approximating the partial differential operators $\partial_x, \partial_y, \partial_{xx}, \partial_{yy}$ with accuracy h^{2m}. Apparently, $\tilde{G}_0^{(m)}$ belongs to the perfect difference ideal generated by $\tilde{G}$:

$$(\forall m \in \mathbb{N}_{\geq 1}) \quad [\tilde{G}_0^{(m)} \in [\![\tilde{G}]\!]].$$

These difference operators are composed of the translations (14). For example,

$$D_i^{(1)} := \frac{\sigma_i - \sigma_i^{-1}}{2h}, \quad D_{i,i}^{(1)} := \frac{\sigma_i - 2 + \sigma_i^{-1}}{h^2}, \quad i \in \{1, 2\}$$

and

$$D_i^{(2)} := \frac{-\sigma_i^2 + 8\sigma_i - 8\sigma_i^{-1} + \sigma_i^{-2}}{12h}, \quad D_{i,i}^{(2)} := \frac{-\sigma_i^2 + 16\sigma_i - 30 + 16\sigma_i^{-1} - \sigma_i^{-2}}{12h^2}$$

with $\sigma_1^{-1} \circ u(j, k) = u(j-1, k)$, $\sigma_2^{-1} \circ u(j, k) = u(j, k-1)$, etc., $\sigma_i^2 = \sigma_i \circ \sigma_i$ and $\sigma_i^{-2} = \sigma_i^{-1} \circ \sigma_i^{-1}$.

From Eqs. (26) and (27), we obtain

$$\tilde{G}_0^{(1)} = F_x^{(2)} + F_y^{(3)} + \frac{1}{\mathrm{Re}}\left(F_{xx}^{(1)} + F_{yy}^{(1)}\right) - F^{(4)} + \mathcal{O}(h^2),$$

$$\Rightarrow F_x^{(2)} + F_y^{(3)} + \frac{1}{\mathrm{Re}}\left(F_{xx}^{(1)} + F_{yy}^{(1)}\right) - F^{(4)} = 0, \tag{28}$$

$$\tilde{G}_0^{(2)} = h^2 \left(\partial_x r_1^{(2)} + \partial_y r_1^{(3)} + \frac{1}{\mathrm{Re}}\left(\partial_{xx} r_1^{(1)} + \partial_{yy} r_1^{(1)}\right)\right) + \mathcal{O}(h^4)$$

$$\Rightarrow \partial_x r_1^{(2)} + \partial_y r_1^{(3)} + \frac{1}{\mathrm{Re}}\left(\partial_{xx} r_1^{(1)} + \partial_{yy} r_1^{(1)}\right) = 0, \tag{29}$$

$$\vdots$$

$$\tilde{G}_0^{(k)} = h^{2k}\left(\partial_x r_k^{(1)} + \partial_y r_k^{(3)} + \frac{1}{\mathrm{Re}}\left(\partial_{xx} r_k^{(1)} + \partial_{yy} r_k^{(2)}\right)\right) + \mathcal{O}(h^{2k+2})$$

$$\Rightarrow \partial_x r_k^{(2)} + \partial_y r_k^{(3)} + \frac{1}{\mathrm{Re}}\left(\partial_{xx} r_k^{(1)} + \partial_{yy} r_k^{(1)}\right) = 0 \ \dots.$$

The implication in Eq. (29) follows from the fact that the normal form of the differential polynomial (29) modulo Eq. (3), if it is nonzero, does not belong to the differential ideal generated by the polynomials in (3) that contradicts the s-consistency of $\tilde{G}$. Because of the same argument, the equality (30) holds for any k.   □

Corollary 2. *A w-consistent difference scheme for system (3) is s-consistent if and only if its set of polynomials is a difference Janet/Gröbner basis for the POT ranking (9).*

Proof. "⇐" Because of our choice (9) of the ranking and the structure (3), differential Janet/Gröbner basis with the underlined leaders, a w-consistent difference scheme composed of four difference polynomials $\{\tilde{G}^{(1)}, \tilde{G}^{(2)}, \tilde{G}^{(3)}, \tilde{G}^{(4)}\}$ has the only difference S-polynomial of the form (27) which approximates the left-hand side of the differential integrability condition (25). Together with the Taylor expansion (26), the relations (28)–(30) imply the reduction of S-polynomial (27) to zero modulo $\{\tilde{G}^{(1)}, \tilde{G}^{(2)}, \tilde{G}^{(3)}, \tilde{G}^{(4)}\}$. Thus, the scheme is a Janet/Gröbner basis.

"⇒" If a w-consistent set $\{\tilde{G}^{(1)}, \tilde{G}^{(2)}, \tilde{G}^{(3)}, \tilde{G}^{(4)}\}$ is a Janet/Gröbner basis, then by Theorem 1 it is s-consistent.   □

We illustrate Proposition 2 and Corollary 2 by the s-inconsistent difference scheme $\{\tilde{F}_1^{(1)}, \tilde{F}_1^{(2)}, \tilde{F}_1^{(3)}, \tilde{F}_1^{(4)}\}$ of Sect. 3 where the first three difference equations coincide with those of the system (10),

$$\tilde{F}_1^{(i)} = \tilde{F}^{(i)} \quad (i = 1, 2, 3),$$

and $\tilde{F}_1^{(4)}$ is given by Eq. (20). Because of the distinction of the last equation from $\tilde{F}^{(4)}$ in (10), the reduced Taylor expansions of equations $\tilde{F}_1^{(1)} = 0$ and $\tilde{F}_1^{(4)} = 0$ are different from $\tilde{F}^{(1)} = 0$ and $\tilde{F}^{(4)} = 0$ in system (24):

$$
\begin{cases}
\tilde{F}_1^{(1)} := u_x + v_y - \dfrac{h^2 \operatorname{Re} p_{yy}}{6} + \dfrac{h^2 v_{yyy}}{3} + \mathcal{O}(h^4) = 0, \\[2mm]
\tilde{F}_1^{(2)} := p_x + \dfrac{1}{\operatorname{Re}} v_{xy} - \dfrac{1}{\operatorname{Re}} u_{yy} - f^{(1)} + \dfrac{h^2 f_{xx}^{(1)}}{6} + \dfrac{h^2 f_{yy}^{(1)}}{4} + \dfrac{h^2 f_{xy}^{(2)}}{4} \\[2mm]
\qquad - \dfrac{h^2 p_{xyy}}{2} + \dfrac{h^2 u_{yyyy}}{6\operatorname{Re}} + \mathcal{O}(h^4) = 0, \\[2mm]
\tilde{F}_1^{(3)} := p_y - \dfrac{1}{\operatorname{Re}} v_{xx} - \dfrac{1}{\operatorname{Re}} v_{yy} - f^{(2)} - \dfrac{h^2 f_{xy}^{(1)}}{12} + \dfrac{h^2 f_{xx}^{(2)}}{12} - \dfrac{h^2 f_{yy}^{(2)}}{6} \\[2mm]
\qquad + \dfrac{h^2 p_{yyy}}{3} - \dfrac{h^2 v_{yyyy}}{6\operatorname{Re}} + \mathcal{O}(h^4) = 0, \\[2mm]
\tilde{F}_1^{(4)} := p_{xx} + p_{yy} - f_x^{(1)} - f_y^{(2)} - \dfrac{h^2 f_{xxx}^{(1)}}{12} - \dfrac{h^2 f_{xyy}^{(1)}}{12} \\[2mm]
\qquad + \dfrac{h^2 f_{xxy}^{(2)}}{12} - \dfrac{h^2 f_{yyy}^{(2)}}{4} + \dfrac{h^2 p_{yyyy}}{6} + \mathcal{O}(h^4) = 0.
\end{cases}
\tag{30}
$$

If we expand $\tilde{F}_1^i$ $(i = 1 \div 4)$ up to the fourth order terms in h and substitute the obtained expansions into the left-hand side of the integrability condition (25), then we obtain

$$\frac{h^2 f_{xxx}^{(1)}}{4} - \frac{h^2 f_{xyy}^{(1)}}{4} + \frac{h^2 f_{xxy}^{(2)}}{4} - \frac{h^2 f_{yyy}^{(2)}}{4} + \frac{h^2 p_{yyyy}}{2} + \mathcal{O}(h^4). \tag{31}$$

Expression (31) contains terms of second order in h. Up to the factor 4, the sum of these terms is the differential polynomial $F^{(6)}$ in Eq. (22). Thus, the presence of the second-order terms in (31) is intimately related to the s-inconsistency of (24) with governing Stokes equations (1). It is clear that the PDE system (30) cannot be considered as a modified Stokes flow.

5   Numerical Simulation

In this section, we present a numerical simulation in order to experimentally validate the s-consistent difference scheme (10) for which we constructed the modified Stokes flow (24). For that, we suppose that the Stokes system (1) is defined in the rectangular domain which is discretized in the x- and y-directions by means of equidistant points. We simulate a fluid flow through porous media which is often mainly caused by the viscous forces, so that its modeling using the Stokes system (1) is reasonable; see Fig. 2. Such a setup has many practical applications in the field of petroleum engineering [5].

We measure the maximum relative error of the average velocities compared to a ground truth result obtained by computing with extremely tiny h-values. From several simulations with varying h-values, we can follow that a maximum relative

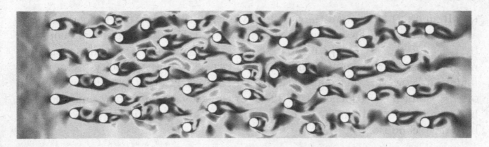

Fig. 2. Visualization of the simulation of a fluid flow through porous media using the s-consistent difference scheme (10).

error of more than 15% in the velocity space compared to the ground truth should not be tolerated in order to ensure for a sufficient degree of global accuracy. Using this restriction we evaluate the performance of the s-consistent difference scheme (10) against the popular classic marker-and-cell (MAC) method [13]. We observe that compared to MAC, using the scheme (10), one can simulate with around a factor of 1.7, i.e., with significantly larger h-values and, at the same time, keep the relative error below the 15%-bar. Moreover, we observe that this factor is only slightly dependent on the Reynolds number.

6 Conclusion

For the two-dimensional incompressible steady Stokes flow (1) and a regular Cartesian solution grid, we presented a computer·algebra-based approach in order to derive the s-consistent difference scheme (10) for which we constructed the modified Stokes flow (24). It shows that the generated scheme has order $\mathcal{O}(h^2)$.

Our computational procedure for the derivation of the modified Stokes flow is based on a combination of differential and difference Gröbner basis techniques. The first is applied to the governing Stokes equations (1) to complete them to the involution form (3) incorporating the pressure Poisson equation $F^{(4)}$, and to verify the s-consistency of the scheme by applying the criterion of s-consistency (Theorem 1) which is fully algorithmic for linear systems of PDEs. The difference Gröbner bases technique is used for the derivation of the scheme on the chosen grid by means of difference elimination.

In addition, we used both techniques to construct a modified Stokes flow (24). Its structure as well as that of the scheme depends on the used difference ranking. We experimented with several rankings and finally preferred the POT ranking satisfying (2) for the differential case and (9) for the difference case as the best suited. To perform the related computations we used the Maple packages JANET [4] and LDA [11].

Since our difference scheme (10) for ranking (9) is obtained from its first three equations $\{\tilde{F}^{(1)}, \tilde{F}^{(2)}, \tilde{F}^{(3)}\}$ by constructing the difference Janet/Gröbner basis (see Remark 2), it is interesting to check via the Gröbner bases whether there are approximations of the continuity equation $F^{(1)}$ in the difference ideal generated by $\tilde{F} := \{\tilde{F}^{(2)}, \tilde{F}^{(3)}, \tilde{F}^{(4)}\}$. In the case of existence of such approximations they might be used for the numerical study of Stokes flow in the velocity-pressure formulation. However, the computation with LDA shows that the discrete version of $F^{(1)}$ is not a consequence of $\tilde{F}$. Thus, in the velocity-pressure formulation one has to add information on the continuity equation to $\tilde{F}$ via the corresponding boundary condition (cf. [18]).

Acknowledgments. The authors are grateful to Daniel Robertz for his help with respect to the use of the packages JANET and LDA and to the anonymous referees for their suggestions. This work has been partially supported by the King Abdullah University of Science and Technology (KAUST baseline funding), the Russian Foundation for Basic Research (16-01-00080) and the RUDN University Program (5-100).

References

1. Adams, W.W., Loustanau, P.: Introduction to Gröbner Bases. Graduate Studies in Mathematics, vol. 3, American Mathematical Society, Providence (1994)
2. Amodio, P., Blinkov, Y., Gerdt, V., La Scala, R.: On consistency of finite difference approximations to the Navier-Stokes equations. In: Gerdt, V.P., Koepf, W., Mayr, E.W., Vorozhtsov, E.V. (eds.) CASC 2013. LNCS, vol. 8136, pp. 46–60. Springer, Cham (2013). https://doi.org/10.1007/978-3-319-02297-0_4
3. Amodio, P., Blinkov, Y.A., Gerdt, V.P., La Scala, R.: Algebraic construction and numerical behavior of a new s-consistent difference scheme for the 2D Navier-Stokes equations. Appl. Math. Comput. **314**, 408–421 (2017)
4. Blinkov, Y.A., Cid, C.F., Gerdt, V.P., Plesken, W., Robertz, D.: The MAPLE package Janet: II. Linear partial differential equations. In: Ganzha, V.G., Mayr, E.W., Vorozhtsov, E.V. (eds.) Proceedings of 6th International Workshop on Computer Algebra in Scientific Computing, CASC 2003, pp. 41–54. Technische Universität München (2003). Package Janet is freely available on the web pagehttp://wwwb.math.rwth-aachen.de/Janet/
5. Fancher, G.H., Lewis, J.A.: Flow of simple fluids through porous materials. Indus. Eng. Chem. Res. **25**(10), 1139–1147 (1933)
6. Ganzha, V.G., Vorozhtsov, E.V.: Computer-Aided Analysis of Difference Schemes for Partial Differential Equations. Wiley, New York (1996)
7. Gerdt, V.P., Blinkov, Y.A., Mozzhilkin, V.V.: Gröbner bases and generation of difference schemes for partial differential equations. SIGMA **2**, 051 (2006)
8. Gerdt, V.P., Blinkov, Y.A.: Involution and difference schemes for the Navier–Stokes equations. In: Gerdt, V.P., Mayr, E.W., Vorozhtsov, E.V. (eds.) CASC 2009. LNCS, vol. 5743, pp. 94–105. Springer, Heidelberg (2009). https://doi.org/10.1007/978-3-642-04103-7_10
9. Gerdt, V.P.: Consistency analysis of finite difference approximations to PDE systems. In: Adam, G., Buša, J., Hnatič, M. (eds.) MMCP 2011. LNCS, vol. 7125, pp. 28–42. Springer, Heidelberg (2012). https://doi.org/10.1007/978-3-642-28212-6_3
10. Gerdt, V.P.: Involutive algorithms for computing Gröbner bases. In: Cojocaru, S., Pfister, G., Ufnarovski, V. (eds.) Computational Commutative and Non-Commutative Algebraic Geometry, NATO Science Series, pp. 199–225. IOS Press (2005)

11. Gerdt, V.P., Robertz, D.: Computation of difference Gröbner bases. Comput. Sci. J. Moldova **20** 2(59), 203–226 (2012). Package LDA is freely available on the web page http://wwwb.math.rwth-aachen.de/Janet/
12. Gerdt, V.P., Robertz, D.: Consistency of finite difference approximations for linear PDE systems and its algorithmic verification. In: Watt, S.M. (ed.) ISSAC 2010, pp. 53–59. Association for Computing Machinery, New York (2010)
13. Harlow, F.H., Welch, J.E.: Numerical calculation of time-dependent viscous incompressible flow of fluid with a free surface. Phys. Fluids **8**, 2182–2189 (1965)
14. Kohr, M., Pop, I.: Viscous Incompressible Flow for Low Reynolds Numbers. Advances in Boundary Elements, vol. 16. WIT Press, Sauthampton (2004)
15. Levin, A.: Difference Algebra. Algebra and Applications, vol. 8. Springer, Heidelberg (2008). https://doi.org/10.1007/978-1-4020-6947-5
16. Milne-Tompson, L.M.: Theoretical Hydrodynamics, 5th edn. Macmillan Education LTD, Banjul (1968)
17. Moin, P.: Fundamentals of Engineering Numerical Analysis, 2nd edn. Cambridge University Press, Cambridge (2010)
18. Petersson, N.A.: Stability of pressure boundary conditions for Stokes and Navier-Stokes equations. J. Comput. Phys. **172**, 40–70 (2001)
19. Seiler, W.M.: Involution: The Formal Theory of Differential Equations and its Applications in Computer Algebra. Algorithms and Computation in Mathematics, vol. 24. Springer, Heidelberg (2010). https://doi.org/10.1007/978-3-642-01287-7
20. Shokin, Y.I.: The Method of Differential Approximation. Springer, Berlin (1983)

Symbolic-Numeric Methods for Nonlinear Integro-Differential Modeling

François Boulier[1]([✉]), Hélène Castel[3], Nathalie Corson[2], Valentina Lanza[2], François Lemaire[1], Adrien Poteaux[1], Alban Quadrat[1], and Nathalie Verdière[2]

[1] Univ. Lille, CNRS, Centrale Lille, Inria,
UMR 9189 - CRIStAL - Centre de Recherche en Informatique
Signal et Automatique de Lille, 59000 Lille, France
`Francois.Boulier@univ-lille.fr`
[2] Normandie Univ, France, UNIHAVRE, LMAH, FR CNRS 3335, ISCN,
76600 Le Havre, France
[3] INSERM, DC2N, Normandie Univ, UNIROUEN, 76000 Rouen, France

Abstract. This paper presents a proof of concept for symbolic and numeric methods dedicated to the parameter estimation problem for models formulated by means of nonlinear integro-differential equations (IDE). In particular, we address: the computation of the model input-output equation and the numerical integration of IDE systems.

1 Introduction

This paper is concerned with the problem of modeling phenomena by systems of nonlinear integro-differential equations (IDE). Motivations for IDE modeling are presented in [14]. In turn, this scientific question raises the two following problems: how to determine the identifiability property of such IDE models? how to estimate parameters from experimental data? We focus on a particular method, called the "input-output (IO) ideal" method, which is available in the nonlinear differential case. The idea of this method consists in computing an equation (called the "IO equation") which is a consequence of the model equations and only depends on the model inputs, outputs and parameters. In the nonlinear differential case, it is known since [27] that it can serve to decide the identifiability property of the model. It is known since [17] that it can also be used to determine a first guess of the parameters from the experimental data. This first guess may then be refined by means of a nonlinear fitting algorithm (of type Levenberg-Marquardt) which requires many different numerical integrations of the model.

Designing analogue theories and algorithms in the IDE case is almost a completely open problem in spite of many recent progresses on the algebraic properties of integro-differential algebras and their operator rings [2–4,19,20,33,36].

This article provides two contributions:

1. a symbolic method for computing an IO equation from a given nonlinear IDE model. This method is incomplete but it is likely to apply over an important class of models that are interesting for modelers;

V. P. Gerdt et al. (Eds.): CASC 2018, LNCS 11077, pp. 82–98, 2018.
https://doi.org/10.1007/978-3-319-99639-4_6

2. an algorithm for the numerical integration of IDE systems, implemented within a new open source C library, endowed with a new MAPLE package called `Blineide`. The library does not seem to have any available equivalent. Our algorithm is an explicit Runge-Kutta method which is restricted to Butcher tableaux specifically designed in order to avoid solving integral equations at each step. In this paper, we provide three such tableaux.

The structure of the paper is as follows. Section 2 provides examples of IDE equations and the symbolic method for computing an IO equation from an IDE model. Section 3 describes our algorithm for the numerical integration of IDE. Section 4 describes its implementation.

2   An IDE Input-Output Equation

This section starts with a short presentation of the Volterra-Kostitzin model, which gives some insight on the point of introducing kernels in models. The second section presents an academic IDE model and explains, over an example, how to compute an IO equation. The last section contains a discussion on how algorithmic the process illustrated by the example is.

2.1   The Volterra-Kostitzin Model

As pointed out in [14], one of the simplest nonlinear integro-differential models studied in the literature is the Volterra-Kostitzin model [26, pp. 66–69] (more recently revisited in [32, Chap. 4]), which may be used for describing the evolution of a population, in a closed environment, intoxicated by its own metabolic products (other applications of the same model are considered in Kostitzin's book). It is an integro-differential equation since the unknown function $y(x)$ appears both differentiated and under some integral sign.

$$\dot{y}(x) = \varepsilon\, y(x) - \lambda\, y(x)^2 - \mu\, y(x) \int_{x-T}^{x} K(x - \xi)\, y(\xi)\, \mathrm{d}\xi. \qquad (1)$$

The independent variable x is time. The dependent variable $y(x)$ is the population, varying with time. The symbols ε, λ, μ and T denote parameters. The *kernel* (or *nucleus*) $K(x, \xi) = K(x - \xi)$ is the *residual action function*. For instance, it could be very similar to a "survival function" in population dynamics [23, p. 3]: a decreasing function, starting at $K(0) = 1$, equal to 0 outside the interval $[0, T]$. Then $K(x - \xi)$ would represent the "toxicity factor" of metabolic products which are the most toxic when produced, at $x = \xi$, become less toxic with time, and have a negligible toxic effect at time $x = \xi + T$.

In the case of models presented by chemical reaction systems, similar kernels could arise from stochastic considerations. Indeed, if the molecularity (the number of reactants) of each reaction is one, then the statistical moments of the random variables which count molecules can be described by ODE [31]. However, if the molecularity of some reactions is greater than one, then the ODE system for the statistical moments becomes infinite and is in general very difficult to approximate by a finite system. A natural idea would then consist in tabulating the density probability of the event under consideration and incorporate the tabulated curve as an integral kernel in some IDE model. See [24, Sect. 3.6].

2.2   A Compartmental IDE Model

The academic two-compartment model depicted in Fig. 1 is a close variant of [40, (1), p. 517] endowed with an input $u(x)$ and an IDE variant of the model studied in [14]. Compartment 1 represents the blood system and compartment 2 represents some organ. Both compartments are supposed to have unit volumes. The function $u(x)$, which has the dimension of a flow, represents a medical drug, injected in compartment 1. The drug diffuses between the two compartments, following linear laws: the proportionality constants are named k_{12} and k_{21}. In this paper, we assume that the drug exits compartment 1, following a law given by an integral term (this model is thus new), depending on a parameter μ (see the Volterra-Kostitzin model for a possible modeling argument). The state variables in this system are $z_1(x)$ and $z_2(x)$. They represent the concentrations of drug in each compartment. This information is sufficient to write the two first equations of the mathematical model (2). The last equation of (2) states that the output, denoted $y(x)$, is equal to $z_1(x)$. This means that only $z_1(x)$ is observed: some numerical data are available for $z_1(x)$ but not for $z_2(x)$. The problem addressed here then consists in estimating the three parameters k_{12}, k_{21} and μ from these data and the knowledge of $u(x)$.

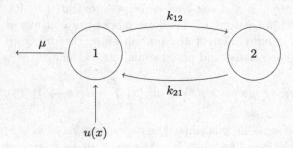

Fig. 1. A two-compartment model featuring three parameters.

In order to estimate the model parameters over such a model, the strategy of the "input-output ideal" method consists in computing from the model equations, an "input-output (IO) equation" featuring only the input $u(x)$, the output $y(x)$ and the unknown parameters. If the model were differential only, the computation of the IO equation, which is an elimination problem, could be handled by means of the elimination theory of differential algebra. See [17,27] and references therein. The IO equation itself could be algebraically described as the single differential polynomial of the regular differential chain are associated to some differential polynomial ideal of some differential polynomial ring. In the IDE case, there does not exist (yet) any integro-differential algebra theory, rich enough to enunciate such a precisely defined statement.

$$\dot{z}_1(x) = -k_{12}\, z_1(x) + k_{21}\, z_2(x) - \mu\, z_1(x) \underbrace{\int_0^x K(x-\xi)\, z_1(\xi)\, \mathrm{d}\xi}_{\text{integral term}} + u(x),$$

$$\dot{z}_2(x) = k_{12}\, z_1(x) - k_{21}\, z_2(x), \tag{2}$$

$$y(x) = z_1(x).$$

2.3   A Work-Around Strategy

It turns out that a work-around strategy is available for a wide class of IDE models. We present it over Model (2).

Renaming Integrals. The idea consists in renaming the integral term using a new unknown function $F(x)$, yielding a polynomial differential model (3), and process this differential model by the classical IO ideal method.

$$\dot{z}_1(x) = -k_{12}\, z_1(x) + k_{21}\, z_2(x) - \mu\, z_1(x)\, F(x) + u(x),$$
$$\dot{z}_2(x) = k_{12}\, z_1(x) - k_{21}\, z_2(x), \tag{3}$$
$$y(x) = z_1(x).$$

Model (3) can be viewed as a polynomial system of the differential polynomial ring $\mathscr{R} = \mathscr{F}\{z_1, z_2, y, u, F\}$, where $\mathscr{F} = \mathbb{Q}(k_{12}, k_{21}, \mu)$. As such, it generates a perfect (even a prime) differential ideal $\mathfrak{A}$. It is even a regular differential chain for $\mathfrak{A}$, with respect to some orderly ranking.

Eliminating State Variables. By an elimination procedure (eliminating z_1 and z_2) one can compute a regular differential chain C_{io} such that $C_{\text{io}} \cap \mathscr{F}\{y, F\}$ is a regular differential chain for the differential ideal $\mathfrak{A} \cap \mathscr{F}\{y, F\}$. The regular differential chain $C_{\text{io}} \cap \mathscr{F}\{y, F\}$ is made of the following single differential polynomial

$$\begin{aligned} D_{\text{io}} = &\; \ddot{y}(x) + \mu\, \dot{y}(x)\, F(x) + k_{12}\, \dot{y}(x) + k_{21}\, \dot{y}(x) - \dot{u}(x) + \mu\, y(x)\, \dot{F}(x) \\ &+ \mu\, k_{21}\, y(x)\, F(x) - k_{21}\, u(x). \end{aligned} \tag{4}$$

Integrating the IO Equation. Applying an integration algorithm for differential fractions, one gets the following reformulation of (4)

$$\begin{aligned} D_{\text{io}} = &\; \mu\, k_{21}\, y(x)\, F(x) - k_{21}\, u(x) \\ &+ \frac{\mathrm{d}}{\mathrm{d}x}\left((k_{12} + k_{21})\, y(x) + \mu\, y(x)\, F(x) - u(x) \right) + \frac{\mathrm{d}^2}{\mathrm{d}x^2}\, y(x), \end{aligned} \tag{5}$$

which can now easily be transformed into an integral equation (integrate twice between 0 and x and use the kernel $x + \xi$ (not to be confused with the kernel $K(x, \xi)$ present in model (3)) to encode double integrals by single ones—see [23, Sect. 1.3.1]):

$$
\begin{aligned}
I_{\text{io}} = {} & \mu \, k_{21} \int_0^x (x - \xi) \, y(\xi) \, F(\xi) \, d\xi - k_{21} \int_0^x (x - \xi) \, u(\xi) \, d\xi \\
& + (k_{12} + k_{21}) \left(\int_0^x y(\xi) \, d\xi - x \, y(0) \right) \\
& + \mu \left(\int_0^x y(\xi) \, F(\xi) \, d\xi - x \, y(0) \, F(0) \right) \\
& - \left(\int_0^x u(\xi) \, d\xi - x \, u(0) \right) + y(x) - y(0) - x \, \dot{y}(0).
\end{aligned}
\tag{6}
$$

Normalizing Integral Terms. It is now time to replace $F(x)$ by its value (and $F(0)$ by 0). However, the expression under the integral sign involves an indeterminate (z_1) which is supposed to be eliminated. Since this expression is a differential polynomial, differential algebra tools can again be applied and we can replace z_1 by its normal form with respect to the regular differential chain C_{io}. Since this chain involves the equation $z_1 = y$, the normal form of z_1 is y and we actually replace $F(x)$ by

$$
\int_0^x K(x - \xi) \, \mathrm{NF}(z_1, C_{\text{io}})(\xi) \, d\xi = \int_0^x K(x - \xi) \, y(\xi) \, d\xi.
$$

The Resulting Equation. After replacement, one eventually gets Eq. 7, given in Fig. 2. In order to establish the global identifiability of model (3), the argument would be the following: Eq. 7 is a linear combination $c_1 m_1 + \cdots + c_4 m_4 = m_0$. In

$$
\begin{aligned}
I_{\text{io}} = {} & \underbrace{\mu \, k_{21}}_{c_4} \underbrace{\left(\int_0^x (x - \xi) \, y(\xi) \int_0^\xi K(\xi - \tau) \, F(\tau) \, d\tau \, d\xi \right)}_{m_4} \\
& \underbrace{- \, k_{21}}_{c_3} \underbrace{\int_0^x (x - \xi) \, u(\xi) \, d\xi}_{m_3} + \underbrace{(k_{12} + k_{21})}_{c_2} \underbrace{\left(\int_0^x y(\xi) \, d\xi - x \, y(0) \right)}_{m_2} \\
& + \underbrace{\mu}_{c_1} \underbrace{\left(\int_0^x y(\xi) \int_0^\xi K(\xi - \tau) \, y(\tau) \, d\tau \, d\xi \right)}_{m_1} \\
& \underbrace{- \left(\int_0^x u(\xi) \, d\xi - x \, u(0) \right) + y(x) - y(0) - x \, \dot{y}(0)}_{m_0}.
\end{aligned}
\tag{7}
$$

Fig. 2. An IO equation for model 3.

principle, the "monomials" m_i can be evaluated at different values of x over the experimental data, yielding a linear system whose unknowns are the blocks of parameters c_i. If the matrix of this linear system has full rank, the system can be solved, providing estimates for the blocks of parameters. Over this system, it is—in principle—easy to recover estimates of the model parameters k_{12}, k_{21}, μ from the estimates of the c_i. These questions are not addressed in this paper.

2.4 Discussion

By many aspects, the computation of Eq. 7 from model (3) suggests algorithms for processing models presented by systems of IDE.

Renaming Integrals. Indeed, it is always possible to rename many different integral terms by new unknown functions $F_i(x)$. The resulting model is a system of differential polynomials (more generally, of differential fractions) in the sense of differential algebra. If the initial IDE model is a dynamical system (i.e. is solved w.r.t. differentiated state variables z_i) then the resulting model defines a prime differential ideal and is a regular differential chain for this ideal, w.r.t. some orderly ranking.

Reference books for differential algebra are [25,35]. Regular differential chains are generalizations of Ritt's characteristic sets. In the non differential context, regular chains provide an alternative to Gröbner bases for describing polynomial ideals and performing some ideal-theoretic constructs. In the differential context, the Gröbner bases theory does not generalize satisfactorily: regular differential chains and other close concepts are the only tools available for investigating properties of differential ideals. See [13] for a recent study of this concept.

Orderly rankings are defined in [25, I, 8, p. 75]. The fact that the differential ideal defined by a dynamical system is prime follows from the fact that each equation of the regular differential chain is linear in its leading derivative, hence cannot be represented as the product of two differential polynomials with positive degree in this leading derivative.

Eliminating State Variables. Eliminating the state variables can be achieved by means of a differential elimination algorithm [1,10,11,28,34], using some specific ranking, leading to some regular differential chain C_{io}. These elimination algorithms can be applied over any system of differential polynomials. They can also be applied over any system of differential fractions, by handling the numerators of the differential fractions as differential polynomial equations and the denominators as differential polynomial inequations (polynomials that are required to be nonzero). See the implementation of [7, RosenfeldGroebner]. Moreover, if the input model already is a regular differential chain w.r.t. some (orderly) ranking, it is possible to apply an improved elimination method [12,30] which avoids splitting cases. Let us conclude this section by a few remarks:

- the state-of-the-art elimination algorithms do not try to minimize the number of times these unknown functions get differentiated, which might be problematic if the integral terms depend on (say) kernels which are not indefinitely differentiable. A similar issue arises in the case of PDE [42];
- if the integral terms satisfy some known differential algebraic relation, it is possible to enlarge the model equations with this relation before the elimination process.

Integrating the IO Equation. For simplicity, let us assume that, among the many different differential polynomials occurring in the regular differential chain C_{io}, a single one (called the IO equation) is free of the state variables.

The integration algorithm [9] may be applied over the IO equation or over any equivalent differential fraction, obtained by dividing the equation by some other differential polynomial, such as the leading coefficient (the initial) of the IO polynomial. The result can then be converted into an IDE (such as (6)) by means of classical techniques. See [8] and [23, Formula (1.45)].

From a theoretical point of view, this integration step is not mandatory. In practice, it leads to formulas which are much more suitable for parameter estimation, as established in [29,39].

Normalizing Integral Terms. Substituting back the unknown functions $F_i(x)$ by the integral terms they represent does not raise any problem. The normalization of the expressions lying under the integral terms leads to a more subtle issue.

In general, an integral term involves, as sub-expressions, many different (e.g. in the case of nested integrals) differential fractions $[f_1, f_2, \ldots, f_r]$. The normal form algorithm presented in [6] can be applied over all these fractions, w.r.t. the whole regular differential chain C_{io}. These normal forms are themselves differential fractions $[\mathrm{NF}_1, \mathrm{NF}_2, \ldots, \mathrm{NF}_r]$. Replacing each f by its normal form, one gets another formulation of the integral term, which is equivalent to the original one.

In full generality, the normal forms may themselves depend on unknown functions $F_i(x)$ and one may consider to iterate this substitution process. If the ranking w.r.t. which C_{io} is defined is not carefully chosen, the substitution process may transform an IO equation into a non-IO equation or (worse) may not terminate at all. A careful study of this issue is left for investigation in another paper.

The Resulting Equation. If the resulting equation does not depend on the state variables at all, it is a candidate for an IO equation. However, in the absence of any sound integro-differential elimination theory, it is not clear that it is minimal. For similar reasons, if the resulting equation depends on state variables so that it is not an IO equation, it is not clear that no IO equation exists at all.

3   Numerical Integration of IDE Systems

According to [41], IDE are a particular case of delay differential equations (DDE) (continuous delay differential equations). However, though there exist numerical solvers for DDE with constant delays [37], there does not seem to be any widely available software for IDE. Within a whole section dedicated to DDE [21, Sect. II.17], a single page is dedicated to the numerical integration problem of IDE in [21, p. 351], which refers to [15] and sketches solutions in particular cases only. In this article, we focus on explicit Runge-Kutta methods. See [18] to a theoretical study of their application to the numerical integration of IDE. The relationship between these early works and our paper still needs some investigation.

3.1   The Method

We are concerned with the numerical integration of IDE of the form

$$\dot{y}(x) = f(x, y(x)), \tag{8}$$

over some integration interval $[x_0, x_{\text{end}}]$. The independent variable x is real. The dependent variable y may actually be a vector of n functions of x. The function f may depend on inputs $u(x)$ and on integral terms of the form

$$\int_{\alpha(x)}^{\beta(x)} K(x, \xi) \, G(y(\xi)) \, d\xi. \tag{9}$$

The inputs $u(x)$ and the kernels $K(x, \xi)$ present within the integral terms (9) are supposed to be C^ρ for some $\rho \geq 0$. For instance, we want to allow inputs to be piecewise defined and kernels to be given by, say, cubic splines. It is required that the integral upper bounds $\beta(x) \leq x$ (typically, $\beta(x) = x$) in order to obtain "causal" systems; various lower bounds are allowed (typically $\alpha(x) = x_0$ or $\alpha(x) = x - T$ for some $T > 0$). Some initial values need also be given. Depending on integral lower bounds $\alpha(x)$, the value of $y(x)$ may need to be prescribed on some sufficiently large interval.

   In this article, we are concerned with the integration problem by means of a numerical integrator derived from explicit Runge-Kutta methods. Moreover, we focus on the study of "fixed step size" integrators. On the one hand, once such an integrator is designed, it is not difficult to design an adaptive step size integrator following the approach which is classical for ODE—since embedded formulas are available. See [21, Sect. II.4]. On the other hand, adaptive step size controllers use the knowledge of the orders of both the principal and the embedded formulas in order to estimate the local error. It is thus important to make sure that the theoretical orders of these formulas correspond to their practical orders—an investigation to be carried out using a "fixed step size" integrator.

   The quotes surrounding the qualifier "fixed step size" are due to the fact that step sizes will actually vary during the integration process. Indeed, assuming some number of steps N is prescribed, one can define a reference step size $h_r = (x_{\text{end}} - x_0)/N$. Assuming moreover that an order p Runge-Kutta method is prescribed, one expects the local error produced by the explicit Runge-Kutta algorithm to be of the order of h_r^{p+1} by [21, Theorem 3.4]. Now, if we had to integrate an ODE, it would be sufficient to perform N steps of size h_r. This strategy does not apply here because we also want to avoid solving integral equations or, more generally, implicit equations involving integrals.

Avoiding Solving Integral Equations. Assume that the current point (x_0, y_0) is known. Consider some integral (9) to be evaluated at $x = x_0$. Assume thus that an approximation of $y(\xi)$ is known over the interval $[\alpha(x_0), x_0]$. Since $\beta(x) \leq x$, we have $[\alpha(x_0), \beta(x_0)] \subset [\alpha(x_0), x_0]$ and the integral (9) can be approximated by a mere quadrature. Thus $f(x_0, y_0)$ also can be approximated by quadratures and, given any step size h, the order 1 Euler method (10) permits to compute an approximation of the next point (x_1, y_1)

$$y_1(h) = y_0 + h\,f(x_0, y_0). \tag{10}$$

This is however not true anymore for order $p > 1$ classical Runge-Kutta methods. Consider Runge midpoint formula, summarized by the following Butcher tableau[1] with $s = 2$ stages [21, Chap. II.1, Table 1.1]

$$
\begin{array}{c|cc}
0 & & \\
c_2 & a_{21} & \\
\hline
& b_1 & b_2
\end{array}
\quad = \quad
\begin{array}{c|cc}
0 & & \\
\frac{1}{2} & \frac{1}{2} & \\
\hline
& 0 & 1
\end{array}
$$

The Runge-Kutta formula [21, II.1, (1.8)] requires s evaluation of the function f of formula (8). These evaluations have the form

$$k_i = f(x_0 + c_i\,h, y_0 + h\,(a_{i,1}\,k_1 + \cdots a_{i,i-1}\,k_{i-1})) \quad (1 \leq i \leq s)$$

Assuming (x_0, y_0) is the current, known, position and the stepsize $h > 0$, we see that negative c_i correspond to an evaluation of f for $x < x_0$ i.e. in the past. *A contrario*, if any c_i is positive (which is the case for all classical tableaux), the evaluation of the formula implies an evaluation in the future which, in the context of IDE, implies solving an integral equation—or worse. To overcome this issue, we have designed the Butcher tableaux of Fig. 3 with negative c_i only. They were obtained, using the MAPLE computer algebra system, by brute force identification of the coefficients of the Taylor series of the exact solution $y(x_0 + h)$ and the ones of the result of the Runge-Kutta formula, denoted $y_1(h)$ in [21, II.1,

---

[1] Butcher tableaux were introduced by Butcher in [16] to provide a compact description of "Runge-Kutta methods". To each tableau is associated a number of stages (customarily denoted s) and an order (customarily denoted p). The computational cost of a Runge-Kutta method increases with the number of stages. The efficiency increases with the order. The coefficients of the tableaux are denoted c_i (the leftmost column), b_j (the bottom row) and $a_{i,j}$ (the matrix).

(1.8)]. The rightmost tableau has 5 stages since a Gröbner basis computation proved that all 4 stages tableaux of order 4 must have $c_4 = 1$ (a result which is known, at least under some simplifying assumptions—see [21, Theorem 1.6]).

$$
\begin{array}{c|cc}
0 & & \\
-\frac{1}{2} & -\frac{1}{2} & \\
\hline
y_1 & 2 & -1 \\
\hat{y}_1 & 1 & 0
\end{array}
\qquad
\begin{array}{c|ccc}
0 & & & \\
-\frac{1}{3} & -\frac{1}{3} & & \\
-\frac{2}{3} & -\frac{4}{9} & -\frac{2}{9} & \\
\hline
y_1 & \frac{19}{4} & -6 & \frac{9}{4} \\
\hat{y}_1 & \frac{5}{2} & -\frac{3}{2} & 0
\end{array}
\qquad
\begin{array}{c|ccccc}
0 & & & & & \\
-\frac{1}{4} & -\frac{1}{4} & & & & \\
-\frac{1}{2} & 0 & -\frac{1}{2} & & & \\
-\frac{3}{4} & 0 & -\frac{3}{8} & -\frac{3}{8} & & \\
-1 & -\frac{23}{123} & \frac{10}{41} & -\frac{38}{41} & \frac{16}{123} & \\
\hline
y_1 & \frac{35}{6} & 0 & -\frac{58}{3} & \frac{64}{3} & -\frac{41}{6} \\
\hat{y}_1 & 1 & 0 & \frac{29}{3} & -\frac{52}{3} & \frac{23}{3}
\end{array}
$$

Fig. 3. The leftmost tableau has $s = 2$ stages, order $p = 2$ and an embedded formula of order $\hat{p} = 1$ (Euler). The tableau in the middle has $s = 3$ stages, order $p = 3$ and an embedded formula of order $\hat{p} = 2$. The rightmost tableau has $s = 5$ stages, order $p = 4$ and an embedded formula of order $\hat{p} = 3$. The coefficients c_i of all tableaux (see the leftmost columns) satisfy $-1 \leq c_i \leq 0$ for $1 \leq i \leq s$.

Stability Analysis. From a theoretical point of view, the stability of Butcher tableaux can be determined by computing the stability function $R(z)$ of each tableau and establishing that its stability domain—which is the subset of the complex plane such that $|R(z)| < 1$—is not empty. Some existing computer algebra software are dedicated to this study [38] but we could not take advantage of them by lack of access to Mathematica. Instead, we directly computed $R(z)$ using [22, IV, (2.8)]. We observed that our two first tableaux, for which $p = s$, exhibit the stability function given in [22, IV, (2.12)]. The leftmost tableau has the following stability function, which admits a non empty stability domain:

$$
R(z) = 1 + z + \frac{z^2}{2} + \frac{z^3}{6} + \frac{z^4}{24} - \frac{z^5}{24}.
$$

Experimental evidence of the existence of non empty stability regions for the tableaux of Fig. 3 is provided in Sect. 4.

Step Size Control. Runge-Kutta methods with $c_i < 0$ have however a drawback when x_0 is the initial value or is on the border of a piecewise defined domain, since the integrator will try to estimate the current derivative of the integral curve on the right hand piece of the domain from derivatives evaluated on the left hand piece. This drawback is certainly a feature for integral terms (by design of the equations). But the terms which lie outside integrals should be evaluated on the right hand piece of the integration domain.

To achieve this goal, our strategy consists in starting with a single Euler step, using a very small step size h_0, then switch to some prescribed more efficient

Runge-Kutta method of Fig. 3 and double the step size at each iteration until the reference step size h_r is reached. Precisely, assume we want to apply some Runge-Kutta method of order $p > 1$. We expect a local error of order h_r^{p+1}. This local error can also be obtained by an Euler step with step size h_0 such that $h_0^2 = h_r^{p+1}$ i.e. such that $h_0 = h_r^{(p+1)/2}$. Let now k be an integer such that $h_0 \simeq h_r/2^k$. Solving, one gets

$$k = \left\lceil -\log_2\left(h_r^{\frac{p-1}{2}}\right)\right\rceil, \quad h_0 = \frac{h_r}{2^k}.$$

Let us assume we are starting the integration with x_0 precisely on the border between two pieces of the integration interval or at its beginning. The first Euler step with step size h_0 does not involve any negative c_i: the terms depending on y and lying outside integrals are evaluated over the border, which may be considered as part of the right hand piece. The second iteration starts at $x_0 + h_0$. Since the coefficients c_i of Fig. 3 satisfy $0 \geq c_i \geq -1$, this step can be performed using the order p Runge-Kutta method, with step size h_0: all terms depending on y and lying outside integrals are thus evaluated within the right hand piece. The third iteration starts at $x_0 + 2h_0$. This step can be performed using the order p Runge-Kutta method, with step size $h = 2h_0$. Continue likewise, doubling the step size at each iteration. At the iteration $k + 2$, the reference step $h = h_r$ is reached (see below) and the integrator may continue with this fixed step size.

Step number	Step size	Method
1	$h_0 = h_r/2^k$	Euler
2	h_0	Order p RK
3	$2h_0$	Order p RK
$\vdots$		
$k + 2$	$2^k h_0 = h_r$	Order p RK

Evaluating Quadratures. In order to evaluate quadratures at any x, it is necessary to be able to evaluate the dependent variable y at any $\xi \in [x_0, x]$.

For this, the whole sequence of points (x_k, y_k) computed by the numerical integrator is recorded as well as the value $f_k = f(x_k, y_k)$ (the derivative of y) whenever it is available. Two methods are implemented for estimating $y(\xi)$:

1. by evaluating the interpolation polynomial defined by a set of points surrounding ξ (the optimal number of points depends on the order of the Butcher tableau), using Newton's divided differences;
2. by evaluating the interpolation polynomial defined by Hermite interpolation i.e. over a dense output of the integrator. See [21, Chap. II.6].

For quadratures, since the orders of our tableaux do not exceed 4, we use basic integration schemes i.e. Newton and Simpson order 4 formulas, with step size equal to the reference step size h_r.

4   Implementation

Our numerical integrator is implemented within an open source C library (about 4000 lines plus 3200 lines for the test suite of version 2.1) available at [5]. It compiles over Linux platforms. It is endowed with a MAPLE library which considerably simplifies the C code generation from mathematical systems.

The C code can be compiled using floating point numbers of various sizes (simple, double, long double and quadruple precision). Its main functionalities are a fixed step size numerical integrator for IDE systems and a function which seeks the best fitting parameters of an IDE system w.r.t. experimental data. This function is mostly a call to the GSL implementation of the Levenberg-Marquardt algorithm, which relies on our numerical integrator in order to compute errors.

4.1   Data Types

Here is a quick review of the main data types. For a better flexibility, most of them are parametrized by functions.

The library has been designed to apply over a piecewise defined integration interval. Pieces may arise from many different sources: inputs may be piecewise defined, delayed evaluations such as $y(x - T)$ may occur from differentiated integral terms ... The boundaries between the pieces of the integration interval are called *critical points*.

A specific data type permits to describe the possibly many different inputs $u(x)$ of the IDE system to be integrated. An input is defined by a name, an evaluation function and a function which permits to enlarge the set of the model critical points with the ones which are due to the input.

A specific data type is dedicated to the model parameters. A parameter is defined by a name, a floating point value, a function which permits to enlarge the set of the model critical points with the ones which are related to the parameter, and two functions providing a transformation and its reciprocal before performing nonlinear fitting methods (an example of such a useful pair of transformations is the pair log / exp to keep positive small parameters which must remain positive).

A specific data type describes the problem i.e. the IDE system to be integrated. A problem is defined by a dimension n, an integration interval $[x_0, x_{\text{end}}]$, an array of n initial value functions (in the general case, the numerical integration of an IDE requires the knowledge of the dependent variable y over an interval, not only a single value at x_0), an array of inputs, an array of parameters and a function *fcn* for evaluating the right hand sides of the IDE equations. A field of the problem data structure contains a description of the problem critical points.

The integral terms (9) occurring in the right hand sides of the IDE equations are described in a separate array of the problem data structure. This permits to evaluate them before calling *fcn* in order to speed up the integrator as follows.

Recall that, at each integration step, the m (say) integral terms have to be evaluated s (the number of stages) times. Thus: (1) by grouping the $m \times s$ quadrature evaluations in the code, it is much easier to compute them in parallel using OpenMP facilities; (2) in some cases, the s evaluations of a given integral term can be computed almost at the cost of a single evaluation, by updating the current result from one stage to the following one.

A last data type contains the whole data needed by the integration process (it is called the "history"). It contains the sequence of all points computed so far (which is the actual history), the problem, the Butcher tableau to be used and a few other fields of minor importance.

4.2   Usefulness of the Computer Algebra Package

A MAPLE package, called `Blineide` and shipped with the C library, permits to handle IDE problems given by mathematical formulas. It permits either to directly perform computations from MAPLE or to generate C code to help programmers who want to work at C level.

Beyond the obvious simplification provided to the user, the idea of generating C code from a computer algebra software provides two important enhancements which are not yet implemented: (1) it should permit to detect linear (algebraic?) dependencies between the integral terms occurring in the IDE model and use this information to reduce the computation cost; (2) it might permit a symbolic study of the location of critical points for a better reliability of the integrator.

4.3   Tests

Checking Convergence Towards Exact Solution. Some tests are designed to check that the numerical integrator converges toward the true solution of a given IDE system, with the expected experimental order. An example of such an IDE is the following one, which admits $y(x) = \cos(x)$ as a solution:

$$\dot{y}(x) = \sin(x) - y^2(x) + 4 \int_0^x (x - \xi)^2 \sin(\xi)\, y(\xi)\, \mathrm{d}\xi - x^2 + 1,$$
$$y(0) = 1.$$

In order to test the experimental order of the numerical integrator over a given example, the test function computes the relative error produced with 2^k integration steps, for many different values of k. The experimental order is then estimated, by linear least squares, as the slope a of the following equation:

$$k\,a + b = -\log_2(\text{relerr}). \tag{11}$$

Other tests check the behaviour of the numerical integrator using various inputs and kernels, including kernels defined by cubic splines.

Checking Experimental Orders. We checked our integrator over Volterra-Kostitzin model (1) and the compartmental IDE model (3). In the case of the Volterra-Kostitzin model, we estimated the practical order of the integrator when used with the Butcher tableaux of Fig. 3. In particular, we addressed the case of non smooth kernels in integrals (see the curves of Fig. 4) and non smooth inputs (curves not shown).

On the left hand picture of Fig. 4, the kernel is a cubic spline (i.e. a C^2 curve). On the right hand picture, the kernel is a smooth curve (on the two pictures, the mathematical problem to be solved is thus not the same). In each picture, there is one curve per Butcher tableau of Fig. 3. Each curve was obtained as follows: a first integration was performed with 2^{15} steps. Its result was then considered as a reference value and compared with the result of other integrations with $2^8, 2^9, \ldots, 2^{14}$ steps, giving 7 points hence a curve, which should be a straight line (see formula (11)). Its slope is a numerical estimate of the order of the numerical integrator. In the case of a C^2 kernel, the integrator has order 2 when used with an order 2 tableau; and a non reliable order close to 2 when used with order 3 and 4 tableaux. In the case of a smooth kernel, the integrator has the same order as the tableau with which it is used (the curve for order 4 is slightly irregular because the order of the quadrature formula does not match the one of the Butcher tableau).

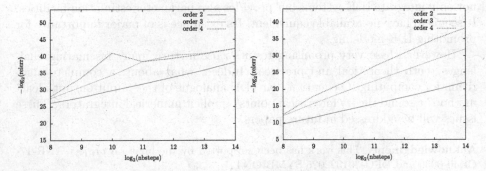

Fig. 4. Experimental evaluation of the order of the IDE numerical integrator over Volterra-Kostitzin model (1), with an integral lower bound equal to zero.

Nonlinear Fit. A test solves the fitting problem addressed by Kostitzin over data obtained on a population of *staphylococcus*, obtaining a much better result than [26, p. 72] which is to be expected since Kostitzin estimated parameters using his mathematical skills, without any computer! See Fig. 5.

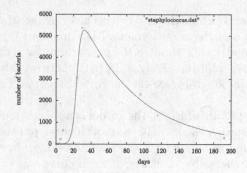

Fig. 5. Best fitting curve for (1) with the trivial kernel $K(x, \xi) = 1$ and an integral lower bound equal to zero, against the *staphylococcus* population reported in [26, p. 72]. Optimal parameters are $\varepsilon = 3.97 \times 10^{-1}$, $\lambda = 6.56 \times 10^{-5}$ and $\mu = 1.02 \times 10^{-6}$.

Conclusion

We have presented and discussed a symbolic method for computing the IO equation of a given IDE system which is likely to apply over an important class of IDE models, together with an open source library dedicated to the numerical integration of such systems, endowed with a new MAPLE package. This library does not only integrate IDE systems but provides also parameter estimation facilities. It seems to have no available equivalent. Its existence is of major importance for promoting IDE modeling.

However, these very promising results raise in turn many fascinating challenges, both theoretical and practical. Indeed, what about: a complete algorithm for computing IO equations? an IDE analogue of the "input-output ideal" method? a sound theory for critical points? implicit numerical integrators? These issues will be addressed in future papers.

Acknowledgment. This work has been supported by the bilateral project ANR-17-CE40-0036 and DFG-391322026 SYMBIONT.

References

1. Bächler, T., Gerdt, V., Lange-Hegermann, M., Robertz, D.: Algorithmic Thomas decomposition of algebraic and differential systems. J. Symb. Comput. **47**(10), 1233–1266 (2012)
2. Bavula, V.V.: The algebra of integro-differential operators on a polynomial algebra. J. Lond. Math. Soc. **83**(2), 517–543 (2011)
3. Bavula, V.V.: The algebra of integro-differential operators on an affine line and its modules. J. Pure Appl. Algebra **17**(3), 495–529 (2013)
4. Bavula, V.V.: The algebra of polynomial integro-differential operators is a holonomic bimodule over the subalgebra of polynomial differential operators. Algebras Represent. Theory **17**(1), 275–288 (2014)
5. Boulier, F., et al.: BLINEIDE. http://cristal.univ-lille.fr/~boulier/BLINEIDE

6. Boulier, F., Lemaire, F.: A normal form algorithm for regular differential chains. Math. Comput. Sci. **4**(2), 185–201 (2010). https://doi.org/10.1007/s11786-010-0060-3

7. Boulier, F., Cheb-Terrab, E.: DifferentialAlgebra. Package of MapleSoft MAPLE Standard Library Since MAPLE 14 (2008)

8. Boulier, F., Korporal, A., Lemaire, F., Perruquetti, W., Poteaux, A., Ushirobira, R.: An algorithm for converting nonlinear differential equations to integral equations with an application to parameter estimation from noisy data. In: Gerdt, V.P., Koepf, W., Seiler, W.M., Vorozhtsov, E.V. (eds.) CASC 2014. LNCS, vol. 8660, pp. 28–43. Springer, Cham (2014). https://doi.org/10.1007/978-3-319-10515-4_3

9. Boulier, F., Lallemand, J., Lemaire, F., Regensburger, G., Rosenkranz, M.: Additive normal forms and integration of differential fractions. J. Symb. Comput. **77**, 16–38 (2016)

10. Boulier, F., Lazard, D., Ollivier, F., Petitot, M.: Representation for the radical of a finitely generated differential ideal. In: ISSAC 1995: Proceedings of the 1995 International Symposium on Symbolic and Algebraic Computation, pp. 158–166. ACM Press, New York (1995). http://hal.archives-ouvertes.fr/hal-00138020

11. Boulier, F., Lazard, D., Ollivier, F., Petitot, M.: Computing representations for radicals of finitely generated differential ideals. Appl. Algebra Eng. Commun. Comput. **20**(1), 73–121 (2009). (1997 Technical report IT306 of the LIFL). https://doi.org/10.1007/s00200-009-0091-7

12. Boulier, F., Lemaire, F., Moreno Maza, M.: Computing differential characteristic sets by change of ordering. J. Symb. Comput. **45**(1), 124–149 (2010). https://doi.org/10.1016/j.jsc.2009.09.04

13. Boulier, F., Lemaire, F., Moreno Maza, M., Poteaux, A.: An equivalence theorem for regular differential chains. J. Symb. Comput. (2018, to appear)

14. Boulier, F., Lemaire, F., Rosenkranz, M., Ushirobira, R., Verdière, N.: On symbolic approaches to integro-differential equations. In: Advances in Delays and Dynamics. Springer (2017). https://hal.archives-ouvertes.fr/hal-01367138

15. Brunner, H., van der Hoeven, P.J.: The Numerical Solution of Volterra Equations. North-Holland, Amsterdam (1986)

16. Butcher, J.C.: On Runge-Kutta processes of high order. J. Austral. Math. Soc. IV, Part 2 **4**, 179–194 (1964)

17. Denis-Vidal, L., Joly-Blanchard, G., Noiret, C.: System identifiability (symbolic computation) and parameter estimation (numerical computation). Numer. Algorithms **34**, 282–292 (2003)

18. Feldstein, A., Sopka, J.R.: Numerical methods for nonlinear Volterra integro-differential equations. SIAM J. Numer. Anal. **11**(4), 826–846 (1974)

19. Gao, X., Guo, L.: Constructions of free commutative integro-differential algebras. In: Barkatou, M., Cluzeau, T., Regensburger, G., Rosenkranz, M. (eds.) AADIOS 2012. LNCS, vol. 8372, pp. 1–22. Springer, Heidelberg (2014). https://doi.org/10.1007/978-3-642-54479-8_1

20. Guo, L., Regensburger, G., Rosenkranz, M.: On integro-differential algebras. J. Pure Appl. Algebra **218**(3), 456–473 (2014)

21. Hairer, E., Norsett, S.P., Wanner, G.: Solving Ordinary Differential Equations I. Nonstiff Problems. Computational Mathematics, vol. 8, 2nd edn. Springer, New York (1993)

22. Hairer, E., Wanner, G.: Solving Ordinary Differential Equations II. Stiff and Differential-Algebraic Problems. Computational Mathematics, vol. 14, 2nd edn. Springer, New York (1996)

23. Jerri, A.J.: Introduction to Integral Equations with Applications. Monographs and Textbooks in Pure and Applied Mathematics, vol. 93. Marcel Dekker Inc., New York (1985)

24. Keener, J., Sneyd, J.: Mathematical Physiology I: Cellular Physiology. Interdisciplinary Applied Mathematics, vol. 8/I, 2nd edn. Springer, New York (2010)

25. Kolchin, E.R.: Differential Algebra and Algebraic Groups. Academic Press, New York (1973)

26. Kostitzin, V.A.: Biologie Mathématique. Armand Colin (1937). (with a foreword by Vito Volterra)

27. Ljung, L., Glad, S.T.: On global identifiability for arbitrary model parametrisations. Automatica **30**, 265–276 (1994)

28. Mansfield, E.L.: Differential Gröbner bases. Ph.D. thesis, University of Sydney, Australia (1991)

29. Moulay, D., Verdière, N., Denis-Vidal, L.: Identifiability of parameters in an epidemiologic model modeling the transmission of the Chikungunya. In: Proceedings of the 9ème Conférence Internationale de Modélisation, Optimisation et SIMulation (2012)

30. Ollivier, F.: Le problème de l'identifiabilité structurelle globale: approche théorique, méthodes effectives et bornes de complexité. Ph.D. thesis, École Polytechnique, Palaiseau, France (1990)

31. Paulsson, J., Elf, J.: Stochastic modeling of intracellular kinetics. In: Szallasi, Z., Stelling, J., Periwal, V. (eds.) System Modeling in Cellular Biology: From Concepts to Nuts and Bolts, pp. 149–175. The MIT Press, Cambridge (2006)

32. Pavé, A.: Modeling Living Systems: From Cell to Ecosystem. ISTE/Wiley, Hoboken (2012)

33. Quadrat, A., Regensburger, G.: Polynomial solutions and annihilators of ordinary integro-differential operators. In: IFAC Proceedings, vol. 46, no. 2, pp. 308–313 (2013)

34. Reid, G.J., Wittkopf, A.D., Boulton, A.: Reduction of systems of nonlinear partial differential equations to simplified involutive forms. Eur. J. Appl. Math. **7**(6), 635–666 (1996)

35. Ritt, J.F.: Differential Algebra, American Mathematical Society Colloquium Publications, vol. 33. American Mathematical Society, New York (1950)

36. Rosenkranz, M., Regensburger, G.: Integro-differential polynomials and operators. In: Jeffrey, D. (ed.) ISSAC 2008: Proceedings of the 2008 International Symposium on Symbolic and Algebraic Computation. ACM Press (2008)

37. Shampine, L.F., Thompson, S.: Solving DDEs in MATLAB. Appl. Numer. Math. **37**, 441–458 (2001)

38. Sofroniou, M.: Order stars and linear stability theory. J. Symb. Comput. **21**, 101–131 (1996)

39. Verdière, N., Denis-Vidal, L., Joly-Blanchard, G.: A new method for estimating derivatives based on a distribution approach. Numer. Algorithms **61**, 163–186 (2012)

40. Verdière, N., Denis-Vidal, L., Joly-Blanchard, G., Domurado, D.: Identifiability and estimation of pharmacokinetic parameters for the ligands of the macrophage mannose receptor. Int. J. Appl. Math. Comput. Sci. **15**(4), 517–526 (2005)

41. Wikipedia, the Free Encyclopedia: Delay Differential Equations. https://en.wikipedia.org/wiki/Delay_differential_equation

42. Zhu, S.: Modeling, identifiability analysis and parameter estimation of a spatial-transmission model of Chikungunya in a spatially continuous domain. Ph.D. thesis, Université de Technologie de Compiègne, Compiègne, France (2017)

A Continuation Method for Visualizing Planar Real Algebraic Curves with Singularities

Changbo Chen and Wenyuan Wu[⊠]

Chongqing Key Laboratory of Automated Reasoning and Cognition,
Chongqing Institute of Green and Intelligent Technology,
Chinese Academy of Sciences, University of Chinese Academy of Sciences,
Beijing, China
changbo.chen@hotmail.com, wuwenyuan@cigit.ac.cn

Abstract. We present a new method for visualizing planar real algebraic curves inside a bounding box based on numerical continuation and critical point methods. Since the topology of the curve near a singular point is not numerically stable, we trace the curve only outside neighborhoods of singular points and replace each neighborhood simply by a point, which produces a polygonal approximation that is ϵ-close to the curve. Such an approximation is more stable for defining the numerical connectedness of the complement of the curve, which is important for applications such as solving bi-parametric polynomial systems.

The algorithm starts by computing three types of key points of the curve, namely the intersection of the curve with small circles centered at singular points, regular critical points of every connected component of the curve, as well as intersection points of the curve with the given bounding box. It then traces the curve starting with and in the order of the above three types of points. This basic scheme is further enhanced by several optimizations, such as grouping singular points in natural clusters and tracing the curve by a try-and-resume strategy. The effectiveness of the algorithm is illustrated by numerous examples.

1 Introduction

Visualizing an implicit plane real algebraic curve is a classical and fundamental problem in computational geometry and computer graphics. There have been many works on this topic [5–7,15,18,20,21,24]. In the literature, a correct visualization usually requires two conditions: (i) the generated polygonal approximation is ϵ-close to the curve, and (ii) the approximation is "topologically correct", which often means that the approximation is isotopic to the curve. There are also many works [12,14,19,27] focusing only on (ii).

Different techniques [16] exist for visualizing plane curves, such as implicit-to-parametric conversion, curve continuation and space subdivision. Symbolic or hybrid symbolic-numeric approaches stand out for being capable of computing the exact topology and many of them are variants of cylindrical algebraic

© Springer Nature Switzerland AG 2018
V. P. Gerdt et al. (Eds.): CASC 2018, LNCS 11077, pp. 99–115, 2018.
https://doi.org/10.1007/978-3-319-99639-4_7

decomposition. For the continuation-based approach, several difficulties arise, such as finding at least one seed point from each connected component, dealing with curve jumping and handling singularities. Each of the three problems has its own interests. For instance, there are several approaches for computing at least one witness point for a real variety, either symbolically [8,13,26] or numerically [17,29,30]. Different techniques for robustly tracing curves are proposed [2,4,25,30]. Techniques for handling singularities also exist [1,11,23].

For curves with singularities, observe that condition (ii) is numerically ill-posed, since a slight perturbation may completely change the topology of the curve near a singular point, see Example 1 for instance. On the other hand, in many applications, such as solving bi-parametric polynomial systems, condition (ii) is unnecessary. Let us illuminate this point now. For a given bi-parametric polynomial system, one can compute a border curve [9,10], or a border polynomial [31] or discriminant variety [22] in general in the parametric space, where the complement of the curve is a disjoint union of connected open cells, such that above each cell the number of solutions of the system is constant and the solutions are continuous functions of parameters with disjoint graphs. Let B be a border curve and $\tilde{B}$ be a polygonal approximation ϵ-close to it. In [10], we introduced the notion of ϵ-connectedness and showed that two points are ϵ-connected w.r.t. $\tilde{B}$ implies that they are connected w.r.t. B, which in turn implies that the parametric system has the same number of solutions at the two points. Thus an ϵ-approximation of the border curve meeting only condition (i) is good enough for the purpose of solving parametric systems. The curve tracing subroutine in [10] relies on perturbation to handle singularities. In this work, we develop a perturbation free algorithm. The algorithm traces the curve only outside neighborhoods of singular points and replaces each neighborhood simply by a point. An approximation produced in this way is more stable for defining the numerical connectedness of the complement of the curve than those approximations preserving the topology around singular points.

The paper is organized as follows. In Sect. 2, we formalize the problem and provide a theoretical base algorithm to guarantee ϵ-closeness based on a robust curve tracing method. Several strategies for improving the numerical stability of tracing is proposed in Sect. 3, such as tracing the curve away from the singular points rather than towards it, tracing the curve by a try-and-resume strategy, and classifying singular points into natural clusters [3]. The theoretical algorithm may require the step size to be very small. In Sect. 4, we present a more practical algorithm based on optimizations in Sect. 3. Instead of preventing curve jumping, it maintains a simple data structure to detect curve jumping. The effectiveness of the algorithm is demonstrated through several nontrivial examples in Sect. 5. Finally, in Sect. 6, we draw the conclusion and point out some possible future directions to improve the current work.

2   A Theoretical Base Algorithm

It is highly nontrivial for continuation methods to guarantee that the polygonal chains are ϵ-close to the curve even when the curve contains no singular points.

Robust tracing without curve jumping must be involved. In the literature, there are several techniques [2, 4, 25] that can solve this. Here we rely on a technique developed in [30], which has been used to estimate the error of numerically computed border curves in [10]. In particular, we have Theorem 1, which provides a way to obtain ϵ-approximation of a regular section of a curve.

2.1   Robust Tracing of Regular Curves

Definition 1. *Let X and Y be two non-empty subsets of a metric space (M, d). The Hausdorff distance $d_H(X, Y)$ is defined by*

$$d_H(X, Y) = \max\{\sup_{x \in X} \inf_{y \in Y} d(x, y), \sup_{y \in Y} \inf_{x \in X} d(x, y)\}.$$

Given a squarefree polynomial $f \in \mathbb{R}[x, y]$, a finite box $B \subset \mathbb{R}^2$ and a given precision $\epsilon \in R$. A set S of polygonal chains contained in B is called an ϵ-approximation of $V_{\mathbb{R}}(f)$ if $d_H(Z_{\mathbb{R}}(S), V_{\mathbb{R}}(f) \cap B) \leq \epsilon$ holds.

Let f be a squarefree polynomial of $\mathbb{R}[x, y]$. Let $\mathcal{J}_f$ be the Jacobian of (f), or simply $\mathcal{J}$ if no confusion arises. Let D be the unit disk centered at the origin. Let B be a bounding box of $\mathbb{R}^2$. W.l.o.g, we assume that $B \subset D$ and that $K(f) = \max(\{\|\nabla \mathcal{J}_{ij}(z)\|) \mid z \in D\}) \leq 1$ holds, which can always be achieved by shifting and rescaling.

Let $\tilde{z}_0$ be an approximate point of $V_{\mathbb{R}}(f)$, such that there exists a τ to make the intersection of $\|z - \tilde{z}_0\| \leq \tau$ and $V_{\mathbb{R}}(f)$ have only one connected component[1] and the line in the gradient direction of f at $\tilde{z}_0$ have only one intersection point z_0 with the component, see Fig. 1. We call z_0 the associated exact point of $\tilde{z}_0$ on $V_{\mathbb{R}}(f)$. Similarly we define $\tilde{z}_1$ and z_1.

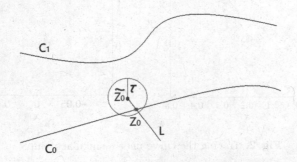

Fig. 1. The associated exact point of an approximate point of the curve.

Let $\tilde{\sigma}_0$ and $\tilde{\sigma}_1$ be respectively the singular value of $\mathcal{J}_f(\tilde{z}_0)$ and $\mathcal{J}_f(\tilde{z}_1)$. Let $\tilde{\sigma} := \max(\tilde{\sigma}_0, \tilde{\sigma}_1)$. Let $\rho \geq 1$ and $\omega = \sqrt{2(2\rho - 1)\left(2\rho - 2\sqrt{\rho(\rho - 1)} - 1\right)}$

---

[1] This component is a subset of a connected component C of $V_{\mathbb{R}}(f)$ and the point $\tilde{z}_0$ belongs to the Voronoi cell of C.

Assume that $2\rho > 3\omega$ holds (which is true for any $\rho \geq 1.6$). Let $s = \frac{\tilde{\sigma} - \sqrt{2}\tau}{2\sqrt{2}\rho}$. Assume that $s > 0$ and $\|\tilde{z}_1 - \tilde{z}_0\| < \omega \cdot s - 2\tau$. We have the following theorem.

Theorem 1. *The points z_0 and z_1 are on the same component of $V_R(f)$. Let $C_{z_0 z_1}$ be the curve segment between z_0 and z_1 in $V_R(f)$. The Hausdorff distance between $C_{z_0 z_1}$ and the segment $\overline{\tilde{z}_0 \tilde{z}_1}$ is at most*

$$\tan\left(2\arccos(\frac{1}{\omega})\right) \frac{\omega}{4\sqrt{2}\rho}(\sqrt{2}\tau + \tilde{\sigma}) + \tau, \text{ or } 1.082\tau + 0.058\tilde{\sigma} \text{ if } \rho = 1.6.$$

Remark 1. *The proof of the theorem is almost the same as Theorem 1 in [10] and thus will not to be replicated here.*

2.2   Handling Singularities

It is a well known fact that tracing a curve near singularities is difficult, as illustrated in Fig. 2. The left subfigure illustrates tracing the zero of $f = y^2 - (-x^2 + x)^3$ starting with a regular point, where the right subfigure zooms in the part of the left subfigure near the origin, which is a singular point. We see that it may be difficult for curve tracing to escape out of the area near the origin, as near the origin, Newton's method requires to solve a linear system $Az = b$ with a very large condition number. As a result, the errors are radically amplified.

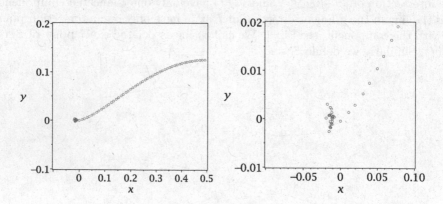

Fig. 2. Tracing the curve near a singular point.

Even worse, the topology of the curve near singularities is not numerically stable, as illustrated by Example 1.

Example 1. *Let $f := x^2 - y^2$. A slight perturbation of its coefficients changes completely the local topology near its singular point $(0,0)$, as depicted in Fig. 3.*

So instead of tracing through a singular point, we bypass it. Before presenting an algorithm, we first use a simple example to illustrate the main idea.

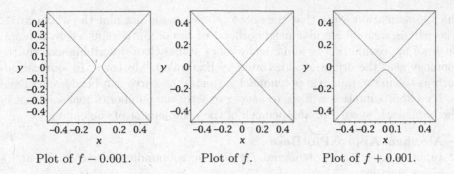

Plot of $f - 0.001$.   Plot of f.   Plot of $f + 0.001$.

Fig. 3. Plot of f and its perturbations.

Example 2. *Consider the polynomial* $f := 6\,xy^7 + 85\,x^4y^3 - 60\,x^2y^5 - 32\,x^2y^3 + 14\,x^4 - 35\,y^4$. *Its real zero set is displayed in Fig. 4 as the red curve.*

It has three connected components inside the box $-3 \le x \le 3, -4 \le y \le 2$. The component on the top has an isolated singular point $(0,0)$, colored in green. To plot this component, we first draw a circle centered at the origin, which has four intersection points with the curve, colored in black. Then we trace the four branches starting with the four black points until meeting the boundary. Next we plot the component at the left bottom corner. To do that, we start with a blue point, which is an intersection point of the curve with a boundary of the box, and trace the curve until meeting a boundary of the box. At last, we plot the closed component at the right bottom corner. To do that, we compute critical points of the curve in x-direction and get two yellow points. Starting with any point of them, trace the curve until meeting the point itself. Finally we plot the singular point. See the right subfigure of Fig. 4 for a visualization of

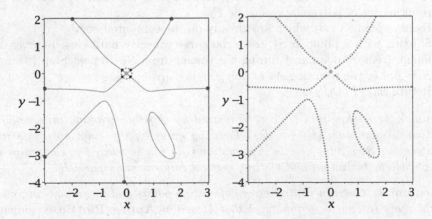

Fig. 4. Left: the curve and key points. Right: an approximation of the curve ($\epsilon = 0.4$). (Color figure online)

the approximation. Note that the above procedure did not plot the whole curve. The part in a small circular neighborhood of the origin is replaced by a point. Such an approximation is numerically more stable than describing exactly the topology near the origin, as illustrated by Example 1. Moreover, in applications such as solving parametric polynomial systems, the curve is a border curve and such an approximation suffices to answer exactly the number of real solutions of the parametric system in an open cell of the complement of the curve.

- Algorithm **ApproxPlotBase**
- Input: a squarefree polynomial $f \in \mathbb{R}[x, y]$; a bounding box $B \subset \mathbb{R}^2$, and a given precision ϵ.
- Output: an ϵ-approximation of $V_{\mathbb{R}}(f)$.
- Assumptions: (i) the singular points are not on the boundary of the box B; (ii) the distance between two singular points is at least ϵ; (iii) $V_R(f)$ has no vertical components or equivalently f has no factors in $\mathbb{R}[x]$.

1. Compute the singular points S_0 (inside B) by solving $\{f, \frac{\partial f}{\partial x}, \frac{\partial f}{\partial y}\}$.
2. Compute the intersection of the curve with circles centered at the singular points with radius less than $\epsilon/2^2$. Set S_1 to be the set of these points, called circular ring points, and C_1 to be the set of these circles.
3. Compute the intersection of the curve with the boundaries. Set S_2 to be the set of these points.
4. Compute the witness points of $V_R(f)$ (inside B) by solving $\{f, \frac{\partial f}{\partial y}\}$. Set S_3 to be the set of these points. Remove from S_3 points that are already inside any circle in C_1.
5. Starting with a point in S_1, trace the curve robustly based on Theorem 1 until meeting (ϵ-close to) a point in S_1 or S_2. Remove the corresponding points met in S_1 or S_2. Repeat Step (5) until $S_1 = \emptyset$. Let the resulting set of polygonal chains be P_1.
6. Starting with a point in S_2, trace the curve robustly until meeting a point in S_2. Remove the point met in S_2. Repeat Step (6) until $S_2 = \emptyset$. Let the resulting set of polygonal chains be P_2.
7. Remove points of S_3 which are already on the computed curve.
8. Starting with a point in S_3, trace the curve robustly until closed curves are found. Remove point met during the tracing from S_3. Repeat Step (8) until $S_3 = \emptyset$. Let the resulting set of polygonal chains be P_3.
9. Return $S_0 \cup P_1 \cup P_2 \cup P_3$.

Remark 2. *Assumption (i) can be relaxed by slightly shrinking or expanding the box. Assumption (ii) can be relaxed by grouping the singular points into clusters. See Sect. 3 for details. Assumption (iii) can be relaxed by computing an irreducible factorization and plotting vertical components separately.*

Theorem 2. *One can control errors of staring points and prediction-correction in the above tracing algorithm, such that Algorithm **ApproxPlotBase** computes an ϵ-approximation of $V_{\mathbb{R}}(f)$.*

---

[2] One could replace the circles with axis aligned boxes inside them.

Proof. We remark that to obtain an ϵ-approximation of the curve, one must have one witness point from each connected component of the curve. If a component is a solitary point, it must be in S_0. For the other components which intersect with the boundary or have singular points, the starting points are in S_2 and S_1 respectively. Note that although S_3 may not contain witness points for every connected component of C_f, it must contain at least one witness point for each smooth closed component of $V_{\mathbb{R}}(f)$, as their extreme points in the direction of x must be inside S_3. By the assumptions, the polynomial systems with zero sets S_i, $i = 0, \ldots, 3$ are all zero-dimensional. If the interval Newton method [28] converges, the error of solving these zero-dimensional systems and the error of Newton iterations (in the corrector step), as well as the distance between the curve and the polygonal chains can be controlled to be much less than ϵ by Theorem 1. Otherwise, one can switch to α-theory [2,4] to guarantee the convergence of Newton iterations. Moreover, by Theorem 1, curve jumping can be avoided. Finally note that in the $\epsilon/2$-neighborhood of the singular points, the distance between the curve and the polygonal chains are less than ϵ. Thus, an ϵ-approximation of $V_{\mathbb{R}}(f)$ can be computed.

3   Improvements

In this section, we propose several strategies for improving the numerical stability of the tracing algorithm in last section.

This first strategy is plotting the curve in the direction away from the singular points rather than towards the singular point. In practice, the former can better avoid curve jumping, as illustrated by Fig. 5. In this figure, the black curve is the locus of $f := x^5 - y^2$. To trace the upper branch, we have two possible starting points, namely the red $\times$ point, say z_0, and the blue $\times$ point, say z_1. If we start from z_0 and move in the tangent direction towards z_1 in step size 0.09, we get a red $\bullet$ point close to the upper branch, with which as an initial point, Newton

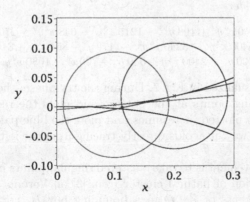

Fig. 5. Jump is more likely to happen when tracing towards singular points. (Color figure online)

iteration converges to a point still in the upper branch. However, if we start from z_1 and move in the tangent direction towards z_0 in step size 0.09, we get a blue
• point close to the lower branch. As a result, Newton iteration converges to a point in the lower branch.

This justifies why we first start with circular ring points instead of the boundary points to trace the curve. However, this first strategy does not consider the situation that there are two singular points in the same component, for which a try-stop-resume strategy is needed, as illustrated by the following example.

Example 3. *Consider again the polynomial* $f := y^2 - \left(-x^2 + x\right)^3$. *It is a closed curve with two singular points* $(0,0)$ *and* $(1,0)$.

In Fig. 6, the algorithm first plots the red points starting from two circular ring points near $(0,0)$ and stops when the singular values drop (at the two × points, which are called front points). It then starts from the two circular ring points near $(1,0)$ and plots the blue points, which happen to meet the front points before singular values drop.

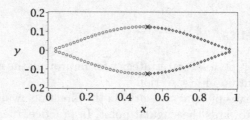

Fig. 6. Try to plot the curve away from the singular points and stop when singular values drop. (Color figure online)

The above example does not need the resuming step. Consider another one.

Example 4. *Consider*

$$f := -3375\,y^{14} - 4050\,x^4y^9 + 108\,y^{13} - 1215\,x^8y^4 - 648\,x^2y^9 + 2700\,y^{11} + 1620\,x^4y^6$$
$$+ 1296\,x^4y^5 - 5400\,x^2y^7 - 3240\,x^6y^2 - 1170\,y^8 - 864\,x^6y - 810\,x^4y^3$$
$$- 720\,x^2y^4 + 4000\,y^6 + 2400\,x^4y + 540\,y^5 + 720\,x^4 - 1080\,x^2y - 135\,y^2 + 800.$$

The locus of f is visualized in Fig. 7. During the try phase, the algorithm starts with the circular ring points at the bottom and plots the red point. After all red parts have been plotted, it resumes and plots the blue parts and finally the green parts. In this way, it avoids directly tracing from the left singular point to the right one.

The third improvement is to take clustered singular points into consideration. We borrow the notion of natural cluster from [3] on Voronoi vertices. Given a set S of singular points of $Z_{\mathbb{R}}(f)$ in a bounding box B. For any disk $D(z,r)$ centered at z of radius r, let $\Delta_S(z,r)$ be the set of points in S contained in $D(z,r)$. If it is not empty, we call it a cluster of S. It is called a natural cluster

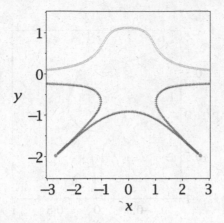

Fig. 7. Try to plot the curve away from the singular points and stop when singular values drop and resume. (Color figure online)

if $D(z,r)$ and $D(z,3r)$ contains exactly the same set of points of S. We call $D(z,r)$ an associated disk of $\Delta_S(z,r)$. Note that the associated disks of two different clusters are also disjoint and the distance between their centers are at least $3r$. For a given S, it is easy to generate a set of disjoint natural clusters and their associated disks. For instance, one can first sort the singular points in an ascending order by the minimal distances between them. One can then check if the points form natural clusters of radius r incrementally. If not, let d be the minimal distances among points in S, one can always obtain natural clusters of radius less than $d/3$. Let's consider an example.

Example 5. *Let* $g := -28\,x^4yz + 58\,xy^5 - 65\,xy^2z^3 + 23\,x^4y + 24\,x^3yz - 64\,x^2z^3 - 32\,xyz^3 - 72\,xy^2z + 6\,z^4 + 56\,xyz + 1$ *and* f *be the discriminant of* g *w.r.t.* z, *which is an irreducible polynomial in* $\mathbb{Q}[x,y]$. *A visualization of it in the box* $-1 \leq x \leq 1, -1 \leq y \leq 1$ *is depicted in Fig. 8. The two points* $(-0.9257645305e-1, 0.7100519895)$ *and* $(-0.6009009066e-1, 0.7790657631)$ *on the top of Fig. 8 form a natural cluster of radius* 0.1.

4   A Practical Algorithm

In Sect. 2, we presented a theoretical algorithm to compute an ϵ-approximation of a curve, which may not be practical due to the small step size chosen. In practice, one has to make a compromise between efficiency and accuracy. Based on the improvement strategies in last section, next we develop a more practical algorithm. Instead of preventing curve jumping, in the algorithms below, we maintain a simple data structure to record if a start point has been visited. If a point is visited more than once, then there is a possible curve jumping.

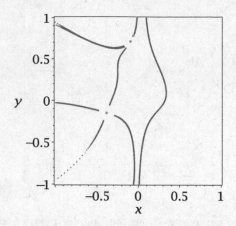

Fig. 8. Plotting the curve with the help of natural clusters.

Algorithm 1. ApproxPlot

Input: A nonconstant squarefree polynomial $f \in \mathbb{Q}[x, y]$. A bounding box
$B \subset \mathbb{R}^2$. A precision $\epsilon > 0$.

Output: An ϵ-approximation of $f^{-1}(0)$ in B.

```
1  begin
2      let S₀ = V_ℝ({f, ∂f/∂x, ∂f/∂y}) ∩ B be the set of singular points of f in B;
3      let δ ≤ ε/2 be the radius of natural clusters;
4      cwp := ∅; bwp := ∅;
5      for each natural cluster C do
6          let p be the center point of C;
7          for each associated circular ring point q of p do
8              let s := ‖∇_f(q)‖₂; // s is the singular value of 𝒥_f(q)
9              let v := q − p; let c := 0; add (q, v, s, c) to cwp;
10     for each point q of f⁻¹(0) ∩ ∂B do
11         let s := ‖∇_f(q)‖₂; let v be the direction towards the interior of B;
12         let c := 0; add (q, v, s, c) to bwp;
13     let Δ be the union of disks associated with the natural clusters; set
       rwp := V_ℝ({f, ∂f/∂y}) ∩ B \ Δ;
14     rescale the coefficients of f if necessary;
       /* Note that below the function PlotMain is called multiple times
          with different arguments and flags.                          */
15     S₁, front := PlotMain(f, B, cwp, bwp, rwp, δ, try);
16     S₂ := PlotMain(f, B, front, bwp, rwp, δ, resume);
17     S₃ := PlotMain(f, B, bwp, cwp, rwp, δ, boundary);
18     S₄ := PlotOval(f, B, rwp, cwp ∪ bwp, δ);
19     return {∪⁴ᵢ₌₀Sᵢ};
```

Algorithm 2. PlotMain($f, B, cwp, bwp, rwp, \delta, tag$)

---

begin

    $S := \emptyset$; $front := \emptyset$;

    for j to $|cwp|$ **do**

        $P := \emptyset$;

        $(q, v, s, c) := cwp[j]$;

        if $cwp[j].c > 0$ **then**

          | next;

        else

          $cwp[i].c := 1$;

        $mb := false$; $mc := false$; $mf := false$;

        while $q \in B$ **do**

          $s' := s$; $q' := q$; $v' := v$;

          $P := P \cup \{q\}$;

          choose step size $h \le \delta/2$ according to δ and s;

          $q := q + hv$;

          with q as initial point, apply Newton iterations to update q;

          let $s := \|\nabla_f(q)\|_2$;

          let $v := (\partial f/\partial y(q), -\partial f/\partial x(q))^T$ and $v := v/\|v\|_2$;

          if $v \bullet v' < 0$ **then** $v := -v$;

          remove any element of rwp on $\overline{q'q}$;

          for i to $|bwp|$ **do**

            if $(bwp[i].v) \bullet v < 0 \wedge bwp[i].q \in \overline{q'q}$ **then**

              if $bwp[i].c > 0$ **then**

                | report curve jump error;

              else

                $P := P \cup \{bwp[i].q\}$;

              $mb := true$; $bwp[i].c := bwp[i].c + 1$; break;

          if mb **then** break;

          for i to $|cwp|$ **do**

            if $i \ne j$ and $(cwp[i].v) \bullet v < 0$ and $cwp[i].q \in \overline{q'q}$ **then**

              if $cwp[i].c > 0$ **then**

                | report curve jump error;

              else if tag is 'resume' or 'try' **then**

                $P := P \cup \{cwp[i].q\}$;

              $mc := true$; $cwp[i].c := cwp[i].c + 1$; break;

          if mc **then** break;

          if tag='try' **then**

            for i to $|front|$ **do**

              if $front[i].v \bullet v < 0$ and $front[i].q \in \overline{q'q}$ **then**

                $P := P \cup \{front[i].q\}$; $mf := true$; remove $front[i]$ from

                $front$;

                break;

            if mf **then** break;

            if $s < s'$ **then**

              then add $(q, v, s, 0)$ to $front$;break;

        $S := S \cup \{P\}$;

    return S;

Remark 3. The main features of Algorithm ApproxPlot, such as tracing the curve away from the singular points, and grouping the singular points into natural clusters and the try-and-resume strategy has been explained in last section. Another feature of the algorithm is to detect curve jumping by counting the number of times that a circular ring or boundary point is visited.

To achieve this, each circular point, boundary point, or new front point generated due to the drop of singular value, is treated as an object with four attributes (q, v, s, c), where q is the point itself, v is the tracing direction, s is the singular value of $J_f(q)$ and c counts the times that q is visited. For an object ob, the notation $ob.q$ means taking the value of the attribute q. Each q should be visited one and only one time. If its visiting time $c > 1$, there is a possible curve jumping at q. It is easy to check that if there is no curve jumping, after executing Algorithm PlotMain, the value of any c (counting visiting times of a circular ring point or boundary point) can not be greater than 1. Moreover, if the numerical errors are well controlled, after executing line 16 of Algorithm ApproxPlot, all the points in rwp will only be on the closed components of the curve. Thus the value of any c can not increase after executing Algorithm PlotOval. Finally we remark that the algorithm may not detect curve jumping errors caused by exchanging branches during tracing.

5   Experimentation

In this section, we provide some nontrivial examples to illustrate the effectiveness of our method. Example 6 is selected from a list of challenges in [21] for plane curve visualization. Example 7 is the discriminant of a random trivariate polynomial. Example 8 is the resultant of two random trivariate polynomials. Example 9 is a discriminant variety of a bi-parametric polynomial system. To make a fair comparison with the Plots:-implicitplot command of Maple 18, all polynomials are plotted using their irreducible factors.

We have implemented our algorithm in Maple. In the algorithms of last section, there are several places where ones needs to solve zero-dimensional polynomial systems, namely computing singular points, computing circular ring points, computing boundary points and computing witness points. For the first three, we find that it is more robust to call a symbolic solver and use RootFinding:-Isolate of Maple. For the last one, we find it is more efficient to use homotopy based methods and we implemented a Maple interface to hom4ps2.

Example 6. *Let $f := 1/4\,x^6 y^2 - 1/2\,x^4 y^4 + 1/4\,x^2 y^6 - \left(x^2 + y^2\right)^7$. Visualizations of it by Plot:-implicitplot in Maple and ApproxPlot are depicted in Fig. 9. No curve jumping is reported by Algorithm ApproxPlot.*

Example 7. *Let f be the same polynomial as in Example 5. Visualizations of it by Plot:-implicitplot in Maple and ApproxPlot are depicted in Fig. 10. The polynomial f has branches very close to each other and the algorithm detects curve jumping.*

Algorithm 3. PlotOval(f, B, rwp, wp, δ)

```
begin
    S := ∅;
    while rwp ≠ ∅ do
        P := ∅;
        choose p ∈ rwp and set rwp := rwp \ {p}; k := 0; q := p;
        let s := ‖∇_f(q)‖₂; let v := (∂f/∂y(q), −∂f/∂x(q))ᵀ and v := v/‖v‖₂;
        mt := false;
        while q ∈ B do
            k := k + 1; q' := q; v' := v; P := P ∪ {q};
            choose step size h ≤ δ/2 according to δ and s;
            q := q + hv;
            with q as initial point, apply Newton iterations to update q;
            let s := ‖∇_f(q)‖₂;
            let v := (∂f/∂y(q), −∂f/∂x(q))ᵀ and v := v/‖v‖₂;
            if v • v' < 0 then  v := −v ;
            if k > 2 and p ∈ q'q then
                break;
            remove any element of rwp other than p on q'q;
            for i to |wp| do
                if (wp[i].v) • v < 0 ∧ wp[i].q ∈ q'q then
                    if wp[i].c > 0 then
                        report curve jump error;
                    mt := true; wp[i].c := wp[i].c + 1; break;
            if mt then break;
        S := S ∪ {P};
    return S;
```

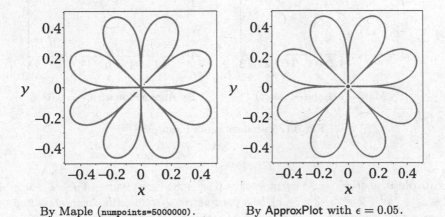

By Maple (`numpoints=5000000`).   By ApproxPlot with $\epsilon = 0.05$.

Fig. 9. Visualization of Example 6.

112   C. Chen and W. Wu

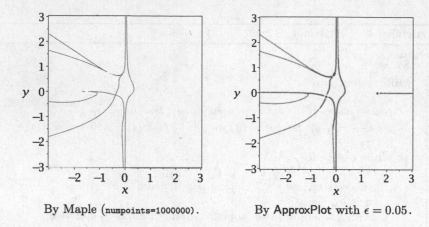

By Maple (numpoints=1000000).   By ApproxPlot with $\epsilon = 0.05$.

Fig. 10. Visualization of Example 7.

Example 8. Let $f_1 := 72\,y^2 z^5 + 26\,x^2 yz^3 - 84\,x^2 y^2 - 73\,xz^2 + 6$, $f_2 := -24\,x^4 z^2 - 35\,yz^3 + 43\,yz^2 - 66\,z^3 + 3$. Let f be the resultant of f_1 and f_2 w.r.t. z. A visualization of it is depicted in Fig. 11. The polynomial f has branches very close to each other and the algorithm detects curve jumping.

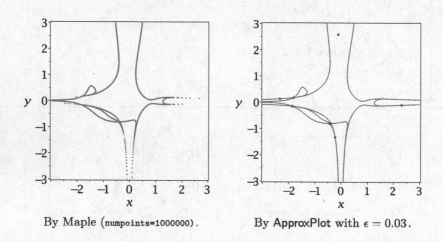

By Maple (numpoints=1000000).   By ApproxPlot with $\epsilon = 0.03$.

Fig. 11. Visualization of Example 8.

Example 9. Let $F := \{8\,wyx^2 + z^4 + 6\,w^3 + 8\,w^2 x - 9\,y^2 - 4\,y + 1, -wx^3 + x^4 + z^4 + 7\,x^3 + 2\,y^2 x + 2\,x^2 + 1\}$ be a parametric system with parameters x, y. A discriminant variety of F is a union of zero sets of two irreducible polynomials in (x, y). A visualization of it is depicted in Fig. 12. No curve jumping is reported by ApproxPlot.

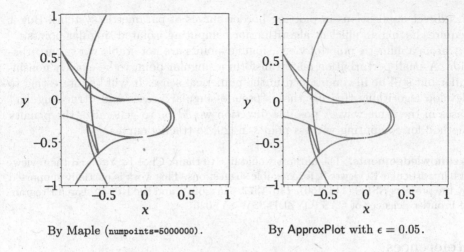

By Maple (numpoints=5000000).   By ApproxPlot with $\epsilon = 0.05$.

Fig. 12. Visualization of Example 9.

Remark 4. *The running time of* implicitplot *largely depends on the value of the option* numpoints. *The running time of* ApproxPlot *depends on the precision* ϵ. *For the options chosen in this paper, here is a summary of the running time (in seconds): Note that for these examples, lifting the value of* numpoints *for Maple (from* numpoints $= 1000000$) *helps little on the quality of the visualization by* implicitplot, *but increases significantly the running time.*

System	implicitplot	ApproxPlot
Example 6	4	26
Example 7	26	47
Example 8	20	47
Example 9	32	6

6 Conclusion and Future Work

In this paper, we presented algorithms for visualizing planar algebraic curves with singularities. The theoretical algorithm guarantees the polygonal approximation ϵ-close to the curve. We introduced several strategies to turn the theoretical algorithm to be practical and illustrate its effectiveness by examples. One bottleneck of the algorithm is the computation of singular points, whose efficiency might be improved if the curve is known to be the resultant or discriminant of two polynomials [19].

The algorithm presented in this paper can be readily generated to tracing space curves with singularities in ambient space with dimension ≥ 3, which has direct applications in plotting border curves of parametric systems. But it requires having an efficient algorithm for computing isolated singular points.

From a numeric point of view, singular points are not stable w.r.t. perturbation. A small perturbation may transform a singular point to be exactly nonsingular but still be ill-conditioned in the numerical sense. It will be interesting to develop algorithms treating these "pseudo-singular" cases and "true-singular" cases in the same way. A possible direction would be to generalize the penalty method for computing witness points in [29] to tracing curves.

Acknowledgements. The authors would like to thank Chee K. Yap and the reviewers, in particular Reviewer 3, for valuable suggestions. This work is partially supported by the projects NSFC (11471307, 11671377, 61572024), and the Key Research Program of Frontier Sciences of CAS (QYZDB-SSW-SYS026).

References

1. Bajaj, C., Xu, G.: Piecewise rational approximations of real algebraic curves. J. Comput. Math. **15**(1), 55–71 (1997)
2. Beltrán, C., Leykin, A.: Robust certified numerical homotopy tracking. Found. Comput. Math. **13**(2), 253–295 (2013)
3. Bennett, H., Papadopoulou, E., Yap, C.: Planar minimization diagrams via subdivision with applications to anisotropic Voronoi diagrams. Comput. Graph. Forum **35** (2016)
4. Blum, L., Cucker, F., Shub, M., Smale, S.: Complexity and Real Computation. Springer, Secaucus (1998). https://doi.org/10.1007/978-1-4612-0701-6
5. Bresenham, J.: A linear algorithm for incremental digital display of circular arcs. Commun. ACM **20**(2), 100–106 (1977)
6. Burr, M., Choi, S.W., Galehouse, B., Yap, C.K.: Complete subdivision algorithms, II: isotopic meshing of singular algebraic curves. J. Symb. Comput. **47**(2), 131–152 (2012)
7. Chandler, R.E.: A tracking algorithm for implicitly defined curves. IEEE Comput. Graph. Appl. **8**(2), 83–89 (1988)
8. Chen, C., Davenport, J., May, J., Moreno Maza, M., Xia, B., Xiao, R.: Triangular decomposition of semi-algebraic systems. J. Symb. Comp. **49**, 3–26 (2013)
9. Chen, C., Wu, W.: A numerical method for analyzing the stability of bi-parametric biological systems. In: SYNASC 2016, pp. 91–98 (2016)
10. Chen, C., Wu, W.: A numerical method for computing border curves of biparametric real polynomial systems and applications. In: Gerdt, V.P., Koepf, W., Seiler, W.M., Vorozhtsov, E.V. (eds.) CASC 2016. LNCS, vol. 9890, pp. 156–171. Springer, Cham (2016). https://doi.org/10.1007/978-3-319-45641-6_11
11. Chen, C., Wu, W., Feng, Y.: Full rank representation of real algebraic sets and applications. In: Gerdt, V.P., Koepf, W., Seiler, W.M., Vorozhtsov, E.V. (eds.) CASC 2017. LNCS, vol. 10490, pp. 51–65. Springer, Cham (2017). https://doi.org/10.1007/978-3-319-66320-3_5
12. Cheng, J., Lazard, S., Peñaranda, L., Pouget, M., Rouillier, F., Tsigaridas, E.: On the topology of real algebraic plane curves. Math. Comput. Sci. **4**(1), 113–137 (2010)

13. Collins, G.E.: Quantifier elimination for real closed fields by cylindrical algebraic decompostion. In: Brakhage, H. (ed.) GI-Fachtagung 1975. LNCS, vol. 33, pp. 134–183. Springer, Heidelberg (1975). https://doi.org/10.1007/3-540-07407-4_17

14. Daouda, D., Mourrain, B., Ruatta, O.: On the computation of the topology of a non-reduced implicit space curve. In: ISSAC 2008, pp. 47–54 (2008)

15. Emeliyanenko, P., Berberich, E., Sagraloff, M.: Visualizing arcs of implicit algebraic curves, exactly and fast. In: Bebis, G., et al. (eds.) ISVC 2009. LNCS, vol. 5875, pp. 608–619. Springer, Heidelberg (2009). https://doi.org/10.1007/978-3-642-10331-5_57

16. Gomes, A.J.: A continuation algorithm for planar implicit curves with singularities. Comput. Graph. **38**, 365–373 (2014)

17. Hauenstein, J.D.: Numerically computing real points on algebraic sets. Acta Applicandae Mathematicae **125**(1), 105–119 (2012)

18. Hong, H.: An efficient method for analyzing the topology of plane real algebraic curves. Math. Comput. Simul. **42**(4), 571–582 (1996)

19. Imbach, R., Moroz, G., Pouget, M.: A certified numerical algorithm for the topology of resultant and discriminant curves. J. Symb. Comput. **80**, 285–306 (2017)

20. Jin, K., Cheng, J.-S., Gao, X.-S.: On the topology and visualization of plane algebraic curves. In: Gerdt, V.P., Koepf, W., Seiler, W.M., Vorozhtsov, E.V. (eds.) CASC 2015. LNCS, vol. 9301, pp. 245–259. Springer, Cham (2015). https://doi.org/10.1007/978-3-319-24021-3_19

21. Labs, O.: A list of challenges for real algebraic plane curve visualization software. In: Emiris, I., Sottile, F., Theobald, T. (eds.) Nonlinear Computational Geometry. The IMA Volumes in Mathematics and Its Applications, vol. 151, pp. 137–164. Springer, New York (2010). https://doi.org/10.1007/978-1-4419-0999-2_6

22. Lazard, D., Rouillier, F.: Solving parametric polynomial systems. J. Symb. Comput. **42**(6), 636–667 (2007)

23. Leykin, A., Verschelde, J., Zhao, A.: Newton's method with deflation for isolated singularities of polynomial systems. TCS **359**(1), 111–122 (2006)

24. Lopes, H., Oliveira, J.B., de Figueiredo, L.H.: Robust adaptive polygonal approximation of implicit curves. Comput. Graph. **26**(6), 841–852 (2002)

25. Martin, B., Goldsztejn, A., Granvilliers, L., Jermann, C.: Certified parallelotope continuation for one-manifolds. SIAM J. Numer. Anal. **51**(6), 3373–3401 (2013)

26. Rouillier, F., Roy, M.F., Safey El Din, M.: Finding at least one point in each connected component of a real algebraic set defined by a single equation. J. Complex. **16**(4), 716–750 (2000)

27. Seidel, R., Wolpert, N.: On the exact computation of the topology of real algebraic curves. In: Proceedings of the Twenty-First Annual Symposium on Computational Geometry, SCG 2005, pp. 107–115. ACM, New York (2005)

28. Shen, F., Wu, W., Xia, B.: Real root isolation of polynomial equations based on hybrid computation. In: ASCM 2012, pp. 375–396 (2012)

29. Wu, W., Chen, C., Reid, G.: Penalty function based critical point approach to compute real witness solution points of polynomial systems. In: Gerdt, V.P., Koepf, W., Seiler, W.M., Vorozhtsov, E.V. (eds.) CASC 2017. LNCS, vol. 10490, pp. 377–391. Springer, Cham (2017). https://doi.org/10.1007/978-3-319-66320-3_27

30. Wu, W., Reid, G., Feng, Y.: Computing real witness points of positive dimensional polynomial systems. Theor. Comput. Sci. **681**, 217–231 (2017)

31. Yang, L., Xia, B.: Real solution classifications of a class of parametric semi-algebraic systems. In: A3L 2005, pp. 281–289 (2005)

From Exponential Analysis to Padé Approximation and Tensor Decomposition, in One and More Dimensions

Annie Cuyt, Ferre Knaepkens, and Wen-shin Lee[(✉)]

Department of Mathematics and Computer Science, Universiteit Antwerpen (CMI),
Middelheimlaan 1, 2020 Antwerp, Belgium
{annie.cuyt,ferre.knaepkens,wen-shin.lee}@uantwerpen.be
http://cma.uantwerpen.be

Abstract. Exponential analysis in signal processing is essentially what is known as sparse interpolation in computer algebra. We show how exponential analysis from regularly spaced samples is reformulated as Padé approximation from approximation theory and tensor decomposition from multilinear algebra.

The univariate situation is briefly recalled and discussed in Sect. 1. The new connections from approximation theory and tensor decomposition to the multivariate generalization are the subject of Sect. 2. These connections immediately allow for some generalization of the sampling scheme, not covered by the current multivariate theory.

An interesting computational illustration of the above in blind source separation is presented in Sect. 3.

Keywords: Exponential analysis · Parametric method · Multivariate
Padé approximation · Tensor decomposition

1 The Univariate Connections

Let us first introduce the problem statement of exponential analysis, which is known in the computer algebra community as sparse interpolation [4,10]. Afterward we rewrite it as a rational approximation problem and as a tensor decomposition problem. In this section, we restrict ourselves to the univariate case.

Let the signal $f(t)$ be given by

$$f(t) = \sum_{j=1}^{n} \alpha_j \exp(\phi_j t), \qquad \alpha_j, \phi_j \in \mathbb{C}, \tag{1}$$

where the objective is to recover the values of the coefficients $\alpha_j, j = 1, \ldots, n$ and the mutually distinct exponents $\phi_j, j = 1, \ldots, n$. Already in 1795, de Prony

© Springer Nature Switzerland AG 2018
V. P. Gerdt et al. (Eds.): CASC 2018, LNCS 11077, pp. 116–130, 2018.
https://doi.org/10.1007/978-3-319-99639-4_8

[14] proved that the problem can be solved from $2n$ equidistant samples if the sparsity n is known, as we assume in the sequel.

In the following, we choose a real $\Delta \neq 0$ such that $|\Im(\phi_j)| < \pi/|\Delta|$, in order to comply with [12,17], where $\Im(\cdot)$ denotes the imaginary part of a complex number. The value Δ denotes the sampling step in the equidistant sampling scheme

$$f_k := f(k\Delta) = \sum_{j=1}^{n} \alpha_j \exp(\phi_j k\Delta) = \sum_{j=1}^{n} \alpha_j \Phi_j^k, \qquad \Phi_j = \exp(\phi_j \Delta). \qquad (2)$$

With the samples $f_k, k = 0, \ldots, 2n - 1$, we fill the Hankel matrices

$$H_n^{(m)} := (f_{m+i+j-2})_{i,j=1}^{n} = \begin{bmatrix} f_m & f_{m+1} & \cdots & f_{m+n-1} \\ f_{m+1} & f_{m+2} & \cdots & f_{m+n} \\ \vdots & \vdots & \ddots & \vdots \\ f_{m+n-1} & f_{m+n} & \cdots & f_{m+2n-2} \end{bmatrix}, \qquad m \geq 0.$$

From the expression (2) for the samples f_k, we immediately find that $H_n^{(m)}$ can be factored as

$$H_n^{(m)} = V_n D_\alpha D_\Phi^m V_n^T,$$

where V_n is the Vandermonde matrix

$$V_n = \left(\Phi_j^{i-1}\right)_{i,j=1}^{n}$$

and D_α and D_Φ are diagonal matrices respectively filled with the vectors $(\alpha_1, \ldots, \alpha_n)$ and $(\Phi_1, \ldots, \Phi_n)$ on the diagonal. So the $\Phi_j, j = 1, \ldots, n$ can be found as the generalized eigenvalues $\lambda_j, j = 1, \ldots, n$ of the problem [11]

$$H_n^{(1)} v_j = \lambda_j H_n^{(0)} v_j, \qquad (3)$$

where the $v_j, j = 1, \ldots, n$ are the right generalized eigenvectors. From the generalized eigenvalues $\Phi_j = \exp(\phi_j \Delta)$, the complex values ϕ_j can be extracted uniquely because $|\Im(\phi_j)\Delta| < \pi$. After recovering the Φ_j, the α_j can be computed from the Vandermonde structured linear system

$$\sum_{j=1}^{n} \alpha_j \Phi_j^k = f_k, \qquad k = 0, \ldots, 2n - 1. \qquad (4)$$

In a noisefree mathematical context, n equations of (4) are linearly dependent because of the relationship (3) between the Φ_j. How to proceed in a noisy context is analyzed in great detail and including several variations in a forthcoming paper and is outside the scope of the current presentation, where we focus on the mathematical interrelationship between seemingly disconnected problem statements.

1.1   From Exponential Analysis to Padé Approximation in 1-D

Instead of filling Hankel matrices with the samples f_k, we construct a formal power series expansion

$$F(z) = \sum_k f_k z^k.$$

The Padé approximant $[m/n]_F$ for $F(z)$ of degree m in the numerator and n in the denominator is defined as the irreducible form of the rational function $p_{m,n}(z)/q_{m,n}(z)$, with

$$p_{m,n}(z) := \sum_{j=0}^{m} a_j z^j,$$

$$q_{m,n}(z) := \sum_{j=0}^{n} b_j z^j,$$

that satisfies

$$F(z)q_{m,n}(z) - p_{m,n}(z) = \sum_{k \geq m+n+1} c_k z^k.$$

The computation of Padé approximants is closely connected to the solution of Toeplitz structured linear systems. The $[m/n]_F$ is computed from putting to zero the terms of degree 0 to $m+n$ in $(Fq_{m,n} - p_{m,n})(z)$:

$$\sum_{j=0}^{n} f_{k-j} b_j = a_k, \qquad k = 0, \ldots, m,$$

where $f_k = 0$ if $k < 0$, and

$$\sum_{j=0}^{n} f_{m+k-j} b_j = 0, \qquad k = 1, \ldots, n.$$

Again using expression (2) for the f_k and under the assumption that the Φ_j are mutually distinct, it is not difficult to see that [2]

$$F(z) = \sum_k f_k z^k$$

$$= \sum_k \left(\sum_{j=1}^{n} \alpha_j \Phi_j^k \right) z^k$$

$$= \sum_{j=1}^{n} \alpha_j \left(\sum_k \Phi_j^k z^k \right)$$

$$= \sum_{j=1}^{n} \frac{\alpha_j}{1 - \Phi_j z}.$$

So the function $F(z)$ is itself a rational function of degree $n - 1$ in the numerator and n in the denominator. The consistency property of Padé approximants guarantees that a rational function like $F(z)$ is reconstructed from its formal series expansion by its $[n - 1/n]_F$ Padé approximant, thereby needing only the series coefficients $f_0, \ldots, f_{2n-1}$. So we can also obtain the Φ_j from the Padé denominator

$$\prod_{j=1}^{n} (1 - \Phi_j z) \tag{5}$$

and the α_j from the partial fraction decomposition of the approximant $[n-1/n]_F$, through

$$P_{n-1,n}(z) = \sum_{j=1}^{n} \alpha_j L_j(z), \qquad L_j(z) = \sum_{\substack{i=1 \\ i \neq j}}^{n} (1 - \Phi_i z).$$

The poles $1/\Phi_j$ of $F(z)$ can even directly be computed from the f_k, in the order of increasing magnitude, using the qd-algorithm [1].

1.2   From Exponential Analysis to Tensor Decomposition in 1-D

With the samples f_k we can also fill an order N tensor $T \in \mathbb{C}^{n_1 \times \cdots \times n_N}$ where

$$2 \leq n_j \leq n, \qquad 1 \leq j \leq N, \qquad 3 \leq N \leq 2n - 1,$$

$$\sum_{j=1}^{N} n_j = 2n + N - 1,$$

and

$$T_{k_1, \ldots, k_N} := f_{k_1 + \cdots + k_N - N}, \qquad 1 \leq k_j \leq n_j. \tag{6}$$

The tensor of smallest order $N = 3$ is, for instance, of size $n \times n \times 2$ [13] and the one of largest order $N = 2n - 1$ is symmetric and of size $2 \times \cdots \times 2$ [6]. For the sequel, we generalize the definition of the square Hankel matrix above to cover rectangular Hankel structured matrices

$$H_{n_1,n_2}^{(m)} = \begin{bmatrix} f_m & f_{m+1} & \cdots & f_{m+n_2-1} \\ f_{m+1} & f_{m+2} & \cdots & f_{m+n_2} \\ \vdots & \vdots & \ddots & \vdots \\ f_{m+n_1-1} & f_{m+n_1} & \cdots & f_{m+n_1+n_2-2} \end{bmatrix}.$$

The tensor slices $T_{\cdot, \cdot, k_3, \ldots, k_N}$ equal

$$T_{\cdot, \cdot, k_3, \ldots, k_N} = H_{n_1,n_2}^{(k_3 + \cdots + k_N - N + 2)}$$

and so are Hankel structured. The tensor T decomposes as

$$T = \sum_{j=1}^{n} \alpha_j \begin{bmatrix} 1 \\ \Phi_j \\ \vdots \\ \Phi_j^{n_1-1} \end{bmatrix} \circ \cdots \circ \begin{bmatrix} 1 \\ \Phi_j \\ \vdots \\ \Phi_j^{n_N-1} \end{bmatrix}, \tag{7}$$

where still the $\Phi_j = \exp(\phi_j \Delta)$ are mutually distinct and $\circ$ denotes the outer product. Decomposition (7) is easily verified by checking the element at position $(k_1, \ldots, k_N)$ in (7):

$$T_{k_1,\ldots,k_N} = \sum_{j=1}^{n} \alpha_j \Phi_j^{k_1-1} \cdots \Phi_j^{k_N-1}$$

$$= \sum_{j=1}^{n} \alpha_j \Phi_j^{k_1+\cdots+k_N-N}$$

$$= f_{k_1+\cdots+k_N-N}.$$

The factor matrices are the rectangular Vandermonde structured matrices

$$V_{n_k,n} = \left(\Phi_j^{i-1}\right)_{i=1,j=1}^{n_k,n}, \qquad 1 \le k \le N.$$

Because of the Vandermonde structure of the factor matrices with $n_k \le n, k = 1,\ldots,N$, their Kruskal rank equals n_k for all k. Since $n_1 + \cdots + n_N = 2n + N - 1$ we find that the sum of the Kruskal ranks of the N factor matrices of the rank n tensor T is bounded below by $2n + N - 1$. Hence the Kruskal condition is satisfied and the unicity of the decomposition is guaranteed.

2   The Multivariate Connections

The result from de Prony that (1) can be solved from only $2n$ samples if the sparsity n is known and that the recovery of the linear coefficients α_j and the nonlinear parameters ϕ_j can be separated, is only recently truly generalized [5] to d-variate functions of the form

$$f(x_1, \ldots, x_d) = \sum_{j=1}^{n} \alpha_j \exp\left(\phi_{j1}x_1 + \cdots + \phi_{jd}x_d\right), \qquad \alpha_j, \phi_{j\ell} \in \mathbb{C}. \qquad (8)$$

In [5], is proved that the $\alpha_j, j = 1,\ldots,n$ and $\phi_{j\ell}, j = 1,\ldots,n, \ell = 1,\ldots,d$ can be recovered from $(d+1)n$ samples in the absence of collisions or cancellations of terms when sampling. In the latter case, the problem is still solvable but requires some additional samples to untangle the collisions or cancellations [5]. For the sequel, we also introduce the vectors $x = (x_1, \ldots, x_d)$ and $\phi_j = (\phi_{j1}, \ldots, \phi_{jd})$ where it is clear from the context whether ϕ_j refers to a complex value as in the previous section or a vector of complex values. Using the vector notation, (8) becomes

$$f(x) = \sum_{j=1}^{n} \alpha_j \exp\left(\langle \phi_j, x \rangle\right).$$

The way to achieve the generalization (8) is by falling back on a one-dimensional projected generalized eigenvalue problem requiring $2n$ samples, complemented with $d-1$ structured linear systems each requiring n samples along linearly independent directions to cover the additional dimensions, and one more structured linear system set up from the first $2n$ samples to recover the linear coefficients α_j.

We introduce the real linearly independent d-dimensional vectors $\Delta_\ell, \ell = 1, \ldots, d$ satisfying $|\Im(\langle \phi_j, \Delta_\ell \rangle)| < \pi, j = 1, \ldots, n, \ell = 1, \ldots, d$. We further collect the samples

$$f_k := f(k\Delta_1), \qquad 0 \le k \le 2n-1,$$
$$f_{k\ell} := f(k\Delta_1 + \Delta_\ell), \qquad 0 \le k \le n-1, \qquad 2 \le \ell \le d$$

and denote $\Phi_{j\ell} := \exp(\langle \phi_j, \Delta_\ell \rangle)$.

We assume now that all the values Φ_{j1} are mutually distinct so that the $\Phi_{j1}, j = 1, \ldots, n$ can be obtained as the generalized eigenvalues of a generalized eigenvalue problem of the form (3) where the Hankel matrices are filled with the samples f_k. Subsequently the α_j are the solution of the Vandermonde linear system

$$\sum_{j=1}^{n} \alpha_j \Phi_{j1}^k = f_k, \qquad k = 0, \ldots, 2n-1. \tag{9}$$

Of course, from $\langle \phi_j, \Delta_1 \rangle$ extracted from Φ_{j1}, the individual $\phi_{j\ell}$ cannot yet be identified. For that purpose, we need the additional $(d-1)n$ samples $f_{k\ell}$ which we reinterpret for each $2 \le \ell \le d$ as

$$\sum_{j=1}^{n} (\alpha_j \Phi_{j\ell}) \Phi_{j1}^k = f_{k\ell}, \qquad k = 0, \ldots, n-1. \tag{10}$$

In other words, with the samples $f_{k\ell}$ as right hand side for ℓ fixed and with the first n rows of the same Vandermonde coefficient matrix as in (9), we obtain the unknown coefficients $\alpha_j \Phi_{j\ell}$ and subsequently the values $\Phi_{j\ell}$ from

$$\frac{\alpha_j \Phi_{j\ell}}{\alpha_j}, \qquad j = 1, \ldots, n, \qquad 2 \le \ell \le d$$

and $\langle \phi_j, \Delta_\ell \rangle$ from $\Phi_{j\ell}$. We remark that $\Phi_{j\ell}$ is easily paired to its associated generalized eigenvalue Φ_{j1} through the structured linear systems (9) and (10), a problem that remained unsolved in various other approaches [9,15].

We now have extracted all the inner products $\langle \phi_j, \Delta_\ell \rangle, j = 1, \ldots, n, \ell = 1, \ldots, d$ for linearly independent vectors Δ_ℓ and so for each $1 \le j \le n$ the individual $\phi_{j\ell}$ can be retrieved as the solution of the following regular linear system:

$$\begin{bmatrix} \Delta_{11} & \cdots & \Delta_{1d} \\ \vdots & & \vdots \\ \Delta_{d1} & \cdots & \Delta_{dd} \end{bmatrix} \begin{bmatrix} \phi_{j1} \\ \vdots \\ \phi_{jd} \end{bmatrix} = \begin{bmatrix} \langle \phi_j, \Delta_1 \rangle \\ \vdots \\ \langle \phi_j, \Delta_d \rangle \end{bmatrix}.$$

In [6], some preliminary work was done leading to a novel technique based on the use of multivariate Padé approximation, but a proper rewrite of the problem statement (8) in terms of Padé approximants was still lacking. We fill this gap here by turning our attention to the concept of simultaneous Padé approximant. We continue along the same lines with a reformulation into a tensor decomposition problem of smaller order than in [6].

2.1   From Exponential Analysis to Padé Approximation in d-D

Instead of one formal power series, we now set up d formal power series, namely

$$F_1(z) = \sum_k f_k z^k,$$

$$F_\ell(z) = \sum_k f_{k\ell} z^k, \qquad 2 \le \ell \le d.$$

Making use of the expressions (9) and (10) for f_k and $f_{k\ell}$, respectively, we again find that the functions

$$F_1(z) = \sum_{j=1}^{n} \frac{\alpha_j}{1 - \Phi_{j1} z},$$

$$F_\ell(z) = \sum_{j=1}^{n} \frac{\alpha_j \Phi_{j\ell}}{1 - \Phi_{j1} z}, \qquad 2 \le \ell \le d$$

are rational functions, each of degree $n-1$ in the numerator and degree n in the denominator. Note that for all $\ell = 1, \ldots, d$, the denominator of $F_\ell(z)$ is the same and reveals the generalized eigenvalues Φ_{j1} which are assumed to be mutually distinct.

The rational functions $F_\ell(z), 1 \le \ell \le d$ can be recovered from the multivariate samples $f_k, 0 \le k \le 2n-1$ and $f_{k\ell}, 0 \le k \le n-1, 2 \le \ell \le d$ by computing the simultaneous Padé approximant $[(n-1, \ldots, n-1)/n]_{(F_1, \ldots, F_d)}$ for the vector of functions $(F_1(z), \ldots, F_d(z))$ [3, pp. 415–417], defined more precisely as the vector of irreducible forms of the rational functions $p_{n-1,n,\ell}(z)/q_{n-1,n}(z), 1 \le \ell \le d$ satisfying

$$(F_\ell q_{n-1,n} - p_{n-1,n,\ell})(z) = \begin{cases} \sum\limits_{k \ge 2n} c_k z^k, & \ell = 1, \\ \sum\limits_{k \ge n} c_{k\ell} z^k, & 2 \le \ell \le d. \end{cases} \tag{11}$$

So the denominator polynomial $q_{n-1,n}(z) = b_0 + \cdots + b_n z^n$ is computed from the Toeplitz structured linear system

$$\sum_{j=0}^{n} f_{n+k-j} b_j = 0, \qquad k = 0, \ldots, n-1,$$

arising from the accuracy-through-order conditions (11) for $F_1(z)$. We remark that again the α_j and $\Phi_{j\ell}, 2 \le \ell \le d$ are naturally paired to the poles $1/\Phi_{j1}$ of each rational function $p_{n-1,n,\ell}(z)/q_{n-1,n}(z)$, which can be computed directly from the samples using the qd-algorithm [1] applied to the formal series $F_1(z)$. This pairing is essential to recover the individual $\phi_{j\ell}$.

It is worthwhile to note that the Padé formulation of (8) allows a slight generalization compared to the generalized eigenvalue formulation of the multivariate problem. The simultaneous Padé approximant $[(n-1,\ldots,n-1)/n]_{(F_1,\ldots,F_d)}$ can also be computed from ν_1 samples f_k and ν_ℓ samples $f_{k\ell}$ for $2 \le \ell \le d$, where the total number of samples equals

$$\sum_{\ell=1}^{d} \nu_\ell = (d+1)n, \qquad \nu_\ell \ge n,$$

instead of $2n$ samples f_k and n samples $f_{k\ell}$ for $2 \le \ell \le d$. In that setting (11) becomes

$$(F_\ell q_{n-1,n} - p_{n-1,n,\ell})(z) = \sum_{k \ge \nu_\ell} c_{k\ell} z^k, \qquad 1 \le \ell \le d,$$

and the common denominator $q_{n-1,n}(z)$ is computed from the linear system

$$\sum_{j=0}^{n} f_{n+k-j} b_j = 0, \qquad k = 0.\ldots,\nu_1 - n - 1,$$

$$\sum_{j=0}^{n} f_{n+k-j,\ell} b_j = 0, \qquad k = 0,\ldots,\nu_\ell - n - 1, \qquad 2 \le \ell \le d.$$

2.2   From Exponential Analysis to Tensor Decomposition in d-D

Along the same lines as above, a connection to a so-called coupled tensor decomposition problem can be made. With the samples $f_k, k = 0,\ldots,2n-1$, a first order N tensor $T^{(1)}$ of dimension $n_1 \times \cdots \times n_N$ is defined as in (6), which decomposes as in (7), but with Φ_j replaced by Φ_{j1}:

$$T^{(1)} = \sum_{j=1}^{n} \alpha_j \begin{bmatrix} 1 \\ \Phi_{j1} \\ \vdots \\ \Phi_{j1}^{n_1-1} \end{bmatrix} \circ \cdots \circ \begin{bmatrix} 1 \\ \Phi_{j1} \\ \vdots \\ \Phi_{j1}^{n_N-1} \end{bmatrix}.$$

As explained in Sect. 1.2, this decomposition is unique as long as the Φ_{j1} are mutually distinct. Remains to recover the $\Phi_{j\ell}, 2 \le \ell \le d$.

To this end, we construct another $d-1$ order N tensors $T^{(\ell)}, 2 \le \ell \le d$ of dimension $n_{1\ell} \times \cdots \times n_{N\ell}$, where

$$2 \le n_{j\ell} \le n, \qquad \sum_{j=1}^{N} n_{j\ell} = n + N - 1,$$

with tensor elements

$$T^{(\ell)}_{k_1,\ldots,k_N} := f_{k_1+\cdots+k_N-N,\ell}, \qquad 2 \le \ell \le d,$$

of which the slices $T^{(\ell)}_{\cdot,\cdot,k_3,\ldots,k_N}$ are still Hankel structured. With

$$
H^{(m,\ell)}_{n_1,n_2} =
\begin{bmatrix}
f_{m,\ell} & f_{m+1,\ell} & \cdots & f_{m+n_2-1,\ell} \\
f_{m+1,\ell} & f_{m+2,\ell} & \cdots & f_{m+n_2,\ell} \\
\vdots & \vdots & \ddots & \vdots \\
f_{m+n_1-1,\ell} & f_{m+n_1,\ell} & \cdots & f_{m+n_1+n_2-2,\ell}
\end{bmatrix},
$$

the tensor slices $T^{(\ell)}_{\cdot,\cdot,k_3,\ldots,k_N}$ equal

$$
T^{(\ell)}_{\cdot,\cdot,k_3,\ldots,k_N} = H^{(k_3+\cdots+k_N-N+2,\ell)}_{n_1,n_2}.
$$

The tensors $T^{(\ell)}$ decompose as

$$
T^{(\ell)} = \sum_{j=1}^{n} \alpha_j \Phi_{j\ell}
\begin{bmatrix} 1 \\ \Phi_{j1} \\ \vdots \\ \Phi_{j1}^{n_{1\ell}-1} \end{bmatrix}
\circ \cdots \circ
\begin{bmatrix} 1 \\ \Phi_{j1} \\ \vdots \\ \Phi_{j1}^{n_{N\ell}-1} \end{bmatrix},
$$

where the entries in the factor matrices from $T^{(\ell)}$ can all be obtained from the decomposition of $T^{(1)}$, hence the term coupled tensor decomposition. Only the sizes $n_{j\ell} \times n$ of the factor matrices may be smaller as the sum of the $n_{j\ell}$ is bounded by $n + N - 1$ instead of $2n + N - 1$. The decomposition of the $T^{(\ell)}$ only serves the purpose of recovering the $\alpha_j \Phi_{j\ell}, j = 1,\ldots,n, 2 \le \ell \le d$. Note again the natural pairing of the α_j and $\alpha_j \Phi_{j\ell}, 2 \le \ell \le d$ to the Φ_{j1}, which is required to recover the individual $\phi_{j\ell}$ in (8).

A similar generalization as in Sect. 2.1 where now

$$
\sum_{j=1}^{N} n_j + \sum_{\ell=2}^{d} \sum_{j=1}^{N} n_{j\ell} = (d+1)n + d(N-1)
$$

is obviously also possible. Then the order N tensor $T^{(1)}$ of dimension $n_1 \times \cdots \times n_N$ is such that

$$
2 \le n_j \le n, \qquad \sum_{j=1}^{N} n_j = \nu_1 + N - 1
$$

and decomposes in the same way as $T^{(1)}$ above (only the sum of the dimensions is bounded differently). Similarly $T^{(\ell)}, 2 \le \ell \le d$ of dimension $n_{1\ell} \times \cdots \times n_{N\ell}$ obeys

$$
2 \le n_{j\ell} \le n, \qquad \sum_{j=1}^{N} n_{j\ell} = \nu_\ell + N - 1
$$

and decomposes as $T^{(\ell)}$ above. Note that Kruskal's condition only guarantees a unique decomposition if $\nu_1 \ge 2n$. However, the unicity is guaranteed through the other formulations of the problem statement, be it as a simultaneous Padé approximation problem or a multivariate exponential analysis.

3   Illustration: Blind Source Separation

We now illustrate the connections between exponential analysis or sparse interpolation with on the one hand Padé approximation and on the other hand tensor decomposition. The emphasis is on the mathematical reformulations of the problem statement and not on the numerical aspects of the various algorithms that can be used in either of the three settings.

We analyze a demo signal consisting of some wild bird chirps mixed with the whistle of a passing train (the original signal is available at our website[1]). The signal is graphed in Fig. 1: it lasts somewhat longer than 1.5 seconds and consists of 12850 samples collected at a rate of 8192 samples per second with a high signal-to-noise ratio. In Fig. 2, the signal's spectrogram is given, put together by applying the short-time Fourier transform to 257 non-overlapping frames of each 50 consecutive samples multiplied by a rectangular window function. It exhibits clearly the Fourier transform's typical leakage. Also the resolution is poor as we consider windows of only 50 samples. The horizontal stripes in the spectrogram represent the train whistle while the bird chirps are found in the higher frequency flame-like elements.

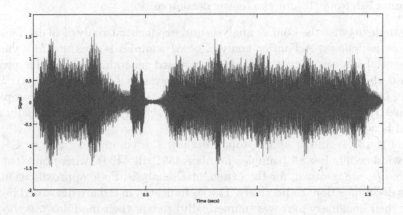

Fig. 1. Real-valued demo signal

The objective now is to identify the bird chirps and the train whistle using a sparse technique instead of the fast Fourier transform, thereby avoiding the leakage and limited resolution. So we recover each contributing α_j and ϕ_j in (1) from the signal samples following the outline of Sect. 1. To this end, we again divide the full data set into 257 non-overlapping windows of 50 samples. In each window, we take the sparsity $n = 20$, meaning that we choose a model consisting of 20 exponential terms, that we fit to 50 samples, in the least squares sense since $50 > 2n$. For the practical computation, use was made of:

---

[1] http://cma.uantwerpen.be/publications.

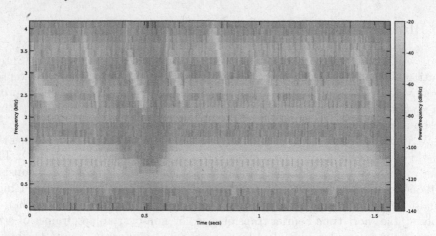

Fig. 2. Spectrogram of the demo signal

- the ESPRIT algorithm from [16] for the exponential analysis,
- the qd-algorithm as in [1] for the rational function reformulation,
- Tensorlab from [18] for the tensor decomposition.

Complexitywise the Fourier analysis and exponential analysis of each window compare as follows. A Fourier analysis of M samples is $O(M \log M)$ while an exponential analysis using the Hankel structured generalized eigenvalue problem (3) and the Vandermonde structured linear system (4) is $O(n^2 \log n)$. When solving (3)–(4) in a least squares sense from $m > 2n$ samples then the complexity increases to $O((m-n)n^2)$ [7,8]. Note that in practical applications usually $M \gg m$ and hence also $M \gg n$.

In Figs. 3, 4, and 5 at the top, we show the computed $\phi_j, j = 1, \ldots, 20$ from window number 88 (samples number 4351 till 4400), where only the blue coloured ϕ_j are retained, for the exponential analysis, Padé approximation, and tensor decomposition, respectively. The ϕ_j indicated in red are discarded because either their imaginary part was (numerically) zero or their modulus was too large ($|\cdot| > 1.05$). The former does not contribute to a sound signal, while the latter may cause ill-conditioning when setting up the Vandermonde matrices involved.

In the same figures at the bottom, the spectrogram results for each of exponential analysis, Padé approximation, and tensor decomposition is shown. It is clear that the sparse technique of exponential analysis and its reformulations do not suffer from the undesirable leakage and limited resolution, as they identify the frequency content in the signal $f(t)$.

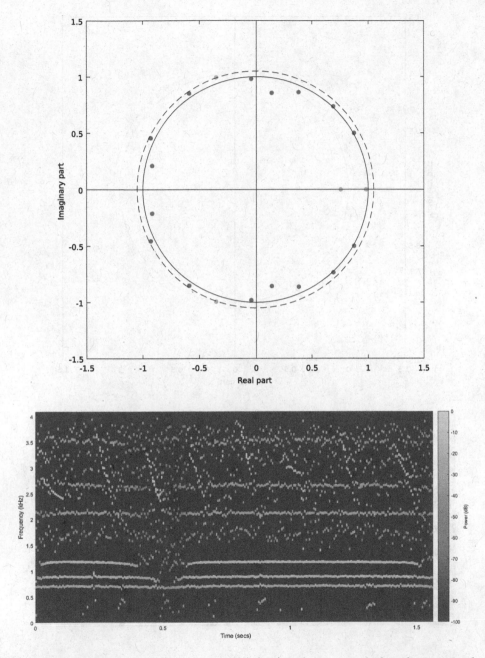

Fig. 3. Extracted $\phi_j, j = 1, \ldots, 20$ using (3) (top) and spectrogram based on retained ϕ_j (bottom) (Color figure online)

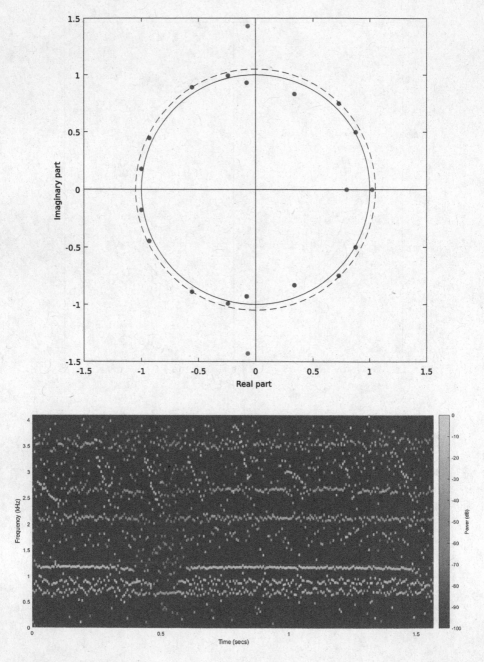

Fig. 4. Extracted $\phi_j, j = 1, \ldots, 20$ using (5) (top) and spectrogram based on retained ϕ_j (bottom) (Color figure online)

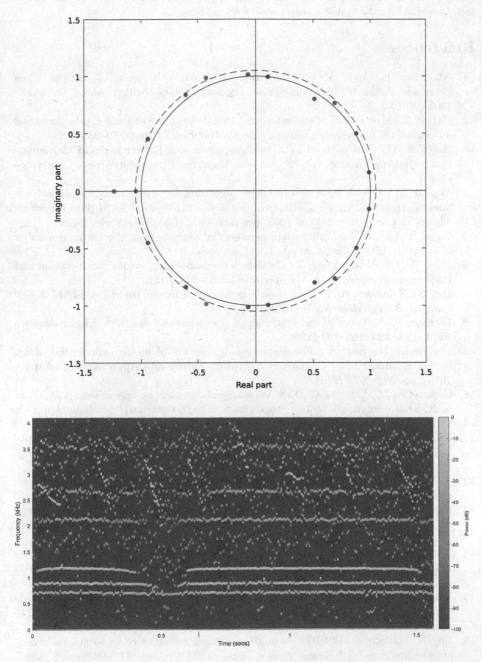

Fig. 5. Extracted $\phi_j, j = 1, \ldots, 20$ using (7) (top) and spectrogram based on retained ϕ_j (bottom) (Color figure online)

Acknowledgment. The authors want to thank George Labahn (University of Waterloo, Canada) for making the dataset available to them.

References

1. Allouche, H., Cuyt, A.: Reliable root detection with the qd-algorithm: when Bernoulli, Hadamard and Rutishauser cooperate. Appl. Numer. Math. **60**, 1188–1208 (2010)
2. Bajzer, Z., Myers, A.C., Sedarous, S.S., Prendergast, F.G.: Padé-Laplace method for analysis of fluorescence intensity decay. Biophys. J. **56**(1), 79–93 (1989)
3. Baker Jr., G., Graves-Morris, P.: Padé Approximants. Encyclopedia of Mathematics and its Applications, vol. 59, 2nd edn. Cambridge University Press, Cambridge (1996)
4. Ben-Or, M., Tiwari, P.: A deterministic algorithm for sparse multivariate polynomial interpolation. In: Proceedings of the Twentieth Annual ACM Symposium on Theory of Computing, STOC 1988, pp. 301–309. ACM, New York (1988)
5. Cuyt, A., Lee, W.-s.: Multivariate exponential analysis from the minimal number of samples. Adv. Comput. Math. (2017, to appear)
6. Cuyt, A., Lee, W.-s., Yáng, X.: On tensor decomposition, sparse interpolation and Padé approximation. Jaén J. Approx. **8**(1), 33–58 (2016)
7. Das, S., Neumaier, A.: Solving overdetermined eigenvalue problems. SIAM J. Sci. Comput. **35**(2), A541–A560 (2013)
8. Demeure, C.J.: Fast QR factorization of Vandermonde matrices. Linear Algebra Appl. **122–124**, 165–194 (1989)
9. Diederichs, B., Iske, A.: Parameter estimation for bivariate exponential sums. In: IEEE International Conference Sampling Theory and Applications (SampTA 2015), pp. 493–497 (2015)
10. Giesbrecht, M., Labahn, G., Lee, W.-s.: Symbolic-numeric sparse interpolation of multivariate polynomials. In: Proceedings of 2006 International Symposium on Symbolic and Algebraic Computation, ISSAC 2006, pp. 116–123 (2006)
11. Hua, Y., Sarkar, T.K.: Matrix pencil method for estimating parameters of exponentially damped/undamped sinusoids in noise. IEEE Trans. Acoust. Speech Sig. Process. **38**, 814–824 (1990)
12. Nyquist, H.: Certain topics in telegraph transmission theory. Trans. Am. Inst. Electr. Eng. **47**(2), 617–644 (1928)
13. Papy, J.M., Lathauwer, L.D., Van Huffel, S.: Exponential data fitting using multilinear algebra: the single-channel and multi-channel case. Numer. Linear Algebra Appl. **12**, 809–826 (2005)
14. de Prony, R.: Essai expérimental et analytique sur les lois de la dilatabilité des fluides élastiques et sur celles de la force expansive de la vapeur de l'eau et de la vapeur de l'alkool, à différentes températures. J. Ec. Poly. **1**, 24–76 (1795)
15. Rouquette, S., Najim, M.: Estimation of frequencies and damping factors by two-dimensional ESPRIT type methods. IEEE Trans. Sig. Process. **49**(1), 237–245 (2001)
16. Roy, R., Kailath, T.: ESPRIT-estimation of signal parameters via rotational invariance techniques. IEEE Trans. Acoust., Speech Sig. Process. **37**(7), 984–995 (1989)
17. Shannon, C.E.: Communication in the presence of noise. Proc. IRE **37**, 10–21 (1949)
18. Vervliet, N., Debals, O., Sorber, L., Van Barel, M., De Lathauwer, L.: Tensorlab 3.0, March 2016. https://www.tensorlab.net. Available online

Symbolic Algorithm for Generating the Orthonormal Bargmann–Moshinsky Basis for SU(3) Group

A. Deveikis[1], A. A. Gusev[2], V. P. Gerdt[2,3], S. I. Vinitsky[2,3(✉)], A. Góźdź[4],
and A. Pędrak[5]

[1] Department of Applied Informatics, Vytautas Magnus University,
Kaunas, Lithuania
algirdas.deveikis@vdu.lt
[2] Joint Institute for Nuclear Research, Dubna, Russia
vinitsky@theor.jinr.ru
[3] RUDN University, 6 Miklukho-Maklaya, 117198 Moscow, Russia
[4] Institute of Physics, Maria Curie-Skłodowska University, Lublin, Poland
[5] National Centre for Nuclear Research, Warsaw, Poland

Abstract. A symbolic algorithm which can be implemented in any computer algebra system for generating the Bargmann–Moshinsky (BM) basis with the highest weight vectors of SO(3) irreducible representations is presented. The effective method resulting in analytical formula of overlap integrals in the case of the non-canonical BM basis [S. Alisauskas, P. Raychev, R. Roussev, J. Phys. G **7**, 1213 (1981)] is used. A symbolic recursive algorithm for orthonormalisation of the obtained basis is developed. The effectiveness of the algorithms implemented in Mathematica 10.1 is investigated by calculation of the overlap integrals for up to $\mu = 5$ with $\lambda > \mu$ and orthonormalization of the basis for up to $\mu = 4$ with $\lambda > \mu$. The action of the zero component of the quadrupole operator onto the basis vectors with $\mu = 4$ is also obtained.

Keywords: SU(3) non-canonical basis · Group theory
Gram-Schmidt orthonormalization · Symbolic algorithms

1 Introduction

The formalism of SU(3) group provides a comprehensive theoretical foundation for understanding this symmetry in nuclear structure [4,6–9]. However, the construction of the SU(3) bases can usually be performed analytically only for some special cases. In this respect, because of mathematical simplicity of its definition, the Bargmann–Moshinsky (BM) basis [3,10] is especially convenient for calculation. However, the necessity to introduce the physically relevant angular momentum observable gives rise to non-canonical group reduction SU(3) ⊃ O(3) ⊃ O(2). The BM vectors may be calculated from the simplest vectors which correspond to the highest angular momentum projection $M = L$,

© Springer Nature Switzerland AG 2018
V. P. Gerdt et al. (Eds.): CASC 2018, LNCS 11077, pp. 131–145, 2018.
https://doi.org/10.1007/978-3-319-99639-4_9

i.e., the highest weight basis vectors with respect to the SO(3) group that was proved in [10]. It should be stressed that the analytical and what is very important an effective algorithm for construction of this basis is required for analysis of some quantum systems.

As an example, one can consider the vibration (in particular, quadrupole) and rotation motions which are the most important low energy nuclear motions. The simplest SU(3) model Hamiltonian consists of the quadrupole-quadrupole interaction, the rotational term, and the other terms constructed from generators of the partner groups $G = \text{SU}(3) \times \overline{\text{SU}(3)}$, see [8] and references therein. A possible Hamiltonian H used in this schematic nuclear model can be written as

$$
\begin{aligned}
H &= \gamma C_2(\text{SU}(3)) - \kappa Q \cdot Q + \beta L \cdot L + H''(\bar{Q}, \bar{L}) \\
&= (\gamma - \kappa)C_2(\text{SU}(3)) + (3\kappa + \beta)L^2 + H''(\bar{Q}, \bar{L}),
\end{aligned} \tag{1}
$$

where the second order Casimir operator $C_2(\text{SU}(3)) = Q \cdot Q + 3L \cdot L$, Q and L are generators of SU(3), i.e., quadrupole and angular momentum, respectively; $\bar{Q}$ and $\bar{L}$ are generators of the intrinsic group $\overline{\text{SU}(3)}$. Some examples of physically interesting forms of the interaction H'' can be written as

$$
H_{3Q} = h_{3Q} \left((Q \otimes Q)_2^3 - (Q \otimes Q)_{-2}^3 \right), \tag{2}
$$

$$
H_{3LQ} = h_{3LQ} \left((L \otimes Q)_2^3 - (L \otimes Q)_{-2}^3 \right), \tag{3}
$$

$$
H_{4Q} = h_{4Q} \left(\sqrt{\frac{14}{5}} (Q \otimes Q)_0^4 + (Q \otimes Q)_{-4}^4 + (Q \otimes Q)_4^4 \right), \tag{4}
$$

where $(T_{\lambda'} \otimes T_{\lambda})_M^L$ denotes the tensor product of two spherical tensors [13]. These interaction terms can simulate either the tetrahedral or octahedral nuclear symmetry now widely considered in nuclear physics [5]. To find the corresponding energies and quantum nuclear states one needs to solve the eigenvalue problem of the Hamiltonian (1).

To solve the eigenvalue problem for H the appropriate basis constructed according to the group chain $\text{SU}(3) \supset \text{SO}(3) \supset \text{SO}(2)$ is required. There were several attempts to construct such bases. They were based on different group theoretical technics, for a short review see the introduction in the paper [11]. In all those cases one obtains the non-orthogonal basis. This increases a complexity of calculations of the reduced matrix elements of different operators, Clebsch–Gordan coefficients, etc. It requires an adaptation of the Gram–Schmidt orthogonalization procedure to be more effective in symbolic calculations.

We start from the BM states which are linearly independent but as in other approaches not orthonormal. However, we develop an effective symbolic algorithm suitable for implementation in computer algebra systems. It is based on the adapted Gram–Schmidt orthonormalization procedure but using the overlap integrals calculated in an analytical form [2]. It provides the analytic construction of the desirable orthonormalized basis. Our adaptation of Gram–Schmidt orthonormalization procedure consists in construction of recursive calculation of the required quantities and the normalization integrals do not involve any square root operation. This distinct feature of the proposed orthonormalization algorithm may make the large scale symbolic calculations feasible.

Then one can calculate in this orthonormalized basis the zero component of the quadrupole operator Q_0 in the analytical form using its simple form given in the non-canonical BM basis [1,12]. The other components of the quadrupole operator Q_k written in the analytical form can be obtained by making use of the Wigner–Eckart theorem with conventional SO(3) Clebsch–Gordan coefficients [13]. Thus, one can construct the above Hamiltonian (1) also in an analytical form.

The paper is organized as follows. In Sect. 2, the *Symbolic Algorithm 1* for calculation of overlap integrals of BM vectors is shown. In Sect. 3, the *Symbolic Algorithm 2* for orthonormalization of BM basis is given. In Sect. 4, an action of the quadrupole operator Q_0 onto the constructed basis is presented.

In the Conclusion, further applications of the elaborated *symbolic algorithms* are pointed out.

2 Overlap Integrals of Bargmann–Moshinsky Basis

The effective method for constructing a non-canonical BM basis with the highest weight vectors of SO(3) irreducible representations corresponding to the group chain $SU(3) \supset O(3) \supset O(2)$ was described in [2]. Let us introduce the notation for the vectors of this basis:

$$\left| \begin{matrix} (\lambda, \mu)_B \\ \alpha, L, L \end{matrix} \right\rangle . \tag{5}$$

Here the quantum numbers λ, μ label irreducible representations (irreps), $\lambda, \mu = 0, 1, 2, \ldots$ and $\lambda > \mu$; L, M are the quantum numbers of angular momentum and its projection (in our case, $M = L$); α is the additional index that is used for unambiguously distinguishing the equivalent SO(3) irreps (L) in a given SU(3) irrep (λ, μ). The dimension of subspace irrep for given λ, μ can be calculated by using the following formula:

$$D_{\lambda\mu} = \frac{1}{2}(\lambda + 1)(\mu + 1)(\lambda + \mu + 2). \tag{6}$$

In order to perform classification of the BM states (5) one should determine the set of allowed values of α and L. It is well known that the ranges of quantum numbers α and L are determined by the values of quantum numbers λ and μ. However, the determination of former quantities is rather cumbersome. The easiest way to get the allowed values of α and L is by using the *Symbolic Algorithm 1* that consists of the following steps:

Step 1. Firstly we should start with choosing some particular value of the quantum number μ. For the following consideration, it is convenient to introduce auxiliary label K [7] which varies in the ranges

$$K = \mu, \mu - 2, \ldots, 1 \text{ or } 0, \quad \text{since } \lambda > \mu. \tag{7}$$

The label K is related to α by

$$\alpha = \frac{1}{2}(\mu - K). \tag{8}$$

So, for every fixed μ, the set of possible values of K can be obtained directly from the definition of K from (7). Now, the set of allowed values of α may be determined from these K values using relation (8).

Step 2. In the case $K = 0$, that may occur only for even values of μ, the allowed values of L are determined by the label λ:

$$L = \lambda, \lambda - 2, \ldots, 1 \text{ or } 0. \tag{9}$$

Step 3. In the case $K \neq 0$, the $L_{\min} = K$. Since for every particular μ, there is a number of possible K numbers, according to (7) there exists a number of the corresponding α numbers. It means that for every particular μ, there will be a number of pairs $(\alpha, L_{\min})$. The maximum value of L is defined by the expression $L_{\max} = \mu - 2\alpha + \lambda - \beta$, where

$$\beta = \begin{cases} 0, & \lambda + \mu - L \text{ even}, \\ 1, & \lambda + \mu - L \text{ odd}. \end{cases} \tag{10}$$

To determine $L_{\max}$ it is convenient to consider two alternatives: $\lambda - L$ is even and $\lambda - L$ is odd. In both cases, the label β is defined by the given μ value, and the number $L_{\max}$ is also determined. An illustrative example for calculation of allowed values of α and L is presented in Table 1. The results for the case $K = 0$ are not included since in this case, their termination is rather straightforward. It should be noted that the set of allowed values of L for overlap integrals is given by the intersection of these sets for the corresponding $<bra|$ and $|ket>$ vectors.

In this paper, we use the following form of the formula for the overlap integral $<\alpha|\alpha'>=<\alpha'|\alpha>$ of the non-canonical BM states presented in [2]:

$$\left\langle \alpha \middle| \alpha' \right\rangle = \left\langle \begin{matrix} (\lambda, \mu)_B \\ \alpha, L, L \end{matrix} \middle| \begin{matrix} (\lambda, \mu)_B \\ \alpha', L, L \end{matrix} \right\rangle = C_1(\lambda, L, \Delta)(\lambda + 2)^\beta (L - \mu + 2\alpha)!$$

$$\times (\lambda - L + \mu - 2\alpha' - \beta)!!(\mu - 2\alpha' - \beta + \Delta - 1)!!$$

$$\times \sum_{l,z} \binom{\alpha'}{\frac{1}{2}(l - \beta - \Delta)} (-1)^{(\mu + 2\alpha - \Delta - \beta)/2 + z} \binom{\frac{1}{2}(\mu - 2\alpha - \Delta - \beta)}{z}$$

$$\times \frac{(\mu - l)!!}{(\mu - l - 2z)!!} \frac{(\mu + \beta + \Delta)!!}{(\mu - 2\alpha' + l)!!}(l - \Delta + \beta - 1)!!(\mu - \Delta - \beta - 2z)!!$$

$$\times \frac{(\lambda - L + \mu - 2\alpha - \beta)!!}{(\lambda - L + \Delta + 2z)!!} \frac{(\lambda + L - \Delta + 2)!!}{(\lambda + L - \mu + 2\alpha + \beta + 2z + 2)!!} \frac{(L + l)!}{L!}$$

$$\times \frac{(\lambda + \mu + L + \beta + 2)!!}{(\lambda + L + l + \beta + 2z + 2)!!} \frac{(\lambda + \beta + 2z + 1)!}{(\lambda + \beta + 1)!} \frac{(\lambda + \mu - l - L + \Delta)!!}{(\lambda - L + \mu - 2\alpha' - \beta)!!}$$

$$\times C_2(\lambda, L, \Delta, z). \tag{11}$$

Table 1. The allowed values of α, $L_{\min}$, and $L_{\max}$ for up to $\mu = 5$ when $K \neq 0$.

μ	α	$L_{\min}$	$L_{\max}(\lambda - L$ even$)$	$L_{\max}(\lambda - L$ odd$)$
1	0	1	λ	$\lambda + 1$
2	0	2	$\lambda + 2$	$\lambda + 1$
3	0	3	$\lambda + 2$	$\lambda + 3$
	1	1	λ	$\lambda + 1$
4	0	4	$\lambda + 4$	$\lambda + 3$
	1	2	$\lambda + 2$	$\lambda + 1$
5	0	5	$\lambda + 4$	$\lambda + 5$
	1	3	$\lambda + 2$	$\lambda + 3$
	2	1	λ	$\lambda + 1$

Here $\alpha \geq \alpha'$ and β from (10) and we use the following notations:

$$\Delta = \begin{cases} 0, & \lambda - L \text{ even}, \\ 1, & \lambda - L \text{ odd}, \end{cases} \qquad \binom{m}{n} = \frac{m!}{n!(m-n)!},$$

$$C_1(\lambda, L, \Delta) = \begin{cases} 1, & L > \lambda + \Delta, \\ \frac{(\lambda+L+\Delta+1)!!}{(2L+1)!!}, & L \leq \lambda + \Delta, \end{cases}$$

$$C_2(\lambda, L, \Delta, z) = \begin{cases} \frac{(\lambda+L+\Delta+1+2z)!!}{(2L+1)!!}, & L > \lambda + \Delta, \\ \frac{(\lambda+L+\Delta+1+2z)!!}{(\lambda+L+\Delta+1)!!}, & L \leq \lambda + \Delta. \end{cases}$$

Remark 1. The upper alternative in definition of the coefficients C_1 and C_2 corresponds to the overlap integrals which contain only λ in their final expression. The summation parameter z runs from 0 to $\frac{1}{2}(-2\alpha - \beta - \Delta + \mu)$ except when $z < \frac{1}{2}(-\Delta - \lambda + L)$. The summation parameter l runs from $\beta + \Delta$ to $2\alpha' + \beta + \Delta$ except when $\mu - l$ or $l - \Delta - \beta$ is odd.

The *Algorithm 1* realized **Steps 1–3** and function (11) was implemented in the Mathematica code. This code was verified by calculating the overlap integrals presented in [2]. We reproduced the results presented there for $\mu = 1, 2, 3, 4$, however, with some exceptions for $\mu = 4$. Our results for $\mu = 4$ are presented in Tables 3 and 4 with specification of indices of the overlap integrals given in Table 2. New corrected expressions of the overlap integrals with respect to the incorrect results from Table 1 of Ref. [2] are marked by asterisk (*).

In this paper, the new results for the overlap integrals for the non-canonical BM basis with the highest weight vectors of the SO(3) group irreps for $\mu = 5$ were calculated and presented in Table 5. Here the more concise notation for the overlap integrals $\langle u_\alpha | u_{\alpha'} \rangle$ of states (5) is introduced. We present the obtained expressions for overlap integrals in Tables 6 and 7. The above algorithm was realized in the form of the program implemented in the computer algebra system Wolfram Mathematica 10.1. The typical running time of calculating the irreducible representations $\mu = 4$ and $\mu = 8$ is 3 and 57 s and memory is 35

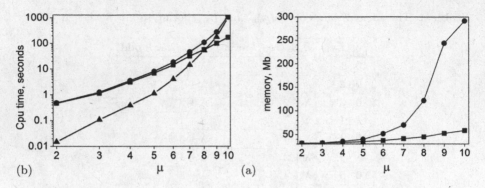

Fig. 1. The CPU time versus parameter μ (a) and MaxMemoryUsed versus parameter μ (b): maximum number of megabytes (Mb) used to store all data for the current Wolfram system session during the calculations of the orthogonal BM basis (circles) consisting of calculation of the overlap integrals by means of Algorithm 1 code (squares) and execution of the othonormalization Gram–Schmidt procedure by means of Algorithm 2 code (triangles).

and 47 Mb, respectively, using the PC Intel Pentium CPU 1.50 GHz 4 GB 64 bit Windows 8. In Fig. 1 we show the CPU time and MaxMemoryUsed during calculations of overlap integrals by Algorithm 1 versus parameter μ.

3   Orthonormalization of Bargmann–Moshinsky Basis

Let us construct the orthonormal basis in the space spanned by the non-canonical BM vectors (5), ($M = L$). For this purpose, we propose a bit more efficient form of the Gram–Schmidt orthonormalisation procedure

$$\left| \begin{matrix} (\lambda, \mu) \\ f_i, L, L \end{matrix} \right\rangle = \sum_{\alpha=0}^{\alpha_{\max}} A_{i,\alpha}^{(\lambda,\mu)}(L) \left| \begin{matrix} (\lambda, \mu)_B \\ \alpha, L, L \end{matrix} \right\rangle. \tag{12}$$

Here multiplicity index i is introduced to differentiate the orthonormalized states and $A_{i,\alpha}^{(\lambda,\mu)}(L)$ are the BM basis orthonormalization coefficients. These coefficients fulfill the following condition

$$A_{i,\alpha}^{(\lambda,\mu)}(L) = 0, \quad \text{if } i > \alpha. \tag{13}$$

Because the BM vectors (5) are linearly independent, one can require the orthonormalization properties for the vectors (12)

$$\left\langle \begin{matrix} (\lambda, \mu) \\ f_i, L, L \end{matrix} \right| \left. \begin{matrix} (\lambda, \mu) \\ f_k, L, L \end{matrix} \right\rangle = \delta_{ik}. \tag{14}$$

In this paper, we developed the analytical orthonormalization procedure based on the Gram–Schmidt orthonormalization algorithm. For explicit construction of orthonormalized BM basis, let us consider step by step the *Symbolic Algorithm 2.*

Table 2. Overlap integrals of non-canonical BM basis for $\mu = 4$.

$(\alpha|\alpha')$	L	$\lambda - L$ even	L	$\lambda - L$ odd
$(2|2)$	$0,\ldots,\lambda$	$\langle u_2|u_2\rangle$		
$(2|1)$	$2,\ldots,\lambda$	$\langle u_2|u_1\rangle$		
$(2|0)$	$4,\ldots,\lambda$	$\langle u_2|u_0\rangle$		
$(1|1)$	$2,\ldots,\lambda$	$\langle u_1|u_1\rangle$	$2,\ldots,\lambda+1$	$\langle \tilde{u}_1|\tilde{u}_1\rangle$
$(1|1)$	$\lambda+2$	$\langle u_1'|u_1'\rangle$		
$(1|0)$	$4,\ldots,\lambda$	$\langle u_1|u_0\rangle$	$4,\ldots,\lambda+1$	$\langle \tilde{u}_1|\tilde{u}_0\rangle$
$(1|0)$	$\lambda+2$	$\langle u_1'|u_0'\rangle$		
$(0|0)$	$4,\ldots,\lambda$	$\langle u_0|u_0\rangle$	$4,\ldots,\lambda+1$	$\langle \tilde{u}_0|\tilde{u}_0\rangle$
$(0|0)$	$\lambda+2$	$\langle u_0'|u_0'\rangle$	$\lambda+3$	$\langle \tilde{u}_0'|\tilde{u}_0'\rangle$
$(0|0)$	$\lambda+4$	$\langle u_0''|u_0''\rangle$		

Step 1. First one needs to organize the loop running over all indices $\alpha = \alpha_{\max}, \alpha_{\max}-1,\ldots,0$ of a given set of the BM states. Then the first orthonormalization coefficients of the orthogonal BM states (i.e., some linear combination of initial states (5)) for a given value of α are calculated by the formula

$$b_{\alpha,\alpha_{\max}} = \frac{\langle u_\alpha|u_{\alpha_{\max}}\rangle}{\langle u_{\alpha_{\max}}|u_{\alpha_{\max}}\rangle^{1/2}}, \tag{15}$$

where the $\langle u_\alpha|u_{\alpha'}\rangle$ denotes the overlap integrals (11).

Step 2. Secondly one needs to organize the inner loop inside the loop defined in Step 1 of this algorithm. This inner loop should run over all indices $\alpha' = \alpha_{\max} - 1, \alpha_{\max} - 2,\ldots,\alpha + 1$ of a given set of BM states. For the following calculations, it is convenient to introduce the intermediate quantity

$$f_{\alpha,\alpha'} = -\langle u_\alpha|u_{\alpha'}\rangle + \frac{\langle u_\alpha|u_{\alpha_{\max}}\rangle\langle u_{\alpha_{\max}}|u_{\alpha'}\rangle}{\langle u_{\alpha_{\max}}|u_{\alpha_{\max}}\rangle}. \tag{16}$$

Now the orthonormalization coefficients for the BM states for any given values of α and α' are calculated by the formula

$$b_{\alpha,\alpha'} = \frac{f_{\alpha,\alpha'}}{\langle \psi_{\alpha'}|\psi_{\alpha'}\rangle^{1/2}}. \tag{17}$$

Here the normalization integral is defined as

$$\langle \psi_\alpha|\psi_\alpha\rangle = \langle u_\alpha|u_\alpha\rangle - \sum_{i=\alpha+1}^{\alpha_{\max}} b_{\alpha,i}^2. \tag{18}$$

Step 3. Now we make the recursive step and calculate the next quantity $f_{\alpha,\alpha'}$ from the results of the previous step:

$$f_{\alpha,\alpha'-1} = f_{\alpha,\alpha'\to\alpha'-1} + \frac{1}{\langle \psi_{\alpha'}|\psi_{\alpha'}\rangle} f_{\alpha\to\alpha'-1,\alpha'} f_{\alpha,\alpha'}. \tag{19}$$

Table 3. Overlap integrals of the non-canonical BM basis. for $\mu = 4$ and $\lambda - L$ even.

$\mu = 4$ and $\lambda - L$ even	
	$\langle u_2 \| u_2 \rangle = 8L!(\lambda - L)!!(\lambda + L + 1)!!(3L^4 + 6L^3$ $-(8\lambda(\lambda + 8) + 135)L^2 - 2(4\lambda(\lambda + 8) + 69)L$ $+8(\lambda + 3)^2(\lambda + 5)^2)/(2L + 1)!!$
	$\langle u_2 \| u_1 \rangle = 8L!(-\lambda + L - 2)(\lambda + L + 6)(\lambda - L)!!(\lambda + L + 1)!!$ $\times (3(L - 1)L - 2(2\lambda(\lambda + 8) + 33))/(2L + 1)!!$
	$\langle u_2 \| u_0 \rangle = 24L!(\lambda - L + 2)(\lambda - L + 4)(\lambda + L + 4)$ $\times (\lambda + L + 6)(\lambda - L)!!(\lambda + L + 1)!!/(2L + 1)!!,$
(*)	$\langle u_1 \| u_1 \rangle = -4(L - 2)!(\lambda - L + 2)!!(\lambda + L + 1)!! \left(6L^5 + 6(\lambda + 5)L^4 \right.$ $-(\lambda(7\lambda + 59) + 150)L^3 - (\lambda + 6)(\lambda(7\lambda + 55) + 118)L^2$ $-(\lambda + 2)(\lambda(5\lambda + 48) + 129)L - 6(\lambda + 2)(\lambda(\lambda + 10) + 27))/(2L + 1)!!$
(*)	$\langle u_1' \| u_1' \rangle = 4(\lambda + 2)(\lambda + 3)(\lambda + 4)(\lambda + 35)\lambda!.$
	$\langle u_1 \| u_0 \rangle = 24(L - 2)!(\lambda - L + 4)(\lambda + L + 6)(\lambda - L + 2)!!$ $\times (\lambda + L(\lambda + L(\lambda + L + 4) + 2) + 2)(\lambda + L + 1)!!/(2L + 1)!!$
	$\langle u_1' \| u_0' \rangle = 96(\lambda + 2)(\lambda + 3)(\lambda + 4)\lambda!$
(*)	$\langle u_0 \| u_0 \rangle = 24(L - 4)!(\lambda - L + 4)!!(\lambda + L + 1)!!(9(\lambda + 2)(\lambda + 4)$ $+L^6 + 2(\lambda + 3)L^5 + 8(\lambda + 2)(\lambda + 3)L + (\lambda(\lambda + 4) - 8)L^4$ $-2(\lambda + 3)(\lambda + 6)L^3 + (\lambda(5\lambda + 38) + 88)L^2)/(2L + 1)!!,$
	$\langle u_0' \| u_0' \rangle = 48(\lambda + 2)(\lambda + 3)(\lambda + 4)(2\lambda^2 + \lambda + 3)(\lambda - 2)!$
	$\langle u_0'' \| u_0'' \rangle = 24(\lambda + 2)(\lambda + 3)(\lambda + 4)(\lambda + 5)\lambda!$

Here the arrows in the right hand side of the (19) indicate that the quantity $f_{\alpha,\alpha'}$ obtained at the previous step is used with the appropriate substitution of indices. Having calculated the quantity $f_{\alpha,\alpha'}$, the expression of the next orthonormalization coefficient $b_{\alpha,\alpha'}$ can be obtained by Eq. (17). The steps of the orthonormalization algorithm defined above are recursively repeated doing the loop over all allowed values of indices α and α'.

Step 4. Finally, we should collect all the coefficients in the recursively obtained analytical expansion representing the orthonormalized state for every independent BM state (5). In this way, we get the required orthonormalization coefficients of expansion (12).

Remark 2. The two advantages of the proposed *Algorithm 2.* First of all, its simplicity: at any recursive step, $f_{\alpha,\alpha'}$ is composed of fragments that are no more complicated than that defined in the right hand side of Eq. (16) and the normalization integrals (18). Secondly, recursive calculation of the quantities $f_{\alpha,\alpha'}$ (19) and the normalization integrals (18) do not involve any square root operation. This distinct feature of the proposed orthonormalization algorithm may make the large scale symbolic calculations feasible.

In this paper, the new results for the orthonormalization coefficients of the non-canonical BM basis with the highest weight vectors of SO(3) irreps for $\mu = 4$ were calculated and presented in Table 8. It should be noted that the orthonormalization coefficients for up to $\mu = 3$ were calculated as well and their values are equal to those presented in Table 2 of Ref. [2]. Let us illustrate the calculation of

Table 4. Overlap integrals of the non-canonical BM basis for $\mu = 4$ and $\lambda - L$ odd.

$\mu = 4$ and $\lambda - L$ odd
$\langle \tilde{u}_1 \vert \tilde{u}_1 \rangle = 6(\lambda + 2)(2(\lambda(\lambda + 10) + 27) - L^2 - L)$ $\times (L + 1)(L + 2)(L - 2)!(\lambda - L + 1)!!(\lambda + L + 2)!!/(2L + 1)!!.$
$\langle \tilde{u}_1 \vert \tilde{u}_0 \rangle = -6(\lambda + 2)(L + 1)(L + 2)(\lambda - L + 3)(\lambda + L + 7)$ $\times (L - 2)!(\lambda - L + 1)!!(\lambda + L + 2)!!/(2L + 1)!!$
$\langle \tilde{u}_0 \vert \tilde{u}_0 \rangle = -6(\lambda + 2)(9(\lambda + 3) + L(\lambda + L(\lambda + L + 5) - 5))$ $\times (L + 1)(L + 2)(L - 4)!(\lambda - L + 3)!!(\lambda + L + 2)!!/(2L + 1)!!$
$(*)$ $\langle \tilde{u}_0' \vert \tilde{u}_0' \rangle = 6(\lambda + 2)(\lambda + 3)(\lambda + 4)^2(\lambda + 5)(\lambda - 1)!$

Table 5. Overlap integrals of non-canonical BM basis for $\mu = 5$.

$(\alpha\vert\alpha')$	L	$\lambda - L$ even	L	$\lambda - L$ odd
$(2\vert 2)$	$1, \ldots, \lambda$	$\langle u_2 \vert u_2 \rangle$	$1, \ldots, \lambda + 1$	$\langle \tilde{u}_2 \vert \tilde{u}_2 \rangle$
$(2\vert 1)$	$3, \ldots, \lambda$	$\langle u_2 \vert u_1 \rangle$	$3, \ldots, \lambda + 1$	$\langle \tilde{u}_2 \vert \tilde{u}_1 \rangle$
$(2\vert 0)$	$5, \ldots, \lambda$	$\langle u_2 \vert u_0 \rangle$	$5, \ldots, \lambda + 1$	$\langle \tilde{u}_2 \vert \tilde{u}_0 \rangle$
$(1\vert 1)$	$3, \ldots, \lambda$	$\langle u_1 \vert u_1 \rangle$	$3, \ldots, \lambda + 1$	$\langle \tilde{u}_1 \vert \tilde{u}_1 \rangle$
$(1\vert 1)$	$\lambda + 2$	$\langle u_1' \vert u_1' \rangle$	$\lambda + 3$	$\langle \tilde{u}_1' \vert \tilde{u}_1' \rangle$
$(1\vert 0)$	$5, \ldots, \lambda$	$\langle u_1 \vert u_0 \rangle$	$5, \ldots, \lambda + 1$	$\langle \tilde{u}_1 \vert \tilde{u}_0 \rangle$
$(1\vert 0)$	$\lambda + 2$	$\langle u_1' \vert u_0' \rangle$	$\lambda + 3$	$\langle \tilde{u}_1' \vert \tilde{u}_0' \rangle$
$(0\vert 0)$	$5, \ldots, \lambda$	$\langle u_0 \vert u_0 \rangle$	$5, \ldots, \lambda + 1$	$\langle \tilde{u}_0 \vert \tilde{u}_0 \rangle$
$(0\vert 0)$	$\lambda + 2$	$\langle u_0' \vert u_0' \rangle$	$\lambda + 3$	$\langle \tilde{u}_0' \vert \tilde{u}_0' \rangle$
$(0\vert 0)$	$\lambda + 4$	$\langle u_0'' \vert u_0'' \rangle$	$\lambda + 5$	$\langle \tilde{u}_0'' \vert \tilde{u}_0'' \rangle$

orthonormalization coefficients and output of their values that are symbolically represented in Table 8. The explicit expressions for these coefficients calculated by the *Algorithm 2* realized **Steps 1–4** was implemented in the Mathematica code in terms of the overlap integrals $<u_i \vert u_j>$ and $<\tilde{u}_i \vert \tilde{u}_j>$ listed in Tables 3 and 4, respectively, are given below. The above algorithm was realized in the form of the program implemented in the computer algebra system Wolfram Mathematica 10.1. The typical running time of calculating the irreducible representations $\mu = 4$ is 30 s and memory is 60 Mb using the PC Intel Pentium CPU 1.50 GHz 4 GB 64bit Windows 8. In Fig. 1, we show the CPU time and MaxMemoryUsed during calculations of the orthonormal BM basis versus parameter μ by means of Algorithms 1 and 2. One can see that the CPU time (in double logarithmic scale) of execution of Algorithm 1 is linearly growing in contradistinction to Algorithm 2, whose execution time is growing faster than linearly due to manipulations with rational expressions.

Table 6. Overlap integrals of the non-canonical BM basis for $\mu = 5$ and $\lambda - L$ even.

$\mu = 5$ and $\lambda - L$ even
$\langle u_2\vert u_2\rangle = 24(\lambda+2)(L+1)(L-1)!(\lambda-L)!!(\lambda+L+1)!!$ $\qquad \times(-(4\lambda(\lambda+10)+109)L^2 - 2(2\lambda(\lambda+10)+55)L$ $\qquad +8(\lambda(\lambda+10)(\lambda(\lambda+10)+49)+603)+L^4+2L^3)/(2L+1)!!$
$\langle u_2\vert u_1\rangle = 24(\lambda+2)(L+1)(-\lambda+L-2)(\lambda+L+8)(L-1)!$ $\qquad \times((L-1)L-2(\lambda(\lambda+10)+27))(\lambda-L)!!(\lambda+L+1)!!/(2L+1)!!$
$\langle u_2\vert u_0\rangle = 24(\lambda+2)(L+1)(-\lambda+L-4)(-\lambda+L-2)(\lambda+L+6)$ $\qquad \times(L-1)!(\lambda+L+8)(\lambda-L)!!(\lambda+L+1)!!/(2L+1)!!$
$\langle u_1\vert u_1\rangle = 12(\lambda+2)(L+1)(L-3)!(\lambda-L+2)!!(\lambda+L+1)!!$ $\qquad \times((\lambda(3\lambda+29)+96)L^3+(\lambda(\lambda(3\lambda+53)+316)+680)L^2$ $\qquad +(\lambda(\lambda(7\lambda+100)+487)+716)L-2L^5-2(\lambda+7)L^4$ $\qquad +2(\lambda(\lambda(7\lambda+102)+491)+684))/(2L+1)!!$
$\langle u_1'\vert u_1'\rangle = 12(\lambda+2)(\lambda+3)^2(\lambda+4)(\lambda+5)(\lambda+20)(\lambda-1)!$
$\langle u_1\vert u_0\rangle = 24(\lambda+2)(L+1)(\lambda-L+4)(\lambda+L+8)(L-3)!(\lambda-L+2)!!$ $\qquad \times(5\lambda+L(3\lambda+L(\lambda+L+6)+8)+12)(\lambda+L+1)!!/(2L+1)!!$
$\langle u_1'\vert u_0'\rangle = 96(\lambda+2)(\lambda+3)^2(\lambda+4)(\lambda+5)(\lambda-1)$
$\langle u_0\vert u_0\rangle = 24(\lambda+2)(L+1)(L-5)!(\lambda-L+4)!!(\lambda+L+1)!!$ $\qquad \times(9(\lambda+4)(9\lambda+22)+(\lambda(\lambda+12)+12)L^4$ $\qquad +2(\lambda-1)(\lambda+8)L^3+(\lambda(17\lambda+114)+248)L^2$ $\qquad +4(\lambda(13\lambda+68)+96)L+L^6+2(\lambda+5)L^5)/(2L+1)!!$
$\langle u_0'\vert u_0'\rangle = 48(\lambda+2)(\lambda+3)^2(\lambda+4)(\lambda+5)(\lambda(2\lambda+3)+10)(\lambda-3)!$
$\langle u_0''\vert u_0''\rangle = 24(\lambda+2)(\lambda+3)(\lambda+4)(\lambda+5)^2(\lambda+6)(\lambda-1)!$

In the case of the subset of three independent BM vectors (5) indicated by the displayed values of labels, expansion (12) takes the form

$$\left\vert \begin{matrix} (\lambda,\mu) \\ f_0, L, L \end{matrix} \right\rangle = A_{0,0}^{(\lambda,\mu)}(L)\left\vert \begin{matrix} (\lambda,\mu)_B \\ 0, L, L \end{matrix} \right\rangle + A_{0,1}^{(\lambda,\mu)}(L)\left\vert \begin{matrix} (\lambda,\mu)_B \\ 1, L, L \end{matrix} \right\rangle + A_{0,2}^{(\lambda,\mu)}(L)\left\vert \begin{matrix} (\lambda,\mu)_B \\ 2, L, L \end{matrix} \right\rangle,$$

$$A_{0,0}^{(\lambda,4)}(L) = -\langle\psi_0\vert\psi_0\rangle^{-1/2},$$

$$A_{0,1}^{(\lambda,4)}(L) = -\frac{\langle\psi_0\vert\psi_0\rangle^{-1/2}}{\langle\psi_1\vert\psi_1\rangle}\left(-\langle u_1\vert u_0\rangle+\frac{\langle u_2\vert u_1\rangle\langle u_2\vert u_0\rangle}{\langle u_2\vert u_2\rangle}\right),$$

$$A_{0,2}^{(\lambda,4)}(L) = \langle\psi_0\vert\psi_0\rangle^{-1/2}\left[\frac{\langle u_2\vert u_0\rangle}{\langle u_2\vert u_2\rangle}+\frac{1}{\langle\psi_1\vert\psi_1\rangle}\left(-\langle u_1\vert u_0\rangle\right.\right.$$

$$\left.\left.+\frac{\langle u_2\vert u_1\rangle\langle u_2\vert u_0\rangle}{\langle u_2\vert u_2\rangle}\right)\frac{\langle u_2\vert u_1\rangle}{\langle u_2\vert u_2\rangle}\right],$$

$$\langle\psi_0\vert\psi_0\rangle = \langle u_0\vert u_0\rangle - \frac{\langle u_2\vert u_0\rangle^2}{\langle u_2\vert u_2\rangle} - \frac{1}{\langle\psi_1\vert\psi_1\rangle}\left(-\langle u_1\vert u_0\rangle+\frac{\langle u_2\vert u_1\rangle\langle u_2\vert u_0\rangle}{\langle u_2\vert u_2\rangle}\right)^2,$$

$$\langle\psi_1\vert\psi_1\rangle = \langle u_1\vert u_1\rangle - \frac{\langle u_2\vert u_1\rangle^2}{\langle u_2\vert u_2\rangle}.$$

Table 7. Overlap integrals of the non-canonical BM basis for $\mu = 5$ and $\lambda - L$ odd.

$\mu = 5$ and $\lambda - L$ odd
$\langle \tilde{u}_2 \vert \tilde{u}_2 \rangle = 24(L+1)(L-1)!(\lambda - L + 1)!!(\lambda + L + 2)!!$ $\qquad \times (-(12\lambda(\lambda + 10) + 317)L^2 - 2(6\lambda(\lambda + 10) + 161)L$ $\qquad + 8(\lambda(\lambda + 10)(\lambda(\lambda + 10) + 49) + 603) + 10L^3 + 5L^4)/(2L+1)!!$
$\langle \tilde{u}_2 \vert \tilde{u}_1 \rangle = 24(L+1)(-\lambda + L - 3)(\lambda + L + 7)(\lambda - L + 1)!!$ $\qquad \times 5(L-1)L - 6(\lambda(\lambda + 10) + 26)(L-1)!(\lambda + L + 2)!!/(2L+1)!!$
$\langle \tilde{u}_2 \vert \tilde{u}_0 \rangle = 120(L+1)(-\lambda + L - 5)(-\lambda + L - 3)(\lambda + L + 5)$ $\qquad \times (L-1)!(\lambda + L + 7)(\lambda - L + 1)!!(\lambda + L + 2)!!/(2L+1)!!$
$\langle \tilde{u}_1 \vert \tilde{u}_1 \rangle = 12(L+1)(L-3)!(\lambda - L + 3)!!(\lambda + L + 2)!!$ $\qquad \times (6(\lambda + 3)(\lambda(5\lambda + 62) + 204) + (\lambda(11\lambda + 135) + 466)L^3$ $\qquad + (\lambda(\lambda(11\lambda + 162) + 781) + 1222)L^2 - 10L^5 - 10(\lambda + 4)L^4$ $\qquad - (\lambda(\lambda(17\lambda + 299) + 1722) + 3414)L)/(2L+1)!!$
$\langle \tilde{u}'_1 \vert \tilde{u}'_1 \rangle = 12(\lambda + 2)(\lambda + 3)(\lambda + 4)(\lambda + 5)(\lambda + 54)\lambda!$
$\langle \tilde{u}_1 \vert \tilde{u}_0 \rangle = 120(L+1)(\lambda - L + 5)(\lambda + L + 7)(L-3)!(\lambda - L + 3)!!$ $\qquad \times (3(\lambda + 3) + L(-\lambda + L(\lambda + L + 3) - 7)(\lambda + L + 2)!!/(2L+1)!!$
$\langle \tilde{u}'_1 \vert \tilde{u}'_0 \rangle = 480(\lambda + 2)(\lambda + 3)(\lambda + 4)(\lambda + 5)\lambda!$
$\langle \tilde{u}_0 \vert \tilde{u}_0 \rangle = 120(L+1)(L-5)!(\lambda - L + 5)!!(\lambda + L + 2)!!$ $\qquad \times \left(L^6 + 2(\lambda + 2)L^5 + (\lambda - 7)(\lambda + 5)L^4 \right.$ $\qquad + 45(\lambda + 3)(\lambda + 5) - 16(\lambda + 4)(\lambda + 11)L$ $\qquad \left. - 2(\lambda(3\lambda + 19) + 10)L^3 + (\lambda(21\lambda + 188) + 439)L^2) \right)/(2L+1)!!$
$\langle \tilde{u}'_0 \vert \tilde{u}'_0 \rangle = 240(\lambda + 2)(\lambda + 3)(\lambda + 4)(\lambda + 5)(\lambda(2\lambda + 1) + 4)(\lambda - 2)!$
$\langle \tilde{u}''_0 \vert \tilde{u}''_0 \rangle = 120(\lambda + 2)(\lambda + 3)(\lambda + 4)(\lambda + 5)(\lambda + 6)\lambda!$

In the case of the subset of two independent BM vectors (5) indicated by the displayed values of labels expansion (12) takes one of three possible linear combinations. The first one is given by expressions

$$\left| \begin{matrix} (\lambda, \mu) \\ \tilde{f}_0, L, L \end{matrix} \right\rangle = \tilde{A}^{(\lambda,\mu)}_{0,0}(L) \left| \begin{matrix} (\lambda, \mu)_B \\ 0, L, L \end{matrix} \right\rangle + \tilde{A}^{(\lambda,\mu)}_{0,1}(L) \left| \begin{matrix} (\lambda, \mu)_B \\ 1, L, L \end{matrix} \right\rangle,$$

$$\tilde{A}^{(\lambda,4)}_{0,0}(L) = -\langle \tilde{\psi}_0 \vert \tilde{\psi}_0 \rangle^{-1/2},$$

$$\tilde{A}^{(\lambda,4)}_{0,1}(L) = \langle \tilde{\psi}_0 \vert \tilde{\psi}_0 \rangle^{-1/2} \frac{\langle \tilde{u}_1 \vert \tilde{u}_0 \rangle}{\langle \tilde{u}_1 \vert \tilde{u}_1 \rangle},$$

$$\langle \tilde{\psi}_0 \vert \tilde{\psi}_0 \rangle = \langle \tilde{u}_0 \vert \tilde{u}_0 \rangle - \frac{\langle \tilde{u}_1 \vert \tilde{u}_0 \rangle^2}{\langle \tilde{u}_1 \vert \tilde{u}_1 \rangle}.$$

Table 8. Transformation coefficients $A_{i,\alpha}^{(\lambda,\mu)}(L)$ for $\mu = 4$.

α	i	L	$\lambda - L$ even	$\lambda - L$ odd
2	2	$0, 1, \ldots, \lambda$	$A_{2,2}^{(\lambda,4)}(L)$	—
	1	$2, 3, \ldots, \lambda$	$A_{1,2}^{(\lambda,4)}(L)$	—
	0	$4, 5, \ldots, \lambda$	$A_{0,2}^{(\lambda,4)}(L)$	—
1	1	$2, 3, \ldots, \lambda + 1$	$A_{1,1}^{(\lambda,4)}(L)$	$\tilde{A}_{1,1}^{(\lambda,4)}(L)$
		$\lambda + 2$	$A_{1,1}^{(\lambda,4)}(\lambda + 2)$	—
	0	$4, 5, \ldots, \lambda + 1$	$A_{0,1}^{(\lambda,4)}(L)$	$\tilde{A}_{0,1}^{(\lambda,4)}(L)$
		$\lambda + 2$	$A_{0,1}^{(\lambda,4)}(\lambda + 2)$	—
0	0	$4, 5, \ldots, \lambda + 1$	$A_{0,0}^{(\lambda,4)}(L)$	$\tilde{A}_{0,0}^{(\lambda,4)}(L)$
		$\lambda + 2$	$A_{0,0}^{(\lambda,4)}(\lambda + 2)$	—
		$\lambda + 3$	—	$\tilde{A}_{0,0}^{(\lambda,4)}(\lambda + 3)$
		$\lambda + 4$	$A_{0,0}^{(\lambda,4)}(\lambda + 4)$	—

The second expression is of the following form:

$$\left| \begin{matrix} (\lambda, \mu) \\ f_0, \lambda+2, \lambda+2 \end{matrix} \right\rangle = A_{0,0}^{(\lambda,\mu)}(\lambda+2) \left| \begin{matrix} (\lambda, \mu)_B \\ 0, \lambda+2, \lambda+2 \end{matrix} \right\rangle + A_{0,1}^{(\lambda,\mu)}(\lambda+2) \left| \begin{matrix} (\lambda, \mu)_B \\ 1, \lambda+2, \lambda+2 \end{matrix} \right\rangle,$$

$$A_{0,0}^{(\lambda,4)}(\lambda+2) = -\left(\frac{\lambda+35}{48(\lambda+2)(\lambda+3)^2(\lambda+4)(\lambda+5)(2\lambda+7)(\lambda-2)!} \right)^{1/2},$$

$$A_{0,1}^{(\lambda,4)}(\lambda+2) = \frac{2\sqrt{3}}{\lambda+3} \left((\lambda+2)(\lambda+4)(\lambda+5)(\lambda+35)(2\lambda+7)(\lambda-2)! \right)^{-1/2}.$$

The third linear combination for two independent BM vectors (5) is read as

$$\left| \begin{matrix} (\lambda, \mu) \\ f_1, L, L \end{matrix} \right\rangle = A_{1,1}^{(\lambda,\mu)}(L) \left| \begin{matrix} (\lambda, \mu)_B \\ 1, L, L \end{matrix} \right\rangle + A_{1,2}^{(\lambda,\mu)}(L) \left| \begin{matrix} (\lambda, \mu)_B \\ 2, L, L \end{matrix} \right\rangle,$$

$$A_{1,1}^{(\lambda,4)}(L) = -\langle \psi_1 | \psi_1 \rangle^{-1/2},$$

$$A_{1,2}^{(\lambda,4)}(L) = \langle \psi_1 | \psi_1 \rangle^{-1/2} \frac{\langle u_2 | u_1 \rangle}{\langle u_2 | u_2 \rangle},$$

$$\left| \begin{matrix} (\lambda, \mu) \\ \tilde{f}_1, L, L \end{matrix} \right\rangle = \tilde{A}_{1,1}^{(\lambda,\mu)}(L) \left| \begin{matrix} (\lambda, \mu)_B \\ 1, L, L \end{matrix} \right\rangle,$$

$$\tilde{A}_{1,1}^{(\lambda,4)}(L) = (\langle \tilde{u}_1 | \tilde{u}_1 \rangle)^{-1/2}.$$

If there is the only one BM vector (5) indicated by the displayed values of labels the orthonormalization coefficient is equal to the reciprocal square root of the corresponding overlap integral. In our case, there are two possible linear combinations. The first one is read as

$$\left|\begin{matrix}(\lambda,\mu)\\\tilde{f}_0,\lambda+3,\lambda+3\end{matrix}\right\rangle = \tilde{A}_{0,0}^{(\lambda,\mu)}(\lambda+3)\left|\begin{matrix}(\lambda,\mu)_B\\0,\lambda+3,\lambda+3\end{matrix}\right\rangle,$$

$$A_{0,0}^{(\lambda,4)}(\lambda+3) = [6(\lambda+2)(\lambda+3)(\lambda+4)^2(\lambda+5)(\lambda-1)!]^{-1/2},$$

$$\left|\begin{matrix}(\lambda,\mu)\\f_0,\lambda+4,\lambda+4\end{matrix}\right\rangle = A_{0,0}^{(\lambda,\mu)}(\lambda+4)\left|\begin{matrix}(\lambda,\mu)_B\\0,\lambda+4,\lambda+4\end{matrix}\right\rangle,$$

$$A_{0,0}^{(\lambda,4)}(\lambda+4) = [24(\lambda+2)(\lambda+3)(\lambda+4)(\lambda+5)\lambda!]^{-1/2}.$$

The second form of the linear combination when there is only one BM vector (5) is read as

$$\left|\begin{matrix}(\lambda,\mu)\\f_1,\lambda+2,\lambda+2\end{matrix}\right\rangle = A_{1,1}^{(\lambda,\mu)}(\lambda+2)\left|\begin{matrix}(\lambda,\mu)_B\\1,\lambda+2,\lambda+2\end{matrix}\right\rangle,$$

$$A_{1,1}^{(\lambda,4)}(\lambda+2) = [4(\lambda+2)(\lambda+3)(\lambda+4)(\lambda+35)\lambda!]^{-1/2},$$

$$\left|\begin{matrix}(\lambda,\mu)\\f_2,L,L\end{matrix}\right\rangle = A_{2,2}^{(\lambda,\mu)}(L)\left|\begin{matrix}(\lambda,\mu)_B\\2,L,L\end{matrix}\right\rangle,$$

$$A_{2,2}^{(\lambda,4)}(L) = (\langle u_2|u_2\rangle)^{-1/2}.$$

4   The Action of the Zero Component of the Quadrupole Operator onto the Orthogonal Basis

Following the paper [12], we determine the action of the zero component of the second order generator of SU(3) group onto the BM basis vectors

$$Q_0\left|\begin{matrix}(\lambda,\mu)_B\\\alpha,L,L\end{matrix}\right\rangle = \sum_{\substack{k=0,1,2\\s=0,\pm1}} a_s^{(k)}\left|\begin{matrix}(\lambda,\mu)_B\\\alpha+s,L+k,L\end{matrix}\right\rangle, \tag{20}$$

where the coefficients $a_s^{(k)}$ can be calculated as in [1] and they have the form given in [12], and the inverse transformation $\tilde{A}_{i,\alpha}^{(\lambda\mu)}(L)$ from formula (12)

$$\left|\begin{matrix}(\lambda,\mu)_B\\\alpha,L,L\end{matrix}\right\rangle = \sum_{i=0}^{\alpha} \tilde{A}_{i,\alpha}^{(\lambda\mu)}(L)\left|\begin{matrix}(\lambda,\mu)\\f_i,L,L\end{matrix}\right\rangle, \tag{21}$$

where conventional relations take place

$$\sum_i \tilde{A}_{i,\alpha'}^{(\lambda\mu)}(L)A_{i,\alpha}^{(\lambda\mu)}(L) = \delta_{\alpha',\alpha} \quad \text{and} \quad \sum_\alpha \tilde{A}_{i',\alpha}^{(\lambda\mu)}(L)A_{i,\alpha}^{(\lambda\mu)}(L) = \delta_{i',i}. \tag{22}$$

Using (20), (21), and (22), one obtains the action of the zero component of the quadrupole operator onto the orthogonal BM basis vectors

$$Q_0\left|\begin{matrix}(\lambda,\mu)\\f_i,L,L\end{matrix}\right\rangle = \sum_{\substack{j=0,\dots,\alpha_{max}\\k=0,1,2}} q_{ijk}^{(\lambda\mu)}(L)\left|\begin{matrix}(\lambda,\mu)\\f_j,L+k,L\end{matrix}\right\rangle, \tag{23}$$

where the coefficients $q_{ijk}^{(\lambda\mu)}(L)$ are calculated by the formula

$$q_{ijk}^{(\lambda\mu)}(L) = \sum_{\substack{\alpha=0,\ldots,\alpha_{\max} \\ s=0,\pm 1}}' A_{i,\alpha}^{(\lambda\mu)}(L) a_s^{(k)} \tilde{A}_{j,(\alpha+s)}^{(\lambda\mu)}(L+k), \qquad (24)$$

and $\tilde{A}_{i,\alpha}^{(\lambda\mu)}(L)$ are elements of the inverse and the transpose of matrix

$$\tilde{A}_{i,\alpha}^{(\lambda\mu)}(L) = (A^{-1})_{\alpha,i}^{(\lambda\mu)}(L). \qquad (25)$$

The matrix elements of the quadrupole operators, generators of the group SU(3) can be reduced to the calculation of the reduced matrix elements by means of the Wigner–Eckart theorem

$$\left\langle \begin{matrix} (\lambda\mu) \\ jL+kM \end{matrix} \middle| Q_m \middle| \begin{matrix} (\lambda\mu) \\ iLM' \end{matrix} \right\rangle = \frac{(LM'\,2m|L+k,M)}{\sqrt{2(L+k)+1}} \left\langle \begin{matrix} (\lambda\mu) \\ j,L+k \end{matrix} \middle\| Q \middle\| \begin{matrix} (\lambda\mu) \\ i,L \end{matrix} \right\rangle. \qquad (26)$$

The corresponding reduced matrix element is determined by formula

$$\left\langle \begin{matrix} (\lambda\mu) \\ j,L+k \end{matrix} \middle\| Q \middle\| \begin{matrix} (\lambda\mu) \\ i,L \end{matrix} \right\rangle = (-1)^k \frac{\sqrt{2L+1}}{(L+k,L,20|LL)} q_{i,j,k}^{(\lambda,\mu)}(L), \qquad (27)$$

where the coefficients $q_{i,j,k}^{(\lambda,\mu)}(L)$ are defined by (24). In this definition, $k \geq 0$. Dimension of subspace of the ket vectors $|(\lambda\mu)iLM\rangle$ at fixed λ and μ are defined by Formula (6). The dimension of this subspace determines the complexity of the above algorithms, i.e., required computer memory and execution time.

In this paper, the new results for the coefficients $q_{ijk}^{(\lambda\mu)}(L)$ in the orthonormal BM basis with the highest weight vectors of SO(3) irreps for $\mu = 4$ were calculated. The results are not presented here because of a restricted volume of the issue and will be published elsewhere. Note that the coefficients $q_{ijk}^{(\lambda\mu)}(L)$ for up to $\mu = 3$ were calculated as well and their values are equal to those presented in Table 1 of Ref. [12].

5 Conclusion

We present the practical symbolic algorithm for constructing the non-canonical Bargmann–Moshinsky (BM) basis with the highest weight vectors of SO(3) irreps which can be implemented in any computer algebra system. This kind of basis is widely used for calculating spectra and electromagnetic transitions in molecular and nuclear physics. The program in the Mathematica language is now prepared for calculating the non-canonical BM basis overlap integrals following the analytical formula given by [2]. The effective symbolic algorithm for orthonormalization of the obtained BM basis based on the Gram–Schmidt orthonormalization procedure is developed. The proposed recursive orthonormalization algorithm allows one to find the analytical expressions of the orthonormalized basis. The distinct advantage of this method is that it does not involve any square root operation on the expressions coming from the previous recursion steps for computation of the orthonormalization coefficients for this basis. This makes the proposed method very suitable for calculations on computer algebra systems. The symbolic nature of the developed algorithms allows one to avoid the numerical round-off errors in calculation of spectral characteristics (especially close to resonances) of quantum systems under consideration and to study their analytical properties for understanding the dominant symmetries. Calculations of spectral characteristics of the above nuclei models and study of their dominant symmetries will be done in our next publications.

Acknowledgment. The work was partially funded by the RFBR grant No. 16-01-00080, the Bogoliubov–Infeld program, and the RUDN University Program 5-100.

References

1. Afanasjev, G.N., Avramov, S.A., Raychev, P.P.: Realization of the physical basis for SU(3) and the probabilities of E2 transitions in the SU(3) formalism. Sov. J. Nucl. Phys. **16**, 53–83 (1973)
2. Alisauskas, S., Raychev, P., Roussev, R.: Analytical form of the orthonormal basis of the decomposition $SU(3) \supset O(3) \supset O(2)$ for some (λ, μ) multiplets. J. Phys. G: Nucl. Phys. **7**, 1213–1226 (1981)
3. Bargmann, V., Moshinsky, M.: Group theory of harmonic oscillators (II). Nucl. Phys. **23**, 177–199 (1961)
4. Cseh, J.: Algebraic models for shell-like quarteting of nucleons. Phys. Lett. B **743**, 213–217 (2015)
5. Dudek, J., Goźdź, A., Schunck, N., Miśkiewicz, M.: Nuclear tetrahedral symmetry: possibly present throughout the periodic table. Phys. Rev. Lett. **88**(25), 252502 (2002)
6. Dytrych, T.: Efficacy of the SU(3) scheme for ab initio large-scale calculations beyond the lightest nuclei. Comp. Phys. Comun. **207**, 202–210 (2016)
7. Elliott, J.P.: Collective motion in the nuclear shell model I. Proc. R. Soc. Lond. A **245**, 128–145 (1958)
8. Goźdź, A., Pędrak, A., Gusev, A.A., Vinitsky, S.I.: Point symmetries in the nuclear SU(3) partner groups model. Acta Phys. Pol. B Proc. Suppl. **11**, 19–27 (2018)
9. Harvey, M.: The nuclear SU_3 model. In: Baranger, M., Vogt, E. (eds.) Advances in Nuclear Physics. Springer, Boston (1968). https://doi.org/10.1007/978-1-4757-0103-6_2
10. Moshinsky, M., Patera, J., Sharp, R.T., Winternitz, P.: Everything you always wanted to know about $SU(3) \supset O(3)$. Ann. Phys. **95**(N.Y.), 139–169 (1975)
11. Pan, F., Yuan, S., Launey, K.D., Draayer, J.P.: A new procedure for constructing basis vectors of $SU(3) \supset SO(3)$. Nucl. Phys. A **743**, 70–99 (2016)
12. Raychev, P., Roussev, R.: Matrix elements of the generators of SU(3) and of the basic O(3) scalars in the enveloping algebra of SU(3). J. Phys. G: Nucl. Phys. **7**, 1227–1238 (1981)
13. Varshalovitch, D.A., Moskalev, A.N., Hersonsky, V.K.: Quantum Theory of Angular Momentum. Nauka, Leningrad (1975). (Also World Scientific (1988))

About Some Drinfel'd Associators

G. H. E. Duchamp[1]([✉]), V. Hoang Ngoc Minh[2], and K. A. Penson[3]

[1] Sorbonne Université, Université Paris XIII, 99 Jean-Baptiste Clément,
93430 Villetaneuse, France
gheduchamp@gmail.com
[2] Université Lille, 1, Place Déliot, 59024 Lille, France
[3] Sorbonne Université, Université Paris VI, 75252 Paris Cedex 05, France

Abstract. We study, by means of a fragment of theory about non-commutative differential equations, existence and unicity of Drinfel'd solutions $G_i, i = 0, 1$ (with asymptotic conditions). From there, we give examples of Drinfel'd series with rational coefficients.

Keywords: Drinfel'd series · Harmonic sums
Knizhnik-Zamolodchikov equations polylogarithms · Polyzetas
Regularization · Renormalization
Noncommutative differential equations
Noncommutative generating series

1 Knizhnik-Zamolodchikov Differential Equations and Coefficients of Drinfel'd Associators

In 1986 [11], in order to study the linear representations of the braid group B_n coming from the monodromy of the Knizhnik-Zamolodchikov differential equations, Drinfel'd introduced a class of formal power series Φ on noncommutative variables over the finite alphabet $X = \{x_0, x_1\}$. Such power series Φ are called *Drinfel'd series* (or *associators*). For $n = 3$, it leads to the following fuchsian differential equation with three regular singularities in $\{0, 1, +\infty\}$:

$$(DE) \quad dG(z) = \left(x_0 \frac{dz}{z} + x_1 \frac{dz}{1-z} \right) G(z).$$

This is connected to the fact that the pure braid group on three strands P_3 is the semi-direct product of the pure braid group on two strands (a copy of $\mathbb{Z}$) with a copy of the free group on two generators. Although this interpretation of (DE) does not play an explicit role below, it can be kept in mind with a view towards applications.

Solutions of (DE) are power series, with coefficients which are mono-valued functions on the simply connected domain $\Omega := \mathbb{C} \setminus (]-\infty, 0] \cup [1, +\infty[)$ and can be seen as multi-valued over[1] $\mathbb{C} \setminus \{0, 1\}$ on noncommutative variables on X.

---

[1] In fact, we have mappings from the universal covering $\widetilde{\mathbb{C} \setminus \{0, 1\}}$.

© Springer Nature Switzerland AG 2018
V. P. Gerdt et al. (Eds.): CASC 2018, LNCS 11077, pp. 146–163, 2018.
https://doi.org/10.1007/978-3-319-99639-4_10

Drinfel'd proved that (DE) admits two particular mono-valued solutions on Ω [12,13]

$$G_0(z) \underset{z \to 0}{\sim} \exp[x_0 \log(z)] \text{ and } G_1(z) \underset{z \to 1}{\sim} \exp[-x_1 \log(1-z)]. \quad (1)$$

and the existence of an associator $\Phi_{KZ} \in \mathbb{R}\langle\langle X \rangle\rangle$ such that $G_0 = G_1 \Phi_{KZ}$ [12,13] but he did not make explicit neither G_0 and G_1 nor Φ_{KZ}. After that, via representations of the chord diagram algebras, Lê and Murakami [26] expressed the coefficients of Φ_{KZ} as *linear* combinations of special values of several complex variables *zeta* functions, $\{\zeta_r\}_{r \in \mathbb{N}_+}$,

$$\zeta_r : \mathcal{H}_r \longrightarrow \mathbb{R}, \ (s_1, \ldots, s_r) \longmapsto \sum_{n_1 > \ldots > n_k > 0} \frac{1}{n_1^{s_1} \ldots n_k^{s_r}}, \quad (2)$$

where $\mathcal{H}_r := \{(s_1, \ldots, s_r) \in \mathbb{C}^r \mid \forall m = 1, .., r, \sum_{i=1}^m \Re(s_i) > m\}$.

For $(s_1, \ldots, s_r) \in \mathcal{H}_r$, one has two ways of thinking $\zeta_r(s_1, \ldots, s_r)$ as limits, fulfilling identities [1,20,21]. Firstly, they are limits of *polylogarithms* and secondly, as truncated sums, they are limits of *harmonic sums*:

$$\text{Li}_{s_1, \ldots, s_k}(z) = \sum_{n_1 > \ldots > n_k > 0} \frac{z^{n_1}}{n_1^{s_1} \ldots n_k^{s_k}}, \text{ for } z \in \mathbb{C}, |z| < 1, \quad (3)$$

$$\text{H}_{s_1, \ldots, s_k}(N) = \sum_{n_1 > \ldots > n_k > 0}^{N} \frac{1}{n_1^{s_1} \ldots n_k^{s_k}}, \text{ for } N \in \mathbb{N}_+. \quad (4)$$

More precisely, if $(s_1, \ldots, s_r) \in \mathcal{H}_r$ then, after a theorem by Abel, one has

$$\lim_{z \to 1} \text{Li}_{s_1, \ldots, s_k}(z) = \lim_{n \to \infty} \text{H}_{s_1, \ldots, s_k}(n) =: \zeta_r(s_1, \ldots, s_k) \quad (5)$$

else it does not hold, for $(s_1, \ldots, s_r) \notin \mathcal{H}_r$, while $\text{Li}_{s_1, \ldots, s_k}$ is well defined over $\{z \in \mathbb{C}, |z| < 1\}$ and so is $\text{H}_{s_1, \ldots, s_k}$, as Taylor coefficients of the following function

$$\frac{\text{Li}_{s_1, \ldots, s_k}(z)}{1 - z} = \sum_{n \geq 1} \text{H}_{s_1, \ldots, s_k}(n) z^n, \text{ for } z \in \mathbb{C}, |z| < 1. \quad (6)$$

For $r = 1$, ζ_1 is nothing else but the famous Riemann zeta function and, for $r = 0$, it is convenient to set ζ_0 to the constant function $s \mapsto 1_{\mathbb{R}}$. In all the sequel, for simplification, we will adopt the notation ζ for $\zeta_r, r \in \mathbb{N}$.

In this work, we will describe the regularized solutions of (DE). Remark also that replacing letters $\{x_i\}_{i=0,1}$ by constant matrices $\{M_i\}_{i=0,1}$ (resp. analytical vector fields $\{A_i\}_{i=0,1}$), one deals with linear (resp. nonlinear) differential equations [3,5,19,25] (resp. [6,9,22]). Hence, (DE) can also be considered as the universal linear and nonlinear differential equation with three singularities. Therefore these computations can undergo an automatic treatment (see, for instance [16] and the subsequent sessions).

For that, we are considering the alphabets $X := \{x_0, x_1\}$ and $Y_0 := \{y_s\}_{s \geq 0}$ equipped with the total ordering $x_0 < x_1$ and $y_0 > y_1 > y_2 > \ldots$, respectively. Let us also consider $Y := Y_0 \setminus \{y_0\}$.

The free monoid generated by X (resp. Y, Y_0) is denoted by X^* (resp. Y^*, Y_0^*) and admits 1_{X^*} (resp. $1_{Y^*}, 1_{Y_0^*}$) as unit. The sets of polynomials and formal power series, with coefficients in a commutative $\mathbb{Q}$-algebra A, over X^* (resp. Y^*, Y_0^*) are denoted by $A\langle X\rangle$ (resp. $A\langle Y\rangle, A\langle Y_0\rangle$) and $A\langle\langle X\rangle\rangle$ (resp. $A\langle\langle Y\rangle\rangle, A\langle\langle Y_0\rangle\rangle$), respectively.

The sets of polynomials are A-modules and endowed with the associative concatenation, the associative commutative shuffle (resp. quasi-shuffle) product, over $A\langle X\rangle$ (resp. $A\langle Y\rangle, A\langle Y_0\rangle$). Their associated coproducts are denoted, respectively, $\Delta_{\shuffle}$ and $\Delta_{\uplus}$.

The shuffle algebra $(A\langle X\rangle, \shuffle, 1_{X^*})$ and quasi-shuffle algebra $(A\langle Y\rangle, \uplus, 1_{Y^*})$ admit the sets of Lyndon words denoted, respectively, by $\mathcal{L}ynX$ and $\mathcal{L}ynY$, as transcendence bases [27] (resp. [22,23]).

Now, for $Z = X$ or Y, denoting $\mathrm{Lie}_A\langle Z\rangle$ and $\mathrm{Lie}_A\langle\langle Z\rangle\rangle$ the sets of, respectively, Lie polynomials and Lie series, the enveloping algebra $\mathcal{U}(\mathrm{Lie}_A\langle Z\rangle)$ is isomorphic to the (Hopf) bialgebra

$$\mathcal{H}_{\shuffle}(Z) := (A\langle Z\rangle, ., 1_{Z^*}, \Delta_{\shuffle}, \mathrm{e}). \tag{7}$$

We get also

$$\mathcal{H}_{\uplus}(Y) := (A\langle Y\rangle, ., 1_{Y^*}, \Delta_{\uplus}, \mathrm{e}) \cong \mathcal{U}(\mathrm{Prim}(\mathcal{H}_{\uplus}(Y))), \tag{8}$$

where $\mathrm{Prim}(\mathcal{H}_{\uplus}(Y)) = \mathrm{span}_A\{\pi_1(w)|w \in Y^*\}$ and, for any $w \in Y^*$ [2,22,23],

$$\pi_1(w) = \sum_{k=1}^{(w)} \frac{(-1)^{k-1}}{k} \sum_{u_1,\dots,u_k \in Y^+} \langle w|u_1 \uplus \dots \uplus u_k\rangle u_1 \dots u_k. \tag{9}$$

The paper is organised as follows: Sect. 1 is devoted to setting the combinatorial framework of noncommutative differential Knizhnik-Zamolodchikov equations and Drinfel'd associators. Afterwards, in Sect. 2, we recall some algebraic structures about polylogarithms and harmonic sums, through their indexing by words. In Sect. 3, we will study, by means of a fragment of theory about noncommutative differential equations[2], existence and unicity of Drinfel'd solutions (1). Finally, in Sect. 4, we will renormalize solutions of (DE) and will regularize them at singularites. Also some examples of Drinfel'd series with rational coefficients are provided. Some results in this paper have been presented in [10], as preprint, but never published before (see also [24]).

2 Indexing Polylogarithms and Harmonic Sums by Words and Their Generating Series

For any $r \in \mathbb{N}$, any combinatorial composition $(s_1, \dots, s_r) \in \mathbb{N}_+^r$ can be associated with words

$$x_0^{s_1-1}x_1 \dots x_0^{s_r-1}x_1 \in X^*x_1 \text{ and } y_{s_1} \dots y_{s_r} \in Y^*. \tag{10}$$

---

[2] The main theorem, although not very difficult once the correct setting has been implemented, is very powerful and new here in its two-sided version (see Subsect. 3.1).

Similarly, any multi-index[3] $(s_1, \ldots, s_r) \in \mathbb{N}^r$ can be associated with words $y_{s_1} \ldots y_{s_r} \in Y_0^*$. Then let us index polylogarithms and harmonic sums by words [2,21]:

$$\mathrm{Li}_{x_0^r}(z) := \frac{(\log(z))^r}{r!}, \; \mathrm{Li}_{x_0^{s_1-1}x_1 \ldots x_0^{s_r-1}x_1} := \mathrm{Li}_{s_1,\ldots,s_r}, \; \mathrm{H}_{y_{s_1} \ldots y_{s_r}} := \mathrm{H}_{s_1,\ldots,s_r}. \quad (11)$$

Similarly, let $\mathrm{Li}_{-s_1,\ldots,-s_k}$ and $\mathrm{H}_{-s_1,\ldots,-s_k}$ be indexed by words[4] as follows [7,8]:

$$\mathrm{Li}_{y_0^r}^-(z) := \left(\frac{z}{1-z}\right)^r, \; \mathrm{Li}_{y_{s_1} \ldots y_{s_r}}^- := \mathrm{Li}_{-s_1,\ldots,-s_r} \text{ and} \quad (12)$$

$$\mathrm{H}_{y_0^r}^-(n) := \binom{n}{r} = \frac{(n)_r}{r!}, \; \mathrm{H}_{y_{s_1} \ldots y_{s_r}}^- := \mathrm{H}_{-s_1,\ldots,-s_r}.$$

There exists a law of algebra, denoted by $\top$, in $\mathbb{Q}\langle\langle Y_0 \rangle\rangle$, such that the morphism (14) of algebras is *surjective*. With this, we get the following [7]

$$\mathrm{H}_\bullet^- : (\mathbb{Q}\langle Y_0 \rangle, \uplus, 1_{Y_0^*}) \longrightarrow (\mathbb{Q}\{\mathrm{H}_w^-\}_{w \in Y_0^*}, \times, 1), \; w \longmapsto \mathrm{H}_w^-, \quad (13)$$

$$\mathrm{Li}_\bullet^- : (\mathbb{Q}\langle Y_0 \rangle, \top, 1_{Y_0^*}) \longrightarrow (\mathbb{Q}\{\mathrm{Li}_w^-\}_{w \in Y_0^*}, \times, 1), \; w \longmapsto \mathrm{Li}_w^-, \quad (14)$$

such that [7]

$$\ker \mathrm{H}_\bullet^- = \ker \mathrm{Li}_\bullet^- = \mathbb{Q}\langle\langle \{w - w \top 1_{Y_0^*} | w \in Y_0^*\} \rangle\rangle. \quad (15)$$

Moreover, the families $\{\mathrm{H}_{y_k}^-\}_{k \geq 0}$ and $\{\mathrm{Li}_{y_k}^-\}_{k \geq 0}$ are $\mathbb{Q}$-linearly independent. On the other hand, the following morphisms of algebras are *injective*

$$\mathrm{H}_\bullet : (\mathbb{Q}\langle Y \rangle, \uplus, 1_{Y^*}) \longrightarrow (\mathbb{Q}\{\mathrm{H}_w\}_{w \in Y^*}, \times, 1), \; w \longmapsto \mathrm{H}_w, \quad (16)$$

$$\mathrm{Li}_\bullet : (\mathbb{Q}\langle X \rangle, \sqcup\!\sqcup, 1_{X^*}) \longrightarrow (\mathbb{Q}\{\mathrm{Li}_w\}_{w \in X^*}, \times, 1), \; w \longmapsto \mathrm{Li}_w. \quad (17)$$

Moreover, the families $\{\mathrm{H}_w\}_{w \in Y^*}$ and $\{\mathrm{Li}_w\}_{w \in X^*}$ are $\mathbb{Q}$-linearly independent and the families $\{\mathrm{H}_l\}_{l \in \mathcal{L}yn Y}$ and $\{\mathrm{Li}_l\}_{l \in \mathcal{L}yn X}$ are $\mathbb{Q}$-algebraically independent. But at singularities of $\{\mathrm{Li}_w\}_{w \in X^*}, \{\mathrm{H}_w\}_{w \in Y^*}$, the following convergent values

$$\forall u \in Y^* - y_1 Y^*, \zeta(u) := \mathrm{H}_u(+\infty) \text{ and } \forall v \in x_0 X^* x_1, \zeta(v) := \mathrm{Li}_v(1) \quad (18)$$

are no longer linearly independent and the values $\{\mathrm{H}_l(+\infty)\}_{l \in \mathcal{L}yn Y \setminus \{y_1\}}$ (resp. $\{\mathrm{Li}_l(1)\}_{l \in \mathcal{L}yn X \setminus X}$) are no longer algebraically independent [21,28]. The graphs of the isomorphisms of algebras, $\mathrm{Li}_\bullet$ and $\mathrm{H}_\bullet$, as generating series, read then [3,4,21]

$$L := \sum_{w \in X^*} \mathrm{Li}_w w = \prod_{l \in \mathcal{L}yn X}^{\searrow} e^{\mathrm{Li}_{s_l} P_l} \text{ and } H := \sum_{w \in Y^*} \mathrm{H}_w w = \prod_{l \in \mathcal{L}yn Y}^{\searrow} e^{\mathrm{H}_{\Sigma_l} \Pi_l}, (19)$$

---

[3] The weight of $(s_1, \ldots, s_r) \in \mathbb{N}_+^r$ (resp. $\mathbb{N}^r$) is defined as the integer $s_1 + \ldots + s_r$ which corresponds to the weight, denoted (w), of its associated word $w \in Y^*$ (resp. Y_0^*) and also (in the case of Y) to the length, denoted $|u|$, of its associated word $u \in X^*$.

[4] Note that, all these $\{\mathrm{Li}_w^-\}_{w \in Y_0^*}$ and $\{\mathrm{H}_w^-\}_{w \in Y_0^*}$ diverge at their singularities.

where the PBW basis $\{P_w\}_{w \in X^*}$ (resp. $\{\Pi_w\}_{w \in Y^*}$) is expanded over the basis of $\mathcal{U}(\mathrm{Lie}_A\langle X\rangle)$ (resp. $\mathcal{U}(\mathrm{Prim}(\mathcal{H}_{\text{⊎}}(Y)))$), $\{P_l\}_{l \in \mathcal{L}ynX}$ (resp. $\{\Pi_l\}_{l \in \mathcal{L}ynY}$), and $\{S_w\}_{w \in X^*}$ (resp. $\{\Sigma_w\}_{w \in Y^*}$) is the basis of the shuffle $(\mathbb{Q}\langle Y\rangle, \text{⧢}, 1_{X^*})$ (resp. the quasi-shuffle $(\mathbb{Q}\langle Y\rangle, \text{⊎}, 1_{Y^*})$) containing the transcendence basis $\{S_l\}_{l \in \mathcal{L}ynX}$ (resp. $\{\Sigma_l\}_{l \in \mathcal{L}ynY}$).

By termwise differentiation, L satisfies the noncommutative differential equation (DE) with the boundary condition $\mathrm{L}(z) \underset{z \to 0^+}{\sim} e^{x_0 \log(z)}$. It is immediate that the power series H and L are group-like, for $\Delta_{\text{⊎}}$ and $\Delta_{\text{⧢}}$, respectively. Hence, the following noncommutative generating series are well defined and are group-like, for $\Delta_{\text{⊎}}$ and $\Delta_{\text{⧢}}$, respectively [21–23]:

$$Z_{\text{⊎}} := \overset{\searrow}{\prod_{l \in \mathcal{L}ynY\setminus\{y_1\}}} e^{\mathrm{H}_{\Sigma_l}(+\infty)\Pi_l} \quad \text{and} \quad Z_{\text{⧢}} := \overset{\searrow}{\prod_{l \in \mathcal{L}ynX\setminus X}} e^{\mathrm{Lis}_l(1)P_l}. \tag{20}$$

Definitions (5) and (18) lead then to the following *surjective* poly-morphism

$$\zeta : \begin{array}{c} (\mathbb{Q}1_{X^*} \oplus x_0\mathbb{Q}\langle X\rangle x_1, \text{⧢}, 1_{X^*}) \\ \overline{(\mathbb{Q}1_{Y^*} \oplus (Y - \{y_1\})\mathbb{Q}\langle Y\rangle, \text{⊎}, 1_{Y^*})} \end{array} \longrightarrow (\mathcal{Z}, \times, 1), \tag{21}$$

$$\begin{array}{c} x_0 x_1^{r_1-1} \ldots x_0 x_1^{r_k-1} \\ \overline{y_{s_1} \ldots y_{s_k}} \end{array} \longmapsto \sum_{n_1 > \ldots > n_k > 0} n_1^{-s_1} \ldots n_k^{-s_k}, \tag{22}$$

where $\mathcal{Z}$ is the $\mathbb{Q}$-algebra generated by $\{\zeta(l)\}_{l \in \mathcal{L}ynX\setminus X}$ (resp. $\{\zeta(S_l)\}_{l \in \mathcal{L}ynX\setminus X}$), or equivalently, generated by $\{\zeta(l)\}_{l \in \mathcal{L}ynY\setminus\{y_1\}}$ (resp. $\{\zeta(\Sigma_l)\}_{l \in \mathcal{L}ynY\setminus\{y_1\}}$).

Now, let $t_i \in \mathbb{C}, |t_i| < 1, i \in \mathbb{N}$. For $z \in \mathbb{C}, |z| < 1$, we have [18]

$$\sum_{n \geq 0} \mathrm{Li}_{x_0^n}(z) \, t_0^n = z^{t_0} \quad \text{and} \quad \sum_{n \geq 0} \mathrm{Li}_{x_1^n}(z) \, t_1^n = \frac{1}{(1-z)^{t_1}}. \tag{23}$$

These suggest to extend the morphism $\mathrm{Li}_\bullet$ over $(\mathrm{Dom}(\mathrm{Li}_\bullet), \text{⧢}, 1_{X^*})$, via *Lazard's elimination*, as follows (subjected to be convergent):

$$\mathrm{Li}_S(z) = \sum_{n \geq 0} \langle S|x_0^n\rangle \frac{\log^n(z)}{n!} + \sum_{k \geq 1} \sum_{w \in (x_0^* x_1)^k x_0^*} \langle S|w\rangle \mathrm{Li}_w(z), \tag{24}$$

with $\mathbb{C}\langle X\rangle \text{⧢} \mathbb{C}^{\mathrm{rat}}\langle\langle x_0\rangle\rangle \text{⧢} \mathbb{C}^{\mathrm{rat}}\langle\langle x_1\rangle\rangle \subset \mathrm{Dom}(\mathrm{Li}_\bullet) \subset \mathbb{C}^{\mathrm{rat}}\langle\langle X\rangle\rangle$ and $\mathbb{C}^{\mathrm{rat}}\langle\langle X\rangle\rangle$ denotes the closure, of $\mathbb{C}\langle X\rangle$ in $\mathbb{C}\langle X\rangle X$, by $\{+, ., *\}$. For example [18,19],

1. For any $x, y \in X$ and for any $i, j \in \mathbb{N}_+, u, v \in \mathbb{C}$ such that $|u| < 1$ and $|v| < 1$, since

$$(ux + vy)^* = (xx)^* \text{⧢} (vy)^* \quad \text{and} \quad (x^*)^{\text{⧢}i} = (ix)^* \tag{25}$$

then

$$\mathrm{Li}_{(x_0^*)^{\text{⧢}i} \text{⧢} (x_1^*)^{\text{⧢}j}}(z) = \frac{z^i}{(1-z)^j}. \tag{26}$$

2. For $a \in \mathbb{C}, x \in X, i \in \mathbb{N}_+$, since

$$(ax)^{*i} = (ax)^* \sqcup (1 + ax)^{i-1} \tag{27}$$

then

$$\mathrm{Li}_{(ax_0)^{*i}}(z) = z^a \sum_{k=0}^{i-1} \binom{i-1}{k} \frac{(a \log(z))^k}{k!},$$

$$\mathrm{Li}_{(ax_1)^{*i}}(z) = \frac{1}{(1-z)^a} \sum_{k=0}^{i-1} \binom{i-1}{k} \frac{(a \log((1-z)^{-1})^k}{k!}. \tag{28}$$

3. Let $V = (t_1 x_0)^{*s_1} x_0^{s_1-1} x_1 \ldots (t_r x_0)^{*s_r} x_0^{s_r-1} x_1$, for $(s_1, \ldots, s_r) \in \mathbb{N}_+^r$. Then

$$\mathrm{Li}_V(z) = \sum_{n_1 > \ldots > n_r > 0} \frac{z^{n_1}}{(n_1 - t_1)^{s_1} \ldots (n_r - t_r)^{s_r}}. \tag{29}$$

In particular, for $s_1 = \ldots = s_r = 1$, one has

$$\mathrm{Li}_V(z) = \sum_{n_1, \ldots, n_r > 0} \mathrm{Li}_{x_0^{n_1-1} x_1 \ldots x_0^{n_r-1} x_1}(z) \, t_0^{n_1-1} \ldots t_r^{n_r-1}$$

$$= \sum_{n_1 > \ldots > n_r > 0} \frac{z^{n_1}}{(n_1 - t_1) \ldots (n_r - t_r)}. \tag{30}$$

4. From the previous points, one gets

$$\{\mathrm{Li}_S\}_{S \in \mathbb{C}\langle X \rangle \sqcup \mathbb{C}[x_0^*] \sqcup \mathbb{C}[(-x_0^*)] \sqcup \mathbb{C}[x_1^*]} = \mathrm{span}_{\mathbb{C}} \left\{ \frac{z^a}{(1-z)^b} \mathrm{Li}_w(z) \right\}_{w \in X^*}^{a \in \mathbb{Z}, b \in \mathbb{N}}$$

$$\subset \mathrm{span}_{\mathbb{C}}\{\mathrm{Li}_{s_1, \ldots, s_r}\}_{s_1, \ldots, s_r \in \mathbb{Z}^r} \oplus \mathrm{span}_{\mathbb{C}}\{z^a | a \in \mathbb{Z}\}, \tag{31}$$

$$\{\mathrm{Li}_S\}_{S \in \mathbb{C}\langle X \rangle \sqcup \mathbb{C}^{\mathrm{rat}}\langle\langle x_0 \rangle\rangle \sqcup \mathbb{C}^{\mathrm{rat}}\langle\langle x_1 \rangle\rangle} = \mathrm{span}_{\mathbb{C}} \left\{ \frac{z^a}{(1-z)^b} \mathrm{Li}_w(z) \right\}_{w \in X^*}^{a, b \in \mathbb{C}}$$

$$\subset \mathrm{span}_{\mathbb{C}}\{\mathrm{Li}_{s_1, \ldots, s_r}\}_{s_1, \ldots, s_r \in \mathbb{C}^r} \oplus \mathrm{span}_{\mathbb{C}}\{z^a | a \in \mathbb{C}\}. \tag{32}$$

3 Noncommutative Evolution Equations

As was previously said, Drinfel'd proved that (DE) admits two particular solutions on Ω. These new tools and results can be considered as pertaining to the domain of *noncommutative evolution equations*. We will, here, only mention what is relevant for our needs.

Even for one sided [5] differential equations, in order to cope with limit initial conditions (see applications below), one needs the two sided version.

---

[5] As the left (DE) for instance (see [6]).

Let then $\Omega \subset \mathbb{C}$ be open simply connected and $\mathcal{H}(\Omega)$ denotes the algebra of holomorphic functions on Ω. We suppose we are given two series (called *multipliers*) without constant term $M_1, M_2 \in \mathcal{H}(\Omega)_+\langle\langle X \rangle\rangle$ (X is an alphabet and the subscript indicates that the series have no constant term). Let then

$$(DE_2) \quad \mathbf{d}S = M_1 S + S M_2.$$

be our two sided differential equation. A solution of it is a series $S \in \mathcal{H}(\Omega)\langle\langle X \rangle\rangle$ such that (DE_2) is satisfied.

In the sequel, we will use of the following lemma.

Lemma 1. *Let $\mathcal{B}$ be a filter basis on Ω and S a solution of (DE_2) such that $\lim_{\mathcal{B}}\langle S(z)|w \rangle = 0$, for all $w \in X^*$, then $S \equiv 0$.*

Proof. Let us suppose $S \not\equiv 0$ and w be a word of minimal length of supp(S). Then for this word, one has

$$\frac{d}{dz}\langle S|w \rangle = \langle M_1 S + S M_2|w \rangle = 0,$$

due to the fact that M_i have no constant term. Then, for this word, $z \mapsto \langle S(z)|w \rangle$ is constant on Ω. But, due to the fact that $\lim_{\mathcal{B}}\langle S|w \rangle = 0$, one must have this constant to be zero in contradiction with the reasoning on the support.

3.1 The Main Theorem

The following theorem, although not very difficult to establish once the correct setting has been implemented, is very powerful and new here in its two-sided version.[6]

Theorem 1. *(i) Solutions of (DE_2) form a $\mathbb{C}$-vector space.*
(ii) Solutions of (DE_2) have their constant term (as coefficient of 1_{X^*}) which are constant functions (on Ω); there exist solutions with constant coefficient 1_Ω (hence invertible).
(iii) If two solutions coincide at one point $z_0 \in \overline{\Omega}$, they coincide everywhere.
(iv) Let be the following one-sided equations

$$(DE^{(1)}) \quad \mathbf{d}S = M_1 S \quad and \quad (DE^{(2)}) \quad \mathbf{d}S = S M_2,$$

and let $S_i, i = 1, 2$ be a solution of $(DE^{(i)})$. Then $S_1 S_2$ is a solution of (DE_2). Conversely, every solution of (DE_2) can be constructed so.
(v) If $M_i, i = 1, 2$ are primitive and if S, a solution of (DE_2), is group-like at one point, (or, even at one limit point) it is globally group-like.

Proof. (i) Straightforward.

---

[6] It implies the previous (one-sided) version [6] which was aimed at the linear independence of coordinate functions.

(ii) One can use Lemma 1 or directly remark that the map $S \mapsto \langle S | 1_{X^*} \rangle = \epsilon(S)$ is a character (of $\mathcal{H}(\Omega)\langle\langle X \rangle\rangle$) which commutes with the derivations, i.e.,

$$\epsilon(\mathbf{d}S) = \frac{d}{dz}\epsilon(S).$$

Hence, as $\epsilon(M_i) = 0$, for every solution of (DE_2), one has $\frac{d}{dz}(\epsilon(S)) = 0$ whence the claim, as Ω is connected.

Now, for each $z_0 \in \Omega$, one can construct the unique solution of (DE_2) such that $S(z_0) = 1_{X^*}$ by the following process (Picard's process)

$$S_0 = 1_{X^*}, \ S_{n+1} = 1_{X^*} + \int_{z_0}^z M_1(s)S_n(s) + S_n(s)M_2(s)ds$$

(term by term integration). Due to the fact that $M_i(s), i = 1,2$ has no constant term, its limit $S_{Pic}^{z_0} := \lim_{n\to\infty} S_n$ exists and is such that $S_{Pic}^{z_0}(z_0) = 1_{X^*}$. Then its constant term is everywhere $1_{\mathbb{C}}$ (i.e. $\langle S_{Pic}^{z_0} | 1_{X^*}\rangle = 1_\Omega$) and therefore $S_{Pic}^{z_0}$ is invertible in $\mathcal{H}(\Omega)\langle\langle X \rangle\rangle$.

(iii) In fact, the previous reasoning can be carried over for any length (in point "ii" it was for length 0). The claim is an easy consequence of Lemma 1.

(vi) The fact that the product S_1S_2 (for $S_i, i = 1,2$ solutions of $(DE^{(i)})$) is a solution of (DE_2) is straightforward. Let us now suppose S to be a solution of (DE_2) and set, here for short, $S_2 := S_{Pic}^{z_0}$, the corresponding Picard solution of $(DE^{(2)})$ (notation as above). We now compute with $T := S(S_2)^{-1}$

$$\mathbf{d}T = \mathbf{d}(S(S_2)^{-1}) = \mathbf{d}S(S_2)^{-1} + S\mathbf{d}(S_2)^{-1}$$
$$= (M_1S + SM_2) + S(-S_2)^{-1}\mathbf{d}S_2(S_2)^{-1} = M_1T,$$

which proves the claim (as $S = TS_2$).

(v) One first remarks that the two preceding points hold if (DE_2) is stated for series over any locally finite monoid [15]. Such a monoid M has the property (and in fact is defined by it) that every element $x \in M$ has a finite number of factorizations

$$x = x_1 \ldots x_n \text{ with } x_i \in M \setminus \{1_M\},$$

and the length above is replaced by $l(x) := \sup(n)$ for all factorisations as above[7]. Series over M are just functions $S \in R^M$ (the ring R here is $R = \mathcal{H}(\Omega)$ and $\langle S|m\rangle$ is another notation for the image of m by S), polynomials are finitely supported series $S \in R^{(M)}$ and the canonical pairing series-polynomials, $\langle S|P\rangle$ reads

$$\langle S|P\rangle := \sum_{m\in M} \langle S|m\rangle\langle m|P\rangle.$$

---

[7] For example $l(1_M) = 0$ and $l(x) = 1$ for $x \in M_+ \setminus (M_+)^2$ (with $M_+ = M \setminus \{1_M\}$) the minimal set of generators of M [15]).

154 G. H. E. Duchamp et al.

Now, we return to the monoid X^* but we will reason on $M = X^* \otimes X^* \simeq X^* \times X^*$ (direct product, thus also locally finite). Let S be a solution of (DE_2) with $M_i, i = 1, 2$ primitive (hence without constant term). One has

$$\mathbf{d}(S \otimes S) = \mathbf{d}S \otimes S + S \otimes \mathbf{d}S$$
$$= (M_1 S + S M_2) \otimes S + S \otimes (M_1 S + S M_2)$$
$$= (M_1 \otimes 1 + 1 \otimes M_1)(S \otimes S) + (S \otimes S)(M_2 \otimes 1 + 1 \otimes M_2),$$
$$\mathbf{d}(\Delta_{\text{ш}}(S)) = (\Delta_{\text{ш}}(\mathbf{d}S)) = \Delta_{\text{ш}}(M_1 S + S M_2) = \Delta_{\text{ш}}(M_1)\Delta_{\text{ш}}(S) + \Delta_{\text{ш}}(S)\Delta_{\text{ш}}(M_2)$$

(again $M = X^* \otimes X^* \subset \mathcal{H}(\Omega)\langle X \rangle \otimes \mathcal{H}(\Omega)\langle X \rangle$, all tensor products are over $\mathcal{H}(\Omega)$). Hence, we see that $S \otimes S$ and $\Delta_{\text{ш}}(S)$ (double series, i.e., series over $X^* \otimes X^*$) satisfy two-sided differential equations with the same multipliers (left$= M_1 \otimes 1 + 1 \otimes M_1 = \Delta_{\text{ш}}(M_1)$ and right$= M_2 \otimes 1 + 1 \otimes M_2 = \Delta_{\text{ш}}(M_2)$), then it suffices that they coincide at one point of $\bar{\Omega}$ in order that $\Delta_{\text{ш}}(S) = S \otimes S$ (the property $\langle S | 1_{X^*} \rangle = 1$ is granted from the fact that S is group-like at one point of $\bar{\Omega}$).

Remark 1. – Every holomorphic series $S(z) \in \mathcal{H}(\Omega)\langle\langle X \rangle\rangle$ which is group-like ($\Delta(S) = S \otimes S$ and $\langle S | 1_{X^*} \rangle$) is a solution of a one-sided dynamics with primitive multiplier (take $M_1 = (\mathbf{d}S)S^{-1}$ and $M_2 = 0$, or $M_2 = S^{-1}(\mathbf{d}S)$ and $M_1 = 0$).
– Invertible solutions of an equation of type $S' = M_1 S$ are on the same orbit by multiplication on the right by *invertible constant series*, i.e., let $S_i, i = 1, 2$ be invertible solutions of $(DE^{(1)})$, then there exists an unique invertible $T \in \mathbb{C}\langle\langle X \rangle\rangle$ such that $S_2 = S_1 T$. From this and point (iv) of the theorem, one can parametrize the set of invertible solutions of (DE_2).

3.2 Application : Unicity of Solutions with Asymptotic Conditions

In a previous work [6], we proved that asymptotic group-likeness, for a series, implies[8] that the series in question is group-like everywhere. The process above (Theorem 1, Picard's process) can still be performed, under certain conditions with improper integrals. We then construct the series L recursively as

$$\langle \mathrm{L}(z) | w \rangle = \begin{cases} \dfrac{\log^n(z)}{n!} & \text{if } w = x_0^n \\ \displaystyle\int_0^z \dfrac{ds}{1-z} \langle \mathrm{L}(z) | u \rangle & \text{if } w = x_1 u \\ \displaystyle\int_0^z \dfrac{ds}{z} \langle \mathrm{L}(z) | u x_1 x_0^n \rangle & \text{if } w = x_0 u x_1 x_0^n. \end{cases} \quad (33)$$

---

[8] Under the condition that the multiplier be primitive, result extended as point (v) of the theorem above.

One can show that (see [6] for details):

– this process is well defined at each step and computes the series L as below;
– L is solution of (DE), is exactly G_0 and is group-like.

We here only prove that G_0 is unique using the theorem above. Consider the series

$$T(z) = \mathrm{L}(z)e^{-x_0 \log(z)}. \tag{34}$$

Then T is solution of an equation of the type (DE_2)

$$T'(z) = \left(\frac{x_0}{z} + \frac{x_1}{1-z}\right)T(z) + T(z)\frac{-x_0}{z}, \tag{35}$$

but

$$\lim_{z \to z_0} G_0(z)e^{-x_0 \log(z)} = 1, \tag{36}$$

so, by Theorem 1, one has

$$G_0(z)e^{-x_0 \log(z)} = \mathrm{L}(z)e^{-x_0 \log(z)} \tag{37}$$

and then[9]

$$G_0 = \mathrm{L}. \tag{38}$$

A similar (and symmetric) argument can be performed for G_1 and then, in this interpretation and context, Φ_{KZ} is unique.

4   Double Global Regularization of Associators

4.1   Global Renormalization by Noncommutative Generating Series

Global singularities analysis leads to to the following global renormalization [3,4]:

$$\lim_{z \to 1} \exp\left(-y_1 \log \frac{1}{1-z}\right) \pi_Y (\mathrm{L}(z)) \tag{39}$$

$$= \lim_{n \to \infty} \exp\left(\sum_{k \geq 1} \mathrm{H}_{y_k}(n)\frac{(-y_1)^k}{k}\right) \mathrm{H}(n) = \pi_Y(Z_{\mathcyr{ш}}).$$

Thus, the coefficients $\{\langle Z_{\mathcyr{ш}}|u\rangle\}_{u \in X^*}$ (i.e. $\{\zeta_{\mathcyr{ш}}(u)\}_{u \in X^*}$) and $\{\langle Z_{\mathcyr{ш}}|v\rangle\}_{v \in Y^*}$ (i.e. $\{\zeta_{\mathcyr{ш}}(v)\}_{v \in Y^*}$) represent the finite part of the asymptotic expansions, in $\{(1-z)^{-a}\log^b(1-z)\}_{a,b \in \mathbb{N}}$ (resp. $\{n^{-a}\mathrm{H}_1^b(n)\}_{a,b \in \mathbb{N}}$) of $\{\mathrm{Li}_w\}_{w \in X^*}$ (resp. $\{\mathrm{H}_w\}_{v \in Y^*}$). On the other way, by a transfer theorem [17], let $\{\gamma_w\}_{v \in Y^*}$ be the

---

[9] See also [24].

finite parts of $\{H_w\}_{v \in Y^*}$, in $\{n^{-a} \log^b(n)\}_{a,b \in \mathbb{N}}$, and let Z_γ be their noncommutative generating series. Hence,

$$\gamma_\bullet : (\mathbb{Q}\langle Y \rangle, \shuffle, 1_{Y^*}) \longrightarrow (\mathcal{Z}, \times, 1), \quad w \longmapsto \gamma_w, \tag{40}$$

is a character and Z_γ is group-like, for $\Delta_{\shuffle}$. Moreover [22,23],

$$Z_\gamma = \exp(\gamma y_1) \overset{\searrow}{\prod_{l \in \mathcal{L}yn Y \setminus \{y_1\}}} \exp(\zeta(\Sigma_l)\Pi_l) = \exp(\gamma y_1) Z_{\shuffle}. \tag{41}$$

The asymptotic behavior leads to the bridge[10] equation [3,4,22,23]

$$Z_\gamma = B(y_1)\pi_Y(Z_{\shuffle}), \text{ or equivalently } Z_{\shuffle} = B'(y_1)\pi_Y(Z_{\shuffle}), \tag{42}$$

where (see [3,4] and [22,23])

$$B(y_1) = \exp\left(\gamma y_1 - \sum_{k \geq 2}(-y_1)^k \frac{\zeta(k)}{k}\right) \text{ and } B'(y_1) = \exp\left(-\sum_{k \geq 2}(-y_1)^k \frac{\zeta(k)}{k}\right). \tag{43}$$

Similarly, there is $C_w^- \in \mathbb{Q}$ and $B_w^- \in \mathbb{N}$, such that [7]

$$H_w^-(N) \underset{N \to +\infty}{\sim} N^{(w)+|w|} C_w^- \text{ and } \text{Li}_w^-(z) \underset{z \to 1}{\sim} (1-z)^{-(w)-|w|} B_w^-. \tag{44}$$

Moreover,

$$C_w^- = \prod_{w=uv, v \neq 1_{Y_0^*}} ((v) + |v|)^{-1} \text{ and } B_w^- = ((w) + |w|)! C_w^-. \tag{45}$$

Now, one can then consider the following noncommutative generating series:

$$L^- := \sum_{w \in Y_0^*} \text{Li}_w^- w, \quad H^- := \sum_{w \in Y_0^*} H_w^- w, \quad C^- := \sum_{w \in Y_0^*} C_w^- w. \tag{46}$$

Then H^- and C^- are group-like for, respectively, $\Delta_{\shuffle}$ and $\Delta_{\shuffle}$ and [7]

$$\lim_{z \to 1} h^{\odot-1}((1-z)^{-1}) \odot L^-(z) = \lim_{N \to +\infty} g^{\odot-1}(N) \odot H^-(N) = C^-, \tag{47}$$

$$h(t) = \sum_{w \in Y_0^*} ((w) + |w|)! t^{(w)+|w|} w \quad \text{and} \quad g(t) = \left(\sum_{y \in Y_0} t^{(y)+1} y\right)^*. \tag{48}$$

4.2   Global Regularization by Noncommutative Generating Series

Next, for any $w \in Y_0^*$, there exists a unique polynomial $p \in (\mathbb{Z}[t], \times, 1)$ of degree $(w) + |w|$ such that [7]

$$\text{Li}_w^-(z) = \sum_{k=0}^{(w)+|w|} \frac{p_k}{(1-z)^k} = \sum_{k=0}^{(w)+|w|} p_k e^{-k \log(1-z)} \in (\mathbb{Z}[(1-z)^{-1}], \times, 1), \tag{49}$$

$$H_w^-(n) = \sum_{k=0}^{(w)+|w|} p_k \binom{n+k-1}{k-1} = \sum_{k=0}^{(w)+|w|} \frac{p_k}{k!}(n)_k \in (\mathbb{Q}[(n)_\bullet], \times, 1), \tag{50}$$

---

[10] This equation is different from Jean Écalle's one [14].

where[11]

$$(n)_\bullet : \mathbb{N} \longrightarrow \mathbb{Q}, \ i \longmapsto (n)_i = n(n-1)\ldots(n-i+1). \tag{51}$$

In other terms, for any $w \in Y_0^*, k \in \mathbb{N}, 0 \leq k \leq (w) + |w|$, one has

$$\langle \mathrm{Li}_w^- | (1-z)^{-k} \rangle = k! \langle \mathrm{H}_w^- | (n)_k \rangle. \tag{52}$$

Hence, denoting $\tilde{p}$ the exponential transform of the polynomial p, one has

$$\mathrm{Li}_w^-(z) = p((1-z)^{-1}) \text{ and } \mathrm{H}_w^-(n) = \tilde{p}((n)_\bullet) \tag{53}$$

with

$$p(t) = \sum_{k=0}^{(w)+|w|} p_k t^k \in (\mathbb{Z}[t], \times, 1) \text{ and } \tilde{p}(t) = \sum_{k=0}^{(w)+|w|} \frac{p_k}{k!} t^k \in (\mathbb{Q}[t], \times, 1). \tag{54}$$

Let us then associate p and $\tilde{p}$ with the polynomial $\check{p}$ obtained as follows:

$$\check{p}(t) = \sum_{k=0}^{(w)+|w|} k! p_k t^k = \sum_{k=0}^{(w)+|w|} p_k t^{\sqcup\!\sqcup k} \in (\mathbb{Z}[t], \sqcup\!\sqcup, 1). \tag{55}$$

Let us recall also that, for any $c \in \mathbb{C}$, one has

$$(n)_c \underset{n \to +\infty}{\widetilde{}} n^c = e^{c\log(n)}$$

and, with the respective scales of comparison, one has the following finite parts

$$\mathrm{f.p.}_{z \to 1} c \log(1-z) = 0, \ \{(1-z)^a \log^b((1-z)^{-1})\}_{a \in \mathbb{Z}, b \in \mathbb{N}}, \tag{56}$$

$$\mathrm{f.p.}_{n \to +\infty} c \log n = 0, \ \{n^a \log^b(n)\}_{a \in \mathbb{Z}, b \in \mathbb{N}}. \tag{57}$$

Hence, using the notations given in (49) and (50), one can see, from (56) and (57), that the values $p(1)$ and $\tilde{p}(1)$ obtained in (54) represent the following finite parts:

$$\mathrm{f.p.}_{z \to 1} \mathrm{Li}_w^-(z) = \mathrm{f.p.}_{z \to 1} \mathrm{Li}_{R_w}(z) = p(1) \in \mathbb{Z}, \tag{58}$$

$$\mathrm{f.p.}_{n \to +\infty} \mathrm{H}_w^-(n) = \mathrm{f.p.}_{n \to +\infty} \mathrm{H}_{\pi_Y(R_w)}(n) = \tilde{p}(1) \in \mathbb{Q}. \tag{59}$$

One can use then these values $p(1)$ and $\tilde{p}(1)$, instead of the values B_w^- and C_w^-, to regularize, respectively, $\zeta_{\sqcup\!\sqcup}(R_w)$ and $\zeta_\gamma(\pi_Y(R_w))$ as showed Theorem 2 below because, essentially, $B_\bullet^-$ and $C_\bullet^-$ do not realize characters for, respectively, $(\mathbb{Q}\langle X \rangle, \sqcup\!\sqcup, 1_{X^*}, \Delta_{\sqcup\!\sqcup}, e)$ and $(\mathbb{Q}\langle Y \rangle, \sqcup\!\sqcup, 1_{Y^*}, \Delta_{\sqcup\!\sqcup}, e)$ [7].

Now, in virtue of the extension of $\mathrm{Li}_\bullet$, defined as in (23) and (24), and of the Taylor coefficients, the previous polynomials $p, \tilde{p}$ and $\check{p}$ given in (54)–(55) can be determined explicitly thanks to

---

[11] Here, it is also convenient to denote $\mathbb{Q}[(n)_\bullet]$ the set of "polynomials" expanded as follows

$$\forall p \in, \mathbb{Q}[(n)_\bullet], \ p = \sum_{k=0}^{d} p_k (n)_k, \ \deg(p) = d.$$

Proposition 1 ([24]).

1. *The following morphisms of algebras are* <u>*bijective*</u>:

$$\lambda : (\mathbb{Z}[x_1^*], \sqcup\!\sqcup, 1_{X^*}) \longrightarrow (\mathbb{Z}[(1-z)^{-1}], \times, 1), \; R \longmapsto \mathrm{Li}_R,$$
$$\eta : (\mathbb{Q}[y_1^*], \uplus, 1_{Y^*}) \longrightarrow \quad (\mathbb{Q}[(n)_\bullet], \times, 1), \quad S \longmapsto \mathrm{H}_S.$$

2. *For any* $w = y_{s_1}, \ldots y_{s_r} \in Y_0^*$, *there exists a unique polynomial* R_w *belonging to* $(\mathbb{Z}[x_1^*], \sqcup\!\sqcup, 1_{X^*})$ *of degree* $(w) + |w|$, *such that*

$$\mathrm{Li}_{R_w}(z) = \mathrm{Li}_w^-(z) = p((1-z)^{-1}) \in (\mathbb{Z}[(1-z)^{-1}], \times, 1),$$
$$\mathrm{H}_{\pi_Y(R_w)}(n) = \mathrm{H}_w^-(n) = \quad \tilde{p}((n)_\bullet) \quad \in (\mathbb{Q}[(n)_\bullet], \times, 1).$$

In particular, via the extension, by linearity, of $R_\bullet$ *over* $\mathbb{Q}\langle Y_0 \rangle$ *and via the linear independent family* $\{\mathrm{Li}_{y_k}^-\}_{k \geq 0}$ *in* $\mathbb{Q}\{\mathrm{Li}_w^-\}_{w \in Y_0^*}$, *one has*

$$\forall k, l \in \mathbb{N}, \; \mathrm{Li}_{R_{y_k} \sqcup\!\sqcup R_{y_l}} = \mathrm{Li}_{R_{y_k}} \mathrm{Li}_{R_{y_l}} = \mathrm{Li}_{y_k}^- \mathrm{Li}_{y_l}^- = \mathrm{Li}_{y_k \top y_l}^- = \mathrm{Li}_{R_{y_k} \top y_l}.$$

3. *For any* w, *one has* $\check{p}(x_1^*) = R_w$.
4. *More explicitly, for any* $w = y_{s_1}, \ldots y_{s_r} \in Y_0^*$, *there exists a unique polynomial* R_w *belonging to* $(\mathbb{Z}[x_1^*], \sqcup\!\sqcup, 1_{X^*})$ *of degree* $(w) + |w|$, *given by*

$$R_{y_{s_1} \ldots y_{s_r}} = \sum_{k_1=0}^{s_1} \sum_{k_2=0}^{s_1+s_2-k_1} \ldots \sum_{k_r=0}^{\substack{(s_1+\ldots+s_r)- \\ (k_1+\ldots+k_{r-1})}} \binom{s_1}{k_1} \ldots \binom{s_1+\ldots+s_r-k_1-\ldots-k_{r-1}}{k_r} \rho_{k_1} \sqcup\!\sqcup \ldots \sqcup\!\sqcup \rho_{k_r},$$

where, for any $i = 1, \ldots, r$, *if* $k_i = 0$ *then* $\rho_{k_i} = x_1^* - 1_{X^*}$ *else, for* $k_i > 0$, *denoting the Stirling numbers of second kind by* $S_2(k,j)$'s, *one has*

$$\rho_{k_i} = \sum_{j=1}^{k_i} S_2(k_i, j)(j!)^2 \sum_{l=0}^{j} \frac{(-1)^l}{l!} \frac{(x_1^*)^{\sqcup\!\sqcup(j-l+1)}}{(j-l)!}.$$

Proposition 2 ([3,4,22,23]). *With notations of* (21), *similar to the character* $\gamma_\bullet$, *the poly-morphism* ζ *can be extended as follows*

$$\zeta_{\sqcup\!\sqcup} : (\mathbb{Q}\langle X \rangle, \sqcup\!\sqcup, 1_{X^*}) \longrightarrow (\mathcal{Z}, \times, 1) \; and \; \zeta_{\uplus} : (\mathbb{Q}\langle Y \rangle, \uplus, 1_{Y^*}) \longrightarrow (\mathcal{Z}, \times, 1),$$

satisfying, for any $\in \mathcal{L}yn Y \setminus \{y_1\}$,

$$\zeta_{\sqcup\!\sqcup}(\pi_X(l)) = \zeta_{\uplus}(l) = \gamma_l = \zeta(l)$$

and, for the generators of length (resp. weight) one, for X^* *(resp.* Y^*),

$$\zeta_{\sqcup\!\sqcup}(x_0) = \zeta_{\sqcup\!\sqcup}(x_1) = \zeta_{\uplus}(y_1) = 0.$$

Now, to regularize $\{\zeta(s_1, \ldots, s_r)\}_{(s_1, \ldots, s_r) \in \mathbb{C}^r}$, we use

Lemma 2 ([7]).

1. *The family* $\{x_0^*, x_1^*\}$ *is algebraically independent over* $(\mathbb{C}\langle X\rangle, ш, 1_{X^*})$ *within* $(\mathbb{C}\langle\langle X\rangle\rangle, ш, 1_{X^*})$.
   In particular, the power series x_0^* *and* x_1^* *are transcendent over* $\mathbb{C}\langle X\rangle$.
2. *The module* $(\mathbb{C}\langle X\rangle, ш, 1_{X^*})[x_0^*, x_1^*, (-x_0)^*]$ *is* $\mathbb{C}\langle X\rangle$-*free and a* $\mathbb{C}\langle X\rangle$-*basis of it is given by the family* $\{(x_0^*)^{шk}ш(x_1^*)^{шl}\}^{(k,l)\in\mathbb{Z}\times\mathbb{N}}$.
   Hence, $\{wш(x_0^*)^{шk}ш(x_1^*)^{шl}\}_{w\in X^*}^{(k,l)\in\mathbb{Z}\times\mathbb{N}}$ *is a* $\mathbb{C}$-*basis of it.*
3. *One has, for any* $x_i \in X$, $\mathbb{C}^{\mathrm{rat}}\langle\langle x_i\rangle\rangle = \mathrm{span}_{\mathbb{C}}\{(tx_i)^*ш\mathbb{C}\langle x_i\rangle | t \in \mathbb{C}\}$.

Since, for any $t \in \mathbb{C}, |t| < 1$, one has $\mathrm{Li}_{(tx_1)^*}(z) = (1-z)^{-t}$ and [3,4]

$$\mathrm{H}_{\pi_Y(tx_1)^*} = \sum_{k\geq 0} \mathrm{H}_{y_1^k} t^k = \exp\left(-\sum_{k\geq 1} \mathrm{H}_{y_k}\frac{(-t)^k}{k}\right) \tag{60}$$

then, in virtue of Proposition 1, we obtain successively

Proposition 3 ([7]). *The characters* $\zeta_ш$ *and* $\gamma_\bullet$ *can be extended as follows:*

$$\zeta_ш : (\mathbb{C}\langle X\rangle ш\mathbb{C}[x_1^*], ш, 1_{X^*}) \longrightarrow (\mathbb{C}, \times, 1_{\mathbb{C}}) \text{ and}$$
$$\gamma_\bullet : (\mathbb{C}\langle Y\rangle ⊎\mathbb{C}[y_1^*], ⊎, 1_{Y^*}) \longrightarrow (\mathbb{C}, \times, 1_{\mathbb{C}}),$$

such that, for any $t \in \mathbb{C}$ *such that* $|t| < 1$, *one has*

$$\zeta_ш((tx_1)^*) = 1_{\mathbb{C}} \text{ and } \gamma_{(ty_1)^*} = \exp\left(\gamma t - \sum_{n\geq 2}\zeta(n)\frac{(-t)^n}{n}\right) = \frac{1}{\Gamma(1+t)}.$$

Theorem 2 ([24]).

1. *For any* $(s_1, \ldots, s_r) \in \mathbb{N}_+^r$ *associated with* $w \in Y^*$, *there exists a unique polynomial* $p \in \mathbb{Z}[t]$ *of valuation 1 and of degree* $(w) + |w|$ *such that*

$$\begin{aligned}
\check{p}(x_1^*) &= R_w & &\in (\mathbb{Z}[x_1^*], ш, 1_{X^*}),\\
p((1-z)^{-1}) &= \mathrm{Li}_{R_w}(z) & &\in (\mathbb{Z}[(1-z)^{-1}], \times, 1),\\
\tilde{p}((n)_\bullet) &= \mathrm{H}_{\pi_Y(R_w)}(n) & &\in (\mathbb{Q}[(n)_\bullet], \times, 1),\\
\zeta_ш(-s_1, \ldots, -s_r) &= p(1) = \zeta_ш(R_w) & &\in (\mathbb{Z}, \times, 1),\\
\gamma_{-s_1,\ldots,-s_r} &= \tilde{p}(1) = \gamma_{\pi_Y(R_w)} & &\in (\mathbb{Q}, \times, 1).
\end{aligned}$$

2. *Let* $\Upsilon(n) \in \mathbb{Q}[(n)_\bullet]\langle\langle Y\rangle\rangle$ *and* $\Lambda(z) \in \mathbb{Q}[(1-z)^{-1}][\log(z)]\langle\langle X\rangle\rangle$ *be the noncommutative generating series of* $\{\mathrm{H}_{\pi_Y(R_w)}\}_{w\in Y^*}$ *and* $\{\mathrm{Li}_{R_{\pi_Y(w)}}\}_{w\in X^*}$:

$$\Upsilon := \sum_{w\in Y^*} \mathrm{H}_{\pi_Y(R_w)} w \text{ and } \Lambda := \sum_{w\in X^*} \mathrm{Li}_{R_{\pi_Y(w)}} w, \text{ with } \langle\Lambda(z)|x_0\rangle = \log(z).$$

Then Υ *and* Λ *are group-like, for respectively* $\Delta_{⊎}$ *and* $\Delta_ш$, *and:*

$$\Upsilon = \prod_{l\in\mathcal{L}ynY}^{\searrow} e^{\mathrm{H}_{\pi_Y(R_{\Sigma_l})}\Pi_l} \text{ and } \Lambda = \prod_{l\in\mathcal{L}ynX}^{\searrow} e^{\mathrm{Li}_{R_{\pi_Y(S_l)}}P_l}.$$

3. Let $Z_{\gamma}^{-} \in \mathbb{Q}\langle\langle Y \rangle\rangle$ and $Z_{\text{Ш}}^{-} \in \mathbb{Z}\langle\langle X \rangle\rangle$ be the noncommutative generating series of $\{\gamma_{\pi_Y(R_w)}\}_{w \in Y^*}$ and[12] $\{\zeta_{\text{Ш}}(R_{\pi_Y(w)})\}_{w \in X^*}$, respectively:

$$Z_{\gamma}^{-} := \sum_{w \in Y^*} \gamma_{\pi_Y(R_w)} w \text{ and } Z_{\text{Ш}}^{-} := \sum_{w \in X^*} \zeta_{\text{Ш}}(R_{\pi_Y(w)}) w.$$

Then Z_{γ}^{-} and $Z_{\text{Ш}}^{-}$ are group-like, for respectively $\Delta_{\uplus}$ and $\Delta_{\text{Ш}}$, and:

$$Z_{\gamma}^{-} = \prod_{l \in \mathcal{L}yn Y}^{\searrow} e^{\gamma_{\pi_Y(R_{\Sigma_l})} \Pi_l} \text{ and } Z_{\text{Ш}}^{-} = \prod_{l \in \mathcal{L}yn X}^{\searrow} e^{\zeta_{\text{Ш}}(\pi_Y(S_l)) P_l}.$$

Moreover,

$$\text{F.P.}_{\cdot n \to +\infty} \Upsilon(n) = Z_{\gamma}^{-} \text{ and F.P.}_{\cdot z \to 1} \Lambda(z) = Z_{\text{Ш}}^{-}, \tag{61}$$

meaning that, for any $v \in Y^*$ and $u \in X^*$, one has

$$\text{f.p.}_{\cdot n \to +\infty} \langle \Upsilon(n) | v \rangle = \langle Z_{\gamma}^{-} | v \rangle \text{ and f.p.}_{\cdot z \to 1} \langle \Lambda(z) | u \rangle = \langle Z_{\text{Ш}}^{-} | u \rangle. \tag{62}$$

To end this section, let us recall that the function Γ is meromorphic, admits no zeroes and simple poles in $-\mathbb{N}$. Hence, Γ^{-1} is entire and admits simple zeros in $-\mathbb{N}$.

Moreover, using the incomplete beta function, i.e., for $z, a, b \in \mathbb{C}$ such that $|z| < 1, \Re a > 0, \Re b > 0$,

$$\begin{aligned} \text{B}(z; a, b) &:= \int_0^z dt\, t^{a-1}(1-t)^{b-1} \\ &= \text{Li}_{x_0[(ax_0)^* \text{Ш} ((1-b)x_1)^*]}(z) \\ &= \text{Li}_{x_1[((a-1)x_0)^* \text{Ш} (-bx_1)^*]}(z), \end{aligned} \tag{63}$$

and setting

$$\begin{aligned} \text{B}(a, b) &:= \text{B}(1; a, b) \\ &= \zeta_{\text{Ш}}(x_0[(ax_0)^* \text{Ш} ((1-b)x_1)^*]) \\ &= \zeta_{\text{Ш}}(x_1[((a-1)x_0)^* \text{Ш} (-bx_1)^*]). \end{aligned} \tag{64}^*$$

we have, on the one hand, the following Euler's formula

$$\text{B}(a, b) \Gamma(a+b) = \Gamma(a) \Gamma(b) \tag{65}$$

---

[12] On the one hand, by Proposition 2, one has $\langle Z_{\text{Ш}}^{-} | x_0 \rangle = \zeta_{\text{Ш}}(x_0) = 0$.
On the other hand, since $R_{y_1} = (2x_1)^* - x_1^*$ then $\text{Li}_{R_{y_1}}(z) = (1-z)^{-2} - (1-z)^{-1}$ and $\text{H}_{\pi_Y(R_{y_1})}(n) = \binom{n}{2} - \binom{n}{1}$. Hence, one also has $\langle Z_{\text{Ш}}^{-} | x_1 \rangle = \zeta_{\text{Ш}}(R_{\pi_Y(y_1)}) = 0$ and $\langle Z_{\gamma}^{-} | x_1 \rangle = \gamma_{\pi_Y(R_{y_1})} = -1/2$.

and, on the other hand[13], in virtue of Proposition 3,

$$\exp\left(\sum_{n\geq 2}\zeta(n)\frac{(u+v)^n-(u^n+v^n)}{n}\right)=\frac{\Gamma(1-u)\Gamma(1-v)}{\Gamma(1-u-v)} \tag{66}$$

$$=\frac{\gamma_{(-(u+v)y_1)^*}}{\gamma_{(-uy_1)^*}\gamma_{(-vy_1)^*}}=\frac{\gamma_{(-(u+v)y_1)^*}}{\gamma_{(-uy_1)^*}\boxplus(-vy_1)^*}.$$

Hence, it follows that

Corollary 1 ([24]). *For any $u,v\in\mathbb{C}$ such that $|u|<1,|v|<1$ and $|u+v|<1$, one has*

$$\gamma_{(-(u+v)y_1)^*}=\gamma_{(-uy_1)^*}\boxplus(-vy_1)^*\cdot\zeta_{\sqcup\!\sqcup}(x_0[(-ux_0)^*\sqcup\!\sqcup(-(1+v)x_1)^*])$$
$$=\gamma_{(-uy_1)^*}\boxplus(-vy_1)^*\cdot\zeta_{\sqcup\!\sqcup}(x_1[(-(1+u)x_0)^*\sqcup\!\sqcup(-vx_1)^*]).$$

Remark 2 By (25), for any $u,v\in\mathbb{C}$ such that $|u|<1,|v|<1$ and $|u+v|<1$, one also has

$$\zeta_{\sqcup\!\sqcup}((-(u+v)x_1)^*)=\zeta_{\sqcup\!\sqcup}((-ux_1)^*\sqcup\!\sqcup(-vx_1)^*)=\zeta_{\sqcup\!\sqcup}((-ux_1)^*)\zeta_{\sqcup\!\sqcup}((-vx_1)^*)=1.$$

References

1. Bui, V.C., Duchamp, G.H.E., Hoang Ngoc Minh, V.: Structure of polyzetas and explicit representation on transcendence bases of shuffle and stuffle algebras. J. Sym. Comput. (2016)
2. Bui, V.C., Duchamp, G.H.E., Hoan Ngô, Q., Hoang Ngoc Minh, V., Tollu, C.: (Pure) Transcendence Bases in -Deformed Shuffle Bialgebras, Séminaire Lotharingien de Combinatoire, B74f (2018). 22 pp
3. Costermans, C., Minh, H.N.: Some results à la Abel obtained by use of techniques à la Hopf. In: Workshop on Global Integrability of Field Theories and Applications, Daresbury, UK, 1–3 November 2006
4. Costermans, C., Minh, H.N.: Noncommutative algebra, multiple harmonic sums and applications in discrete probability. J. Sym. Comput. 801–817 (2009)

---

[13] The first equality of (66) is already presented in [13].

Since $(-uy_1)^*\boxplus(uy_1)^*=(-u^2y_2)^*$ then, letting $v=-u$ in (66), we have

$$\exp\left(-\sum_{n\geq 1}\zeta(2n)\frac{u^{2n}}{n}\right)=\Gamma(1-u)\Gamma(1+u)=\frac{1}{\gamma_{(-uy_1)^*}\boxplus(uy_1)^*}=\frac{1}{\gamma_{(-u^2y_2)^*}}.$$

It is also a consequence obtained by expanding identities like (60) [3,4]

$$\forall y_r\in Y,\ y_r^k=\frac{(-1)^k}{k!}\sum_{\substack{s_1,\ldots,s_k>0\\s_1+\ldots+ks_k=k}}\frac{(-y_r)^{\boxplus s_1}}{1^{s_1}}\boxplus\ldots\boxplus\frac{(-y_{kr})^{\boxplus s_k}}{k^{s_k}}.$$

5. Deligne, P.: Equations Différentielles à Points Singuliers Réguliers. Lecture Notes in Mathematics, vol. 163. Springer, Heidelberg (1970)
6. Deneufchâtel, M., Duchamp, G.H.E., Hoang Ngoc Minh, V., Solomon, A.I.: Independence of hyperlogarithms over function fields via algebraic combinatorics. In: Winkler, F. (ed.) CAI 2011. LNCS, vol. 6742, pp. 127–139. Springer, Heidelberg (2011). https://doi.org/10.1007/978-3-642-21493-6_8
7. Duchamp, G.H.E., Minh, H.N., Ngo, Q.H.: Harmonic sums and polylogarithms at negative multi-indices. J. Sym. Comput. (2016)
8. Duchamp, G.H.E., Minh, H.N., Ngo, Q.H.: Double regularization of polyzetas at negative multiindices and rational extensions (en préparation)
9. Duchamp, G.H.E., Hoang Ngoc Minh, V., Ngo, Q.H., Penson, K.A., Simonnet, P.: Mathematical renormalization in quantum electrodynamics via noncommutative generating series. In: Kotsireas, I., Martínez-Moro, E. (eds.) ACA 2015. Springer Proceedings in Mathematics & Statistics, vol. 198, pp. 59–100. Springer, Heidelberg (2017). https://doi.org/10.1007/978-3-319-56932-1_6
10. Duchamp, G.H.E., Minh, H.N., Penson, K.: On Drinfel'd associators. arXiv:1705.01882 (2017)
11. Drinfel'd, V.: Quantum group. In: Proceedings of International Congress of Mathematicians, Berkeley (1986)
12. Drinfel'd, V.: Quasi-Hopf algebras. Len. Math. J. **1**, 1419–1457 (1990)
13. Drinfel'd, V.: On quasitriangular quasi-Hopf algebra and a group closely connected with gal($\bar{\mathbb{Q}}/\mathbb{Q}$). Len. Math. J. **4**, 829–860 (1991)
14. Écalle, J.: L'équation du pont et la classification analytique des objets locaux, In: Les fonctions résurgentes, 3, Publications de l'Université de Paris-Sud, Département de Mathématique (1985)
15. Eilenberg, S.: Automata, Languages and Machines, vol. A. Academic Press, New York (1974)
16. Jacob, G., Reutenauer, C. (eds.) The 1st International Conference on Formal Power Series and Algebraic Combinatorics, FPSAC 1988, University of Lille, France, December 1988
17. Flajolet, P., Odlyzko, A.: Singularity analysis of generating functions. SIAM J. Discrete Math. **3**(2), 216–240 (1982)
18. Minh, H.N.: Summations of polylogarithms via evaluation transform. Math. Comput. Simul. **1336**, 707–728 (1996)
19. Minh, H.N., Jacob, G.: Symbolic integration of meromorphic differential equation via Dirichlet functions. Discrete Math. **210**, 87–116 (2000)
20. Minh, H.N., Jacob, G., Oussous, N.E., Petitot, M.: De l'algèbre des ζ de Riemann multivariées à l'algèbre des ζ de Hurwitz multivariées. J. électronique du Séminaire Lotharingien de Combinatoire **44** (2001)
21. Minh, H.N., Petitot, M.: Lyndon words, polylogarithmic functions and the Riemann ζ function. Discrete Math. **217**, 273–292 (2000)
22. Hoang Ngoc Minh, V.: On a conjecture by Pierre Cartier about a group of associators. Acta Math. Vietnamica **38**(3), 339–398 (2013)
23. Minh, H.N.: Structure of polyzetas and Lyndon words. Vietnamese Math. J. **41**(4), 409–450 (2013)
24. Minh, H.N.: On solutions of KZ_3 (to appear)
25. Lappo-Danilevsky, J.A.: Théorie des systèmes des équations différentielles linéaires. Chelsea, New York (1953)
26. Lê, T.Q.T., Murakami, J.: Kontsevich's integral for Kauffman polynomial. Nagoya Math. J. 39–65 (1996)

27. Reutenauer, C.: Free Lie Algebras. London Mathematical Society Monographs (1993)

28. Zagier, D.: Values of zeta functions and their applications. In: Joseph, A., Mignot, F., Murat, F., Prum, B., Rentschler, R. (eds.) First European Congress of Mathematics, vol. 120, pp. 497–512. Birkhäuser, Basel (1994). https://doi.org/10.1007/978-3-0348-9112-7_23

On a Polytime Factorization Algorithm
for Multilinear Polynomials over $\mathbb{F}_2$

Pavel Emelyanov[1,2]($\boxtimes$) and Denis Ponomaryov[1,2]

[1] A.P. Ershov Institute of Informatics Systems,
Lavrentiev av. 6, 630090 Novosibirsk, Russia
{emelyanov,ponom}@iis.nsk.su
[2] Novosibirsk State University, Pirogov st. 1, 630090 Novosibirsk, Russia

Abstract. In 2010, Shpilka and Volkovich established a prominent result on the equivalence of polynomial factorization and identity testing. It follows from their result that a multilinear polynomial over the finite field of order 2 can be factored in time cubic in the size of the polynomial given as a string. Later, we have rediscovered this result and provided a simple factorization algorithm based on computations over derivatives of multilinear polynomials. The algorithm has been applied to solve problems of compact representation of various combinatorial structures, including Boolean functions and relational data tables. In this paper, we describe an improvement of this factorization algorithm and report on preliminary experimental analysis.

1 Introduction

Polynomial factorization is a classic algorithmic problem in algebra [14], whose importance stems from numerous applications. The computer era has stimulated interest to polynomial factorization over finite fields. For a long period of time, Theorem 1.4 in [8] (see also [12, Theorem 1.6]) has been the main source of information on the complexity of this problem: a (densely represented) polynomial $F_{p^r}(x_1, \ldots, x_m)$ of the total degree $n > 1$ over all its variables can be factored in time that is polynomial in n^m, r, and p. In addition, practical probabilistic factorization algorithms have been known.

In 2010, Shpilka and Volkovich [13] established a connection between polynomial factorization and polynomial identity testing. The result has been formulated in terms of the arithmetic circuit representation of polynomials. It follows from these results that a multilinear polynomial over $\mathbb{F}_2$ (the finite field of the order 2) can be factored in the time that is cubic in the size of the polynomial given as a symbol sequence.

Multilinear polynomials over $\mathbb{F}_2$ are well known in the scope of mathematical logic (as Zhegalkine polynomials [15] or Algebraic Normal Form) and in circuit synthesis (Canonical Reed-Muller Form [10]). Factorization of multilinear

---

This work is supported by the Ministry of Science and Education of the Russian Federation under the 5-100 Excellence Program and the grant of Russian Foundation for Basic Research No. 17-51-45125.

V. P. Gerdt et al. (Eds.): CASC 2018, LNCS 11077, pp. 164–176, 2018.
https://doi.org/10.1007/978-3-319-99639-4_11

polynomials is a particular case of decomposition (so-called conjunctive or AND-decomposition) of logic formulas and Boolean functions. By the idempotence law in the algebra of logic, multilinearity (all variables occur in degree 1) is a natural property of these polynomials, which makes the factors have disjoint sets of variables $F(X, Y) = F_1(X)F_2(Y)$, $X \cap Y = \varnothing$. In practice, this property allows for obtaining a factorization algorithm by variable partitioning (see below).

Among other application domains, such as game and graph theory, the most attention has been given to decomposition of Boolean functions in logic circuit synthesis, which is related to the algorithmic complexity and practical issues of electronic circuits implementation, their size, time delay, and power consumption (see [9,11], for example). One may note the renewed interest in this topic, which is due to the novel technological achievements in circuit design.

The logic interpretation of multilinear polynomials over $\mathbb{F}_2$ admits another notion of factorization, which is commonly called Boolean factorization (finding Boolean divisors). For example, there are Boolean polynomials, which have decomposition components sharing some common variables. Their product/conjunction does not produce original polynomials in the algebraic sense but it gives the same functions/formulas in the logic sense. In general, logic-based approaches to decomposition are more powerful than algebraic ones: a Boolean function can be decomposable logically, but not algebraically [9, Chap. 4].

In 2013, the authors have rediscovered the result of Shpilka and Volkovich under simpler settings and in a simpler way [5,7]. A straightforward treatment of sparsely represented multilinear polynomials over $\mathbb{F}_2$ gave the same worst-case cubic complexity of the factorization algorithm. Namely, the authors provided two factorization algorithms based, respectively, on computing the greatest common divisor (GCD) and formal derivatives (FD) for polynomials obtained from the input one.

The algorithms have been used to obtain a solution to the following problems of compact representation of different combinatorial structures (below we provide examples, which intuitively explain their relation to the factorization problem).

– Conjunctive disjoint decomposition of monotone Boolean functions given in positive DNF [5,7]. For example, the following DNF

$$\varphi = (x \wedge u) \vee (x \wedge v) \vee (y \wedge u) \vee (y \wedge v) \vee (x \wedge u \wedge v) \qquad (1)$$

is equivalent to

$$\psi = (x \wedge u) \vee (x \wedge v) \vee (y \wedge u) \vee (y \wedge v), \qquad (2)$$

since the last term in φ is redundant, and we have

$$\psi \equiv (x \vee y) \wedge (u \vee v) \qquad (3)$$

and the decomposition components $x \vee y$ and $u \vee v$ can be recovered from the factors of the polynomial

$$F_\psi = xu + xv + yu + yv = (x + y) \cdot (u + v) \qquad (4)$$

constructed for ψ.

– Conjunctive disjoint decomposition of Boolean functions given in full DNF [5,7]. For example, the following full DNF

$$\varphi = (x \wedge \neg y \wedge u \wedge \neg v) \vee (x \wedge \neg y \wedge \neg u \wedge v) \vee$$
$$\vee (\neg x \wedge y \wedge u \wedge \neg v) \vee (\neg x \wedge y \wedge \neg u \wedge v)$$

is equivalent to

$$(x \wedge \neg y) \vee (\neg x \wedge y) \bigwedge (u \wedge \neg v) \vee (\neg u \wedge v), \tag{5}$$

and the decomposition components of φ can be recovered from the factors of the polynomial

$$F_\varphi = x\bar{y}u\bar{v} + x\bar{y}\bar{u}v + \bar{x}yu\bar{v} + \bar{x}y\bar{u}v = (x\bar{y} + \bar{x}y) \cdot (u\bar{v} + \bar{u}v) \tag{6}$$

constructed for φ.
– Non-disjoint conjunctive decomposition of multilinear polynomials over $\mathbb{F}_2$, in which components can have common variables from a given set. In [3], a fixed-parameter polytime decomposition algorithm has been proposed, for the parameter being the number of the shared variables between components.
– Cartesian decomposition of data tables (i.e., finding tables such that their unordered Cartesian product gives the source table) [4,6] and generalizations thereof for the case of a non-empty subset of shared attributes between the tables. For example, the following table has a decomposition of the form:

B	E	D	A	C
z	q	u	x	y
y	q	u	x	y
y	r	v	x	z
z	r	v	x	z
y	p	u	x	x
z	p	u	x	x

$=$

A	B
x	y
x	z

$\times$

C	D	E
x	u	p
y	u	q
z	v	r

which can be obtained from the factors of the polynomial

$$z_B \cdot q \cdot u \cdot x_A \cdot y_C + y_B \cdot q \cdot u \cdot x_A \cdot y_C +$$
$$y_B \cdot r \cdot v \cdot x_A \cdot z_C + z_B \cdot r \cdot v \cdot x_A \cdot z_C +$$
$$y_B \cdot p \cdot u \cdot x_A \cdot x_C + z_B \cdot p \cdot u \cdot x_A \cdot x_C$$
$$= (x_A \cdot y_B + x_A \cdot z_B) \cdot (q \cdot u \cdot y_C + r \cdot v \cdot z_C + p \cdot u \cdot x_C)$$

constructed for the table's content.
In terms of SQL, Cartesian decomposition means reversing the first operator and the second operator represents some feasible generalization of the problem:

```
T1 CROSS JOIN T2              SELECT T1.*, T2.* EXCEPT(Attr2)
                              FROM T1 INNER JOIN T2
                              ON T1.Attr1 = T2.Attr2
```

where EXCEPT(list) is an informal extension of SQL used to exclude list from the resulting attributes. This approach can be applied to other table-based structures (for example, decision tables or datasets appearing in the K&DM domain, as well as the truth tables of Boolean functions).

Shpilka and Volkovich did not address the problems of practical implementations of the factorization algorithm. However, the applications above require a factorization algorithm to be efficient enough on large polynomials. In this paper, we propose an improvement of the factorization algorithm from [4,6], which potentially allows for working with larger inputs. An implementation of this version of the algorithm in Maple 17 outperforms the native Maple's `Factor(poly)` `mod 2` factorization, which in our experiments failed to terminate on input polynomials having 10^3 variables and 10^5 monomials.

2 Definitions and Notations

A polynomial $F \in \mathbb{F}_2[x_1, \ldots, x_n]$ is called *factorable* if $F = F_1 \cdot \ldots \cdot F_k$, where $k \geq 2$ and $F_1, \ldots, F_k$ are non-constant polynomials. The polynomials $F_1, \ldots, F_k$ are called *factors* of F. It is important to realize that since we consider multilinear polynomials (every variable can occur only in the power of ≤ 1), the factors are polynomials *over disjoint sets of variables*. In the following sections, we assume that the polynomial F does not have *trivial divisors*, i.e., neither x, nor $x + 1$ divides F. Clearly, trivial divisors can easily be recognized.

For a polynomial F, a variable x from the set of variables $Var(F)$ of F, and a value $a \in \{0, 1\}$, we denote by $F_{x=a}$ the polynomial obtained from F by substituting x with a. For multilinear polynomials over $\mathbb{F}_2$, we define a *formal derivative* as $\frac{\partial F}{\partial x} = F_{x=0} + F_{x=1}$, but for non-linear ones, we use the definition of a "standard" formal derivative for polynomials. Given a variable z, we write $z|F$ if z divides F, i.e., z is present in every monomial of F (note that this is equivalent to the condition $\frac{\partial F}{\partial z} = F_{z=1}$).

Given a set of variables Σ and a monomial m, the *projection* of m onto Σ is 1 if m does not contain any variable from Σ, or is equal to the monomial obtained from m by removing all the variables not contained in Σ, otherwise. The *projection* of a polynomial F onto Σ, denoted as $F|_{\Sigma}$, is the polynomial obtained as sum of monomials from the set S, where S is the set of the monomials of F projected onto Σ.

$|F|$ is the *length* of the polynomial F given as a symbol sequence, i.e., if the polynomial over n variables has M monomials of lengths $m_1, \ldots, m_M$ then $|F| = \sum_{i=1}^{M} m_i = O(nM)$.

We note that the correctness proofs for the algorithms presented below can be found in [5,7].

3 GCD-Algorithm

Conceptually, this algorithm is the simplest one. It outputs factors of an input polynomial whenever they exist.

1. Take an arbitrary variable x from $Var(F)$
2. $G := \gcd(F_{x=0}, \frac{\partial F}{\partial x})$
3. If $G = 1$ then stop

4. Output factor $\frac{F}{G}$
5. $F := G$
6. Go to 1

Here the complexity of factorization is hidden in the algorithm for finding the greatest common divisor of polynomials.

Computing GCD is known as a classic algorithmic problem in algebra [14], which involves computational difficulties. For example, if the field is not too rich ($\mathbb{F}_2$ is an example) then intermediate values vanish quite often, which essentially affects the computation performance. In [2], Wittkopf et al. developed the LINZIP algorithm for the GCD-problem. Its complexity is $O(|F|^3)$, i.e., the complexity of the GCD-algorithm is asymptotically the same as for Shpilka and Volkovich's result for the case of multilinear polynomials (given as strings).

4 FD-Algorithm

In the following, we assume that the input polynomial F contains at least two variables. The basic idea of FD-Algorithm is to partition a variable set into two sets with respect to a selected variable:

- the first set Σ_{same} contains the selected variable and corresponds to an irreducible polynomial;
- the second set Σ_{other} corresponds to the second polynomial that can admit further factorization.

As soon as Σ_{same} and Σ_{other} are computed (and $\Sigma_{other} \neq \varnothing$), the corresponding factors can be easily obtained as projections of the input polynomial onto these sets.

1. Take an arbitrary variable x occurring in F
2. Let $\Sigma_{same} := \{x\}, \Sigma_{other} := \varnothing, F_{same} := 0, F_{other} := 0$
3. Compute $G := F_{x=0} \cdot \frac{\partial F}{\partial x}$
4. For each variable $y \in Var(F) \setminus \{x\}$:
   If $\frac{\partial G}{\partial y} = 0$ then $\Sigma_{other} := \Sigma_{other} \cup \{y\}$
            else $\Sigma_{same} := \Sigma_{same} \cup \{y\}$
5. If $\Sigma_{other} = \varnothing$ then report "F is non-factorable" and stop
6. Return polynomials F_{same} and F_{other} obtained as projections onto Σ_{same} and Σ_{other}, respectively

The factors F_{same} and F_{other} have the property mentioned above and hence, the algorithm can be applied to obtain factors for F_{other}.

Note that FD-algorithm takes $O(|F|^2)$ steps to compute the polynomial $G = F_{x=0} \cdot \frac{\partial F}{\partial x}$ and $O(|G|)$ time to test whether the derivative $\frac{\partial G}{\partial y}$ equals zero. As we have to verify this for every variable $y \neq x$, we have a procedure that computes a variable partition in $O(|F|^3)$ steps. The algorithm allows for a straightforward parallelization on the selected variable y: the loop over the variable y (selected in line 4) can be performed in parallel for all the variables.

One can readily see that the complexity of factorization is hidden in the computation of the product G of two polynomials and testing whether a derivative of this product is equal to zero. In the worst case, the length of $G = F_{x=0} \cdot \frac{\partial F}{\partial x}$ equals $\Omega(|F|^2)$, which makes computing this product expensive for large input polynomials. In the next section, we describe a modification of the FD-algorithm, which implements the test above in a more efficient recursive fashion, without the need to compute the product of polynomials explicitly.

5   Modification of FD-Algorithm

Assume the polynomials $A = \frac{\partial F}{\partial x}$ and $B = F_{x=0}$ are computed. By taking a derivative of $A \cdot B$ on y (a variable different from x) we have $D = \frac{\partial F_{x=0}}{\partial y}$ and $C = \frac{\partial^2 F}{\partial x \partial y}$. We need to test whether $AD + BC = 0$, or equivalently, $AD = BC$. The main idea is to reduce this test to four tests involving polynomials of smaller sizes. Proceeding recursively in this way, we obtain smaller, or even constant, polynomials for which identity testing is simpler. Yet again, the polynomial identity testing demonstrates its importance, as Shpilka and Volkovich have readily established.

Steps 3–4 of FD-algorithm are modified as follows:

```
Let A = ∂F/∂x , B = F_{x=0}
For each variable y ∈ Var(F) \ {x}:
   Let D = ∂B/∂y , C = ∂A/∂y
   If IsEqual(A,D,B,C) then   Σ_other := Σ_other ∪ {y},
                       else    Σ_same := Σ_same ∪ {y}
```

where (all the above mentioned variables are chosen from the set of variables of the corresponding polynomials).

```
Define IsEqual(A,D,B,C) returning Boolean
```

1. If $A = 0$ or $D = 0$ then return $(B = 0$ or $C = 0)$
2. If $B = 0$ or $C = 0$ then return FALSE
3. For all variables z occurring in at least one of A, B, C, D :
4.   If $(z|A$ or $z|D)$ xor $(z|B$ or $z|C)$ then return FALSE
5.   Replace every $X \in \{A, B, C, D\}$ with $X := \frac{\partial X}{\partial z}$, provided $z|X$
6. If $A = 1$ and $D = 1$ then return $(B = 1$ and $C = 1)$
7. If $B = 1$ and $C = 1$ then return FALSE
8. If $A = 1$ and $B = 1$ then return $(D = C)$
9. If $D = 1$ and $C = 1$ then return $(A = B)$
10. Pick a variable z
11. If not IsEqual$(A_{z=0}, D_{z=0}, B_{z=0}, C_{z=0})$ then return FALSE
12. If not IsEqual$(\frac{\partial A}{\partial z}, \frac{\partial D}{\partial z}, \frac{\partial B}{\partial z}, \frac{\partial C}{\partial z})$ then return FALSE
13. If IsEqual$(\frac{\partial A}{\partial z}, B_{z=0}, A_{z=0}, \frac{\partial B}{\partial z})$ then return TRUE
14 Return IsEqual$(\frac{\partial A}{\partial z}, C_{z=0}, A_{z=0}, \frac{\partial C}{\partial z})$

```
End Definition
```

Several comments on IsEqual are in order:

- Lines 1–9 implement processing of trivial cases, when the condition $AD = BC$ can easily be verified without recursion. For example, when line 2 is executed, it is already known that neither A nor D equals zero and hence, AD can not be equal to BC. Similar tests are implemented in lines 6–9.
- At line 5 it is known that z divides both, AD and BC and thus, the problem $AD = BC$ can be reduced to the polynomials obtained by eliminating z.
- Finally, lines 11–14 implement recursive calls to IsEqual. Observe that the parameter polynomials are obtained from the original ones by evaluating a variable z to zero or by computing a derivative. Both of the operations yield polynomials of a smaller size than the original ones and can give constant polynomials in the limit. To determine the parameters of IsEqual we resort to a trick that transforms one identity into two smaller ones. This transformation uses a multiplier, which is not unique. Namely, we can select 16 variants among 28 possible ones (see comments in Sect. 5.1 below) and this gives 16 variants of lines 13–14.

5.1   Complete List of Possible Parameters

If A, D, B, C are the parameters of IsEqual, we denote for a $Q \in \{A, D, B, C\}$ the derivative on a variable z and evaluation $z = 0$ as Q_1 and Q_2, respectively.

$$AD = BC \quad \text{iff} \quad (A_1 z + A_2)(D_1 z + D_2) = (B_1 z + B_2)(C_1 z + C_2),$$

$$A_1 D_1 z^2 + (A_1 D_2 + A_2 D_1)z + A_2 D_2 = B_1 C_1 z^2 + (B_1 C_2 + B_2 C_1)z + B_2 C_2.$$

The equality holds iff the corresponding coefficients are equal:

$$\begin{cases} A_1 D_1 = B_1 C_1 & (1) \\ A_2 D_2 = B_2 C_2 & (2) \\ A_1 D_2 + A_2 D_1 = B_1 C_2 + B_2 C_1 & (3) \end{cases}$$

If at least one of the identities (1), (2) does not hold then $AD \neq BC$. Otherwise, we can use these identities to verify (3) in the following way.

By the rule of choosing z, we can assume $A_1, A_2 \neq 0$. Multiplying both sides of (3) by $A_1 A_2$ gives

$$A_1^2 A_2 D_2 + A_1 A_2^2 D_1 = A_1 A_2 B_1 C_2 + A_1 A_2 B_2 C_1.$$

Next, by using the identities (1) and (2),

$$A_1^2 B_2 C_2 + A_1 A_2 B_2 C_1 = A_2^2 B_1 C_1 + A_1 A_2 B_1 C_2,$$

$$A_1 B_2 (A_1 C_2 + A_2 C_1) = A_2 B_1 (A_2 C_1 + A_1 C_2).$$

Hence, it suffices to check $(A_1 B_2 + A_2 B_1)(A_1 C_2 + A_2 C_1) = 0$, i.e., at least one of these factors equals zero. It turns out that we need to test at most 4 polynomial identities, and each of them is smaller than the original identity $AD = BC$.

Notice that the multiplier $A_1 A_2$ is used to construct the version of IsEqual given above.

By the rule of choosing z, we can take different multiplier's combinations of the pairs of 8 elements. Only 16 out of 28 pairs are appropriate:

$$
\begin{aligned}
A_1 A_2 &\rightarrow A_1 C_2 = A_2 C_1, \quad A_1 B_2 = A_2 B_1 \\
A_1 B_2 &\rightarrow A_1 D_2 = B_2 C_1, \quad A_1 B_2 = A_2 B_1 \\
A_1 C_2 &\rightarrow A_1 D_2 = B_1 C_2, \quad A_1 C_2 = A_2 C_1 \\
A_1 D_2 &\rightarrow A_1 D_2 = B_2 C_1, \quad A_1 D_2 = B_1 C_2 \\
A_2 B_1 &\rightarrow A_2 D_1 = B_1 C_2, \quad A_1 B_2 = A_2 B_1 \\
A_2 C_1 &\rightarrow A_2 D_1 = B_2 C_1, \quad A_1 C_2 = A_2 C_1 \\
A_2 D_1 &\rightarrow A_2 D_1 = B_2 C_1, \quad A_2 D_1 = B_1 C_2 \\
B_1 B_2 &\rightarrow B_1 D_2 = B_2 D_1, \quad A_1 B_2 = A_2 B_1 \\
B_1 C_2 &\rightarrow A_2 D_1 = B_1 C_2, \quad A_1 D_2 = B_1 C_2 \\
B_1 D_2 &\rightarrow B_1 D_2 = B_2 D_1, \quad A_1 D_2 = B_1 C_2 \\
B_2 C_1 &\rightarrow A_2 D_1 = B_2 C_1, \quad A_1 D_2 = B_2 C_1 \\
B_2 D_1 &\rightarrow B_1 D_2 = B_2 D_1, \quad A_2 D_1 = B_2 C_1 \\
C_1 C_2 &\rightarrow C_1 D_2 = C_2 D_1, \quad A_1 C_2 = A_2 C_1 \\
C_1 D_2 &\rightarrow C_1 D_2 = C_2 D_1, \quad A_1 D_2 = B_2 C_1 \\
C_2 D_1 &\rightarrow C_1 D_2 = C_2 D_1, \quad A_2 D_1 = B_1 C_2 \\
D_1 D_2 &\rightarrow C_1 D_2 = C_2 D_1, \quad B_1 D_2 = B_2 D_1
\end{aligned}
$$

5.2 Analysis of ModFD-Algorithm for Random Polynomials

We now provide a theoretical analysis of ModFD-algorithm. The complexity estimations we describe here are conservative and, therefore, they give an upper bound greater than $O(|F|^3)$ of the original FD-algorithm. However, the approach presented here could serve as a basis to obtain a more precise upper bound, which would explain the gain in performance in practice; we report on a preliminary experimental evaluation in Sect. 6.

Our estimation is based on

Theorem 1 (Akra and Bazzi, [1]). *Let the recurrence*

$$
T(x) = g(x) + \sum_{i=1}^{k} \lambda_i T(\omega_i x + h_i(x)) \quad \text{for } x \geq C
$$

satisfy the following conditions:

1. *$T(x)$ is appropriately defined for $x < C$;*
2. *$\lambda_i > 0$ and $0 < \omega_i < 1$ are constants for all i;*
3. *$|g(x)| = O(x^c)$; and*
4. *$|h_i(x)| = O\left(\frac{x}{(\log x)^2}\right)$ for all i.*

Then

$$
T(x) = \Theta\left(x^p \left(1 + \int_1^x \frac{g(t)}{t^{p+1}} dt \right) \right),
$$

where p is determined by the characteristic equation $\sum_{i=1}^{k} \lambda_i \omega_i^p = 1$.

For the complexity estimations, we assume that polynomials are represented by alphabetically sorted lists of bitscales corresponding to indicator vectors for the variables of monomials. Hence, to represent a polynomial F over n variables with M monomials $|F| = nM + cM$ bits are required, where c is a constant overhead to maintain the list structure. This guarantees the linear time complexity for the following operations:

- computing a derivative with respect to a variable (the derived polynomial also remains sorted);
- evaluating to zero for a variable with removing the empty bitscale representing the constant 1 if it occurs (the derived polynomial also remains sorted);
- identity testing for polynomials derived from the original sorted polynomial by the two previous operations.

For IsEqual we have

1. $x = |A| + |B| + |C| + |D|$. By taking into account the employed representation of monomials (the bitscale is not shortened when a variable is removed), we may also assume that $|Q| = |Q_1| + |Q_2|$.
2. $\forall i, \lambda_i = 1$.
3. $\forall i, h_i(x) = 0$.
4. $g(x) = O(nx)$. Therefore, the total time for lines 1–10 consists of the constant numbers of linear (with respect to the input of IsEqual) operations executed at most n times. Apparently, n is quite a conservative assumption, because at a single recursion step, at least one variable is removed from the input set of variables.
5. We need to estimate $\omega_1, \omega_2, \omega_3, \omega_4$.

   Among all the possible choices of the multipliers mentioned in Sect. 5.1, let us consider those of the form $Q_1 Q_2$. They induce two equations that do not contain one of the input parameters of IsEqual: A, B, C, D result in the absence of the parts of D, C, B, A, respectively, among the parameters of IsEqual in lines 13 and 14. Hence, the largest parameter can be excluded by taking an appropriate Q; lines 13–14 of ModFD-algorithm are to be rewritten with the help of this observation.

   Without loss of generality, we may assume that the largest parameter is D and thus, we can take Q equal to A. In this case, $\omega_1, \omega_2, \omega_3, \omega_4$ represent the relative lengths of the parameters $|A_1| + |B_1| + |C_1| + |D_1|$, $|A_2| + |B_2| + |C_2| + |D_2|$, $|A| + |B|$, $|A| + |C|$ for the recursive calls to IsEqual with respect to $|A| + |B| + |C| + |D|$.

   Since $|A|, |B|, |C| \leq |D|$, we obtain $|A| + |B|$, $|A| + |C|$, $|B| + |C| \leq 2|D|$. Then the lengths $|A| + |B|$ and $|A| + |C|$, respectively, can be estimated in the following way:

$$|A| + |B| = x - |C| - |D| \leq x - 0 - \frac{|A| + |B|}{2},$$

hence, $|A| + |B|, |A| + |C| \leq \frac{2}{3}$.

Let F be a multilinear polynomial over n variables with M monomials such that no variable divides F. A random polynomial consists of monomials randomly chosen from the set of all monomials over n variables. Variables appear in monomials independently. For each variable x from $\mathrm{var}(F)$, we can consider the following quantity $\mu_x = \frac{\partial F}{\partial x}$ (i.e. the part of monomials containing this variable). We want to estimate the probability that among μ_x there exist at least one, which is (approximately) equal to $\frac{M}{2}$. Hence

$$
\begin{aligned}
\mathbb{P}\left[\text{there exists } x \text{ such that } \mu_x \text{ is a median}\right] &= 1 - \mathbb{P}\left[\bigwedge_x \mu_x \text{ is not median}\right] \\
&= 1 - \mathbb{P}\left[\mu_1 \text{ is not median}\right]^n \\
&= 1 - \left(1 - \mathbb{P}\left[\mu_1 \text{ is a median}\right]\right)^n \\
&= 1 - \left(1 - \tfrac{1}{2}\right)^n \\
&= 1 - \tfrac{1}{2^n}
\end{aligned}
$$

Thus, with a high probability one can pick from a large polynomial (in our case, from D) a variable such that $|D_1| \approx |D_2|$.

Let us consider the following multicriteria linear program:

$$
\text{maximize} \left\{
\begin{aligned}
a_1 + b_1 + c_1 + d_1 \\
a_2 + b_2 + c_2 + d_2 \\
a + b \\
a + c
\end{aligned}
\right\}
\text{ subject to }
\left\{
\begin{aligned}
a + b + c + d &= 1 \\
d_1 &= d_2 \\
a \le d, \ b \le d, \ c &\le d \\
a + b &\le \tfrac{2}{3} \\
a + c &\le \tfrac{2}{3} \\
\text{all nonnegative} &
\end{aligned}
\right.
$$

Since the objective functions and constraints are linear and the optimization domain is bounded, we can enumerate all the extreme points of the problem and select those points that give the maximum solution of the characteristic equation of Theorem 1. By taking into account the symmetries between the first and the second objective functions and between the third and fourth ones, we obtain that

$$
\omega_1 = \frac{3}{4}, \quad \omega_2 = \frac{1}{4}, \quad \omega_3 = \frac{1}{2}, \quad \omega_4 = \frac{1}{2}. \qquad (*)
$$

Hence, the characteristic equation is

$$
\left(\frac{3}{4}\right)^p + \left(\frac{1}{4}\right)^p + \left(\frac{1}{2}\right)^p + \left(\frac{1}{2}\right)^p = 1.
$$

Its unique real solution is $p \approx 2.226552$. Finally, the total time for the ModFD-algorithm obtained this way is

$$
T = O(n^2 |F|^{2.226552}).
$$

6 Preliminary Experiments and Discussion

For a computational evaluation of the developed factorization algorithms, we used Maple 17 for Windows run on 3.0 GHz PC with 8 GB RAM. The factorization algorithm implemented in Maple `Factor(poly) mod 2` can process multilinear polynomials over $\mathbb{F}_2$ with hundreds of variables and several thousands of monomials in several hours. But many attempts of factorization of polynomials with 10^3 variables and 10^5 monomials were terminated by the time limit of roughly one week of execution. In general, a disadvantage of all Maple implementations is that they are memory consuming. For example, the algorithm that requires computing products of polynomials fails to work even for rather small examples (about 10^2 variables and 10^3 monomials). Although GCD-algorithm is conceptually simple, it involves computing the greatest common divisor for polynomials over the "poor" finite field $\mathbb{F}_2$. A practical implementation of LINZIP is not that simple. An older version of Maple reports on some inputs that "LINZIP/not implemented yet". We did not observe this issue in Maple 17. It would be important to conduct an extensive comparison of the performance of GCD- and FD-algorithm implemented under similar conditions. The factorization algorithm (FD-based) for sparsely represented multilinear polynomials over $\mathbb{F}_2$ demonstrates reasonable performance. BDD/ZDD can be considered as some kind of the black box representation. We are going to implement factorization based on this representation and to compare these approaches.

A careful study of the solution (*) given at the end of Sect. 5.2 shows that it describes the case when $|A| \approx |D| \approx \frac{x}{2}$ and $|B| \approx |C| \approx 0$. This means that at the next steps the maximal parameter is A: $|A| \approx \frac{x}{2}$, while the remaining parameters are smaller. Thus, one can see that the lengths of the inputs to the recursive calls of IsEqual are reduced at least twice in at most two levels of the recursion. This allows for obtaining a more precise complexity bound, which will be further studied.

Yet another property is quite important for the performance of the algorithm. Evaluating the predicate IsEqual for the variables from the same factor requires significantly less time compared with evaluation for other variables. For polynomials with 50 variables and 100 monomials in the both components, the speed-up achieves 10–15 times. The reason is evident and it again confirms the importance of (Zero) Polynomial Identity Testing, as shown by Shplika and Volkovich. Testing that the polynomial $AD + BC$ is not zero requires less reduction steps in contrast with the case when it does equal zero. The latter requires reduction to the constant polynomials. Therefore, we used the following approach: if the polynomials A, D, B, C are "small" enough then the polynomial $AD + BC$ was checked to be zero directly via multiplication. For the polynomials with the above mentioned properties, this allows to save about 3–5% of the execution time. The first practical conclusion is that in general, the algorithm works faster for non-factorable polynomials than for factorable ones. The second is that we need to investigate new methods to detect variables from the "opposite" component (factor). Below we give an idea of a possible approach.

It is useful to detect cases of irreducibility before launching the factorization procedure. Using simple necessary conditions for irreducibility, as well as testing simple cases of variable classification for variable partition algorithms, can substantially improve performance. Let F be a multilinear polynomial over n variables with M monomials such that no variable divides F. For each variable x, recall that the value μ_x corresponds to the number of monomials containing

x, i.e. the number of monomials in $\frac{\partial F}{\partial x}$. Then a necessary condition for F to be factorable is

$$\forall x \quad \gcd(\mu_x, M) > 1.$$

In addition, we have deduced several properties, which are based on analyzing occurrences of pairs of variables in the given polynomial (for example, if there is no monomial in which two variables occur simultaneously then these variables can not belong to different factors). Of course, the practical usability of these properties depends on how easily they can be tested.

Finally, we note an important generalization of the factorization problem, which calls for efficient implementations of the factorization algorithm. To achieve a deeper optimization of logic circuits we asked in [5,7] how to find a representation of a polynomial in the form $F(X,Y) = G(X)H(Y) + D(X,Y)$, where a "relatively small" defect" $D(X,Y)$ extends or shrinks the pure disjoint factors. Yet another problem is to find a representation of the polynomial in the form

$$F(X,Y) = \sum_k G_k(X)H_k(Y), \quad X \cap Y = \varnothing,$$

i.e., a complete decomposition without any "defect", which (along with the previous one) has quite interesting applications in the knowledge and data mining domain. Clearly, such decompositions (for example, the trivial one, where each monomial is treated separately) always exist, but not all of them are meaningful from the K&DM point of view. For example, one might want to put a restriction on the size of the "factorable part" of the input polynomial (e.g., by requiring the size to be maximal), which opens a perspective into a variety of optimization problems. Formulating additional constraints targeting factorization is an interesting research topic. One immediately finds a variety of the known computationally hard problems in this direction and it is yet to be realized how the computer algebra and theory of algorithms can mutually benefit from each other along this way.

References

1. Akra, M., Bazzi, L.: On the solution of linear recurrence equations. Comput. Optim. Appl. **10**(2), 195–210 (1998). https://doi.org/10.1023/A:1018373005182
2. de Kleine, J., Monagan, M.B., Wittkopf, A.D.: Algorithms for the non-monic case of the sparse modular GCD algorithm. In: Proceedings of 2005 International Symposium on Symbolic and Algebraic Computation (ISSAC 2005), pp. 124–131. ACM, New York (2005). https://doi.org/10.1145/1073884.1073903
3. Emelyanov, P.: AND–decomposition of boolean polynomials with prescribed shared variables. In: Govindarajan, S., Maheshwari, A. (eds.) CALDAM 2016. LNCS, vol. 9602, pp. 164–175. Springer, Cham (2016). https://doi.org/10.1007/978-3-319-29221-2_14
4. Emelyanov, P.: On two kinds of dataset decomposition. In: Shi, Y., et al. (eds.) ICCS 2018. LNCS, vol. 10861, pp. 171–183. Springer, Cham (2018). https://doi.org/10.1007/978-3-319-93701-4_13

5. Emelyanov, P., Ponomaryov, D.: Algorithmic issues of AND-decomposition of Boolean formulas. Program. Comput. Softw. **41**(3), 162–169 (2015). https://doi.org/10.1134/S0361768815030032. Trans. by: Programmirovanie **41**(3), 62–72 (2015)
6. Emelyanov, P., Ponomaryov, D.: Cartesian decomposition in data analysis. In: Proceedings of Siberian Symposium on Data Science and Engineering (SSDSE 2017), Novosibirsk, Russia, pp. 55–60 (2017). https://doi.org/10.1109/SSDSE.2017.8071964
7. Emelyanov, P., Ponomaryov, D.: On tractability of disjoint AND-decomposition of Boolean formulas. In: Voronkov, A., Virbitskaite, I. (eds.) PSI 2014. LNCS, vol. 8974, pp. 92–101. Springer, Heidelberg (2015). https://doi.org/10.1007/978-3-662-46823-4_8
8. Grigoriev, D.: Theory of Complexity of Computations. II. Notes of Scientific Seminars of LOMI (Zapiski Nauchn. Semin. Leningr. Otdel. Matem. Inst. Acad. Sci. USSR). Nauka, Leningrad, vol. 137, pp. 20–79 (1984). (in Russian)
9. Khatri, S.P., Gulati, K. (eds.): Advanced Techniques in Logic Synthesis, Optimizations and Applications. Springer, New York (2011). https://doi.org/10.1007/978-1-4419-7518-8
10. Muller, D.E.: Application of Boolean algebra to switching circuit design and to error detection. IRE Trans. Electron. Comput. **EC-3**, 6–12 (1954)
11. Perkowski, M.A., Grygiel, S.: A survey of literature on function decomposition, Version IV. PSU Electrical Engineering Department Report, Portland State University, Portland, Oregon, USA, November 1995
12. Shparlinski, I.E.: Computational and Algorithmic Problems in Finite Fields. Springer, New York (1992). https://doi.org/10.1007/978-94-011-1806-4
13. Shpilka, A., Volkovich, I.: On the relation between polynomial identity testing and finding variable disjoint factors. In: Abramsky, S., Gavoille, C., Kirchner, C., Meyer auf der Heide, F., Spirakis, P.G. (eds.) ICALP 2010. LNCS, vol. 6198, pp. 408–419. Springer, Heidelberg (2010). https://doi.org/10.1007/978-3-642-14165-2_35
14. von zur Gathen, J., Gerhard, J.: Modern Computer Algebra, 3rd edn. Cambridge University Press, New York (2013). https://doi.org/10.1017/CBO9781139856065
15. Zhegalkin, I.: Arithmetization of symbolic logics. Sbornik Mathematics **35**(1), 311–377 (1928). (in Russian)

Tropical Newton–Puiseux Polynomials

Dima Grigoriev$^{(\boxtimes)}$

CNRS, Mathématiques, Université de Lille, 59655 Villeneuve d'Ascq, France
Dmitry.Grigoryev@math.univ-lille1.fr
http://en.wikipedia.org/wiki/Dima_Grigoriev

Abstract. We introduce tropical Newton–Puiseux polynomials admitting rational exponents. A resolution of a tropical hypersurface is defined by means of a tropical Newton–Puiseux polynomial. A polynomial complexity algorithm for resolubility of a tropical curve is designed. The complexity of resolubility of tropical prevarieties of arbitrary codimensions is studied.

Keywords: Tropical Newton–Puiseux polynomials
Resolution of tropical hypersurfaces

Introduction

Recall (see e. g. [6]) that in the tropical semiring, $\oplus$ denotes min and $\otimes$ denotes the (classical) addition $+$. As examples of tropical semirings one can take $\mathbb{Z}, \mathbb{R}$. A *tropical* (respectively, *tropical Laurent) monomial* has the form

$$a \otimes x^{\otimes I} := a \otimes x_1^{\otimes i_1} \otimes \cdots \otimes x_n^{\otimes i_n}$$

where $a \in \mathbb{R}$ and $0 \leq i_1, \ldots, i_n \in \mathbb{Z}$ (respectively, $i_1, \ldots, i_n \in \mathbb{Z}$). Thus, classically $a \otimes x^{\otimes I}$ equals a linear function $a + \sum_{1 \leq j \leq n} i_j \cdot x_j$. A *tropical polynomial f* has the form $\bigoplus_I a_I \otimes x^{\otimes I}$, being classically a convex piece-wise linear function.

A vector $\mathbf{x} = (x_1, \ldots, x_n) \in \mathbb{R}^n$ is a *tropical root* of f if the minimum of $a_I \otimes \mathbf{x}^{\otimes I}$ is attained at least for two different tropical monomials of f. The set of all tropical roots of f constitutes a *tropical hypersurface* $T(f) \subset \mathbb{R}^n$ being a finite union of polyhedra of dimensions $n - 1$.

We extend the concept of a tropical polynomial by allowing the exponents $i_1, \ldots, i_n$ to be rational calling it a *tropical Newton–Puiseux polynomial*. Assume that

$$f = \bigoplus_{0 \leq i \leq d} f_i \otimes y^{\otimes i} \tag{1}$$

for some tropical polynomials $f_0, \ldots, f_d$ in the variables $x_1, \ldots, x_n$. We call a Newton–Puiseux polynomial y a *resolution* of f (or of the tropical hypersurface $T(f)$) if for any point $\mathbf{x} \in \mathbb{R}^n$ the point $(\mathbf{x}, y(\mathbf{x})) \in \mathbb{R}^{n+1}$ provides a tropical root of f (one can find the formal definitions below in Sect. 1).

© Springer Nature Switzerland AG 2018
V. P. Gerdt et al. (Eds.): CASC 2018, LNCS 11077, pp. 177–186, 2018.
https://doi.org/10.1007/978-3-319-99639-4_12

This resembles Newton–Puiseux series from algebraic geometry with the difference that we consider finite supports since in the tropical semiring, one takes min. One can view tropical Newton–Puiseux polynomials as a tropical analog of algebraic functions.

In Sect. 1, we show that the set of all the resolutions of a tropical hypersurface is finite and closed under taking min. Thus, there exists a minimal resolution, and in case of a monic tropical polynomial

$$f = y^{\otimes d} \oplus \bigoplus_{0 \leq i < d} f_i \otimes y^{\otimes i}$$

we provide a simple formula for the minimal resolution. In addition, a geometric description of resolutions is given. Also we show that the resolubility of a tropical hypersurface belongs to the complexity class NP.

In Sect. 2, a polynomial (bit-size) complexity algorithm is exhibited for resolving degree 1 tropical polynomials of the form $f_1 \otimes y \oplus f_0$, which is equivalent to the divisibility of f_0 by f_1.

In Sect. 3, we design a polynomial (bit-size) complexity algorithm for testing resolubility of a tropical curve in a real space of a fixed dimension, moreover, the algorithm provides a succinct description of the set of all the resolutions.

In Sect. 4, we study the problem of resolubility of a system of tropical polynomials in a single variable x and in several indeterminates $y_1, \ldots, y_s$ and establish its NP-hardness.

In Sect. 5, we study tropical Newton–Puiseux rational functions, being tropical quotients (or in other words, the classical subtraction) of pairs of tropical Newton-Puiseux polynomials. An algorithm is suggested which tests resolubility of a tropical curve by means of tropical Newton–Puiseux rational functions. The complexity of the algorithm is polynomial for a fixed dimension of the ambient space.

1   Resolution of a Tropical Hypersurface

Let an algebraic (classical) equation

$$F := \sum_{0 \leq i \leq d} F_i \cdot Y^i = 0 \tag{2}$$

where the coefficients $F_i \in K[X_1, \ldots, X_n]$ for the field $K = \mathbb{C}((t^{1/\infty}))$ of Newton–Puiseux series, have a Laurent polynomial solution

$$Y = \sum_I A_I \cdot X^I \tag{3}$$

with a finite sum over multiindices $I \in \mathbb{Z}^n$ and the coefficients $A_I \in K$.

Denote the *tropicalization*

$$Trop(Y) := \bigoplus_I Trop(A_I) \otimes X^{\otimes I} \tag{4}$$

where for a Newton–Puiseux series $A_I = \sum_{0 \leq j < \infty} b_j \cdot t^{s_j/q}$ with $b_j \in \mathbb{C}$, $b_0 \neq 0$ and increasing integers $s_0 < s_1 < \dots$ its tropicalization $Trop(A_I) := s_0/q \in \mathbb{Q}$.

Remark 1. $Trop(Y)$ is a solution of the tropical equation

$$\bigoplus_{0 \leq i \leq d} Trop(F_i) \otimes (Trop(Y))^{\otimes i} \tag{5}$$

This means that for any point $\mathbf{x} = (x_1, \dots, x_n) \in \mathbb{R}^n$, the minimal value of $Trop(F_i) \otimes (Trop(Y))^{\otimes i}$ at $\mathbf{x}$ for $0 \leq i \leq d$ is attained at least for two different $0 \leq i_1 < i_2 \leq d$.

Remark 2. Observe that the validity of (5) does not change if one multiplies all the rational coefficients in $Trop(F_i)$, $0 \leq i \leq d$ by their common denominator m and simultaneously all $Trop(A_I)$ (see (4)) by m to make all the coefficients in $Trop(F_i)$, $0 \leq i \leq d$ integers.

Remark 1 motivates the following definition.

Definition 1. *A tropical hypersurface $T(f) \subset \mathbb{R}^{n+1}$ defined by a tropical polynomial (1) where f_i are tropical polynomials in the variables $x_1, \dots, x_n$ with integer coefficients (cf. Remark 2) has a resolution being a tropical Newton–Puiseux polynomial*

$$y = \bigoplus_I a_I \otimes x^{\otimes I} \tag{6}$$

for a finite sum over multiindices $I \in \mathbb{Q}^n$ and $a_I \in \mathbb{Q}$, if for any point $\mathbf{x} = (x_1, \dots, x_n) \in \mathbb{R}^n$, the minimal value among $f_i \otimes y^{\otimes i}$, $0 \leq i \leq d$ (treated as piecewise linear functions) at $\mathbf{x}$ is attained at least for two different $0 \leq i_1 < i_2 \leq d$.

Denote by N the common denominator of all the rational coordinates of multiindices I from (6). Then $y^{\otimes N}$ is a tropical (Laurent) polynomial which is equal classically to $N \cdot \min_I \{a_I + i_1 x_1 + \dots + i_n x_n\}$.

Proposition 1. *Let y be a resolution of f (see (1), (6)), then $(\mathbf{x}, y(\mathbf{x})) \in T(f)$.*

Example 1. The tropical polynomial $f = y \oplus x \oplus 0$ has a resolution $y = x \oplus 0$. Its graph $\{(\mathbf{x}, y(\mathbf{x})) : \mathbf{x} \in \mathbb{R}\} \subset T(f) \subset \mathbb{R}^2$ consists of two half-lines, while the tropical curve $T(f)$ consists of three half-lines.

Proposition 2. *Let y (see (6)) and $\bigoplus_I b_I \otimes x^{\otimes I}$ be resolutions of (1). Then $\bigoplus_I (a_I \oplus b_I) \otimes x^{\otimes I}$ is also a resolution of (1).*

The proof follows from an observation that for any point $\mathbf{x} \in \mathbb{R}^n$, the minimum on the tropical monomials after opening the parenthesis in a power $y^{\otimes i}$ (see (6)) is attained on the powers of the kind $(a_I \otimes \mathbf{x}^{\otimes I})^{\otimes i}$. □

Below in Remark 4, we show that there is at most a finite number of resolutions of (1). Hence according to Proposition 2, there exists a minimal resolution.

Proposition 3. *If*

$$f = y^{\otimes d} \oplus \bigoplus_{0 \le i < d} f_i \otimes y^{\otimes i}$$

(see (1)) is monic then

$$y = \bigoplus_{1 \le i \le d} f_{d-i}^{\otimes(1/i)}$$

is the minimal resolution.

Proof. For any point $\mathbf{x} \in \mathbb{R}^n$, the minimal $y_0 \in \mathbb{R}$ such that $(\mathbf{x}, y_0)$ belongs to the tropical hypersurface $T(f) \subset \mathbb{R}^{n+1}$ satisfies a (classical) equation $d \cdot y_0 = f_i(\mathbf{x}) + i \cdot y_0$ for suitable $0 \le i < d$ (due to analyzing the Newton polygon). □

Note also that if $f_{d-i} = \bigoplus_J c_J \otimes x^{\otimes J}$ then $f_{d-i}^{\otimes(1/i)} = \bigoplus_J (c_J/i) \otimes x^{\otimes(J/i)}$.

Remark 3. When f is not monic, a resolution does not necessarily exist as in the example $f = (x \oplus 0) \otimes y \oplus 0$. On the other hand, one can write a similar formula

$$y = \bigoplus_{1 \le i \le d} (f_{d-i} \oslash f_d)^{\otimes(1/i)}$$

where $\oslash$ stands for the tropical division, i.e., the classical subtraction. In this case, y is not necessarily a convex function, while being piecewise linear (cf. Sect. 5).

Now we proceed to a geometric description of resolutions. Let (6) be a resolution of (1). Assume that for some I, the (convex) polyhedron $M_I \subset \mathbb{R}^n$ of points at which the (tropical) monomials $\{a_J \otimes x^{\otimes J}\}_J$ of y attain the minimum for $a_I \otimes x^{\otimes I}$, has the full dimension n. Observe that if M_I has a dimension less than n one can discard the monomial $a_I \otimes x^{\otimes I}$ from y.

Assume that for some $0 \le i_1 < i_2 \le d$ and a pair of monomials $c_{i_1,I_1} \otimes x^{\otimes I_1}$, $c_{i_2,I_2} \otimes x^{\otimes I_2}$ from the polynomials f_{i_1}, f_{i_2}, respectively, it holds

$$I_1 + i_1 \cdot I = I_2 + i_2 \cdot I; \quad c_{i_1,I_1} + i_1 \cdot a_I = c_{i_2,I_2} + i_2 \cdot a_I, \tag{7}$$

in other words, the monomials

$$(c_{i_1,I_1} \otimes x^{\otimes I_1}) \otimes (a_I \otimes x^{\otimes I})^{\otimes i_1} = (c_{i_2,I_2} \otimes x^{\otimes I_2}) \otimes (a_I \otimes x^{\otimes I})^{\otimes i_2}$$

coincide. Consider the convex polyhedron $M_{I,i_1,I_1,i_2,I_2} \subset M_I$ of the points from M_I at which the minimum of the monomials $(c_{i,I} \otimes x^{\otimes I}) \otimes (a_I \otimes x^{\otimes I})^{\otimes i}$ for the monomials $c_{i,I} \otimes x^{\otimes I}$ from f_i, $0 \le i \le d$ is attained for $(c_{i_1,I_1} \otimes x^{\otimes I_1}) \otimes (a_I \otimes x^{\otimes I})^{\otimes i_1}$. We get the following lemma.

Lemma 1. *Let (6) be a resolution of (1) and the polyhedron $M_I \subset \mathbb{R}^n$ have the full dimension n. Then the polyhedra M_{I,i_1,I_1,i_2,I_2} having the full dimension n constitute a partition of M_I, i. e., every two elements of the partition either coincide or intersect by a set (face) of dimension less than n.*

It would be interesting to clarify, how many resolutions a tropical hypersurface might have?

Let the tropical degrees $trdeg(f_i) \leq D$, $0 \leq i \leq d$.

Remark 4. The problem of resolving a tropical polynomial (1) belongs to the complexity class NP. This follows from the observation that each coefficient a_I satisfies (7) (or equals infinity), and therefore, there are at most $d^2 \cdot \binom{D+n}{n}$ possibilities for a_I, taking into account that $\binom{D+n}{n}$ bounds the number of monomials in each f_i.

Note that when f_i, $0 \leq i \leq d$ are in sparse encoding, in the latter bound one can replace $\binom{D+n}{n}$ by the number of monomials in f_i, $0 \leq i \leq d$. Thus, the problem of resolubility of (1) belongs to NP for both dense and sparse encodings of (1).

It would be interesting to say more about the complexity of resolubility of (1).

Remark 5. One can extend the results of this section to an input tropical Newton–Puiseux polynomials in place of (1).

2 Polynomial Complexity Testing Divisibility of Tropical Polynomials

If (1) has degree 1, i.e., $f = f_1 \otimes y \oplus f_0$, then according to (7), a resolution (6) is equivalent to the divisibility $f_1 \otimes y = f_0$ with y being a tropical Laurent polynomial. We agree that two tropical (Laurent) polynomials are equal if they are equal as (convex piece-wise linear) functions.

We describe an algorithm for testing divisibility within polynomial complexity. First the algorithm deletes from f_0 all the monomials of the form $b \otimes x_1^{b_1} \otimes \cdots \otimes x_n^{b_n}$ which do not change f_0 as a function. Geometrically, it means that the hyperplane defined as the graph

$$\{(x_1, \ldots, x_n, \sum_{1 \leq j \leq n} b_j \cdot x_j + b) : (x_1, \ldots, x_n) \in \mathbb{R}^n\}$$

of this monomial in $\mathbb{R}^{n+1}$ is higher (with respect to the last coordinate) than the polyhedron P defined by the other monomials of f_0 (observe that P is the graph of f_0 as a function). The latter is a problem of linear programming. Thus, one can suppose f_0 to be reduced, i.e., do not contain unnecessary monomials. Also we suppose that f_1 is reduced.

For every candidate $I = (i_1, \ldots, i_n) \in \mathbb{Z}^n$, $\sum_{1 \leq j \leq n} |i_j| \leq D$ (see (7)) to be in the support of a resolution y the algorithm calculates (again involving linear programming) the minimal a_I such that for each monomial $c \otimes x^{\otimes C}$ of f_1, the hyperplane in $\mathbb{R}^{n+1}$ defined by the monomial $(c \otimes x^{\otimes C}) \otimes (a_I \otimes x^{\otimes I})$ is (non-strictly) higher than P.

Then $y = \bigoplus_I a_I \otimes x^{\otimes I}$ is a resolution of $f_1 \otimes y \oplus f_0$ iff for each monomial $b \otimes x^{\otimes B}$ of f_0, there exists I and a monomial $c \otimes x^{\otimes C}$ of f_1 such that $(a_I \otimes x^{\otimes I}) \otimes$

$(c \otimes x^{\otimes C}) = b \otimes x^{\otimes B}$. Reducing further y as described above, we conclude that there is a unique reduced resolution y (provided that it does exist).

Summarizing, we obtain the following proposition.

Proposition 4. *One can test resolubility of degree 1 tropical polynomial $f_1 \otimes y \oplus f_0$ (or equivalently, the divisibility $f_1 \otimes y = f_0$) within polynomial complexity. In case of the divisibility, the algorithm yields the unique reduced resolution y.*

3   Polynomial Complexity Algorithm for Resolving Tropical Curves

Let a system of tropical polynomials

$$f_i,\ 1 \le i \le k \tag{8}$$

in n variables $x, y_1, \ldots, y_{n-1}$ with integer coefficients determine a tropical prevariety $T := T(f_1, \ldots, f_k) \subset \mathbb{R}^n$. Let the tropical degrees $trdeg(f_i) \le d$, $1 \le i \le k$ and the bit-sizes of the coefficients of f_i, $1 \le i \le k$ do not exceed L.

First, the algorithm constructs T as a union of polyhedra (see e.g. [6]). Each of these polyhedra (including faces of all the dimensions) is defined by specifying the monomials of f_i, $1 \le i \le k$ (treated as linear functions) on which the minima are attained (cf. e.g. [4]). The algorithm can find the partition of $\mathbb{R}^n$ into polyhedra defined by given feasible tuples of signs (i.e., either the positive, either the negative or zero) of all the differences of the monomials of f_i, $1 \le i \le k$ (in other words, by all the feasible orderings of the monomials of f_i, $1 \le i \le k$). Namely, the algorithm finds the partition by recursion on the number of the differences. If for a current subset of the differences, the partition of $\mathbb{R}^n$ w.r.t. this subset is already constructed, the algorithm picks up the next difference and for each element (being a polyhedron) of the current partition verifies which signs of the picked up difference are feasible on this polyhedron (with the help of linear programming). Thereupon, the algorithm discards the unfeasible tuples of signs, which completes the recursive step.

The number of the elements of a current partition at every step of the recursion is bounded by

$$n^2 \cdot 2^n \cdot \binom{k \cdot \binom{d+n}{n}^2}{n} < k^n \cdot d^{2 \cdot n^2}$$

due to the Buck's formula on the number of faces in an arrangement of hyperplanes [3]. Hence the complexity of the recursion is bounded by a polynomial in L, k^n, d^{n^2}, taking into account that the algorithm invokes linear programming $(k^n \cdot d^{n^2})^{O(1)}$ times.

Since the tropical prevariety T is a union of appropriate subset of the elements of the constructed partition of $\mathbb{R}^n$, we get the following proposition.

Proposition 5. *There is an algorithm which constructs the tropical prevariety $T(f_1, \ldots, f_k) \subset \mathbb{R}^n$ determined by (8) within the complexity polynomial in L, k^n, d^{n^2}.*

Now we assume that $\dim T = 1$, thus T is a tropical curve. We design an algorithm which verifies the resolubility of T, i.e., whether there exist tropical Newton–Puiseux polynomials $y_1(x), \ldots, y_{n-1}(x)$ assuring a resolution of (8). The latter is equivalent to the condition that every piecewise linear function y_j, $1 \leq j \leq n - 1$ is convex.

The algorithm produces a directed graph G whose vertices are the edges of T (including the unbounded ones) not lying in a hyperplane of the form $x = c$. Two edges $e^{(-)}$, $e^{(+)}$ of T (being vertices of G) with the same endpoint of the kind

$$e^{(-)} = ((x^{(-)}, y_1^{(-)}, \ldots, y_{n-1}^{(-)}), (x, y_1, \ldots, y_{n-1})),$$

$$e^{(+)} = ((x, y_1, \ldots, y_{n-1}), (x^{(+)}, y_1^{(+)}, \ldots, y_{n-1}^{(+)}))$$

are linked by an edge directed from $e^{(-)}$ to $e^{(+)}$ in G if

$$x^{(-)} < x < x^{(+)}; \quad \frac{y_j - y_j^{(-)}}{x - x^{(-)}} \geq \frac{y_j - y_j^{(+)}}{x - x^{(+)}}, 1 \leq j \leq n - 1. \tag{9}$$

when $e^{(-)}$ (respectively, $e^{(+)}$) is unbounded with an endpoint $(x, y_1, \ldots, y_{n-1})$ (so, is a half-line), which we call unbounded from the left, we take an arbitrary point of $e^{(-)}$ with $x^{(-)} < x$ (respectively, if $e^{(+)}$ is a half-line, we take a point of $e^{(+)}$ with $x^{(+)} > x$, and we call $e^{(+)}$ unbounded from the right). When an edge of T has no endpoints, so is a line, it provides a resolution of T.

After that the algorithm produces a subset S of the vertices of G. It starts with including into S all the edges of T (so, the vertices of G) unbounded from the left (denote this set by S_0). Thereupon, the algorithm includes into S all the vertices of G reachable from S_0. If a vertex of G corresponding to an edge of T unbounded from the right, belongs to S, a path in G leading to such a vertex from S_0 provides a resolution of T (i.e., each piece-wise linear function $y_j(x)$, $1 \leq j \leq n - 1$ corresponding to the path is convex due to (9)). Moreover, the paths in G from S_0 to the vertices corresponding to the edges of T unbounded from the right, are in a bijective correspondence with the resolutions of T.

Summarizing and taking into account Proposition 5, we obtain the following theorem.

Theorem 1. *There is an algorithm which tests resolubility of a tropical curve $T \subset \mathbb{R}^n$ determined by (8), and in case of the resolubility yields a resolution. The complexity of the algorithm is polynomial in L, k^n, d^{n^2}. In particular, the complexity is polynomial for a fixed ambient dimension n.*

Remark 6. Let a system of tropical polynomials of the form (8) depend on the variables $x_1, \ldots, x_m, y_1, \ldots, y_n$ and the tropical prevariety $T \subset \mathbb{R}^{m+n}$ have dimension m. Then one can try different subsets of all m-dimensional faces of T as candidates to constitute a graph of a resolution

$$(x_1, \ldots, x_m) \to (x_1, \ldots, x_m, y_1(x_1, \ldots, x_m), \ldots, y_n(x_1, \ldots, x_m)) \in T$$

of T similar to Remark 4. The latter, in fact, means that firstly, the projections of the chosen m-dimensional faces on $\mathbb{R}^m$ with the coordinates $x_1, \ldots, x_m$ form a partition of $\mathbb{R}^m$ and secondly, that each piece-wise linear function $y_j(x_1, \ldots, x_m)$, $1 \leq j \leq n-1$ is convex.

4   Resolution of Systems of Tropical Polynomials with Several Indeterminates

In this section, we consider the systems of tropical polynomials (instead of a single polynomial (1)) in one variable x and several indeterminates $y_1, \ldots, y_s$. Thus, in a resolution (cf. (6)), each y_i is a tropical Newton–Puiseux polynomial. We show the following proposition.

Proposition 6. *The problem of resolubility of a system of tropical polynomials in a single variable and in several indeterminates is NP-hard.*

Proof. We reduce 3-SAT to the problem under consideration, so we construct a system R of tropical polynomials. For an instance of 3-SAT problem in n variables $u_1, \ldots, u_n$, we introduce indeterminates $y_1, \ldots, y_n, z_1, \ldots, z_n$ and add to R tropical polynomials

$$y_i \otimes z_i \oplus x, \ 1 \leq i \leq n \tag{10}$$

Formula (10) means that the resolutions of y_i and of z_i are both monomials in x. Informally, $0 = x^{\otimes 0}$ encodes the truth and $x = x^{\otimes 1}$ encodes the falsity, y_i corresponds to u_i and z_i corresponds to $\neg u_i$.

For every jth 3-clause of the 3-SAT formula, say, $u_m \vee \neg u_k \vee u_l$, we add to R the following tropical (linear) polynomials

$$y_m \oplus z_k \oplus y_l \oplus v_j; \tag{11}$$

$$v_j \oplus x^{\otimes 1} \oplus w_j; \tag{12}$$

$$w_j \oplus x^{\otimes 1} \oplus 0 \tag{13}$$

with indeterminates v_j, w_j. Note that (13) ensures that in a resolution, the reduced $w_j = x^{\otimes 1} \oplus 0$, then (12) ensures that the reduced v_j contains the constant monomial 0 (and possibly, monomials of the form $c \otimes x^{\otimes b}$ with $0 < b \leq 1$, $c \geq 0$). Finally, (11) ensures that one of the resolutions of y_m, z_k, y_l equals 0.

Thus, existence of a resolution of the system R for all j implies the solvability of the initial 3-SAT formula.

The converse is obvious: for a Boolean vector $(u_1, \ldots, u_n)$ providing a solution of the initial 3-SAT formula put $y_i = 0$, $z_i = x^{\otimes 1}$ when u_i is true and $y_i = x^{\otimes 1}$, $z_i = 0$ for u_i being false. Thereupon put $v_j = y_m \oplus z_k \oplus y_l$. □

We mention that the problem of solvability of a system of tropical polynomials is NP-complete [8].

It would be interesting to understand more about the complexity of the problem under consideration in this section.

5 Tropical Newton–Puiseux Rational Functions

Any tropical Newton–Puiseux rational function $f_1 \oslash f_2$, where f_1, f_2 are tropical Newton–Puiseux polynomials, is a piece-wise linear (continuous) function (cf. Remark 3). The converse is also true (see e.g. [1], [5]): any piece-wise linear continuous function is a difference of two piece-wise linear convex functions. In [7], an algorithm is suggested which represents a piece-wise linear function as a difference of piece-wise linear convex functions with the complexity bound being exponential. In case of one-variable functions, a polynomial complexity algorithm for this problem is exhibited in [2].

Let a tropical curve $T \subset \mathbb{R}^n$ be determined by system (8). As in Sect. 3, the algorithm finds T. Thereupon, similar to Sect. 3 it constructs a graph which comprises all the paths consisting of the edges of T of the form

$$\{(x_l, y_1^{(l)}, \dots, y_{n-1}^{(l)}), (x_{l+1}, y_1^{(l+1)}, \dots, y_{n-1}^{(l+1)}) : 0 \le l \le s\}$$

where $x_0 := -\infty < x_1 < \cdots < x_s < x_{s+1} := \infty$, thus, this path contains $s + 1$ edges. The difference with Sect. 3 is that now we do not impose a requirement on convexity.

The algorithm can pick up any such path (provided that it does exist), then this path yields $n - 1$ piece-wise linear functions $y_i(x)$, $1 \le i < n$. Making use of [2] the algorithm represents $y_i(x) = g_i(x) - h_i(x)$ with piece-wise linear convex functions g_i, h_i. This produces a tropical Newton–Puiseux rational function resolution of T.

Summarizing and invoking the complexity bounds from Sect. 3, we get the following proposition.

Proposition 7. *There is an algorithm which tests resolubility of a tropical curve determined by (8) by means of tropical Newton–Puiseux rational functions within the complexity polynomial in L, k^n, d^{n^2}. The algorithm yields a resolution, provided that it does exist. Therefore, the complexity is polynomial for a fixed dimension n of the ambient space.*

It would be interesting to estimate the complexity of resolubility of tropical prevarieties or arbitrary dimensions by means of tropical Newton–Puiseux rational functions.

Acknowledgments. The author is grateful to the grant RSF 16-11-10075 and to MCCME for inspiring atmosphere.

References

1. Bittner, L.: Some representation theorems for functions and sets and their application to nonlinear programming. Numer. Math. **16**, 32–51 (1970)
2. Dobkin, D., Guibas, L., Hershberger, J., Snoeyink, J.: An efficient algorithm for finding the CSG representation of a simple polygon. Algorithmica **10**, 1–23 (1993)

3. Fukuda, K., Saito, S., Tamura, A.: Combinatorial face enumeration in arrangements and oriented matroids. Discret. Appl. Math. **31**, 141–149 (1991)
4. Grigoriev, D., Podolskii, V.: Complexity of tropical and min-plus linear prevarieties. Comput. Complex. **24**, 31–64 (2015)
5. Kripfganz, A., Schulze, R.: Piecewise affine functions as a difference of two convex functions. Optimization **18**, 23–29 (1987)
6. Maclagan, D., Sturmfels, B.: Introduction to Tropical Geometry. Graduate Studies in Mathematics, vol. 161. AMS, Providence (2015)
7. Ovchinnikov, S.: Max-min representation of piecewise linear functions. Beitr. Algebra Geom. **43**, 297–302 (2002)
8. Theobald, T.: On the frontiers of polynomial computations in tropical geometry. J. Symb. Comput. **41**, 1360–1375 (2006)

Orthogonal Tropical Linear Prevarieties

Dima Grigoriev[1] and Nicolai Vorobjov[2]([✉])

[1] CNRS, Mathématiques, Université de Lille, 59655 Villeneuve d'Ascq, France
Dmitry.Grigoriev@math.univ-lille1.fr
http://en.wikipedia.org/wiki/Dima_Grigoriev
[2] Department of Computer Science, University of Bath, Bath BA2 7AY, England, UK
nnv@cs.bath.ac.uk

Abstract. We study the operation $A^\perp$ of tropical orthogonalization, applied to a subset A of a vector space $(\mathbb{R} \cup \{\infty\})^n$, and iterations of this operation. Main results include a criterion and an algorithm, deciding whether a tropical linear *prevariety* is a tropical linear *variety* formulated in terms of a duality between $A^\perp$ and $A^{\perp\perp}$. We give an example of a countable family of tropical hyperplanes such that their intersection is not a tropical prevariety.

Keywords: Tropical linear prevarieties · Tropical linear varieties · Orthogonalization

Introduction

We study some aspects of tropical linear prevarieties and varieties. General concepts of tropical algebra can be found in [17,18]. Specific questions of tropical linear algebra were considered in [5,6,19,20].

We introduce the operation $A^\perp$ of tropical orthogonalization applied to a subset A of a vector space $(\mathbb{R} \cup \{\infty\})^n$. The special interest to us presents an interplay between the tropical linear prevariety $A^\perp$ and its orthogonalization $A^{\perp\perp}$. Our main results include a criterion and a deciding algorithm for a tropical linear prevariety to be a tropical linear *variety* formulated in terms of a duality between $A^\perp$ and $A^{\perp\perp}$. In this note, we present our results without proofs which can be found in [13].

In Sect. 1 we list basic definitions, including the concept of a *tropical hull* of a subset in $(\mathbb{R} \cup \{\infty\})^n$. We recall a fundamental theorem proved in [2,8], stating that any tropical linear prevariety is the tropical hull of a finite set of vectors. We give an example showing that the restriction of a tropical linear prevariety to $\mathbb{R}^n$ may not be representable in this way. We describe a simple algorithm, with polynomial complexity, testing the membership of a vector in a tropical hull.

In Sect. 2, we study some properties of double orthogonalization. In particular, we state that $A^{\perp\perp}$ is the minimal tropical linear prevariety containing a finite set A and that dimensions of tropical hulls of A and of $A^{\perp\perp}$ coincide.

© Springer Nature Switzerland AG 2018
V. P. Gerdt et al. (Eds.): CASC 2018, LNCS 11077, pp. 187–196, 2018.
https://doi.org/10.1007/978-3-319-99639-4_13

In Sect. 3, we present a theorem stating that, given two mutually complementary and orthogonal linear subspaces P and Q of the vector space $(\mathbb{C}((t^{1/\infty})))^n$ over Puiseux series, there exists a finite $A \subset (\mathbb{R} \cup \{\infty\})^n$ such that tropicalizations of P and Q coincide with $A^{\perp}$ and $A^{\perp\perp}$, respectively.

Section 4 contains a criterion and a deciding algorithm for a tropical linear prevariety to be a tropical linear variety. The algorithm has a doubly exponential complexity. We also give a brief description of algorithms which for a given tropical linear variety $A^{\perp}$ produces a linear subspace P whose tropicalization coincides with $A^{\perp}$.

Finally, in Sect. 5, we give an example of a countable family of tropical hyperplanes in $(\mathbb{R} \cup \{\infty\})^6$ such that their intersection is not a tropical prevariety. This strengthens examples in [7] (example of T. Theobald) and [12] about countable intersections of nonlinear tropical hypersurfaces.

1   Tropical Hull

We use the notation $\mathbb{R}_{\infty}$ for $\mathbb{R} \cup \{\infty\}$. We assume that for all $a \in \mathbb{R}$ the rules $a < \infty$, $a + \infty = \infty$, $\infty + \infty = \infty$, and, for positive a, $a \cdot \infty = \infty$ hold. The element ∞ is a "tropical zero", being the neutral element with respect to taking minimum.

Definition 1. *For a given* $(a_1, \ldots, a_n) \in \mathbb{R}_{\infty}^n$, *a* tropical hyperplane *in* $\mathbb{R}_{\infty}^n$ *is the set of all points* $(x_1, \ldots, x_n) \in \mathbb{R}_{\infty}^n$ *at which a set* $\{x_1 + a_1, \ldots, x_n + a_n\}$ *has at least two minimal elements. A* tropical linear prevariety *in* $\mathbb{R}_{\infty}^n$ *is the intersection of a finite number of tropical hyperplanes.*

Remark 1. The point $(\infty, \ldots, \infty)$ belongs to every tropical linear prevariety. A tropical hyperplane according to Definition 1 corresponds to the notion of a codimension one linear subspace in classical linear algebra. It can be identified with a special case, when $a_{n+1} = \infty$, of a more general notion of a tropical hyperplane defined as a set of all points $(x_1, \ldots, x_n) \in \mathbb{R}_{\infty}^n$ at which a set $\{x_1 + a_1, \ldots, x_n + a_n, a_{n+1}\}$, where $a_i \in \mathbb{R}_{\infty}$, $1 \leq i \leq n+1$, has at least two minimal elements.

Definition 2. *Vectors* $\mathbf{v} = (v_1, \ldots, v_n)$, $\mathbf{a} = (a_1, \ldots, a_n) \in \mathbb{R}_{\infty}^n$ *are called* tropically orthogonal *if among numbers* $v_i + a_i$, $1 \leq i \leq n$ *there are at least two minimal. Note that* $(\infty, \ldots, \infty)$ *is tropically orthogonal to every vector* $\mathbf{a} \in \mathbb{R}_{\infty}^n$. *For a set of vectors* $A = \{\mathbf{a}_1, \ldots, \mathbf{a}_k\} \subset \mathbb{R}_{\infty}^n$ *denote by* $A^{\perp}$ *the set of all vectors in* $\mathbb{R}_{\infty}^n$ *tropically orthogonal to each* $\mathbf{a}_i$, $1 \leq i \leq k$.

It is clear that $\{\mathbf{a}\}^{\perp}$ is a tropical hyperplane for a vector $\mathbf{a} \in \mathbb{R}_{\infty}^n$, while $A^{\perp}$ is a tropical linear prevariety when $A \subset \mathbb{R}_{\infty}^n$ is finite. Conversely, every tropical linear prevariety in $\mathbb{R}_{\infty}^n$ coincides with $A^{\perp}$ for a suitable finite set of vectors $A \subset \mathbb{R}_{\infty}^n$.

Definition 3. *For a finite set of vectors $A = \{\mathbf{a}_1, \ldots, \mathbf{a}_k\} \subset \mathbb{R}_\infty^n$ define its tropical hull* Trophull(A) *as the set of all vectors in $\mathbb{R}_\infty^n$ of the kind*

$$\min_{1 \le i \le k} \{t_i \mathbf{1}_n + \mathbf{a}_i\},$$

where $t_1, \ldots, t_k$ are arbitrary elements in $\mathbb{R}_\infty$, $\min_{1 \le i \le k}$ denotes the component-wise minimum of a set of vectors, and $\mathbf{1}_n = (1, \ldots, 1)$ is the unit vector in $\mathbb{R}^n$. For an arbitrary subset $X \subset \mathbb{R}_\infty^n$ define Trophull(X) *as the union of sets* Trophull(A) *over all finite subsets $A \subset X$.*

Note that Trophull(A) always contains the point $(\infty, \ldots, \infty) \in \mathbb{R}_\infty^n$, because all t_i, $1 \le i \le k$ can be chosen to be ∞.

Lemma 1 ([5,6]). *If B is finite and $B \subset A^\perp$, then* Trophull$(B) \subset A^\perp$.

Definition 4. *For every partition $\{i_1, \ldots, i_p\} \cup \{i_{p+1}, \ldots, i_n\}$ of $\{1, \ldots, n\}$, a chart is an open convex polyhedron*

$$C_{i_1, \ldots, i_p} := \{x_{i_1} = \cdots = x_{i_p} = \infty\} \cap \{x_{i_{p+1}} < \infty\} \cap \cdots \cap \{x_{i_n} < \infty\} \subset \mathbb{R}_\infty^n.$$

Clearly, $\mathbb{R}_\infty^n$ is the union of all 2^n pairwise disjoint charts.

One can extend the standard concepts of a convex polyhedron and a finite polyhedral complex to the case of the subsets of the space $\mathbb{R}_\infty^n$ (see [8]). Restriction of a convex polyhedron $P \subset \mathbb{R}_\infty^n$ to a chart $C_{i_1, \ldots, i_p}$ coincides with a usual convex polyhedron in $\mathbb{R}^{n-p}$ translated by a vector in $\{0, \infty\}^{n-p}$ with ∞ in positions $i_1, \ldots, i_p$. Hence, P is a finite union of translated usual convex polyhedra, and we define the dimension $\dim(P)$ as the maximum of the dimensions of restrictions of P to all charts. The dimension of a finite polyhedral complex is defined as the maximum of dimensions of its convex polyhedra.

The following theorem directly follows from [8, Theorem 1] (part (2) of the theorem was proved earlier in [2, Proposition 2]).

Let $A = \{\mathbf{a}_1, \ldots, \mathbf{a}_k\} \subset \mathbb{R}_\infty^n$ be a set of vectors.

Theorem 1 ([2,8]).

1. *The set* Trophull(A) *is a union of all convex polyhedra of a polyhedral complex in $\mathbb{R}_\infty^n$.*
2. *For any tropical linear prevariety $A^\perp \subset \mathbb{R}_\infty^n$, there exists a finite set of vectors $\{\mathbf{b}_1, \ldots, \mathbf{b}_N\} \subset \mathbb{R}_\infty^n$ such that $A^\perp =$ Trophull$(\{\mathbf{b}_1, \ldots, \mathbf{b}_N\})$.*

Corollary 1. *Any tropical linear prevariety $A^\perp \subset \mathbb{R}_\infty^n$ is a union of all convex polyhedra of a polyhedral complex in $\mathbb{R}_\infty^n$.*

Remark 2. There is an algorithm, with polynomial complexity, which for a given set $A = \{\mathbf{a}_1, \ldots, \mathbf{a}_k\} \subset \mathbb{R}_\infty^n$ and a vector $\mathbf{x} = (x_1, \ldots, x_n) \in \mathbb{R}_\infty^n$ tests the inclusion $\mathbf{x} \in$ Trophull(A). We assume bit complexity, if all input vectors are in $(\mathbb{Z} \cup \{\infty\})^n$, or the complexity of BSS model, if the input vectors are in $\mathbb{R}_\infty^n$. Let, for definiteness, input vectors be in $\mathbb{R}_\infty^n$.

The algorithm attempts to find a vector $(t_1, \ldots, t_k) \in \mathbb{R}_\infty^k$ such that

$$\mathbf{x} = \min_{1 \leq i \leq k} \{t_i \mathbf{1}_n + \mathbf{a}_i\}, \tag{1}$$

where minimum is componentwise.

Fix i, $1 \leq i \leq k$. If $\mathbf{a}_i = (\infty, \ldots, \infty)$, then choose $t_i \in \mathbb{R}_\infty^k$ arbitrarily. Otherwise, let, for definiteness, elements $a_{i1}, \ldots, a_{is}$ be all different from ∞ for $1 \leq s \leq n$. Choose $t_i \in \mathbb{R}_\infty$ as $\max_{1 \leq j \leq s}(x_j - a_{ij})$ (note that t_i will turn out to be ∞ if at least one of $x_j = \infty$, where $1 \leq j \leq s$). Chosen t_i is the minimal such that $t_i(1, \ldots, 1) + \mathbf{a}_i \geq \mathbf{x}$, where the inequality is componentwise.

Repeating the procedure for each i, $1 \leq i \leq k$, we obtain the vector $(t_1, \ldots, t_k)$. The vector $\mathbf{x} \in \mathrm{Trophull}(A)$ iff the equality (1) takes place.

Theorem 1, (2) states that any linear tropical prevariety $A^\perp = \{\mathbf{a}_1, \ldots, \mathbf{a}_n\}^\perp \subset \mathbb{R}_\infty^n$ coincides with the tropical hull of a finite subset of its vectors. The following example shows that this fact is not necessarily true for the restriction $A^\perp \cap \mathbb{R}^n$.

Example 1 (cf. [11]). Let $A_0 = \{\mathbf{a}_1, \ldots, \mathbf{a}_{n-1}\} \subset \mathbb{R}^n$, where

$$\mathbf{a}_i = (\underbrace{1, \ldots, 1}_{i}, 0, 1, \ldots, 1, 0, 0) \text{ for } 1 \leq i \leq n-2 \text{ and } \mathbf{a}_{n-1} = (1, \ldots, 1, 0, 0).$$

It is easy to see that every vector $\mathbf{x} = (x_1, \ldots, x_n) \in A_0^\perp \subset \mathbb{R}_\infty^n$ should have minimal elements x_{n-1}, x_n, and, conversely, every vector $\mathbf{x}$ with minimal elements x_{n-1}, x_n is in $A_0^\perp$. Therefore,

$$A_0^\perp = \{t\mathbf{1}_n + (c_1, \ldots, c_{n-2}, 0, 0) | \text{ for all } 0 \leq c_i \in \mathbb{R}_\infty \text{ and } t \in \mathbb{R}_\infty\}.$$

Directly from definitions it follows that $A_0^\perp = \mathrm{Trophull}(\{\mathbf{b}_1, \ldots, \mathbf{b}_{n-1}\})$, where

$$\mathbf{b}_i = (\underbrace{\infty, \ldots, \infty}_{i}, 0, \infty, \ldots, \infty, 0, 0) \text{ for } 1 \leq i \leq n-2 \text{ and } \mathbf{b}_{n-1} = (\infty, \ldots, \infty, 0, 0).$$

On the other hand, for restrictions $A_0^\perp \cap \mathbb{R}^n$, such a representation, as a tropical hull, is generally not true already when $n = 3$.

Proposition 1. *For any finite set* $\{\mathbf{v}_1, \ldots, \mathbf{v}_N\} \subset A_0^\perp \cap \mathbb{R}^3$ *we have*

$$A_0^\perp \cap \mathbb{R}^3 \not\subset \mathrm{Trophull}(\{\mathbf{v}_1, \ldots, \mathbf{v}_N\}).$$

2   Dual Tropical Linear Prevarieties

We extend the operation $X^\perp$ introduced in Definition 2 so that it can be applied to arbitrary (not necessarily finite) subsets $X \subset \mathbb{R}_\infty^n$. Namely, denote by $X^\perp$ the set of all vectors in $\mathbb{R}_\infty^n$ orthogonal to each $\mathbf{a} \in X$. We will use notations $X^{\perp\perp} := (X^\perp)^\perp$ and $X^{\perp\perp\perp} := (X^{\perp\perp})^\perp$.

Remark 3. Observe that by the definition, for a finite subset $A \subset \mathbb{R}^n_\infty$, the set $A^{\perp\perp}$ is an intersection of an infinite number of tropical hyperplanes in $\mathbb{R}^n_\infty$. As we will show in Sect. 5 below, not every intersection of even a countable number of tropical hyperplanes is a union of cells of a finite polyhedral complex, let alone linear tropical prevariety. However, in the special case of a finite A, the set $A^{\perp\perp}$ is a linear tropical prevariety (Proposition 2).

Lemma 2. *For any subset $X \subset \mathbb{R}^n_\infty$ we have:*

1. $\mathrm{Trophull}(X) \subset X^{\perp\perp}$;
2. $X^\perp = X^{\perp\perp\perp}$.

Proposition 2. *Let A be a finite set of vectors in $\mathbb{R}^n_\infty$. Then $A^{\perp\perp}$ is the minimal tropical linear prevariety containing A.*

For the proof of the next theorem, we recall the following definition.

Let $A = \{\mathbf{a}_1, \dots, \mathbf{a}_k\}$. Since, by Proposition 2 and Corollary 1, both $A^{\perp\perp}$ and $\mathrm{Trophull}(A)$ have the structure of polyhedral complexes in $\mathbb{R}^n_\infty$, dimensions of these sets are defined. By Lemma 2, (1), $\dim(\mathrm{Trophull}(A)) \leq \dim(A^{\perp\perp})$.

Theorem 2. $\dim(\mathrm{Trophull}(A)) = \dim(A^{\perp\perp})$.

Remark 4. Lemma 12 in [11] implies that $\mathrm{trk}(A) + \dim(A^\perp) \geq n$ for a finite $A \subset \mathbb{R}^n_\infty$.

Example 2. Consider the set of vectors $A_0 \subset \mathbb{R}^n$ from Example 1. Arguing as in that example, we see that every vector $\mathbf{y} = (y_1, \dots, y_n) \in A_0^{\perp\perp} \subset \mathbb{R}^n_\infty$ should have minimal elements y_{n-1}, y_n, and, conversely, every vector $\mathbf{y}$ with minimal elements y_{n-1}, y_n is in $A_0^{\perp\perp}$. It follows that $A_0^{\perp\perp} = A_0^\perp$. Observe that $\dim(A_0^\perp) = \dim(A_0^{\perp\perp}) = n - 1$.

By Lemma 2, (1), $\mathrm{Trophull}(A_0) \subset A_0^{\perp\perp}$. On the other hand, $\mathrm{Trophull}(A_0) \neq A_0^{\perp\perp}$. Indeed, we have $A_0 \subset A_0^{\perp\perp} = A_0^\perp$, hence, by Lemma 1, $\mathrm{Trophull}(A_0) \subset A_0^\perp$. Then, by Proposition 1, $A_0^{\perp\perp} \cap \mathbb{R}^n = A_0^\perp \cap \mathbb{R}^n \not\subset \mathrm{Trophull}(A_0)$. In particular, $\mathrm{Trophull}(A_0)$ is not a tropical linear prevariety, by Proposition 2.

3   Tropicalization of Linear Subspaces

Let $\mathbb{F}$ denote the field $\mathbb{C}((t^{1/\infty}))$ of Puiseux series over $\mathbb{C}$. For an element $y \in \mathbb{F}$ different from 0, let $\mathrm{val}(y) \in \mathbb{Q}$ denote the *valuation* of the element y in $\mathbb{F}$, i.e., the power in the lowest term of the Puiseux series y. Separately define $\mathrm{val}(0) = \infty$.

Definition 5 (cf. [17]). *Consider a linear form $f := a_1 x_1 + \cdots + a_n x_n$, where $0 \neq a_i \in \mathbb{F}$ for all $1 \leq i \leq n$. The formal tropicalization of the hyperplane $\{f = 0\} \subset \mathbb{F}^n$ is the tropical hyperplane, $\mathrm{Tropf}(\{f = 0\}) \subset \mathbb{R}^n$, defined by the set $\{y_1 + \mathrm{val}(a_1), \dots, y_n + \mathrm{val}(a_n)\}$ (see Definition 1).*

By Kapranov's Theorem [17, Theorem 3.1.3], $\mathrm{Tropf}(\{f = 0\})$ coincides with the (Euclidean) closure in $\mathbb{R}^n$ of the countable set

$$\{(\mathrm{val}(x_1), \ldots, \mathrm{val}(x_n)) | (x_1, \ldots, x_n) \in \{f = 0\}\}.$$

The following definition is "dual" to Definition 4.

Definition 6. *For every partition* $\{i_1, \ldots, i_r\} \cup \{i_{r+1}, \ldots, i_n\}$ *of* $\{1, \ldots, n\}$, *a* chart *in* $\mathbb{F}^n$ *is an open set*

$$D_{i_1, \ldots, i_r} := \{x_{i_1} = \cdots = x_{i_r} = 0\} \cap \{x_{i_{r+1}} \neq 0\} \cap \cdots \cap \{x_{i_n} \neq 0\} \subset \mathbb{F}^n.$$

Let $P \subset \mathbb{F}^n$ be a linear subspace of arbitrary dimension. Clearly,

$$P = \bigcup_{\{i_1, \ldots, i_r\}} (P \cap D_{i_1, \ldots, i_r}),$$

where the union is taken over all subsets $\{i_1, \ldots, i_r\}$ of $\{1, \ldots, n\}$.

Definition 7 (cf. [17]). *The* tropicalization $\mathrm{Trop}(P \cap D_{i_1, \ldots, i_r})$ *of* $P \cap D_{i_1, \ldots, i_r}$ *is the set of all points* $(y_1, \ldots, y_n) \in \mathbb{R}^n_\infty$ *such that* $y_{i_1} = \cdots = y_{i_r} = \infty$ *and* $(y_{i_{r+1}}, \ldots, y_{i_n})$ *belongs to the Euclidean closure in* $\mathbb{R}^{n-r}$ *of the set*

$$\{(\mathrm{val}(x_{i_{r+1}}), \ldots, \mathrm{val}(x_{i_n})) | (x_{i_{r+1}}, \ldots, x_{i_n}) \in P \cap D_{i_1, \ldots, i_r}\}.$$

The tropicalization $\mathrm{Trop}(P)$ *of* P *is defined as* $\bigcup_{\{i_1, \ldots, i_r\}} \mathrm{Trop}(P \cap D_{i_1, \ldots, i_r})$.

Remark 5. Definition 7 immediately implies that for any two sets $A, B \subset \mathbb{F}^n$, there is the inclusion $\mathrm{Trop}(A \cap B) \subset \mathrm{Trop}(A) \cap \mathrm{Trop}(B)$. The inverse inclusion $\supset$ is not generally true even for linear subspaces.

Let $\dim(P) = d$, and $\mathbf{z}_1, \ldots, \mathbf{z}_d \in P$ be a basis in P. Recall that *Plücker coordinates* of P in the Grassmanian $\mathrm{Gr}(d, \mathbb{F}^n)$ are all $(d \times d)$-minors $p_{j_1, \ldots, j_d}$ of the matrix with rows $\mathbf{z}_1, \ldots, \mathbf{z}_d$, corresponding to columns $1 \le j_1 < \cdots < j_d \le n$. Any $\mathbf{z} = (z_1, \ldots, z_n) \in P$ satisfies *Plücker relation*

$$\sum_{1 \le i \le d+1} (-1)^i p_{j_1, \ldots, j_{i-1}, j_{i+1}, \ldots, j_{d+1}} z_{j_i} = 0, \qquad (2)$$

for every subset of columns $1 \le j_1 < \cdots < j_{d+1} \le n$. Note that relations (2) do not depend on the choice of the basis in P.

Denote the set of points $\mathbf{z}$ satisfying (2) by $P_{j_1, \ldots, j_{d+1}}$.

The following statement is a strengthening of [19, Proposition 4.2].

Lemma 3.

$$\mathrm{Trop}(P) = \bigcap_{j_1, \ldots, j_{d+1}} \mathrm{Trop}(P_{j_1, \ldots, j_{d+1}}). \qquad (3)$$

Let $Q \subset \mathbb{F}^n$ be a linear subspace orthogonal to P with $\dim(Q) = n - d$. According to [1,19], the tropicalizations $\mathrm{Trop}(P)$ and $\mathrm{Trop}(Q)$ are orthogonal, with $\dim(\mathrm{Trop}(P)) = d$ and $\dim(\mathrm{Trop}(Q)) = n - d$.

Theorem 3. *There is a finite subset $A \subset \mathbb{R}_\infty^n$ such that* $\mathrm{Trop}(P) = A^\perp$ *and* $\mathrm{Trop}(Q) = A^{\perp\perp}$.

Corollary 2. *Let* $\mathrm{Trop}(P) = A^\perp$ *for a linear subspace* $P \subset \mathbb{F}^n$ *and* $Q \subset \mathbb{F}^n$ *be the subspace which is complementary orthogonal to* P. *Then* $\mathrm{Trop}(Q) = A^{\perp\perp}$.

4 Criterion and Deciding Algorithm for Being a Tropical Linear Variety

Let $A \subset (\mathbb{Q} \cup \{\infty\})^n$ be a finite set of vectors. Since $A^\perp$ and $A^{\perp\perp}$ are tropical linear prevarieties (see Proposition 2), Theorem 1 implies that

$$A^\perp = \mathrm{Trophull}(\{\mathbf{x}_1, \ldots, \mathbf{x}_p\}) \quad \text{and} \quad A^{\perp\perp} = \mathrm{Trophull}(\{\mathbf{y}_1, \ldots, \mathbf{y}_q\})$$

for some vectors $\mathbf{x}_1, \ldots, \mathbf{x}_p, \mathbf{y}_1, \ldots, \mathbf{y}_q \in (\mathbb{Q} \cup \{\infty\})^n$.

Definition 8. *Let* $\mathbf{x} \in (\mathbb{Q} \cup \{\infty\})^n$ *and* $\mathbf{v} \in \mathbb{F}^n$ *such that* $\mathrm{Trop}(\mathbf{v}) = \mathbf{x}$. *Then* $\mathbf{v}$ *is called the* lifting *of* $\mathbf{x}$.

Theorem 4. *The following three statements are equivalent.*

1. *There exist mutually complementary and orthogonal linear subspaces* P, Q *in* $\mathbb{F}^n$ *such that* $A^\perp = \mathrm{Trop}(P)$, *and* $A^{\perp\perp} = \mathrm{Trop}(Q)$ *(in particular,* $A^\perp$, $A^{\perp\perp}$ *are tropical linear varieties).*
2. *There exist liftings*

   $$\mathbf{v}_1, \ldots \mathbf{v}_p, \mathbf{w}_1, \ldots, \mathbf{w}_q \in \mathbb{F}^n \quad \text{of vectors} \quad \mathbf{x}_1, \ldots \mathbf{x}_p, \mathbf{y}_1, \ldots, \mathbf{y}_q \in (\mathbb{Q} \cup \{\infty\})^n,$$

   respectively, such that $(\mathbf{v}_i, \mathbf{w}_j) = 0$ *for all* $1 \le i \le p,\ 1 \le j \le q$.
3. $A^\perp$ *is a tropical linear variety.*

Corollary 3. *There is an algorithm which for a given tropical linear prevariety* $A^\perp = \mathrm{Trophull}(\{\mathbf{x}_1, \ldots, \mathbf{x}_p\})$ *where* $\mathbf{x}_1, \ldots, \mathbf{x}_p \in (\mathbb{Q} \cup \{\infty\})^n$, *decides whether it is a tropical linear variety.*

Proof. The input of the algorithm under construction is the set $\{\mathbf{x}_1, \ldots, \mathbf{x}_p\} \subset (\mathbb{Q} \cup \{\infty\})^n$. Using the algorithm from the proof of Theorem 1 in [8], construct the set $\{\mathbf{y}_1, \ldots, \mathbf{y}_q\} \subset (\mathbb{Q} \cup \{\infty\})^n$ such that $A^{\perp\perp} = \mathrm{Trophull}(\{\mathbf{y}_1, \ldots, \mathbf{y}_q\})$.

Consider, over $\mathbb{F}^* \cong \mathbb{F} \setminus \{0\}$, the system of equations

$$\sum_{1 \le \nu \le n} V_{i\nu} W_{j\nu} = 0 \tag{4}$$

for all $1 \le i \le p,\ 1 \le j \le q,\ 1 \le \nu \le n$ such that $x_{i\nu} \ne \infty$ and $y_{j\nu} \ne \infty$, where $\mathbf{V}_i = (V_{i1}, \ldots, V_{in}), \mathbf{W}_j = (W_{j1}, \ldots, W_{jn})$ are vectors of variables. Using [1,14], or [16], the algorithm constructs the tropical basis of the system of Eq. (4), which is a finite set of polynomials $H_\ell,\ 1 \le \ell \le N$, where $N \le (p+q)n$, with integer coefficients and variables $V_{i\nu}, W_{j\nu}$ such that $x_{i\nu} \ne \infty$ and $y_{j\nu} \ne \infty$.

(See a detailed definition and properties of a tropical basis in [17, Sect. 2.6].) The algorithm checks whether, for all $1 \leq \ell \leq N$, vectors $\mathbf{x}_1, \ldots \mathbf{x}_p, \mathbf{y}_1, \ldots, \mathbf{y}_q$, from which coordinates $x_{i\nu} = \infty$ and $y_{j\nu} = \infty$ are removed, satisfy tropicalizations $\mathrm{Trop}(H_\ell)$. If yes, then, by the definition of tropical basis, there exist liftings $v_{i\nu}, w_{j\nu} \in \mathbb{F}^*$ of all $x_{i\nu} \neq \infty, y_{j\nu} \neq \infty$, respectively, which satisfy system (4). Let P be the linear hull of vectors $\mathbf{v}_1, \ldots \mathbf{v}_p$ such that in every $\mathbf{v}_i$, each coordinate corresponding to $x_{i\nu} \neq \infty$ is the lifting $v_{i\nu}$, while each coordinate corresponding to $x_{i\nu} = \infty$ is 0. Similarly, let Q be the linear hull of vectors $\mathbf{w}_1, \ldots \mathbf{w}_q$ such that in every $\mathbf{w}_j$, each coordinate corresponding to $y_{j\nu} \neq \infty$ is the lifting $w_{j\nu}$, while each coordinate corresponding to $y_{j\nu} = \infty$ is 0. Then, by Theorem 4, P and Q are mutually complementary and orthogonal linear subspaces of $\mathbb{F}^n$, while $A^\perp = \mathrm{Trop}(P)$ and $A^{\perp\perp} = \mathrm{Trop}(Q)$. In particular, $A^\perp$ and $A^{\perp\perp}$ are tropical linear varieties. If vectors $\mathbf{x}_1, \ldots \mathbf{x}_p, \mathbf{y}_1, \ldots, \mathbf{y}_q$ from which coordinates $x_{i\nu} = \infty$ and $y_{j\nu} = \infty$ are removed do not satisfy $\mathrm{Trop}(H_\ell)$ for all $1 \leq \ell \leq N$, then $A^\perp$ is not a tropical linear variety, by Theorem 4.

The complexity of the algorithm is polynomial in the maximum of absolute values of numerators and denominators of rational coordinates of vectors $\mathbf{x}_i$, $1 \leq i \leq p$, and doubly exponential in n. In this regard, we note that the complexity of the algorithm for computing the tropical basis in [16] is doubly exponential.

Remark 6. There is an algorithm which for a tropical linear *prevariety* $A^\perp = \mathrm{Trophull}(\{\mathbf{x}_1, \ldots, \mathbf{x}_p\})$, where $\mathbf{x}_1, \ldots, \mathbf{x}_p \in (\mathbb{Q} \cup \{\infty\})^n$, produces bases of linear subspaces P and Q, such that $A^\perp = \mathrm{Trop}(P)$ and $A^{\perp\perp} = \mathrm{Trop}(Q)$, in case these subspaces exist.

More precisely, as in the proof of Corollary 2, construct, using the algorithm from the proof of Theorem 1 in [8], the set $\{\mathbf{y}_1, \ldots, \mathbf{y}_q\} \subset (\mathbb{Q} \cup \{\infty\})^n$ such that $A^{\perp\perp} = \mathrm{Trophull}(\{\mathbf{y}_1, \ldots, \mathbf{y}_q\})$. Apply the algorithm from [15] to vectors $\mathbf{x}_i, \mathbf{y}_j$, $1 \leq i \leq p$, $1 \leq j \leq q$, from which coordinates $x_{i\nu} = \infty$ and $y_{j\nu} = \infty$ are removed, and to system (4). The algorithm from [15] will either produce liftings $v_{i\nu}, w_{j\nu} \in \mathbb{F}^*$ of all $x_{i\nu} \neq \infty, y_{j\nu} \neq \infty$, respectively, which satisfy (4), or will indicate that liftings don't exist, i.e., vectors $\mathbf{x}_i, \mathbf{y}_j$ with removed coordinates do not belong to the tropicalization of (4). If liftings exist, then P is the linear hull of vectors $\mathbf{v}_1, \ldots \mathbf{v}_p$ such that in every $\mathbf{v}_i$, each coordinate corresponding to $x_{i\nu} \neq \infty$ is the lifting $v_{i\nu}$, while each coordinate corresponding to $x_{i\nu} = \infty$ is 0. Similarly, Q is the linear hull of vectors $\mathbf{w}_1, \ldots \mathbf{w}_q$ such that in every $\mathbf{w}_j$, each coordinate corresponding to $y_{j\nu} \neq \infty$ is the lifting $w_{j\nu}$, while each coordinate corresponding to $y_{j\nu} = \infty$ is 0. Herewith, all coordinates of vectors $\mathbf{v}_i, \mathbf{w}_j$ are Puiseux series in $\overline{\mathbb{Q}}((t^{1/\infty}))$ and the algorithm computes their expansions up to a given power of t, representing complex algebraic coefficients as zeroes of irreducible univariate polynomials with integer coefficients.

We sketch briefly an alternative algorithm for producing P and Q. A procedure in [15] reduces the construction of liftings to finding all points in a zero-dimensional algebraic set in $(\overline{\mathbb{Q}}((t^{1/\infty})))^N$, by adding generic linear forms to the tropical basis. All such points (absolutely irreducible components of the zero-dimensional algebraic set) can be computed with singly exponential complexity using algorithms from [3,9] (these algorithms have much more general scope). Zero-dimensional components are represented as elements of a finite extension of the field $\mathbb{Q}(t)$ via a primitive element. Then, using a procedure from [4] for Puiseux extension of solutions of polynomial equations over $\overline{\mathbb{Q}}((t^{1/\infty}))$, the algorithm checks whether the tropicalization of a solution coincides with $\mathbf{x}_1, \ldots \mathbf{x}_p, \mathbf{y}_1, \ldots, \mathbf{y}_q$. Because this algorithm uses the procedure from [15], its complexity is doubly exponential in n. The digest of symbolic manipulation technique with Puiseux series can be found in [10].

5 Infinite Intersections of Tropical Linear Prevarieties

By the definition, the intersection of a finite number of tropical hyperplanes is a tropical linear prevariety. In this section, we give an example of a *countable* family of tropical hyperplanes in $\mathbb{R}_\infty^6$ such that their intersection is not a finite union of convex polyhedra, in particular, not a tropical prevariety. This strengthens examples in [7] (example of T. Theobald) and [12], in which intersections of a countable families of tropical (non-linear) prevarieties were shown not to be finite unions of convex polyhedra.

Choose a sequence $\{\varepsilon_i\}_{i=1}^n$ of pairwise distinct real numbers ε_i such that $0 < \varepsilon_i < 1/4$, and consider a tropical hyperplane $L_i \subset \mathbb{R}_\infty^6$ defined by the set

$$\{-i + x_1, -i + x_2, -i/2 - \varepsilon_i + y_1, -i/2 + y_2, z_1, z_2\}$$

(see Definition 1).

Let

$$M := \bigcap_{1 \leq i < \infty} L_i \subset \mathbb{R}_\infty^6.$$

Proposition 3. *The set $M \cap \mathbb{R}^6$ is not a finite union of convex polyhedra. In particular, M is not a tropical prevariety.*

Acknowledgments. We thank M. Joswig, N. Kalinin, H. Markwig, and T. Theobald for useful discussions. Part of this research was carried out during our joint visit in September 2017 to the Hausdorff Research Institute for Mathematics at Bonn University, under the program Applied and Computational Algebraic Topology, to which we are very grateful. D. Grigoriev was partly supported by the RSF grant 16-11-10075.

References

1. Bogart, T., Jensen, A.N., Speyer, D., Sturmfels, B., Thomas, R.R.: Computing tropical varieties. J. Symb. Comput. **42**(1–2), 54–73 (2007)
2. Butkovic, P., Hegedüs, G.: An elimination method for finding all solutions of the system of linear equations over an extremal algebra. Ekon. Mat. Obzor **20**, 203–214 (1984)
3. Chistov, A.: An algorithm of polynomial complexity for factoring polynomials, and determination of the components of a variety in a subexponential time. J. Sov. Math. **34**, 1838–1882 (1986)

4. Chistov, A.L.: Polynomial complexity of the Newton-Puiseux algorithm. In: Gruska, J., Rovan, B., Wiedermann, J. (eds.) MFCS 1986. LNCS, vol. 233, pp. 247–255. Springer, Heidelberg (1986). https://doi.org/10.1007/BFb0016248
5. Develin, M., Santos, F., Sturmfels, B.: On the rank of a tropical matrix. In: Combinatorial and Computational Geometry, vol. 52. MSRI Publications (2005)
6. Develin, M., Sturmfels, B.: Tropical convexity. Doc. Math. **9**, 1–27 (2004)
7. Dress, A., Wenzel, W.: Algebraic, tropical, and fuzzy geometry. Beitr. Algebra Geom. **52**(2), 431–461 (2011)
8. Gaubert, S., Katz, R.D.: Minimal half-spaces and external representation of tropical polyhedra. J. Algebraic Comb. **33**(3), 325–348 (2011)
9. Grigoriev, D.: Polynomial factoring over a finite field and solving systems of algebraic equations. J. Sov. Math. **34**, 1762–1803 (1986)
10. Grigoriev, D.: Polynomial complexity recognizing a tropical linear variety. In: Gerdt, V.P., Koepf, W., Seiler, W.M., Vorozhtsov, E.V. (eds.) CASC 2015. LNCS, vol. 9301, pp. 152–157. Springer, Cham (2015). https://doi.org/10.1007/978-3-319-24021-3_11
11. Grigoriev, D., Podolskii, V.: Complexity of tropical and min-plus linear prevarieties. Comput. Complex. **24**(1), 31–64 (2015)
12. Grigoriev, D., Vorobjov, N.: Upper bounds on Betti numbers of tropical prevarieties. Arnold Math. J. **4**(1), 127–136 (2018)
13. Grigoriev, D., Vorobjov, N.: Orthogonal tropical linear prevarieties. arXiv: 1803.01068
14. Hept, K., Theobald, T.: Tropical bases by regular projections. Proc. Amer. Math. Soc. **137**(7), 2233–2241 (2009)
15. Jensen, A.N., Markwig, H., Markwig, T.: An algorithm for lifting points in a tropical variety. Collect. Math. **59**(2), 129–165 (2008)
16. Kazarnovskii, Y., Khovanskii, A.G.: Tropical noetherity and Gröbner bases. St. Petersburg Math. J. **26**(5), 797–811 (2015)
17. Maclagan, D., Sturmfels B.: Introduction to Tropical Geometry. Graduate Studies in Mathematics, vol. 161. American Mathematical Society, Providence (2015)
18. Richter-Gebert, J., Sturmfels, B., Theobald, T.: First steps in tropical geometry. In: Litvinov, G., Maslov, V. (eds.) Idempotent Mathematics and Mathematical Physics (Proceedings Vienna 2003), Contemporary Mathematics, vol. 377, pp. 289–317. American Mathematical Society (2005)
19. Speyer, D.: Tropical linear spaces. SIAM J. Discret. Math. **22**(4), 1527–1558 (2008)
20. Yu, J., Yuster, D.S.: Representing tropical linear spaces by circuits. In: The 19th International Conference on Formal Power Series and Algebraic Combinatorics (2006). arXiv:0611579

Symbolic-Numerical Algorithms for Solving Elliptic Boundary-Value Problems Using Multivariate Simplex Lagrange Elements

A. A. Gusev[1](✉), V. P. Gerdt[1,2], O. Chuluunbaatar[1,3], G. Chuluunbaatar[1,2], S. I. Vinitsky[1,2], V. L. Derbov[4], A. Góźdź[5], and P. M. Krassovitskiy[1,6]

[1] Joint Institute for Nuclear Research, Dubna, Russia
gooseff@jinr.ru
[2] RUDN University, 6 Miklukho-Maklaya St., Moscow 117198, Russia
[3] Institute of Mathematics, National University of Mongolia, Ulaanbaatar, Mongolia
[4] N.G. Chernyshevsky Saratov National Research State University, Saratov, Russia
[5] Institute of Physics, University of M. Curie–Skłodowska, Lublin, Poland
[6] Institute of Nuclear Physics, Almaty, Kazakhstan

Abstract. We propose new symbolic-numerical algorithms implemented in Maple-Fortran environment for solving the self-adjoint elliptic boundary-value problem in a d-dimensional polyhedral finite domain, using the high-accuracy finite element method with multivariate Lagrange elements in the simplexes. The high-order fully symmetric PI-type Gaussian quadratures with positive weights and no points outside the simplex are calculated by means of the new symbolic-numerical algorithms implemented in Maple. Quadrature rules up to order 8 on the simplexes with dimension $d = 3 - 6$ are presented. We demonstrate the efficiency of algorithms and programs by benchmark calculations of a low part of spectra of exactly solvable Helmholtz problems for a cube and a hypercube.

Keywords: Elliptic boundary-value problem · Finite element method
Multivariate simplex lagrange elements
High-order fully symmetric Gaussian quadratures
Helmholtz equation for cube and hypercube

1 Introduction

The progress of modern computing power offers more possibilities for setting and numerical solution of multidimensional elliptic boundary-value problems (BVPs) with high accuracy. 3D BVPs have wide applications in such areas as vibrating membrane, electromagnetic radiation, motion of thermal neutrons in the reactor, seismology, and acoustics, see, e.g., [4], while multidimensional BVPs have applications in nuclear physics, see, e.g., [7]. For this purpose, novel numerical

© Springer Nature Switzerland AG 2018
V. P. Gerdt et al. (Eds.): CASC 2018, LNCS 11077, pp. 197–213, 2018.
https://doi.org/10.1007/978-3-319-99639-4_14

methods of high accuracy order are being developed. When reducing the boundary value problem to an algebraic one in the finite element method (FEM) of the order p, one of the problems is the calculation of integrals on a finite element (we consider only simplicial finite elements) containing the products of two basis functions of Lagrange or Hermite interpolation polynomials of the order p by the coefficients for the unknown functions [5,9]. There are three possible ways to calculate the integrals:

(i) using analytical calculation, which is possible for a limited number of cases;
(ii) using quadrature formulas with products of two basic functions used as a weight function;
(iii) using quadrature formulas with a single weight function.

It is well known [20] that as a result of applying the pth order FEM to the solution of the discrete spectrum problem for the elliptic (Schrödinger) equation, the eigenfunction and the eigenvalue are determined with an accuracy of the order $p + 1$ and $2p$ provided that all intermediate quantities are calculated with sufficient accuracy. It follows that for the realization of the FEM of the order p in the third case, the integrals must be computed at least with an accuracy of the order $2p$, depending on the problem considered. The most economical calculation of such integrals is achieved using the quadratures of Gaussian type. In the one-dimensional case, the nodes and the quadrature Gaussian weights are expressed analytically; in the two-, three- and four-dimensional case, the high-order quadrature formulas are determined numerically [2,6,8,10,17–19,21]. Note that for multidimensional integrals, numerous quadrature formulas of the Newton–Cotes and third-order Gaussian type are known, too (see Ref. [1]).

The paper presents a new method for constructing fully symmetric multidimensional Gaussian-type quadratures on a standard simplex. The main idea of the method is replacing the coordinates of nodes with their symmetric combinations obtained using the Vieta theorem, which simplifies the system of nonlinear algebraic equations. The construction of the desired systems of equations is performed analytically using an original algorithm implemented in Maple [13]. The derived systems up to the sixth order are solved using the built-in procedure PolynomialSystem, implementing the technique of Gröbner bases, and the systems of higher order are solved using the developed symbolic-numerical algorithm based on numerical methods, implemented in Maple-Fortran environment. We demonstrate the efficiency of algorithms and programs by benchmark calculations of the lower part of spectra in exactly solvable Helmholtz problems for a cube and a hypercube.

The paper is structured as follows. In Sects. 2 and 3, the FEM schemes and algorithms for solving the d-dimensional BVP are presented. In Sect. 4, the algorithms for constructing the d-dimensional fully symmetric Gaussian quadratures are presented. In Sect. 5, the benchmark calculations of the exactly solvable Helmholtz problems for the cube and hypercube are presented. In Conclusion, we discuss the results and perspectives.

2   Setting of the Problem

Consider a self-adjoint boundary-value problem for the elliptic differential equation of the second order:

$$(D - E)\Phi(z) \equiv \left(-\frac{1}{g_0(z)} \sum_{ij=1}^{d} \frac{\partial}{\partial z_i} g_{ij}(z) \frac{\partial}{\partial z_j} + V(z) - E \right) \Phi(z) = 0. \qquad (1)$$

For the principal part coefficients of Eq. (1), the condition of uniform ellipticity holds in the bounded domain $z = (z_1, \ldots, z_d) \in \Omega$ of the Euclidean space $\mathcal{R}^d$, i.e., the constants $\mu > 0$, $\nu > 0$ exist such that $\mu \xi^2 \leq \sum_{ij=1}^{d} g_{ij}(z) \xi_i \xi_j \leq \nu \xi^2$, $\xi^2 = \sum_{i=1}^{d} \xi_i^2$, $\forall \xi_i \in \mathcal{R}$. The left-hand side of this inequality expresses the requirement of ellipticity, while the right-hand side expresses the boundedness of the coefficients $g_{ij}(z)$. It is also assumed that $g_0(z) > 0$, $g_{ji}(z) = g_{ij}(z)$ and $V(z)$ are real-valued functions, continuous together with their generalized derivatives to a given order in the domain $z \in \bar{\Omega} = \Omega \cup \partial\Omega$ with the piecewise continuous boundary $S = \partial\Omega$, which provides the existence of nontrivial solutions obeying the boundary conditions [5] of the first kind

$$\Phi(z)|_S = 0, \qquad (2)$$

or the second kind

$$\frac{\partial \Phi(z)}{\partial n_D} \Big|_S = 0, \qquad \frac{\partial \Phi(z)}{\partial n_D} = \sum_{ij=1}^{d} (\hat{n}, \hat{e}_i) g_{ij}(z) \frac{\partial \Phi(z)}{\partial z_j}, \qquad (3)$$

where $\frac{\partial \Phi_m(z)}{\partial n_D}$ is the derivative along the conformal direction, $\hat{n}$ is the outer normal to the boundary of the domain $S = \partial\Omega$, $\hat{e}_i$ is the unit vector of $z = \sum_{i=1}^{d} \hat{e}_i z_i$, and $(\hat{n}, \hat{e}_i)$ is the scalar product in $\mathcal{R}^d$.

For a discrete spectrum problem, the functions $\Phi_m(z)$ from the Sobolev space $H_2^{s \geq 1}(\Omega)$, $\Phi_m(z) \in H_2^{s \geq 1}(\Omega)$, corresponding to the real eigenvalues E: $E_1 \leq E_2 \leq \ldots \leq E_m \leq \ldots$ satisfy the conditions of normalization and orthogonality

$$\langle \Phi_m(z) | \Phi_{m'}(z) \rangle = \int_{\Omega} dz g_0(z) \Phi_m(z) \Phi_{m'}(z) = \delta_{mm'}, \quad dz = dz_1 \ldots dz_d. \quad (4)$$

The FEM solution of the boundary-value problems (1)–(4) is reduced to the determination of stationary points of the variational functional [3,5]

$$\Xi(\Phi_m, E_m) \equiv \int_{\Omega} dz g_0(z) \Phi_m(z) (D - E_m) \Phi(z) = \Pi(\Phi_m, E_m), \qquad (5)$$

where $\Pi(\Phi, E)$ is the symmetric quadratic functional

$$\Pi(\Phi, E) = \int_{\Omega} dz \left[\sum_{ij=1}^{d} g_{ij}(z) \frac{\partial \Phi(z)}{\partial z_i} \frac{\partial \Phi(z)}{\partial z_j} + g_0(z) \Phi(z)(V(z) - E)\Phi(z) \right].$$

3 FEM Calculation Scheme

In FEM, the domain $\Omega = \Omega_h(z) = \bigcup_{q=1}^{Q} \Delta_q$, specified as a polyhedral domain, is covered with finite elements, in the present case, the simplexes Δ_q with $d+1$ vertices $\hat{z}_i = (\hat{z}_{i1}, \hat{z}_{i2}, \ldots, \hat{z}_{id})$, $i = 0, \ldots, d$. Each edge of the simplex Δ_q is divided into p equal parts, and the families of parallel hyperplanes $H(i,k)$ are drawn, numbered with the integers $k = 0, \ldots, p$, starting from the corresponding face (see also [5]). The equation of the hyperplane is $H(i,k)$: $H(i;z) - k/p = 0$, where $H(i;z)$ is a linear function of z.

The node points of hyperplanes crossing A_r are enumerated with the sets of integers $[n_0, \ldots, n_d]$, $n_i \geq 0$, $n_0 + \ldots + n_d = p$, where n_i, $i = 0, 1, \ldots, d$ are the numbers of hyperplanes, parallel to the simplex face, not containing the ith vertex $\hat{z}_i = (\hat{z}_{i1}, \ldots \hat{z}_{id})$. The coordinates $\xi_r = (\xi_{r1}, \ldots, \xi_{rd})$ of the node point $A_r \in \Delta_q$ are calculated using the formula

$$(\xi_{r1}, \ldots, \xi_{rd}) = (\hat{z}_{01}, \ldots, \hat{z}_{0d})\frac{n_0}{p} + (\hat{z}_{11}, \ldots, \hat{z}_{1d})\frac{n_1}{p} + \ldots + (\hat{z}_{d1}, \ldots, \hat{z}_{dd})\frac{n_d}{p} \quad (6)$$

from the coordinates of the vertices $\hat{z}_j = (\hat{z}_{j1}, \ldots, \hat{z}_{jd})$. Then the Lagrange interpolation polynomials (LIP) $\varphi_r^p(z)$ are equal to one at the point A_r with the coordinates $\xi_r = (\xi_{r1}, \ldots, \xi_{rd})$, characterized by the numbers $[n_0, n_1, \ldots, n_d]$, and equal to zero at the remaining points $\xi_{r'}$, i.e., $\varphi_r^p(\xi_{r'}) = \delta_{rr'}$, have the form

$$\varphi_r^p(z) = \prod_{i=0}^{d} \prod_{n_i'=0}^{n_i-1} \frac{H(i;z) - n_i'/p}{H(i;\xi_r) - n_i'/p}. \quad (7)$$

As shape functions in the simplex Δ_q we use the multivariate Lagrange interpolation polynomials $\varphi_l^p(z)$ of the order p that satisfy the condition $\varphi_l^p(x_{1l'}, x_{2l'}) = \delta_{ll'}$, i.e., equal 1 at one of the points A_l and zero at the other points. In this method, the piecewise polynomial functions $N_l^p(z)$ in the domain Ω are constructed by joining the shape functions $\varphi_l^p(z)$ in the simplex Δ_q:

$$N_l^p(z) = \{\varphi_l^p(z), A_l \in \Delta_q; 0, A_l \notin \Delta_q\}$$

and possess the following properties: the functions $N_l^p(z)$ are continuous in the domain Ω; the functions $N_l^p(z)$ equal 1 at one of the points A_l and zero at the rest of the points; $N_l^p(z_{l'}) = \delta_{ll'}$ in the entire domain Ω. Here l takes the values $l = 1, \ldots, N$.

The functions $N_l^p(z)$ form a basis in the space of polynomials of the pth order. Now, the function $\Phi(z) \in \mathcal{H}^1(\Omega)$ is approximated by a finite sum of piecewise basis functions $N_l^p(z)$:

$$\Phi^h(z) = \sum_{l=1}^{N} \Phi_l^h N_l^p(z). \quad (8)$$

Table 1. The orbits and their number of permutations for $d = 3, 4, 5, 6$.

$d = 3$		$d = 4$		$d = 5$				$d = 6$			
Orbits	Perm.	Orbits	Perm.	Orbits	Perm.	Orbits	Perm.	Orbits	Perm.	Orbits	Perm.
S_4	1	S_5	1	S_6	1	S_{3111}	120	S_7	1	S_{4111}	210
S_{31}	4	S_{41}	5	S_{51}	6	S_{2211}	180	S_{61}	7	S_{3211}	420
S_{22}	6	S_{32}	10	S_{42}	15	S_{21111}	360	S_{52}	21	S_{2221}	630
S_{211}	12	S_{311}	20	S_{33}	20	S_{111111}	720	S_{43}	35	S_{31111}	840
S_{1111}	24	S_{221}	30	S_{411}	30			S_{511}	42	S_{22111}	1260
		S_{2111}	60	S_{321}	60			S_{421}	105	S_{211111}	2520
		S_{11111}	120	S_{222}	90			S_{331}	140	$S_{1111111}$	5040
								S_{322}	210		

After substituting expansion (8) into the variational functional (5) and minimizing it [3,20], we obtain the generalized eigenvalue problem

$$\mathbf{A}^p \boldsymbol{\Phi}^h = \varepsilon^h \mathbf{B}^p \boldsymbol{\Phi}^h. \tag{9}$$

Here $\mathbf{A}^p$ is the symmetric stiffness matrix; $\mathbf{B}^p$ is the symmetric positive definite mass matrix; $\boldsymbol{\Phi}^h$ is the vector approximating the solution on the finite-element grid; and ε^h is the corresponding eigenvalue. The matrices $\mathbf{A}^p$ and $\mathbf{B}^p$ have the form:

$$\mathbf{A}^p = \{a^p_{ll'}\}^N_{ll'=1}, \mathbf{B}^p = \{b^p_{ll'}\}^N_{ll'=1}, \tag{10}$$

where the matrix elements $a^p_{ll'}$ and $b^p_{ll'}$ are calculated for simplex elements as

$$a^p_{ll'} = \sum_{ij=1}^d \int_{\Delta_q} g_{ij}(z) \frac{\partial \varphi^p_l(z)}{\partial z_i} \frac{\partial \varphi^p_{l'}(z)}{\partial z_j} \, dz + \int_{\Delta_q} g_0(z) \varphi^p_l(z) \varphi^p_{l'}(z) V(z) \, dz,$$

$$b^p_{ll'} = \int_{\Delta_q} g_0(z) \varphi^p_l(z) \varphi^p_{l'}(z) dz. \tag{11}$$

The economical implementation of FEM is the following.

The calculations, including those of FEM integrals for mass and stiffness matrices at each subdomain Δ_q are performed in the local (reference) system of coordinates x, in which the coordinates of the simplex vertices are the following: $\hat{x}_j = (\hat{x}_{j1}, \ldots, \hat{x}_{jd})$, $\hat{x}_{jk} = \delta_{jk}$, $j = 0, \ldots, d$, $k = 1, \ldots, d$.

Let us construct the Lagrange interpolation polynomial (LIP) on an arbitrary d-dimensional simplex Δ_q with vertices $\hat{z}_i = (\hat{z}_{i1}, \hat{z}_{i2}, \ldots, \hat{z}_{id})$, $i = 0, \ldots, d$. For this purpose, we introduce the local system of coordinates $x = (x_1, x_2, \ldots, x_d) \in \mathcal{R}^d$, in which the coordinates of the simplex vertices are $\hat{x}_i$. The relation between the coordinates is given by the formula:

$$z_i = \hat{z}_{0i} + \sum_{j=1}^d \hat{J}_{ij} x_j, \quad \hat{J}_{ij} = \hat{z}_{ji} - \hat{z}_{0i}, \quad i = 1, \ldots, d. \tag{12}$$

Table 2. Quadrature rule on tetrahedra.

Orbit	Weight	Abscissas
14-points 4-order rule		
S_{31}	0.0801186758957551214557967806191	0.0963721076152827180679867982109
S_{31}	0.1243674424942431317471251193937	0.3123064218132941261147265437508
S_{22}	0.0303425877400011645313853999915	0.0274707886853344957750132954191
14-points 5-order rule		
S_{31}	0.0734930431163619495437102054863	0.0927352503108912264023239137370
S_{31}	0.1126879257180158507991856523333	0.3108859192633006097973457337635
S_{22}	0.0425460207770814664380694281203	0.0455037041256496494918805262793
24-points 6-order rule		
S_{31}	0.0399227502581674920996906275575	0.2146028712591520292888392193863
S_{31}	0.0100772110553206429480132374459	0.0406739585346113531155794489564
S_{31}	0.0553537181543654722095153277853 7	0.3223378901422755103439944707625
S_{211}	0.0482142857142857142857142857143	0.0636610018750175252992355276057 0.6030056647916491413674311390609
35-points 7-order rule		
S_4	0.0954852894641308488605784361172	0.2500000000000000000000000000000
S_{31}	0.0423295812099670290762861707986	0.3157011497782027994234299995933
S_{22}	0.0318969278328575799342748240829	0.0504898225983963687630538229866
S_{211}	0.0372071307283346213696155611915	0.1888338310260010477364311038546 0.5751716375870000234832415770223
S_{211}	0.0081107708299033415661034334911	0.0212654725414832459883361014998 0.8108302410985485611181053798482
46-points 8-order rule		
S_{31}	0.0063972777406656176515049738764	0.0396757518582111225277078936298
S_{31}	0.0401906214382288067086698161802	0.3144877686588789672386516888007
S_{31}	0.0243081692121760770795396363192	0.1019873469010702748038937565346
S_{31}	0.0548586277637264928464254253584	0.1842037697228154771186065671874
S_{22}	0.0357196747563309013579348149829	0.0634363951662790318385035375295
S_{211}	0.0071831862652404057248973769332	0.0216901288123494021982001218658 0.7199316530057482532021892796203
S_{211}	0.0163720776383284788356885983306	0.2044800362678728018101543629799 0.5805775568740886759781950895673

The inverse transformation and the relation between the differentiation operators are given by the formulas

$$x_i = \sum_{j=1}^{d} (\hat{J}^{-1})_{ij}(z_j - \hat{z}_{0j}), \frac{\partial}{\partial x_i} = \sum_{j=1}^{d} \hat{J}_{ji}\frac{\partial}{\partial z_j}, \frac{\partial}{\partial z_i} = \sum_{j=1}^{d} (\hat{J}^{-1})_{ji}\frac{\partial}{\partial x_j}.$$

Table 3. Quadrature rule on $d = 4$ dimensional simplex.

Orbit	Weight	Abscissas
20-points 4-order rule		
S_{41}	0.03795392242065396108315111760634	0.078422464532008441270186009537
S_{41}	0.068138449514096507307237418942	0.244992500251650682974726724199
S_{32}	0.046953814032624765804805702497	0.065780705401760442932665992362
30-points 5-order rule		
S_{41}	0.049251680175315740938395667283	0.085346630830859408251632945252
S_{41}	0.032511460658739364936949373887	0.236960011661460705646083216339
S_{32}	0.017532710995800450876663590892	0.041298014131848401048205215945
S_{32}	0.041585718587171996185663888521	0.299744338479035286296335489564
56-points 6-order rule		
S_5	0.073279236743554772188440808855	0.200000000000000000000000000000
S_{41}	0.004742912171318373911790594179	0.041703381748481614470367973524
S_{32}	0.037167112402533006986944882925	0.295622797147098049191196334346
S_{311}	0.013336248018481771716654774405	0.154394924873116842736992119567
		0.522750646227696832515158469571
S_{311}	0.013230505900244392702503095144	0.047815675137827492151514862425
		0.281973941992880602871627877781
76-points 7-order rule		
S_5	0.028272766759793510146165467413	0.200000000000000000000000000000
S_{41}	0.017163792015553795559126596836	0.249402089309377969567400055747
S_{32}	0.008426290417736873748764156645	0.039027995660106969047822346802
S_{32}	0.015163362756045314580986291487	0.128311404463812192159465856927
S_{311}	0.004109934841481556020447802548	0.033847470986564263527996961838
		0.746262428681339061102062480377
S_{221}	0.018927101486499483611724700536	0.044833796455796184976390008452
		0.209871085716232476426298177816
110-points 8-order rule		
S_{41}	0.020988963106203348828447185874	0.106416063260142058846827434852
S_{41}	0.002556930429961908711113352905	0.040543282412661311354934088265
S_{32}	0.015336414023745230822528153201	0.055320520485979115777864856400
S_{32}	0.014341370355404557767979712361	0.132984924720748876527117239830
S_{32}	0.021983906357169179701387411959	0.292164962367903993351239086340
S_{311}	0.003699835117610442071728496938	0.033339878866874728719032798603
		0.696028477914025484511728247325
S_{311}	0.010287515395496733244605083680	0.174905546599082503418947240638
		0.471358339480343408015545132262
S_{221}	0.002863553823128017435221922684	0.213995556297885214765130285694
		0.005579447145523524409701578704

Table 4. Quadrature rule on $d = 5$ dimensional simplex.

Orbit	Weight	Abscissas
27-points 4-order rule		
S_6	0.238095238095238095238095238095	0.166666666666666666666666666667
S_{51}	0.047619047619047619047619047619	0.083333333333333333333333333333
S_{33}	0.023809523809523809523809523809	0.311004233964073107793538617922
37-points 5-order rule		
S_6	0.153720220308429361772712636247	0.166666666666666666666666666667
S_{51}	0.028910622449315161561592816285	0.075000000000000000000000000000
S_{42}	0.027230105329857854702523915896	0.062093117793768044826243647512
S_{42}	0.017624297669854123221324781634	0.249411306984993017120659007161
102-points 6-order rule		
S_{51}	0.022060977769991841638517180921	0.093678479665790717950788318494
S_{51}	0.001028893984029374775200119260	0.027056643434076662571355869857
S_{42}	0.015626417261871945741838008061	0.065395098603733917972269240480
S_{42}	0.027828249444582554626634192403	0.229884418162665890105121333939
S_{321}	0.003494012814650919933176886532	0.018286803692430566770820358571
		0.196342639261513886645835928285
137-points 7-order rule		
S_{51}	0.025107991299585124669056837993	0.196250599802720238630278483591
S_{51}	0.026818177307254632568824859414	0.107306452949479294888911283341
S_{42}	0.008885610639738100803748773255	0.049969346573416854851613066075
S_{33}	0.015596510553760956859649640907	0.281229405057665572544934165951
S_{411}	0.001321513025263388127349264056	0.028735658249241368381255596936
		0.724302579453474918796971677329
S_{321}	0.003393053782162819391716791281	0.157327086232615167689860126229
		0.003654828611574876914707129176
257-points 8-order rule		
S_{51}	0.017630371189522179835961517082	0.106207926944053142785182181823
S_{51}	0.002226121210387036603556382974	0.044512875393854674753930540301
S_{42}	0.016674730579721612702949367108	0.221527165448792194555643607607
S_{33}	0.003966020462620965451627027936	0.028736243970238229827352135430
S_{411}	0.001371276128902419350510203067	0.030280731662816118424551232724
		0.574262524074710111906196422273
S_{321}	0.000926197175246393629294125774	0.017865374241004182434331661713
		0.159948503554659605076809985667
S_{321}	0.004831192109776069322662120503	0.097117546422468953758619774787
		0.350913592003902556659821964299
S_{321}	0.002747300611398014069223844427	0.154259841783653690445787981895
		0.017530190266106349578962599571

Table 5. Quadrature rule on $d = 6$ dimensional simplex.

Orbit	Weight	Abscissas
43-points 4-order rule		
S_7	0.16689962424064264240655553487802	0.14285714285714285714285714285714
S_{61}	0.02716619815142700769036736200866	0.07120154347010901732552543625004
S_{43}	0.01836962824855338010745351763331	0.03787627104219600219620536572980
64-points 5-order rule		
S_7	0.10556089403200693223264178793046	0.14285714285714285714285714285714
S_{61}	0.02429904195320186500137946120510	0.07155392508439903058574731017070
S_{52}	0.01171346168792031576174415885910	0.05067728321030771781231841506430
S_{43}	0.01366751762426438233603070421680	0.23043585212440365120245662379560
175-points 6-order rule		
S_{61}	0.00046104931565255285484082283370	0.02509909604870815447009085165340
S_{61}	0.01301991674586050465013068956160	0.16408824850302388029905815038860
S_{52}	0.00203064971090217995679119523050	0.02784407850016651933540912122510
S_{43}	0.01622220926263431272900952737070	0.05427117388472234767215445663260
S_{421}	0.00281158430208050822113571174900	0.12031965897287419105268484181550
		0.00375498171181802169768851192860
266-points 7-order rule		
S_{61}	0.01035837264537888252615510306590	0.16555370691703407135736243874300
S_{52}	0.01279465427717344053399913268920	0.08004169174138494538281587908680
S_{52}	0.00386665797691560684680540249746	0.04620602073726548357076393562060
S_{43}	0.00684822737381594150629804039420	0.22516267723705716736524194439130
S_{43}	0.00130065466676527607925405064060	0.01402083836117134817473437605620
S_{511}	0.00053218990985704857284890002180	0.02466780636399904904470747767340
		0.17596361300651512394911832179360
S_{421}	0.00257183456071513788304591409970	0.12428318118671194564818424084700
		0.00637231310142874735591922490677
553-points 8-order rule		
S_{61}	0.01195769984391890953221406683800	0.16467687533234213409428704255510
S_{61}	0.01700338552088890217399887775380	0.10107026106277182500519132582750
S_{61}	0.00157632710208893572203094203000	0.04450133014588455711806772835280
S_{43}	0.00299601348511639014786666776980	0.04442595335054347436540693296550
S_{43}	0.00578102644320970733099508033590	0.22110512716074526607395675836530
S_{511}	0.00070969810729333061947960575180	0.03038422111823568037998492356500
		0.25759784196158417691648228708090
S_{421}	0.00031727721601467282707436680400	0.01266863837585566447361723432550
		0.21017701247934510298958115975030
S_{421}	0.00152765862898539069491639528510	0.12326753489923003279547226294360
		0.00503160098647695485919297306620
S_{322}	0.00121674348099515619245218166200	0.09558682973748164107782263108660
		0.33778856869063836579701555683620

The integrals that enter the variational functional (5) on the domain $\Omega_h(z) = \bigcup_{q=1}^{Q} \Delta_q$, are expressed via the integrals, calculated on the element Δ_q, and recalculated to the local coordinates x on the element Δ,

$$\int_{\Delta_q} dz g_0(z) \varphi_r^p(z) \varphi_{r'}^p(z) V(z) = J \int_{\Delta} dx g_0(z(x)) \varphi_r^p(x) \varphi_{r'}^p(x) V(z(x)), \quad (13)$$

$$\int_{\Delta_q} dz g_{s_1 s_2}(z) \frac{\partial \varphi_r^p(z)}{\partial z_{s_1}} \frac{\partial \varphi_{r'}^p(z)}{\partial z_{s_2}}$$

$$= J \sum_{t_1,t_2=1}^{d} \hat{J}_{s_1 s_2; t_1 t_2}^{-1} \int_{\Delta} dx g_{s_1 s_2}(z(x)) \frac{\partial \varphi_r^p(x)}{\partial x_{t_1}} \frac{\partial \varphi_{r'}^p(x)}{\partial x_{t_2}},$$

where $J = \det \hat{J} > 0$ is the determinant of the matrix $\hat{J}$ from Eq. (12), $\hat{J}_{s_1 s_2; t_1 t_2}^{-1} = (\hat{J}^{-1})_{t_1 s_1} (\hat{J}^{-1})_{t_2 s_2}$, $dx = dx_1 \ldots dx_d$.

In the local coordinates, the LIP $\varphi_r^p(x)$ is equal to one at the node point ξ_r characterized by the numbers $[n_0, n_1, \ldots, n_d]$, and zero at the remaining node points ξ_r', i.e., $\varphi_r(\xi_r') = \delta_{rr}$ are determined by Eq. (7) at $H(0; x) = 1 - x_1 - \ldots - x_d$, $H(i; z) = x_i$, $i = 1, \ldots, d$:

$$\varphi_r(x) = \prod_{i=1}^{d} \prod_{n_i=0}^{n_i-1} \frac{x_i - n_i/p}{n_i/p - n_i/p} \prod_{n_0=0}^{n_0-1} \frac{1 - x_1 - \ldots - x_d - n_0/p}{n_0/p - n_0/p}. \quad (14)$$

Integrals (13) are evaluated using the Gaussian quadrature of the order $2p$.

Let ε_m and $\Phi_m(z)$ be exact solutions of Eq. (9) and ε_m^h and Φ_m^h be the corresponding numerical solutions. Then the following estimations are valid [20]

$$|\varepsilon_m - \varepsilon_m^h| \leq c_1 |\varepsilon_m| h^{2p}, \quad \|\Phi_m(z) - \Phi_m^h\|_0 \leq c_2 |E_m| h^{p+1}, \quad (15)$$

where $\|a(z)\|_0^2 = \langle a(z)|a(z) \rangle$, h is the maximal step of the finite-element grid, m is the number of the corresponding solution, and the positive constants c_1 and c_2 do not depend on the step h.

To solve the generalized eigenvalue problem (9), we choose the subspace iteration method [3, 20] elaborated by Bathe [3] for the solution of large symmetric banded-matrix eigenvalue problems. This method uses the skyline storage mode which stores the components of the matrix column vectors within the banded region of the matrix, and is ideally suited for banded finite-element matrices.

4   Construction of the d-dimensional Quadrature Formulas

Let us construct the d-dimensional p-ordered quadrature formula

$$\int_{\Delta_q} dz V(z) = |\Delta_q| \sum_{j=1}^{n_t} w_j V(z_j), \quad z = (z_1, \ldots, z_d), \quad dz = dz_1 \ldots dz_d, \quad (16)$$

for integration over the d-dimensional simplex Δ_q with vertices $\hat{z}_i = (\hat{z}_{i1}, \hat{z}_{i2}, \ldots,$ $\hat{z}_{id})$, $i = 0, \ldots, d$, which is exact for all polynomials of the variables $z_1, \ldots, z_d$ of degree not exceeding p, where n_t is the number of nodes that is determined during the calculation process. In Eq. (16), w_j, $j = 1, \ldots, n_t$ are the weights and $z_j = (z_{j1}, z_{j2}, \ldots, z_{jd})$ are the coordinates of nodes. $|\Delta_q|$ denotes the volume of Δ_q. For each node z_j, instead of sets of d coordinates we use the sets of $d+1$ barycentric coordinates (BC) $(x_{j0}, x_{j1}, \ldots, x_{jd})$:

$$z_j = x_{j0}\hat{z}_0 + \ldots + x_{jd}\hat{z}_d, \quad x_{j0} + \ldots + x_{jd} = 1. \tag{17}$$

For this purpose, we introduce the local coordinate system $x = (x_1, x_2, \ldots, x_d)$ and (12). Therefore, without loss of generality, we construct the d-dimensional p-ordered quadrature formula (16) on the standard simplex Δ with vertices $\hat{x}_j = (\hat{x}_{j1}, \ldots, \hat{x}_{jd})$, $\hat{x}_{jk} = \delta_{jk}$, $j = 0, \ldots, d$, $k = 1, \ldots, d$, which is exact for all polynomials of the variables $x_1, \ldots, x_d$ of degree not exceeding p:

$$\int_\Delta dx V(x) = \frac{1}{d!} \sum_{j=1}^{n_t} w_j V(x_{j0}, \ldots, x_{jd}). \tag{18}$$

Since the following formula is valid for all permutations $(l_0, \ldots, l_d)$ of $(k_0, \ldots, k_d)$:

$$\int_\Delta dx x_1^{l_1} \ldots x_d^{l_d} (1 - x_1 - \ldots - x_d)^{l_0} = \frac{\prod_{i=0}^d k_i!}{\left(d + \sum_{i=0}^d k_i\right)!},$$

we consider the fully symmetric Gaussian quadratures

$$\int_\Delta dx V(x) = \frac{1}{d!} \sum_{j=1}^a w_j \sum_{j_0, \ldots, j_d} V(x_{j_0 0}, x_{j_1 1}, \ldots, x_{j_d d}), \tag{19}$$

where the internal summation by $j_0, \ldots, j_d$ is carried out over the different permutations of $(x_{j0}, x_{j1}, \ldots, x_{jd})$. Table 1 presents the orbits and the corresponding number of different permutations for $d = 3, 4, 5, 6$. Here, for example, the orbit S_{331} at $d = 6$ contains BC $(\alpha, \alpha, \alpha, \beta, \beta, \beta, \gamma)$, $\alpha \neq \beta \neq \gamma$, $\alpha \neq \gamma$, $3\alpha + 3\beta + \gamma = 1$ and their different 140 permutations.

Substituting a monomial of the order not exceeding p in Eq. (19) instead of $V(x)$, we arrive at a system of nonlinear algebraic equations, that using the Vieta theorem reduces to the form:

$$\int_\Delta dx s_2^{l_2} s_3^{l_3} \times \ldots \times s_{d+1}^{l_{d+1}} = \frac{1}{d!} \sum_{j=1}^a w_j Q_j s_{j2}^{l_2} s_{j3}^{l_3} \times \ldots \times s_{jd+1}^{l_{d+1}}, \tag{20}$$

$$2l_2 + 3l_3 + \ldots + (d+1)l_{d+1} \leq p, \tag{21}$$

where

$$s_2 = \sum_{i=0, j\neq i}^d x_i x_j, \quad \ldots, \quad s_{d+1} = \prod_{i=0}^d x_i, \tag{22}$$

s_{ji}, $i = 2, \ldots, d+1$, are their values in the BC $(x_{j0}, x_{j1}, \ldots, x_{jd})$, and Q_j is the number of different permutation of the BC. As in Ref. [15], instead of Eq. (22), we can use

$$s_j = \sum_{i=0}^{d} x_i^j, \quad j = 2, \ldots, d+1. \tag{23}$$

The number of all $l_j \geq 0$ solutions of Eq. (21) provides the minimal number of independent nonlinear equations for the quadrature formula of the order p. It means that we can obtain a set of independent polynomials by adding new polynomials when increasing the order p. Below the first few independent polynomials of the order not exceeding $p \leq 6$ for $d \geq 5$ are presented:

$$\begin{aligned} V_1(x) &= s_1, & \text{for } p = 1, \\ V_2(x) &= s_2, & \text{for } p = 2, \\ V_3(x) &= s_3, & \text{for } p = 3, \\ V_4(x) &= s_2^2, V_5(x) = s_4, & \text{for } p = 4, \\ V_6(x) &= s_2 s_3, V_7(x) = s_5, & \text{for } p = 5, \\ V_8(x) &= s_2^3, V_9(x) = s_3^2, V_{10}(x) = s_2 s_4, V_{11}(x) = s_6, & \text{for } p = 6. \end{aligned} \tag{24}$$

We consider fully symmetric rules with positive weights, and no points are outside the simplex (the so-called PI-type).

The n_p-points p-order quadrature rules are constructed with Algorithm 1 [21] implemented by us in Maple and Fortran:

– **for** each decomposition n_p **do**

  repeat
  1. Randomly choose an initial guess for the unknowns n_t.
  2. Find a least square solution to Eqs. (20), (21) using a quasi-Newton algorithm.
  3. If a PI-type solution is found satisfying Eqs. (20), (21), with sufficient accuracy, go to Step 4.
  until maximum number of initial guesses tried.
– **end for**
– **Stop.**
– 4. Minimize the nonlinear equation for unknowns n_t using the Levenberg–Marquardt algorithm with high accuracy [12,14].

The Levenberg–Marquardt Algorithm 2:
Let $f(\mathbf{x})$ be twice differentiable with respect to the variable $\mathbf{x} = (x_1, \ldots, x_n)$. We consider the minimization

$$\min_{\mathbf{x} \in R^n} f(\mathbf{x}). \tag{25}$$

1. Start with an initial value $\mathbf{x}_0$, in S, an initial damping parameter λ_0, and a scaling parameter ρ. For $k \geq 0$ do the following:

2. Determine a trial iterate $\mathbf{y}$, using

$$\mathbf{y} = \mathbf{x}_k - (H_f(\mathbf{x}_k) + \lambda \, \mathrm{diag}(H_f(\mathbf{x}_k)))^{-1} \nabla f(\mathbf{x}_k), \qquad (26)$$

with $\lambda = \lambda_k \rho^{-1}$.

3. If $f(\mathbf{y}) < f(\mathbf{x}_k)$, where $\mathbf{y}$ is determined in Step 2, then set $\mathbf{x}_{k+1} = \mathbf{y}$ and $\lambda_{k+1} = \lambda_k \rho^{-1}$. Return to Step 2, replace k with $k+1$, and compute a new trial iterate.

4. If $f(\mathbf{y}) \geq f(\mathbf{x}_k)$ in Step 3, determine a new trial iterate, $\mathbf{y}$, using (26) with $\lambda = \lambda_k$.

5. If $f(\mathbf{y}) < f(\mathbf{x}_k)$, where $\mathbf{y}$ is determined in Step 4, then set $\mathbf{x}_{k+1} = \mathbf{y}$ and $\lambda_{k+1} = \lambda_k$. Return to Step 2, replace k with $k+1$, and compute a new trial iterate.

6. If $f(\mathbf{y}) \geq f(\mathbf{x}_k)$ in Step 5, then determine the smallest value of m so that when a trial iterate $\mathbf{y}$ is computed using (26) with $\lambda = \lambda_k \rho^m$, then $f(\mathbf{y}) < f(\mathbf{x}_k)$. Set $\mathbf{x}_{k+1} = \mathbf{y}$ and $\lambda_{k+1} = \lambda_k \rho^m$. Return to Step 2, replace k with $k+1$, and compute a new trial iterate.

7. Terminate the algorithm when $\|\nabla f(\mathbf{x}_k)\| < \epsilon$, where ϵ is the specified tolerance.

In the above Algorithm 2, $\nabla f(\mathbf{x})$, $H_f(\mathbf{x})$ are the gradient vector and the Hessian matrix functions of $f(\mathbf{x})$, respectively. $\mathrm{diag}(H_f(\mathbf{x}))$ is the diagonal matrix of the Hessian matrix function $H_f(\mathbf{x})$.

The weights (W) and the BC of PI-type rules of order p are presented in Tables 2, 3, 4 and 5. Here, for example, for the orbit S_{421} at $d = 6$ contains the BC $(\alpha, \alpha, \alpha, \alpha, \beta, \beta, \gamma)$, $\alpha \neq \beta \neq \gamma$, $\alpha \neq \gamma$ and their different 105 permutations. We present α in the first line and β in the second line, since γ is expressed in terms of α, β, i.e., $\gamma = 1 - 4\alpha - 2\beta$. The rules of the fifth and sixth order on tetrahedra coincide with the results of Ref. [2]. We believe that at least some of the rules presented in this paper are new. But we can not guarantee that the presented numbers of points of high-order quadrature rules are minimal. Note that up to the order $p = 6$ W and BC were calculated using Maple with 32 significant digits. For $p > 6$, W and BC were calculated using Fortran with 10 significant digits (the first three steps of Algorithm 1). These calculations were performed using the Central Information and Computer Complex, and HybriLIT heterogeneous computing cluster at JINR. Starting from the approximate values found with the Fortran code, W and BC were then calculated in Maple with 32 significant digits.

5 BVP for Helmholtz Equation in a d-dimensional Hypercube

For benchmark calculations, we use the BVP for the Helmholtz equation (HEQ) with the boundary condition (II) in a d-dimensional hypercube with the edge length π. Since the variables are separated, the eigenvalues $E_m = E_{m_1,\ldots,m_d}$ are sums of squared integers, $E_m = E_{m_1,\ldots,m_d} = m_1^2 + \ldots + m_d^2$, $m_k = 0, 1, \ldots$, $k = 1, \ldots, d$.

210   A. A. Gusev et al.

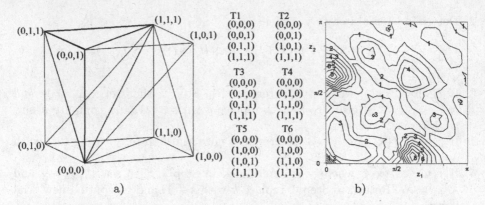

Fig. 1. (a) Division of a 3D cube into $3! = 6$ equal tetrahedrons (T1,...,T6). (b) The error $\Delta\Phi_8(z_1, z_2, z_3) = |\Phi_8^h(z_1, z_2, z_3) - \Phi_8(z_1, z_2, z_3)|$ for the eighth eigenfunction $\Phi_8^h(z_1, z_2, z_3)$ at fixed $z_3 = \pi/9$, calculated using FEM with third-order LIPs versus the exact eigenfunction $\Phi_8(z_1, z_2, z_3)$ corresponding to the eigenvalue $E_8 = 3$. Here the cube is divided into 2^3 cubes, each comprised of 6 tetrahedrons. The isolines marked 1 correspond to the values of $\Delta\Phi_8(z_1, z_2, z_3) = \Delta\Phi_8^{\max}/10$, the isolines marked 2 correspond to the values of $\Delta\Phi_8(z_1, z_2, z_3) = 2\Delta\Phi_8^{\max}/10, \ldots$, at $\Delta\Phi_8^{\max} \approx 0.018$.

Assertion (see also [16]). The hypercube is divided into $d!$ equal simplices. The vertices of each simplex are located on broken lines composed of d mutually perpendicular edges, and the extreme vertices of all polygons are located on one of the diagonals of the hypercube (for $d = 3$ see Fig. 1a).

Algorithm 3.

Input. A single d-dimensional hypercube with vertices the coordinates of which are either 0 or 1 in the Euclidean space $\mathcal{R}^d$. The chosen diagonal of the hypercube connects the vertices with the coordinates $(0, \ldots, 0)$ and $(1, \ldots, 1)$.

Output. $z_k^{(i)} = (z_{k1}^{(i)}, \ldots, z_{kd}^{(i)})$, the coordinates of the ith simplex.

Local. The coordinates of the vertices of the polygonal line are $z_k = (z_{k1}, \ldots, z_{kd})$, $k = 0, \ldots, d$.

1. For all $i = (i_1, \ldots, i_d)$, the permutations of the numbers $(1, \ldots, d)$:

1.1. For all $k = 0, \ldots, d$ and $s = 1, \ldots, d$: $z_{k,s}^{(i)} = \{1, i_s \leq k,; 0, i_s > k\}$

1.2. If $\det(z_{ks}^{(i)})_{ks-1}^d = -1$ then $z_{kd}^{(i)} \leftrightarrow z_{kd-1}^{(i)}$.

3D HEQ for the cube. In Fig. 1b, we show the error $\Delta\Phi_8(z_1, z_2, z_3)$ for the eighth eigenfunction $\Phi_8^h(z_1, z_2, z_3)$ at fixed $z_3 = \pi/9$, calculated using FEM with third-order LIPs versus the exact eigenfunction $\Phi_8(z_1, z_2, z_3)$ corresponding to the eigenvalue $E_8 = 3$. In Fig. 2a, we also show the maximal error $\Delta\Phi_8^{\max}$ for the exact eighth eigenfunction $\Phi_8(z_1, z_2, z_3)$ calculated using FEM with LIPs of the orders $p = 3, 4, 5$ versus the number N of piecewise basis functions $N_l^p(z)$ in the expansion (8). In Fig. 2b, we show the error of eigenvalues of the 3D BVP for the HEQ at $d = 3$ with the boundary condition (II) using the FEM scheme with 3D LIP of the order $p = 6$. As seen from Fig. 2, the errors of the eigenfunctions and eigenvalues lie on parallel lines in the double logarithmic scale

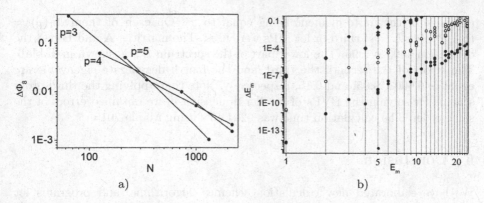

a)   b)

Fig. 2. (a) The maximal error $\Delta\Phi_8^{\max} = \max_{z_1} \in (0,\pi), z_2 \in (0,\pi), z_3 \in (0,\pi)|$ $\Phi_8^h(z_1,z_2,z_3) - \Phi_8(z_1,z_2,z_3)|$ for the exact eighth eigenfunction $\Phi_8(z_1,z_2,z_3)$ calculated using FEM with LIPs of the orders $p = 3, 4, 5$ versus the number N of piecewise basis functions $N_i^p(z)$ in the expansion (8). (b)The error $\Delta E_m = E_m^h - E_m$ calculated using FEM with sixth-order LIPs versus the exact eigenvalue E_m. Squares: the cube divided into 6 tetrahedrons. Circles: the cube divided into 2^3 cubes, each comprised of 6 tetrahedrons. Solid circles: the cube divided into 4^3 cubes, each comprised of 6 tetrahedrons.

Table 6. The lower part of the exact spectrum E_m and the calculated spectrum E_m^h for the 6D hypercube.

E_m	E_m^h
0	0.183360983479286 e−10
1	1.00023, 1.00034, 1.00034, 1.00034, 1.00034, 1.00034
2	2.04760, 2.04760, 2.04760, 2.04760, 2.04760, 2.04760, 2.04760, 2.04760, 2.04760, 2.07391, 2.08478, 2.08478, 2.08478, 2.08478, 2.08478
3	3.15060, 3.15196, 3.15196, 3.15196, 3.15196, 3.15196, 3.15780, 3.15780, 3.15780, 3.15780, 3.15780, 3.16319, 3.16319, 3.16319, 3.16319, 3.16319, 3.16319, 3.16319, 3.16319

which agrees with the theoretical error estimates (15) for the eigenfunctions and eigenvalues depending on the maximal size of the finite element. For a cube with the edge π divided into 4^3 cubes, each of them comprising 6 tetrahedrons, the matrices $\mathbf{A}$ and $\mathbf{B}$ had the dimension 15625×15625. The matrices $\mathbf{A}$ and $\mathbf{B}$ were calculated in two ways: analytically or with Gaussian quadratures from Sect. 4 using Maple 2015, 2x 8-core Xeon E5-2667 v2 3.3 GHz, 512 GB RAM, GPU Tesla 2075. For the considered task, the values of matrix elements agree with Gaussian quadratures up to the order 10 with given accuracy. The generalized algebraic eigenvalue problem (9) was solved during 20 min using Intel Fortran.

6D HEQ for the hypercube. We solved HEQ at $d = 6$ with the boundary condition (II) using FEM scheme with 6D LIP of the order $p = 3$. The 6D hypercube having the edge π was divided into $n = d! = 6! = 720$ simplexes (the size of the finite element being equal to π). On each of them $N_1(p) = (p + d)!/(d!p!) = 84$ third-order LIPs were used. The matrices $\mathbf{A}$ and $\mathbf{B}$ had the dimension 4096×4096. The lower part of the spectrum E_m is shown in Table 6. The errors of the second, the third, and the fourth degenerate eigenvalue are equal to 0.0003, 0.05, and 0.15, respectively. Note that applying the third-order scheme for solving the BVPs of smaller dimension d, we obtained errors of the same order. The calculation time was 9234.46 s using Maple 2015.

6   Conclusion

We have elaborated new calculation schemes, algorithms, and programs for solving the multidimensional elliptic BVP using the high-accuracy FEM with simplex elements. The elaborated symbolic-numerical algorithms and programs implemented in Maple-Fortran environment calculate multivariate finite elements in the simplex and the fully symmetric PI Gaussian quadrature rules. We demonstrated the efficiency of the proposed finite element schemes, algorithms, and codes by benchmark calculations of BVPs for Helmholtz equation of cube and hypercube. The developed approach is aimed at calculations of the spectral characteristics of nuclei models and electromagnetic transitions [7, 11]. This will be done in our next publications.

Acknowledgment. The work was partially supported by the RFBR (grant No. 16-01-00080 and 18-51-18005), the MES RK (Grant No. 0333/GF4), the Bogoliubov-Infeld program, the Hulubei–Meshcheryakov program, the RUDN University Program 5-100 and grant of Plenipotentiary of the Republic of Kazakhstan in JINR. The authors are grateful to prof. R. Enkhbat for useful discussions.

References

1. Abramowitz, M., Stegun, I.A.: Handbook of Mathematical Functions. Dover, New York (1965)
2. Akishin, P.G., Zhidkov, E.P.: Some symmetrical numerical integration formuas for simplexes. Communications of the JINR 11–81-395, Dubna (1981). (in Russian)
3. Bathe, K.J.: Finite Element Procedures in Engineering Analysis. Prentice Hall, Englewood Cliffs (1982)
4. Bériot, H., Prinn, A., Gabard, G.: Efficient implementation of high-order finite elements for Helmholtz problems. Int. J. Numer. Meth. Eng. **106**, 213–240 (2016)
5. Ciarlet, P.: The Finite Element Method for Elliptic Problems. North-Holland Publishing Company, Amsterdam (1978)
6. Cui, T., Leng, W., Lin, D., Ma, S., Zhang, L.: High order mass-lumping finite elements on simplexes. Numer. Math. Theor. Meth. Appl. **10**(2), 331–350 (2017)
7. Dobrowolski, A., Mazurek, K., Góźdź, A.: Consistent quadrupole-octupole collective model. Phys. Rev. C **94**, 054322-1–054322-20 (2017)
8. Dunavant, D.A.: High degree efficient symmetrical Gaussian quadrature rules for the triangle. Int. J. Numer. Meth. Eng. **21**, 1129–1148 (1985)

9. Gusev, A.A., et al.: Symbolic-numerical algorithm for generating interpolation multivariate hermite polynomials of high-accuracy finite element method. In: Gerdt, V.P., Koepf, W., Seiler, W.M., Vorozhtsov, E.V. (eds.) CASC 2017. LNCS, vol. 10490, pp. 134–150. Springer, Cham (2017). https://doi.org/10.1007/978-3-319-66320-3_11

10. Gusev, A.A., et al.: Symbolic-numerical algorithms for solving the parametric self-adjoint 2D elliptic boundary-value problem using high-accuracy finite element method. In: Gerdt, V.P., Koepf, W., Seiler, W.M., Vorozhtsov, E.V. (eds.) CASC 2017. LNCS, vol. 10490, pp. 151–166. Springer, Cham (2017). https://doi.org/10.1007/978-3-319-66320-3_12

11. Gusev, A.A., et al.: Symbolic algorithm for generating irreducible rotational-vibrational bases of point groups. In: Gerdt, V.P., Koepf, W., Seiler, W.M., Vorozhtsov, E.V. (eds.) CASC 2016. LNCS, vol. 9890, pp. 228–242. Springer, Cham (2016). https://doi.org/10.1007/978-3-319-45641-6_15

12. Levenberg, K.: A method for the solution of certain non-linear problems in least squares. Q. Appl. Math. **2**, 164–168 (1944)

13. www.maplesoft.com

14. Marquardt, D.: An algorithm for least squares estimation of parameters. J. Soc. Ind. Appl. Math. **11**, 431–441 (1963)

15. Maeztu, J.I., Sainz de la Maza, E.: Consistent structures of invariant quadrature rules for the n-simplex. Math. Comput. **64**, 1171–1192 (1995)

16. Mead, D.G.: Dissection of the hypercube into simplexes. Proc. Am. Math. Soc. **76**, 302–304 (1979)

17. Mysovskikh, I.P.: Interpolation Cubature Formulas. Nauka, Moscow (1981). (in Russian)

18. Papanicolopulos, S.-A.: Analytical computation of moderate-degree fully-symmetric quadrature rules on the triangle. arXiv:1111.3827v1 [math.NA] (2011)

19. Sainz de la Maza, E.: Fórmulas de cuadratura invariantes de grado 8 para el simplex 4-dimensional. Revista internacional de métodos numéricos para cálculo y diseño en ingeniería **15**(3), 375–379 (1999)

20. Strang, G., Fix, G.J.: An Analysis of the Finite Element Method. Prentice-Hall, Englewood Cliffs (1973)

21. Zhang, L., Cui, T.: Liu. H.: A set of symmetric quadrature rules on triangles and tetrahedra. J. Comput. Math. **27**, 89–96 (2009)

Symbolic-Numeric Simulation of Satellite Dynamics with Aerodynamic Attitude Control System

Sergey A. Gutnik[1($\boxtimes$)] and Vasily A. Sarychev[2]

[1] Moscow State Institute of International Relations (University),
76, Prospekt Vernadskogo, Moscow 119454, Russia
s.gutnik@inno.mgimo.ru
[2] Keldysh Institute of Applied Mathematics (Russian Academy of Sciences),
4, Miusskaya Square, Moscow 125047, Russia
vas31@rambler.ru

Abstract. The dynamics of the rotational motion of a satellite, subjected to the action of gravitational, aerodynamic and damping torques in a circular orbit is investigated. Our approach combines methods of symbolic study of the nonlinear algebraic system that determines equilibrium orientations of a satellite under the action of the external torques and numerical integration of the system of linear ordinary differential equations describing the dynamics of the satellite. An algorithm for the construction of a Gröbner basis was implemented for determining the equilibria of the satellite for specified values of the aerodynamic torque, damping coefficients, and principal central moments of inertia. Both the conditions of the satellite's equilibria existence and the conditions of asymptotic stability of these equilibria were obtained. The transition decay processes of the spatial oscillations of the satellite for various system parameters have also been studied.

1 Introduction

The study of the satellite dynamics under the influence of gravitational and aerodynamic torques is an important topic for practical implementation of attitude control systems of the artificial satellites. The gravity orientation systems are based on the result that a satellite with unequal moments of inertia in the central Newtonian force field in a circular orbit has stable equilibrium orientations [1–3]. An important property of the gravity orientation systems is that these systems can operate for a long time without fuel consumption. However, at altitudes from 250 up to 500 km, the rotational motion of a satellite is subjected to aerodynamic torque too. Therefore, it is necessary to study the joint action of gravitational and aerodynamic torques and, in particular, to analyze the possible satellite equilibria and conditions of stability of these equilibria in a circular orbit. The dynamics of a satellite subjected to gravitational and aerodynamic torques was considered in many papers indicated in [2]. The problem of determining the classes of equilibrium orientations for general values of aerodynamic

© Springer Nature Switzerland AG 2018
V. P. Gerdt et al. (Eds.): CASC 2018, LNCS 11077, pp. 214–229, 2018.
https://doi.org/10.1007/978-3-319-99639-4_15

torque was considered in [4–6]. In [7,8], some equilibrium orientations were found in special cases, when the center of pressure is located on a satellite's principal central axis of inertia and on a satellite's principal central plane of inertia. In [9], all equilibrium orientations were found in the case of axisymmetric satellite. In [10], all cases when the center of pressure is located in the satellite's principal central plane of inertia were considered using Computer Algebra methods. The basic problems of the satellite dynamics with an aerodynamic attitude control system have been presented in [2,6,11]. In [11], necessary and sufficient conditions for the stability of the aligned equilibrium position of the satellite with the aerodynamic orientation system using the damping moments of the gyroscopes were obtained.

In this paper, we consider a new problem, when the satellite is subjected to aerodynamic, gravitational, and active damping torques. The dynamics of the gravitationally-oriented satellite under the action of the damping torque, without taking into account the influence of the atmosphere on the motion of the satellite, was studied in detail in [12]. The main extension here, in comparison with [12], is the consideration of the additional influence of the atmosphere on the dynamics of the satellite under the action of the damping torque. Adding the action of the aerodynamic moment to the satellite leads to the appearance of new parameters in the equations of motion, which complicates their solution, but at the same time, it allows us to obtain new equilibrium solutions. In particular, the appearance of an additional aerodynamic parameter in the algebraic equations determining the stationary motions of the satellite seriously affects the runtime and memory requirements of symbolic computations for solving these equations.

We assume that the center of pressure of aerodynamic forces is located on one of the principal central axes of inertia of the satellite and the damping torque depends on the projections of the angular velocity of the satellite. This damping torque may be provided by using the angular velocity sensors. The action of damping torques can ensure the asymptotic stability of the equilibria of the satellite with aerodynamic attitude control system. The investigation of satellite equilibria was performed by using the Computer Algebra Gröbner basis methods. The regions with an equal number of equilibria were specified by using the Meiman theorem [19] for the construction of discriminant hypersurfaces. The conditions of equilibria stability are determined as a result of an analysis of the linearized equations of motion using the Routh–Hurwitz criterion [20]. The types of transition decay processes of spatial oscillations of the satellite at different aerodynamic and damping parameters have been investigated numerically.

The question of finding regions of parameter space with certain equilibria properties also occurred in relevance to a biology problem and was presented at the CASC 2017 Workshop [21].

2 Equations of Motion

Consider the attitude motion of the satellite subjected to gravitational, aerodynamic, and damping torques in a circular orbit. We assume that the satellite is a triaxial rigid body, and active damping torques depend on the projections of the angular velocity of the satellite.

To write the equations of motion we introduce two right-handed Cartesian coordinate systems with origin at the satellite's center of mass O. The orbital coordinate system is $OXYZ$, where the OZ axis is directed along the radius vector from the Earth center of mass to the satellite center of mass; the OX axis is in the direction of the satellite orbital motion. Then, the OY axis is directed along the normal to the orbital plane. The satellite body coordinate system is $Oxyz$, where Ox, Oy, and Oz are the principal central axes of inertia of the satellite. The orientation of the satellite body coordinate system $Oxyz$ with respect to the orbital coordinate system is determined by means of the aircraft angles of pitch (α), yaw (β), and roll (γ) (Fig. 1), and the direction cosines in the transformation matrix between the orbital coordinate system $OXYZ$ and $Oxyz$ are expressed in terms of aircraft angles using the relations [2]:

$$
\begin{aligned}
a_{11} &= \cos(x, X) = \cos\alpha\cos\beta, \\
a_{12} &= \cos(y, X) = \sin\alpha\sin\gamma - \cos\alpha\sin\beta\cos\gamma, \\
a_{13} &= \cos(z, X) = \sin\alpha\cos\gamma + \cos\alpha\sin\beta\sin\gamma, \\
a_{21} &= \cos(x, Y) = \sin\beta, \\
a_{22} &= \cos(y, Y) = \cos\beta\cos\gamma, \\
a_{23} &= \cos(z, Y) = -\cos\beta\sin\gamma, \\
a_{31} &= \cos(x, Z) = -\sin\alpha\cos\beta, \\
a_{32} &= \cos(y, Z) = \cos\alpha\sin\gamma + \sin\alpha\sin\beta\cos\gamma, \\
a_{33} &= \cos(z, Z) = \cos\alpha\cos\gamma - \sin\alpha\sin\beta\sin\gamma.
\end{aligned}
\tag{1}
$$

For small oscillations of the satellite, the angles of pitch, yaw, and roll correspond to the rotations around the OY, OZ, and OX axes, respectively.

In the derivation of the equations of motion, we will make the following assumptions [2]:

(1) the atmospheric effect on the satellite is reduced to the drag force applied at the center of pressure and directed against the velocity of the satellite center of mass relative to the air; the pressure center is located on the axis Ox of the satellite. This assumption is fulfilled accurately enough for the shape of the satellite close to the spherical;

(2) the atmospheric effect on the translational motion of the satellite is negligible;

(1) the atmospheric drag by the rotating Earth is neglected.

These assumptions make it possible to simplify the mathematical model of the effect of the atmosphere on the rotational motion of the satellite and neglect its influence on the parameters of the circular orbit.

Let the damping torque, in addition to the aerodynamic torque, act on the satellite. Their integral vector projections on the axis Ox, Oy, and Oz are equal to the following values: $M_x = \bar{k}_1 p_1$, $M_y = \bar{k}_2 q_1$, and $M_z = \bar{k}_3 r_1$. Here $\bar{k}_1$, $\bar{k}_2$, and

$\bar{k}_3$ are the damping coefficients; p_1, q_1, and r_1 are the projections of the satellite angular velocity vector onto the axes Ox, Oy, and Oz; ω_0 is the angular velocity of the orbital motion of the satellite's center of mass. Then the equations of satellite attitude motion can be written in the Euler form:

$$Ap_1' + (C - B)q_1r_1 - 3\omega_0^2(C - B)a_{32}a_{33} + \bar{k}_1p_1 = 0,$$
$$Bq_1' + (A - C)r_1p_1 - 3\omega_0^2(A - C)a_{31}a_{33} + \omega_0^2H_1a_{13} + \bar{k}_2q_1 = 0,$$
$$Cr_1' + (B - A)p_1q_1 - 3\omega_0^2(B - A)a_{31}a_{32} - \omega_0^2H_1a_{12} + \bar{k}_3r_1 = 0, \qquad (2)$$

where

$$p_1 = (\alpha' + \omega_0)a_{21} + \gamma',$$
$$q_1 = (\alpha' + \omega_0)a_{22} + \beta'\sin\gamma, \qquad (3)$$
$$r_1 = (\alpha' + \omega_0)a_{23} + \beta'\cos\gamma.$$

Moreover, here A, B, and C are the principal central moments of inertia of the satellite. And $H_1 = -Qa/\omega_0^2$, Q is the drag force acting on the satellite, and $(a, 0, 0)$ are the coordinates of the satellite center of pressure in the reference frame $Oxyz$. For the aerodynamically stable construction of the satellite, the center of pressure lies behind its center of gravity and, therefore, $a < 0$. The prime denotes the differentiation with respect to time t.

Over the systems (2) and (3) applying the change of variables $(p, q, r) = (p_1/\omega_0, q_1/\omega_0, r_1/\omega_0)$ and after this introducing dimensionless parameters $\theta_A = A/B$, $\theta_C = C/B$, $\tilde{k}_1 = \bar{k}_1/B\omega_0$, $\tilde{k}_2 = \bar{k}_2/B\omega_0$, $\tilde{k}_3 = \bar{k}_3/B\omega_0$, $h_1 = H_1/B$, and $\tau = \omega_0 t$, we can rewrite (2) and (3), and finally put respectively (because it is transforming (2) and (3))

$$\theta_A\dot{p} + (\theta_C - 1)qr - 3(\theta_C - 1)a_{32}a_{33} + \tilde{k}_1p = 0,$$
$$\dot{q} + (\theta_A - \theta_C)rp - 3(\theta_A - \theta_C)a_{31}a_{33} + h_1a_{13} + \tilde{k}_2q = 0, \qquad (4)$$
$$\theta_C\dot{r} + (1 - \theta_A)pq - 3(1 - \theta_A)a_{31}a_{32} - h_1a_{12} + \tilde{k}_3r = 0,$$

where

$$p = (\dot{\alpha} + 1)a_{21} + \dot{\gamma},$$
$$q = (\dot{\alpha} + 1)a_{22} + \dot{\beta}\sin\gamma, \qquad (5)$$
$$r = (\dot{\alpha} + 1)a_{23} + \dot{\beta}\cos\gamma.$$

The dot denotes the differentiation with respect to τ.

3   Equilibrium Orientations of Satellite

Assuming the initial condition $(\alpha, \beta, \gamma) = (\alpha_0 = \text{const}, \beta_0 = \text{const}, \gamma_0 = \text{const})$ and also $A \neq B \neq C$ ($\theta_A \neq \theta_C \neq 1$), we obtain from (4) and (5) the equations

$$a_{22}a_{23} - 3a_{32}a_{33} + ka_{21} = 0,$$
$$(1 - \nu)(a_{21}a_{23} - 3a_{31}a_{33}) + h(a_{21}a_{32} - a_{22}a_{31}) + ka_{22} = 0, \qquad (6)$$
$$\nu(a_{21}a_{22} - 3a_{31}a_{32}) - h(a_{23}a_{31} - a_{21}a_{33}) + ka_{23} = 0,$$

which allow us to determine the satellite equilibria in the orbital coordinate system. Here we consider the special case when $\tilde{k}_1/(\theta_C - 1) = \tilde{k}_2/(1 - \theta_C) = \tilde{k}_3/(1 - \theta_C) = k$. This reduction in the number of parameters makes it possible . to simplify the system of equations and solve the problem. In (6), $h = h_1/(1 - \theta_C)$ and $\nu = (1 - \theta_A)/(1 - \theta_C)$.

Substituting the expressions for the direction cosines from (1) in terms of the aircraft angles into Eq. (6), we obtain three equations with three unknowns α, β, and γ. Another way of closing Eq. (6) is to add the following three conditions for the orthogonality of direction cosines:

$$a_{21}^2 + a_{22}^2 + a_{23}^2 - 1 = 0,$$
$$a_{31}^2 + a_{32}^2 + a_{33}^2 - 1 = 0, \qquad (7)$$
$$a_{21}a_{31} + a_{22}a_{32} + a_{23}a_{33} = 0.$$

Equations (6) and (7) form a closed system of equations with respect to the six direction cosines identifying the satellite equilibrium orientations. For this system of equations, we formulate the following problem: for given values of h, k, and ν, it is required to determine all the nine directional cosines, i.e., all satellite equilibrium orientations in the orbital coordinate system. After a_{21}, a_{22}, a_{23}, a_{31}, a_{32}, and a_{33} are found, the direction cosines a_{11}, a_{12}, and a_{13} can be determined from the conditions of orthogonality.

To find solutions of the algebraic system (6), (7) we used the algorithm for constructing the Gröbner bases [13]. The method for constructing a Gröbner basis is an algorithmic procedure for complete reduction of the problem involving systems of polynomials in many variables to consideration of a polynomial in one variable.

In our study, for Gröbner bases construction, we applied the command Groebner[Basis] from the package Groebner implemented in the computer algebra system Maple 15 [14]. We constructed the Gröbner basis of the system of six second-order polynomials (6), (7) with six variables a_{ij} ($i = 2, 3; j = 1, 2, 3$), with respect to the lexicographic ordering of variables by using option plex. In the list of polynomials F:=[f_i, $i = 1, 2, 3, 4, 5, 6$], f_i are the left–hand sides of the algebraic equations (6), (7). Thus, the Maple command used was as follows:

```
G:=map(factor,Groebner[Basis](F,plex(a31,a32,a33,a21,a22,a23)));
```

Here, calculating the Gröbner basis over the field of rational functions in h, k, and ν we compute the generic solutions of our problem only. In our task from the area of the satellite dynamics with aerodynamic attitude control system, the main goal of the study is to estimate a range of system parameters for which the satellite's equilibria exist.

It should be taking into account that in practice, it is difficult to ensure a constant value of the aerodynamic moment on the orbit and there are errors of the angular velocity sensors and the errors of the signals, which generate damping torques, the exact bifurcation values of the coefficients are very difficult to obtain. We are interested in estimating the size of regions in the space of system parameters where equilibria exist. In the case of parametric dynamical system solving, when the parameters reach non-generic solutions, the symbolic application based on comprehensive Gröbner bases [15], discriminant varieties [16], and comprehensive triangular decomposition [17,18] methods are used. In our task, we did not use these methods because we did not consider the cases of bifurcation values of the parameters, and for our problem, these methods are rather complicated.

Here we write down the polynomial in the Gröbner basis that depends only on one variable $x = a_{23}$. This polynomial has the form

$$P(x) = P_1(x)P_2(x) = 0, \tag{8}$$

where

$$
\begin{aligned}
P_1(x) &= x(x^2 - 1), \\
P_2(x) &= p_0 x^4 + p_1 x^2 + p_2 = 0, \\
p_0 &= \big(16(k^2 + (1-\nu)^2)(k^2 + \nu^2)h^4 \\
&\quad - 24(k^2 + \nu(1-\nu))\big(k^2 - 2\nu(1-\nu)\big)^2 h^2 \\
&\quad + 9(k^2 - 2\nu(1-\nu))^4\big)^2, \\
p_1 &= -h^2\big(64(k^2 + 4\nu^2)(k^2 + (1-\nu)^2)^2 h^8 \\
&\quad + 16\big((2+8\nu)k^8 + (72\nu^3 - 50\nu^2 + 8\nu + 7)k^6 \\
&\quad - 4(1-\nu)(48\nu^4 - 58\nu^3 + 20\nu^2 - 8\nu + 1)k^4 \\
&\quad + 4\nu(1-\nu)^2(32\nu^4 - 104\nu^3 + 100\nu^2 - 25\nu + 6)k^2 \\
&\quad + 192\nu^3((1-\nu)^5)h^6 + 12(k^2 - 2\nu(1-\nu))^2((40\nu - 21)k^6 \\
&\quad + 4(32\nu^3 - 28\nu^2 + 5\nu + 6)k^4 \\
&\quad + 4(1-\nu)(56\nu^4 - 78\nu^3 + 24\nu^2 + 13\nu + 3)k^2 \\
&\quad + 288\nu^2(1-\nu)^4)h^4 \\
&\quad - 36(k^2 - 2\nu(1-\nu))^4\big(2(8\nu - 5)k^4 + (16\nu^3 - 24\nu + 17) \\
&\quad + 48\nu(1-\nu)^3)h^2 + 27\big(k^2 - 2\nu(1-\nu)\big)^6((8\nu - 5)k^2 \\
&\quad + 12(1-\nu)^2)), \\
p_2 &= p_{21}p_{22}, \\
p_{21} &= -h^4 k^2 (k^2 + 4\nu^2 - 6\nu)^2 \\
p_{22} &= 4(k^2 + 4\nu^2)h^6 - 4(4k^4 + (14\nu^2 - 2\nu + 1)k^2 \\
&\quad + 4\nu^2(1 + 4\nu - 5\nu^2)h^4 \\
&\quad + 3\big(k^2 - 2\nu(1-\nu)\big)^2(7k^2 + 8\nu + 4\nu^2)h^2 - 9\big(k^2 - 2\nu(1-\nu)\big)^4.
\end{aligned}
$$

The left-hand side of (8) becomes zero under the conditions $P_1(x) = 0$, $P_2(x) = 0$. Whence follows that in order to determine the equilibria it is required to consider separately the three cases: the first $a_{23}^2 = 1$, the second $a_{23} = 0$, and the third $P_2(a_{23}) = 0$. It should also be taken into account that equilibrium solutions are determined only by such real roots (8) whose absolute values should be less than or equal to 1.

In the first case, when $a_{23} = \pm 1$, $(a_{21} = a_{22} = 0)$, system (6), (7) takes the form

$$
\begin{aligned}
-3\nu a_{31}a_{32} - ha_{31}a_{23} + ka_{23} &= 0, \\
a_{31}^2 + a_{32}^2 &= 1, \\
a_{23}^2 &= 1, \\
a_{33} = a_{21} = a_{22} &= 0.
\end{aligned}
\tag{9}
$$

The first two equations of system (9) can be written in a simpler form

$$
P_3(a_{32}) = 9\nu^2 a_{32}^4 \pm 6\nu ha_{32}^3 + (h^2 - 9\nu^2)a_{32}^2 \mp 6\nu ha_{32} + k^2 - h^2 = 0, \tag{10}
$$

$$
a_{31} = \pm \frac{k}{(3\nu a_{32} \pm h)}.
$$

Having solved system (10), one can determine all six direction cosines of system (9). The number of real roots of equations (10) does not exceed 8. It is possible to show that each real root a_{32} of equations (10) corresponds to one equilibrium solution of the original system (6), (7).

In studying the satellite equilibrium orientations in the first case, we determine the conditions for the existence of real roots of equations (10). To identify these conditions, we use the Meiman theorem [19], which yields that the decomposition of the space of parameters into domains with equal number of real roots is determined by the discriminant hypersurface.

In our case, the discriminant hypersurface is given by the discriminant of polynomial $P_3(a_{32})$. This hypersurface contains a component of codimension 1, which is the boundary of domains with equal number of real roots. The set of singular points of the discriminant hypersurface in the space of parameters k, h, and ν is given by the following system of algebraic equations:

$$
P_3(y) = 0, \quad P_3'(y) = 0. \tag{11}
$$

Here $y = a_{32}$, and the prime denotes the differentiation with respect to y.

We eliminate the variable y from system (11) by calculating the determinant of the resultant matrix of Eq. (11) and obtain an algebraic equation of the discriminant hypersurface as

$$
P_4(k, h, \nu) = h^6 - (k^2 + 27\nu^2)h^4 + 9\nu^2(20k^2 + 27\nu^2)h^2 - 9\nu^2(4k^2 - 9\nu^2)^2 = 0. \tag{12}
$$

Now we should check the change in the number of equilibria when the surface (12) is intersected. This can be done numerically by determining the number of equilibria at a point of each domain $P_4(k, h, \nu) = 0$ in the space of parameters k, h, and ν.

Figure 2 presents an example of the properties and form of the discriminant hypersurface $P_4(k, h, \nu) = 0$, which are two-dimensional cross sections of the surface in the plane (k, h) at the fixed value of parameter $\nu = 0.5$. Figure 2 shows the distributions of domains with equal number of real roots of Eq. (10) and indicates the domains where four and two real solutions exist as well as the domains where no real solutions exist (marked by 0).

In the second case, when $a_{23} = 0$, system (6), (7) takes the form

$$ka_{21} - 3a_{32}a_{33} = 0,$$
$$ka_{22} - 3(1 - \nu)a_{31}a_{33} + h(a_{21}a_{32} - a_{22}a_{31}) = 0,$$
$$\nu(a_{21}a_{22} - 3a_{31}a_{32}) + ha_{21}a_{33} = 0, \tag{13}$$
$$a_{21}^2 + a_{22}^2 = 1, \quad a_{21}a_{31} + a_{22}a_{32} = 0,$$
$$a_{31}^2 + a_{32}^2 + a_{33}^2 - 1 = 0.$$

From (13) we can obtain the following solutions:

$$a_{21} = a_{23} = a_{32} = 0, \quad a_{22}^2 = 1,$$
$$P_5(a_{33}) = 9(1 - \nu)^2 a_{33}^4 \pm 6(1 - \nu)ha_{33}^3$$
$$+ (h^2 - 9(1 - \nu)^2)a_{33}^2 \mp 6(1 - \nu)ha_{33} + k^2 - h^2 = 0, \tag{14}$$
$$a_{31} = \pm \frac{k}{3(1 - \nu)a_{33} \pm h}.$$

Note that if in the expressions for the coefficients P_5 from (14) the term $(1 - \nu)$ is replaced by ν, we obtain the form of the coefficients of the polynomial P_3 from (10). Therefore, the conditions for the existence of real roots of Eq. (14) will be determined by the discriminant (12), in which the term ν is replaced by $(1 - \nu)$. For example, for the value $\nu = 0.5$, the conditions for the existence of real roots of Eqs. (10) and (14) will be the same (see Fig. 2).

Now let us consider the third case for which the satellite equilibria are determined by the real roots of the biquadratic equation $P_2(a_{23}) = 0$ from (8). The number of real roots of the biquadratic equation $P_2(a_{23}) = 0$ is even and not greater than 4. For each solution, one can find from the second polynomial from the constructed Gröbner basis two values of a_{22} and, then, their respective values a_{21}. For each set of values a_{21}, a_{22}, and a_{23}, one can unambiguously define from original system (6), (7) the respective values of the direction cosines a_{31}, a_{32}, and a_{33}. Thus, each real root of the biquadratic Eq. (6) is matched with two sets of values a_{ij} (two equilibrium orientations). Since the number of real roots of biquadratic Eq. (6) does not exceed 4, the satellite at the third case can have no more than 8 equilibrium orientations.

Real solutions of the biquadratic equation from (8) exist in the case when the discriminant

$$D(k, h, \nu) = p_1^2 - 4p_0p_2 \tag{15}$$

is non-negative. Using symbolic computations, it is possible to factorize the discriminant (15) in rather simple form

$$D(k, h, \nu) = h^4 D_1(k, h, \nu)\big(D_2(k, h, \nu)\big)^2, \tag{16}$$

where

$$D_1(k, h, \nu) = 4h^4 + 4\big(k^4 - (1 + 4\nu(1 - \nu))k^2 - 6\nu(1 - \nu)\big]h^2$$
$$- (4k^2 - 9)[k^2 - 2\nu(1 - \nu))^2,$$
$$D_2(k, h, \nu) = 27\big(k^2 - 4(1 - \nu)^2\big)\big(k^2 - 2\nu(1 - \nu)\big)^5$$
$$- 32\big((k^2 + (1 - \nu)^2)^2(k^2 + 4\nu^2)h^6$$
$$+ 24\big(4k^8 + (22\nu^2 - 12\nu - 1)k^6$$
$$+ 2(1 - \nu)^2(1 + 4\nu + \nu^2)k^4$$
$$- 4\nu(1 - \nu)^2(6\nu - 21\nu^2 + 19\nu^3 - 1)k^2$$
$$+ 48\nu^3(1 - \nu)^5\big)h^4$$
$$- 18\big(k^2 - 2\nu(1 - \nu)\big)^3\big(5k^4 + 2(\nu^2 + 7\nu - 5)k^2$$
$$- 24\nu(1 - \nu)^3\big)h^2.$$

For the existence of real roots of biquadratic equation from (8), it is necessary to satisfy the inequality $D(k, h, \nu) \geq 0$ ($D_1(k, h, \nu) \geq 0$). In case of the $D_1(k, h, \nu) > 0$ ($D_2(k, h, \nu) \neq 0$) and $0 \leq a_{23}^2 \leq 1$ inequalities fulfillment, biquadratic Eq. (8) has four real roots a_{23}. The boundary of the regions of the necessary conditions for the existence of these solutions is the curve $D_1(k, h, \nu) = 0$.

The regions of the necessary conditions for the existence of the real solutions of biquadratic Eq. (8) on the plane (k, h) are presented in Figs. 3 and 4 for $\nu = 0.2$ and $\nu = 0.5$. For the values ν and $(1 - \nu)$ these regions coincide.

Thus, from Eq. (8), we can obtain all possible values of the direction cosine a_{23} and corresponding values a_{21}, a_{22}, a_{31}, a_{32}, and a_{33} satisfying the initial system (6), (7). Once the set of six values a_{21}, a_{22}, a_{23}, a_{31}, a_{32}, and a_{33} is found, the remaining three values a_{11}, a_{12}, and a_{13} can be uniquely determined from the conditions of the orthogonality of the directional cosines. So we can determine all the equilibrium orientations of the satellite under the influence of aerodynamic, gravitational, and damping torques.

4   Necessary and Sufficient Conditions of Asymptotic Stability Of the Equilibrium Orientations of Satellite

In order to study the necessary and sufficient conditions of asymptotic stability of the equilibrium orientations of System (6) and (7), let us linearize the system of differential Eqs. (4) and (5) in the vicinity of the specific equilibrium solution, from the case 2 ($a_{22}^2 = 1, a_{21} = a_{23} = 0$):

$$\alpha = \alpha_0, \quad \beta_0 = \gamma_0 = 0. \tag{17}$$

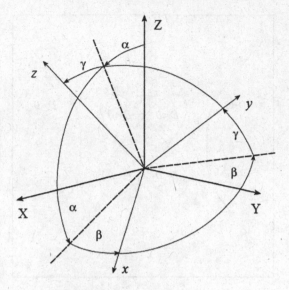

Fig. 1. Orientation of body–fixed axes with respect to the orbital coordinate system

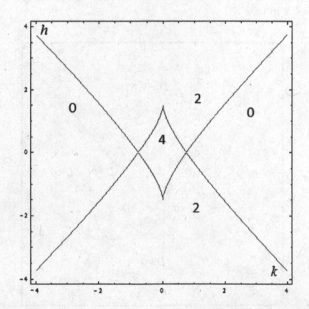

Fig. 2. The regions with the fixed number of equilibria for $\nu = 0.5$ for the cases 1, 2

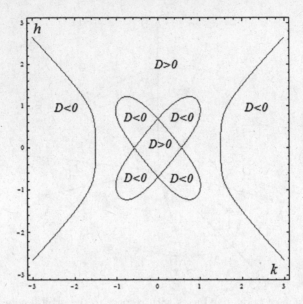

Fig. 3. The regions where the necessary conditions for the existence of equilibria are satisfied for $\nu = 0.2$ in case 3

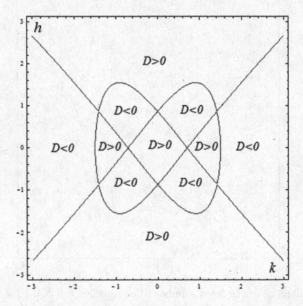

Fig. 4. The regions where the necessary conditions for the existence of equilibria are satisfied for $\nu = 0.5$ in case 3

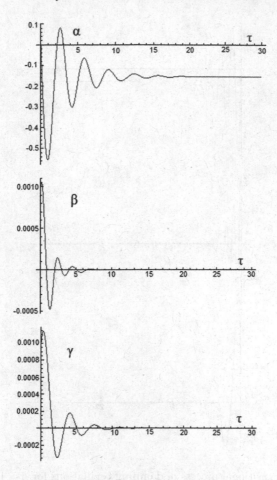

Fig. 5. The transitional process of damping oscillations for $k = 1.0; h = 5.0$

Here $\alpha_0 = \arccos(a_{33})$, where a_{33} is a real root of algebraic Eq. (14). We represent α, β, and γ in the form $\alpha = \alpha_0 + \bar{\alpha}$, $\beta = \beta_0 + \bar{\beta}$, $\gamma = \gamma_0 + \bar{\gamma}$, where $\bar{\alpha}$, $\bar{\beta}$ and $\bar{\gamma}$ are small deviations from the equilibrium orientation (17) of the satellite. The linearized system of equations of motion takes the following form:

$$\ddot{\bar{\alpha}} + (1 - \theta_C)k\dot{\bar{\alpha}} + \big((1 - \theta_C)h\cos\alpha_0 + 3(\theta_A - \theta_C)\cos 2\alpha_0\big)\bar{\alpha} = 0,$$

$$\theta_C\ddot{\bar{\beta}} + (1 - \theta_C)k\dot{\bar{\beta}} - (\theta_A + \theta_C - 1)\dot{\bar{\gamma}} + \big((1 - \theta_C)h\cos\alpha_0$$
$$+ (1 - \theta_A)(1 + 3\sin^2\alpha_0)\big)\bar{\beta} + \big(1.5(1 - \theta_A)\sin 2\alpha_0$$
$$- (1 - \theta_C)((1 - \theta_A)k + h\sin\alpha_0)\big)\bar{\gamma} = 0, \qquad (18)$$

$$\theta_A\ddot{\bar{\gamma}} - (1 - \theta_C)k\dot{\bar{\gamma}} + (\theta_A + \theta_C - 1)\dot{\bar{\beta}} + (1 - \theta_C)(1 + 3\cos^2\alpha_0)\bar{\gamma}$$
$$+ (1 - \theta_C)(1.5\sin 2\alpha_0 - k)\bar{\beta} = 0.$$

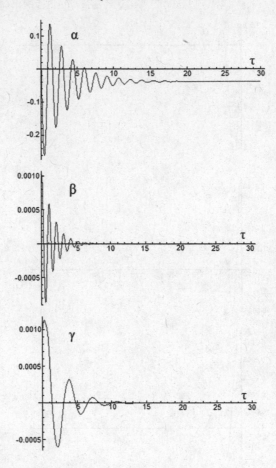

Fig. 6. The transitional process of damping oscillations for $k = 1.0; h = 25.0$

The characteristic equation of system (18)

$$(\lambda^2 + A_{01}\lambda + A_{02})(A_0\lambda^4 + A_1\lambda^3 + A_2\lambda^2 + A_3\lambda + A_4) = 0 \qquad (19)$$

decomposes into quadratic and 4th degree equations. Here the following notations are introduced:

$$A_{01} = (1 - \theta_C)k, \quad A_{02} = (1 - \theta_C)h\cos\alpha_0 + 3(\theta_A - \theta_C)\cos 2\alpha_0,$$
$$A_0 = \theta_A\theta_C, \quad A_1 = (1 - \theta_C)(\theta_A - \theta_C)k,$$
$$A_2 = (\theta_A + \theta_C - 1)^2 - (1 - \theta_C)^2k^2 + (1 - \theta_C)(\theta_A h + \theta_C(1 + 3\cos^2\alpha_0))$$
$$\qquad + \theta_A(1 - \theta_A(1 + 3\sin^2\alpha_0),$$
$$A_3 = k(1 - \theta_C)\big((1 - \theta_C)(1 + 3\cos^2\alpha_0 - h\cos\alpha_0) - (1 - \theta_A)(1 + 3\sin^2\alpha_0)\big)$$

$$+ (\theta_A + \theta_C - 1)\big((1 - \theta_C)(h\sin\alpha_0 + 1.5\sin 2\alpha_0) - 1.5(1 - \theta_A)\sin 2\alpha_0\big],$$
$$A_4 = (1 - \theta_C)(1 + 3\cos^2\alpha_0)\big((1 - \theta_C)h\cos\alpha_0 + (1 - \theta_A)(1 + 3\sin^2\alpha_0)\big)$$
$$+ (\theta_A + \theta_C - 1)\big((1 - \theta_C)(k + h\sin\alpha_0 - 1.5(1 - \theta_A)\sin 2\alpha_0\big).$$

The necessary and sufficient conditions for asymptotic stability (Routh–Hurwitz criterion) of the equilibrium solution (17) take the following form:

$$(1 - \theta_C)k > 0, \quad (1 - \theta_C)h\cos\alpha_0 + 3(\theta_A - \theta_C)\cos 2\alpha_0 > 0,$$
$$\Delta_1 = A_1 > 0,$$
$$\Delta_2 = A_1 A_2 - A_0 A_3 > 0, \tag{20}$$
$$\Delta_3 = A_1 A_2 A_3 - A_0 A_3^2 - A_1^2 A_4 > 0,$$
$$\Delta_4 = \Delta_3 A_4 > 0.$$

The detailed analysis of the fulfillment of inequalities (20), under which necessary and sufficient conditions for stability are satisfied was performed numerically for fixed values of the parameters θ_A, θ_C, k, and h. One should take into account also the following triangle inequalities for the real bodies, which parameters (θ_A and θ_C) should fulfill: $\theta_A + \theta_C > 1$, $\theta_C + 1 > \theta_A$, $\theta_A + 1 > \theta_C$. The triangle conditions isolate the infinite half-band in the (θ_A, θ_C) plane.

The numerical integration of system (4) and (5) was carried out for the fixed values of the parameters θ_A, θ_C, k, and h where the conditions of asymptotic stability (20) and the triangle inequalities hold. The different types of transition decay processes of spatial oscillations of the satellite at different inertial, aerodynamic, and damping parameters are presented in Figs. 5 and 6. The initial values of variables in the calculations were taken to be equal to 0.001.

Figure 5 shows that for rather small values of the damping coefficient and for small values of the aerodynamic torque ($k = 1, h = 5$; $\theta_A = 0.7, \theta_C = 0.4$), the system reaches the equilibrium solution (18) for α angle, when the τ value exceeds 15, and for β and γ angles, when the τ values are equal to about 10. Here equilibrium value $\alpha_0 = \arccos(a_{33}) = -0.155$ and $a_{33} = 0.988$ is the real root of algebraic Eq. (14).

When the value of the aerodynamic torque h increases the satellite oscillation frequency increases in angles α and β and the time of the transient process for $h = 25$, $k = 1.0$, ($\theta_A = 0.7, \theta_C = 0.4$) (Fig. 6) is close to 15 for α angle and less than 10 for β and γ angles. In Fig. 6, $\alpha_0 = -0.0377$. The value of the α angle approaches zero when the aerodynamic moment significantly increases.

In the case of the satellite with an aerogyroscopic stabilization system, when studying the dynamics of this system in [11] it was also shown that the satellite oscillation frequency increased in angles α and β when the magnitude of aerodynamic moment increased.

When the value of the damping coefficient increases, the time of the transient process of the system to the equilibrium solution decreases, for example when $k = 1.5$, $h = 25$ ($\theta_A = 0.7, \theta_C = 0.4$), the time of the transient process is less than 10 for all three angles. For $k = 2.0$, $h = 25$ ($\theta_A = 0.7, \theta_C = 0.4$), the transition time becomes less than 7 also for all three angles, which corresponds to one satellite turnover in the orbit.

5 Conclusion

In this paper, we present the study of the dynamics of the rotational motion of the satellite subject to the gravitational, aerodynamic, and active damping torques, which depend on the projections of satellite angular velocity.

The computer algebra method (based on the construction of Gröbner basis) of determining all equilibrium orientations of the satellite in the orbital coordinate system with given values of aerodynamic torque, damping coefficients and principal central moments of inertia was presented. The conditions for existence of these equilibria were obtained. We have made a detailed analysis of the evolution of domains of existence of equilibrium orientations in the plane of system parameters h and k for the fixed values of parameter ν.

For the special equilibrium orientation, when two axes of the satellite-centered coordinate system coincide with two axes of the orbital coordinate system, the necessary and sufficient conditions for asymptotic stability are obtained.

The numerical study of the character of transient processes of system, entering the special equilibrium orientation, has been carried out for various values of aerodynamic and damping parameters. It has been shown that there is a wide range of values of aerodynamic and damping parameters from which, choosing the required values of parameters, one can provide the asymptotic stability of the equilibrium orientation. The obtained results can be used to design aerodynamic attitude control systems for the artificial Earth satellites.

Acknowledgments. The authors thank the reviewers for very useful remarks and suggestions.

References

1. Beletsky, V.V.: Attitude Motion of Satellite in Gravitational Field. MGU Press, Moscow (1975)
2. Sarychev, V.A.: Problems of orientation of satellites, Itogi Nauki i Tekhniki. Ser. Space Research, vol. 11. VINITI, Moscow (1978)
3. Likins, P.W., Roberson, R.E.: Uniqueness of equilibrium attitudes for earth-pointing satellites. J. Astronaut. Sci. **13**(2), 87–88 (1966)
4. Gutnik, S.A.: Symbolic-numeric investigation of the aerodynamic forces influence on satellite dynamics. In: Gerdt, V.P., Koepf, W., Mayr, E.W., Vorozhtsov, E.V. (eds.) CASC 2011. LNCS, vol. 6885, pp. 192–199. Springer, Heidelberg (2011). https://doi.org/10.1007/978-3-642-23568-9_15
5. Sarychev, V.A., Gutnik, S.A.: Dynamics of a satellite subject to gravitational and aerodynamic torques. Investigation of equilibrium positions. Cosm. Res. **53**, 449–457 (2015)
6. Sarychev, V.A., Gutnik, S.A.: Satellite dynamics under the influence of gravitational and aerodynamic torques. A study of stability of equilibrium positions. Cosm. Res. **54**, 388–398 (2016)

7. Sarychev, V.A., Mirer, S.A.: Relative equilibria of a satellite subjected to gravitational and aerodynamic torques. Cel. Mech. Dyn. Astron. **76**(1), 55–68 (2000)
8. Sarychev, V.A., Mirer, S.A., Degtyarev, A.A.: Equilibria of a satellite subjected to gravitational and aerodynamic torques with pressure center in a principal plane of inertia. Cel. Mech. Dyn. Astron. **100**, 301–318 (2008)
9. Sarychev, V.A., Gutnik, S.A.: Dynamics of an axisymmetric satellite under the action of gravitational and aerodynamic torques. Cosm. Res. **50**, 367–375 (2012)
10. Gutnik, S.A., Sarychev, V.A.: A symbolic investigation of the influence of aerodynamic forces on satellite equilibria. In: Gerdt, V.P., Koepf, W., Seiler, W.M., Vorozhtsov, E.V. (eds.) CASC 2016. LNCS, vol. 9890, pp. 243–254. Springer, Cham (2016). https://doi.org/10.1007/978-3-319-45641-6_16
11. Sarychev, V.A., Sadov, Yu.A.: Analysis of a satellite dynamics with an gyrodamping orientation system. In: Obukhov, A.M., Kovtunenko, V.M. (eds.) Space Arrow. Optical Investigations of an Atmosphere, Nauka, Moscow, pp. 71–88 (1974)
12. Gutnik, S.A., Sarychev, V.A.: A symbolic study of the satellite dynamics subject to damping torques. In: Gerdt, V.P., Koepf, W., Seiler, W.M., Vorozhtsov, E.V. (eds.) CASC 2017. LNCS, vol. 10490, pp. 167–182. Springer, Cham (2017). https://doi.org/10.1007/978-3-319-66320-3_13
13. Buchberger, B.: A theoretical basis for the reduction of polynomials to canonical forms. SIGSAM Bull. **10**(3), 19–29 (1976)
14. Char, B.W., Geddes, K.O., Gonnet, G.H., Monagan, M.B., Watt, S.M.: Maple Reference Manual. Watcom Publications Limited, Waterloo (1992)
15. Weispfenning, V.: Comprehensive Gröbner bases. J. Symb. Comp. **14**(1), 1–30 (1992)
16. Lazard, D., Rouillier, F.: Solving parametric polynomial systems. J. Symb. Comp. **42**(6), 636–667 (1992)
17. Chen, C., Maza, M.M.: Semi-algebraic description of the equilibria of dynamical systems. In: Gerdt, V.P., Koepf, W., Mayr, E.W., Vorozhtsov, E.V. (eds.) CASC 2011. LNCS, vol. 6885, pp. 101–125. Springer, Heidelberg (2011). https://doi.org/10.1007/978-3-642-23568-9_9
18. Chen, C., Golubitsky, O., Lemaire, F., Maza, M.M., Pan, W.: Comprehensive triangular decomposition. In: Ganzha, V.G., Mayr, E.W., Vorozhtsov, E.V. (eds.) CASC 2007. LNCS, vol. 4770, pp. 73–101. Springer, Heidelberg (2007). https://doi.org/10.1007/978-3-540-75187-8_7
19. Meiman, N.N.: Some problems on the distribution of the zeros of polynomials. Uspekhi Mat. Nauk **34**, 154–188 (1949)
20. Gantmacher, F.R.: The Theory of Matrices. Chelsea Publishing Company, New York (1959)
21. England, M., Errami, H., Grigoriev, D., Radulescu, O., Sturm, T., Weber, A.: Symbolic versus numerical computation and visualization of parameter regions for multistationarity of biological networks. In: Gerdt, V.P., Koepf, W., Seiler, W.M., Vorozhtsov, E.V. (eds.) CASC 2017. LNCS, vol. 10490, pp. 93–108. Springer, Cham (2017). https://doi.org/10.1007/978-3-319-66320-3_8

Finding Multiple Solutions in Nonlinear Integer Programming with Algebraic Test-Sets

M. I. Hartillo[1](✉), J. M. Jiménez-Cobano[2](✉), and J. M. Ucha[1,2](✉)

[1] Departamento de Matemática Aplicada I., Universidad de Sevilla, Sevilla, Spain
{hartillo,ucha}@us.es
[2] Insituto de Matemáticas de la Universidad de Sevilla Antonio de Castro Brzezicki, Universidad de Sevilla, Sevilla, Spain
josjimcob@alum.us.es

Abstract. We explain how to compute all the solutions of a nonlinear integer problem using the algebraic test-sets associated to a suitable linear subproblem. These test-sets are obtained using Gröbner bases. The main advantage of this method, compared to other available alternatives, is its exactness within a quite good efficiency.

1 Introduction

In many real-life combinatorial optimization problems it is of great interest for the decision-maker to have not only one solution, but the set of *all* optimal solutions (see [15] or [11], for example). The information provided by this set can give some unexpected insights about the features of the solutions, and sometimes stands as a first step for multi-objective optimization as well.

On the other hand, sometimes these problems require nonlinear constraints to be modeled properly. In [14] a method for problems of the form

$$
\begin{aligned}
\min\ & cx^t \\
\text{s.t.}\ & Ax^t \le b^t \\
& x \in \Omega \\
& x \in \mathbb{N}^n
\end{aligned}
\tag{1}
$$

where $A \in \mathbb{Z}^{m \times n}, c \in \mathbb{Z}^n, b \in \mathbb{Z}^m$ (operator t stands for transposition) and the region Ω is finite and defined by linear and nonlinear constraints, was proposed. This method makes use of the so called *test-sets* associated to the linear subproblem

$$
\begin{aligned}
\min\ & cx^t \\
\text{s.t.}\ & Ax^t \le b^t \\
& x \in \mathbb{N}^n
\end{aligned}
\tag{2}
$$

First author is partially supported by MTM2016-75024-P and MTM2016-74983-C2-1-R. Third author is partially supported by MTM2016-75024-P, P12-FQM-2696 and MTM2016-74983-C2-1-R.

© Springer Nature Switzerland AG 2018
V. P. Gerdt et al. (Eds.): CASC 2018, LNCS 11077, pp. 230–237, 2018.
https://doi.org/10.1007/978-3-319-99639-4_16

Definition 1. *A set $T \subset \mathbb{Z}^n$ is a test-set associated to problem 2 if $T \subset \ker(A)$, and for any non optimal x feasible for 2 there exists a $t \in T$ such that $x - t$ is feasible and $c(x - t) < c(x)$.*

As a consequence of this definition, starting from an optimal point $\hat{x}$ of problem 2 you can *recover* the set of all the feasible points, adding elements of the test-set until you eventually complete all the feasible region. In this way, you can obtain the optimal points of problem 1 *walking back* from the linear optimal point until you reach the region Ω. Technical details can be found in [14]. The feasible region will be supposed finite, although this is not strictly necessary.

There are several ways of computing test-sets (as a matter of fact, they can be computed depending only on the cost and the matrix of constraints, not taking into account the right hand side, for example). One of the most efficient and manageable ways is using Gröbner bases (see for example [7]) with the software 4ti2 (see [9]). This approach is based in the classical paper [4], that shows how to solve a Linear Integer problem obtaining Gröbner bases of a suitable *binomial* ideal with respect to an ordering compatible with the cost function.

In [3,10] the method of [14] is applied to real-life size problems with very competitive results. In this work, we (1) explain how to modify the walk-back method to obtain all the optimal points, and (2) compare its performance with the natural generalization of the algorithm presented in [15] and with the commercial software BARON (see [1]).

Remark 1. The ideals corresponding to the method that we present in this work are not zero-dimensional, so some efficient strategies as Triangular Decomposition can not be applied for the Gröbner bases computations. In principle, the method proposed in [2] would be an alternative to treat some Nonlinear Integer Optimization problems with zero-dimensional ideals, but as soon as some constraints can not be expressed in terms of polynomials or the rank of values of the variables is big the method is useless.

2   Finding All the Optimal Points with a General Integer Cut

In [15] a method is introduced to show how to compute all the optimal points in Integer Linear Problems. Once an optimal point is obtained, the idea is to add some conditions to make it unfeasible and solve again. More precisely, if you are solving the problem 1 and have obtained an optimal solution $(x_1^0, \ldots, x_n^0)$ you can add the constraint

$$\sum_{i=1}^n |x_i - x_i^0| \geq 1$$

to assure that $(x_1^0, \ldots, x_n^0)$ is unfeasible for this new problem. As you obtain new optimal solutions you have to add a similar constraint for each solution. This formulation can be linearized, as it is explained in [15, Prop. 1]. This method can be used for Nonlinear Integer problems simply considering problems of the form

$$\min cx^t$$
$$\text{s.t. } Ax^t \leq b^t$$
$$\sum_{i=1}^{n} |x_i - x_i^j| \geq 1, j = 1, \ldots, N$$
$$x \in \Omega$$
$$x \in \mathbb{N}^n$$

with N constraints to try to obtain the $(N+1)$-th optimal point. One of the aims of this paper is to compare this natural approach to an alternative algebraic method.

3 Finding All the Optimal Points with Test-Sets

Our algorithm is based on the algebraic algorithm of [14] that provides one solution for a given nonlinear integer problem of the form

$$\min cx^t$$
$$\text{s.t. } Ax^t \leq b^t$$
$$x \in \Omega \quad (P)$$
$$x \in \mathbb{N}^n$$

where $A \in \mathbb{Z}^{m \times n}$, $c \in \mathbb{Z}^n$, and $b \in \mathbb{Z}^m$. Let us describe the steps of our method:

– We start, as in the algorithm proposed in [14], in the optimal point for a suitable linear subproblem

$$\min cx^t$$
$$\text{s.t. } Ax^t \leq b^t \; (P_L)$$
$$x \in \mathbb{N}^n$$

Remark 2. The selection of the subproblem has to do first with the computability of the test-set, that can be a bottleneck as it is a computation of a Gröbner basis of a certain ideal (see [7]). Moreover, although the test-set of the whole linear part of problem 1 is available, sometimes it is better to compute the test-set associated to a submatrix of A that gives us a more manageable number of directions to be considered at any point during the walk-back. The constraints that are not included in the submatrix are simply added to the description of Ω.

– Then we systematically add the elements of the corresponding test-set, thus worsening the cost function trying to obtain in return feasible points for problem 1, until we get into Ω.
– The difference to the original method (that was designed to find *only one optimal point*) is that now we have to manage the searching of new possible optimal points inside Ω, once we have reached a candidate. While in the algorithm of [14] you discard new points with the same optimal value, we instead stock them in a provisional list. This list will be the set of optimal points as long as we find a new better value inside the region Ω.

– If we eventually find an improvement in the cost we delete the provisional list. Otherwise, we already have the set of optimal points.

The pseudocode of our algorithm is the following one:

INPUT: $c, A, b; \Omega$; optimal point β of 2; T associated test-set of problem 2.

$Opt := \emptyset$;
$Leaves := \{\beta + t | \forall t \in T\} \cap \mathbb{N}^n$
$costOpt = \infty$

IF $\beta \in \Omega$
THEN $Opt := \{\beta\}$;
$\qquad costOpt := c\beta^t$

WHILE $(Leaves \neq \emptyset)$ DO
$\quad$ FOR $h \in Leaves$ DO
$\qquad$ IF $c(h) < costOpt$
$\qquad\qquad Leaves = (Leaves \setminus \{h\}) \cup (\{h + t | \forall t \in T\} \cap \mathbb{N}^n)$
$\qquad\qquad$ IF $h \in \Omega$
$\qquad\qquad$ THEN $Opt = \{h\}$;
$\qquad\qquad\qquad costOpt = ch^t$;
$\qquad\qquad\qquad Leaves = (Leaves \setminus \{h\}) \cup (\{h + t | \forall t \in T\} \cap \mathbb{N}^n)$

$\qquad\qquad\qquad$ ♯ the list of old candidates is deleted
$\qquad\qquad\qquad$ ♯ and updated with a new candidate

$\qquad$ ELSE IF $c(h) > costOpt$
$\qquad$ THEN $Leaves = Leaves \setminus \{h\}$
$\qquad$ ♯ these branches are discarded

$\qquad$ ELSE IF $c(h) = costOpt$
$\qquad$ THEN $Leaves = (Leave \setminus \{h\}) \cup (\{h + t | \forall t \in T\} \cap \mathbb{N}^n)$
$\qquad\qquad$ IF $h \in \Omega$
$\qquad\qquad$ THEN $Opt = Opt \cup \{h\}$;
$\qquad\qquad$ ♯ a new candidate to be an optimal point has been obtained

END WHILE

OUTPUT: Opt the set of all optimal points of problem 1 with cost $costOpt$

Remark 3. It is straightforward to modify this algorithm to obtain the K best optimal points for a given K, as in [11].

4 Computational Experiments

We have run all the examples to test our algorithm coded in Python in a computer with an Intel Core i5, 3.5 Ghz, 8 Gb of RAM, under Ubuntu. The examples with BARON [1] and COUENNE [6] have been sent to neos-server.org.

We have studied two families of examples: the integer portfolio problem and the problem of reliability in series-parallel systems.

4.1   Integer Portfolio Problem

In the integer portfolio problem (see [3]) we have to solve problems of the form

$$
\begin{aligned}
\max \ & \sum_{i=1}^{n} c_i x_i \\
s.t. \ & xCx^t \le R_0 \\
& \sum_{i=1}^{n} b_i x_i \le B \\
& x \in \mathbb{N}^n
\end{aligned}
\tag{3}
$$

where b_i are the prices today of n alternative investments or assets; c_i a forecast for their future prices; the matrix C has to do with the covariance matrix of the historical returns of the assets and it is a way of measuring the risk of a portfolio; R_0 stands for the maximum admissible risk; B is the available budget.

This so called *mean-variance portfolio model* was introduced by Markowitz with continuous variables (see [12] for the model and [5] or [16] for the mixed-integer case), but it is interesting to consider the case of integer variables: first to take into account the finite divisibility of the assets and second to consider some logical conditions that appear in these problems.

If you consider tailored examples for which two or more variables have the same price and risk you obtain many different optima and can compare our method to the generalization of [15]. As a general outcome we have obtained that as the number of optimal points increases our method overcomes by far the general cut method (coded in GAMS for COUENNE). Thereby, you can consider for example the simple case

$$
\begin{aligned}
\max \ & 2x_1 + x_2 + x_3 + \cdots + x_{n-1} + x_n \\
s.t. \ & x_1 + x_2 + x_3 + \cdots + x_{n-1} + x_n \le B \\
& \left(x_1 \ \cdots \ x_n \right) C \begin{pmatrix} x_1 \\ \vdots \\ x_n \end{pmatrix} \le R_0 \\
& x \in \mathbb{N}^n
\end{aligned}
\tag{4}
$$

with C defined by $c_{ii} = 0.05, c_{ij} = -0.01$ if $i \ne j$ for $n = 10, 15, 20, 50, 100$, $B = 10$ and a not very tight R_0 to include many points. In this family of examples you can obtain thousands of optimal points. In average our method is more than a hundred times faster for big numbers of optimal points.

Remark 4. Comparing with the commercial software BARON, that has the option of computing the K best optimal solutions, we obtain better running times only in 15% of the cases. Nevertheless, our method obtains exactly the complete set of optimal points in all cases. BARON, in contrast to our method, fails in 11% of the cases: in 5% of the cases it does not obtain the optimal cost and in 6% does not find the complete set, due to rounding problems.

4.2   Reliability Problems

The *redundancy allocation problem* can be formulated as the minimization of the design cost of a series-parallel system with multiple component choices, while ensuring a given system reliability level. The obtained model is a nonlinear integer programming problem with a nonlinear, nonseparable constraint (see [8,13] or [10]). It has the form

$$\min \sum_{i=0}^{n} \sum_{j=1}^{k} c_{ij} x_{ij}$$
$$s.a. \ R(x) \geq R_0$$
$$\sum_{j=1}^{k} x_{ij} \geq 1, \ \forall i = 1, \ldots, n \qquad (5)$$
$$0 \leq x_{ij} \leq u_{ij}, \ \forall i = 1, \ldots, n, \ j = 1, \ldots, k$$
$$x_{ij} \in \mathbb{Z}^+$$

with $R(x) = \prod_{i=1}^{n} \left(1 - \prod_{j=1}^{k} (1 - r_{ij})^{x_{ij}} \right)$. In this problem n is the number of subsystems (in series); k_i the number of different types of available components (in parallel) for the i- th subsystem, $1 \leq i \leq n$; r_{ij} the reliability of the j-th component for the i-th subsystem, $1 \leq i \leq n, 1 \leq j \leq k_i$; c_{ij}, the cost of the j-th component for the i-th subsystem, $1 \leq i \leq n, 1 \leq j \leq k_i$; l_{ij}, u_{ij}, lower/upper bounds of number of j components for the i-th subsystem, $1 \leq i \leq n, 1 \leq j \leq k_i$; R_0, an admissible level of reliability of the whole system; x_{ij}, number of j components used in the i-th subsystem, $1 \leq i \leq n, 1 \leq j \leq k_i$.

We have studied about 100 examples with 2, 3 and 4 subsystems and with 2 or 3 components in each subsystem (the costs and reliabilities generated randomly, $r_{ij} \in [0.90, 0.99]$ and $R_0 = 0.90$). The summary is in Tables 1, 2 and 3 and contains only the information about the examples with multiple number of solutions.

Table 1. Reliability examples $n = 3$, $k = 2, 3$.

	Average CPU Time	% Complete set of optimal solutions found
Test-set	0.2	100 %
BARON K-best	0.2	100 %
General cut	0.54	100 %

Table 2. Reliability examples $n = 4$, $k = 2$.

	Average CPU Time	% Complete set of optimal solution found
Test-set	0.46	100 %
BARON K-best	0.21	72.72 %
General cut	0.98	100 %

Table 3. Reliability examples $n = 4$; $k = 3$.

	Average CPU Time	% Complete set of optimal solution found
Test-set	1.1	100 %
BARON K-best	0.19	61.54 %
General cut	1.29	100 %

We can observe that:

- The test-set method and the general cut method are exact. BARON, on the contrary, does not give the complete set of optimal points or even one (in fact, usually provides only one although you ask for the best 3 or 4 best) in about 30 % of the cases.
- BARON is better in CPU time than the test-set method, and this is better than the general cut method, and much better as the number of optimal points increases substantially. If, for example, you treat an example with a hundred optimal points you have to solve 99 problems with 1, 2, ..., 99 new constraints, respectively.

5   Conclusions

We have presented an exact method to obtain the set of all optimal points for a given Nonlinear Integer problem as problem 1. A convenient linear subproblem is selected and then, walking back from an optimal point of this linear subproblem with the help of a test-set, the feasible region of the original problem is reached in all different ways, updating a list of optimal points. In this work we have studied two families of examples:

- Portfolio integer problems that can produce a huge number of optimal points and for which our algorithm overcomes a general cut approach as the one proposed in [15].
- Reliability problems, in which we point out problems of lack of exactness in the commercial software BARON compared with our approach.

References

1. Sahinidis, N. V.: BARON 14.4.0: Global Optimization of Mixed-Integer Nonlinear Programs. User's Manual (2014)
2. Bertsimas, D., Perakis, G., Tayur, S.: A new algebraic geometry algorithm for integer programming. Manag. Sci. **46**(7), 999–1008 (2000)
3. Castro, F.J., Gago, J., Hartillo, I., Puerto, J., Ucha, J.M.: An algebraic approach to integer portfolio problems. Eur. J. Oper. Res. **210**(3), 647–659 (2011)
4. Conti, P., Traverso, C.: Buchberger algorithm and integer programming. In: Mattson, H.F., Mora, T., Rao, T.R.N. (eds.) AAECC 1991. LNCS, vol. 539, pp. 130–139. Springer, Heidelberg (1991). https://doi.org/10.1007/3-540-54522-0_102

5. Corazza, M., Favaretto, D.: On the existence of solutions to the quadratic mixed-integer mean-variance portfolio selection problem. Eur, J. Oper. Res. **176**, 1947–1960 (2007)
6. Belotti, P.: COUENNE: a user's manual. Department of Mathematics Sciences, Clemson University. https://www.coin-or.org/Couenne/
7. Cox, D.A., Little, J., O'Shea, D.: Using Algebraic Geometry. Graduate Texts in Mathematics, vol. 185, 2nd edn. Springer, Hiedelberg (2005). https://doi.org/10.1007/978-1-4757-6911-1
8. Djerdjour, M., Rekab, K.: A branch and bound algorithm for designing reliable systems at a minimum cost. Appl. Math. Comput. **118**, 247–59 (2001)
9. 4ti2 team: 4ti2 - a software package for algebraic, geometric and combinatorial problems on linear spaces (2013). https://www.4ti2.de
10. Gago, J., Hartillo, I., Puerto, J., Ucha, J.M.: Exact cost minimization of a series-parallel reliable system with multiple component choices using an algebraic method. Comput. Oper. Res. **40**(11), 2752–2759 (2013)
11. Leão, A.A.S., Cherri, L.H., Arenales, M.N.: Determining the K-best solutions of knapsack problems. Comput. Oper. Res. **49**, 71–82 (2014)
12. Markowitz, H.: Portfolio selection. J. Finance **7**, 77–91 (1952)
13. Ruan, N., Sun, X.L.: An exact algorithm for cost minimization in series reliability systems with multiple component choices. Appl. Math. Comput. **181**, 732–41 (2006)
14. Tayur, S.R., Thomas, R.R., Natraj, N.R.: An algebraic geometry algorithm for scheduling in presence of setups and correlated demands. Math. Program. Ser. A **69**(3), 369–401 (1995). https://doi.org/10.1007/BF01585566
15. Tsai, J.-F., Lin, M.-H., Hu, Y.-C.: Finding multiple solutions to general integer linear programs. Eur. J. Oper. Res. **184**(2), 802–809 (2008)
16. Li, H.-L., Tsai, J.-F.: A distributed computation algorithm for solving portfolio problems with integer variables. Eur. J. Oper. Res. **186**(2), 882–891 (2008)

Positive Solutions of Systems of Signed Parametric Polynomial Inequalities

Hoon Hong[1] and Thomas Sturm[2,3]($\boxtimes$) (iD)

[1] North Carolina State University, Raleigh, NC, USA
hong@ncsu.edu
[2] CNRS, Inria, and the University of Lorraine, Nancy, France
thomas.sturm@loria.fr
[3] MPI Informatics and Saarland University, Saarbrücken, Germany
sturm@mpi-inf.mpg.de

Abstract. We consider systems of strict multivariate polynomial inequalities over the reals. All polynomial coefficients are parameters ranging over the reals, where for each coefficient we prescribe its sign. We are interested in the existence of positive real solutions of our system for all choices of coefficients subject to our sign conditions. We give a decision procedure for the existence of such solutions. In the positive case our procedure yields a parametric positive solution as a rational function in the coefficients. Our framework allows to reformulate heuristic subtropical approaches for non-parametric systems of polynomial inequalities that have been recently used in qualitative biological network analysis and, independently, in satisfiability modulo theory solving. We apply our results to characterize the incompleteness of those methods.

1 Introduction

We investigate the problem of finding a *parametric positive* solution of a system of *signed parametric* polynomial inequalities, if exists. We illustrate the problem by means of two toy examples:

$$f(x) = c_2 x^2 - c_1 x + c_0, \quad g(x) = -c_2 x^2 + c_1 x - c_0,$$

where c_2, c_1, c_0 are parameters. An expression $z(c)$ is called a parametric positive solution of $f(x) > 0$ if for all $c > 0$ we have $z(c) > 0$ and $f(z(c)) > 0$. One easily verifies that $z(c) = \frac{c_1}{c_2}$ is a parametric positive solution of $f(x)$. However, $g(x) > 0$ does not have any parametric positive solution since $g(x) > 0$ has no positive solution when, e.g., $c_2 = c_1 = c_0 = 1$. Of course, we are interested in tackling much larger cases with respect to numbers of variables, monomials, and polynomials.

The problem is important as systems of polynomial inequalities often arise in science and engineering applications, including, e.g., the qualitative analysis of biological or chemical networks [7,20,21,40] or *Satisfiability Modulo Theories (SMT) solving* [1,22,32]. In both these areas, one is indeed often interested in

V. P. Gerdt et al. (Eds.): CASC 2018, LNCS 11077, pp. 238–253, 2018.
https://doi.org/10.1007/978-3-319-99639-4_17

positive solutions. For instance, unknowns in the biological and chemical context of [7,20,21,40] are positive concentrations of species or reaction rates, where the direction of the reaction is known. In SMT solving, positivity is often not required but, in the satisfiable case, benchmarks typically have also positive solutions; comprehensive statistical data for several thousand benchmarks can be found in [22]. In many areas systems have parameters and one desires to have parametric solutions. Hence, an efficient and reliable tool for finding parametric positive solutions can aid scientists and engineers in developing and investigating their mathematical models.

The problem of finding parametric positive solutions is essentially that of quantifier elimination over the first order theory of real closed fields. In 1930, Tarski [38] showed that real quantifier elimination can be carried out algorithmically. Since then, there has been intensive research, producing profound theories with dramatically improved asymptotic complexity, e.g., [5,10,14,24,33]. Practical complexity was improved as well, often in combination with highly refined implementations, e.g., [2,8,11,13,17,23,25–28,30,35,36,41]. Today several implementations of real quantifier elimination are available in well-supported computer algebra software such as Maple [11,43], Mathematica [42, later editions online], Qepcad B [9], or Reduce [18,28]. However, existing general quantifier elimination software is still too inefficient for finding parametric positive solutions with relevant problem sizes in our above-mentioned fields of applications.

The main contribution of this paper is to provide simple and practically efficient algorithmic criteria for deciding whether or not a given signed parametric system has a parametric positive solution. To be precise, we reduce the problem to SMT solving over quantifier-free linear real arithmetic (QF_LRA). In case of existence we provide an explicit formula (rational function) for a parametric positive solution. The main challenge was eliminating many universal quantifiers in the problem statement. We tackled that challenge by, firstly, carefully approximating/bounding polynomials by suitable multiple of monomials and, secondly, tropicalizing, i.e., linearizing monomials by taking logarithms in the style of [39]. However, unlike standard tropicalization approaches, we determine sufficiently large *finite bases* for our logarithms, in order to get an explicit formula for parametric positive solutions.

Our main result also shines a new light on recent heuristic subtropical methods [22,37]: We provide a precise characterization of their incompleteness in terms of the existence of parametric positive solutions for the originally nonparametric input problems considered there. Furthermore our approach is applicable to generalized polynomials with real exponents. Such polynomials have been studied for related but different questions, also in the context of chemical reaction networks, in [31].

The paper is structured as follows. In Sect. 2, we motivate and present a compact and convenient notation for a system of multivariate polynomials, which will be used throughout the paper. In Sect. 3, we precisely define the key notions of *signed parametric systems* and *parametric positive solutions*. Then we present and prove the main result of this paper, which shows how to check the existence of a parametric positive solution and, in the positive case, how to find one. In Sect. 4, we apply our framework and our result to re-analyze and improve the above-mentioned subtropical methods [22,37].

2 Notation

The principal mathematical object studied in this paper are systems of multi-variate polynomials over the real numbers. In order to minimize cumbersome indices, we are going to introduce some compact notations. Let us start with a motivation by means of a simple example. We are going to use hat accents, like $\hat{f}$, for naming polynomials and systems with concrete coefficients in contrast to parametric ones, which we will introduce and discuss in the next section.

Example 1. Consider the following system of three polynomials in two variables:

$$\hat{f}_1 = -x_1^5 + 4x_1^2 x_2 - 2x_1^2 + x_2^2$$
$$\hat{f}_2 = 6x_1^5 + x_1^2 x_2 + 7x_1^2 - 3x_2^3$$
$$\hat{f}_3 = 4x_1^5 + x_1^2 x_2 - 2x_1^2 - 5x_2^3.$$

We rewrite those polynomials by aligning their signs, coefficients, and monomial support:

$$\hat{f}_1 = -1 \cdot 1 \cdot x_1^5 x_2^0 + 1 \cdot 4 \cdot x_1^2 x_2^1 + -1 \cdot 2 \cdot x_1^3 x_2^0 + \quad 0 \cdot 1 \cdot x_1^2 x_2^0 + 1 \cdot 1 \cdot x_1^0 x_2^2$$
$$\hat{f}_2 = \quad 1 \cdot 6 \cdot x_1^5 x_2^0 + 1 \cdot 1 \cdot x_1^2 x_2^1 + \quad 1 \cdot 7 \cdot x_1^3 x_2^0 + -1 \cdot 3 \cdot x_1^2 x_2^0 + 0 \cdot 1 \cdot x_1^0 x_2^2$$
$$\hat{f}_3 = \quad 1 \cdot 4 \cdot x_1^5 x_2^0 + 1 \cdot 1 \cdot x_1^2 x_2^1 + -1 \cdot 2 \cdot x_1^3 x_2^0 + -1 \cdot 5 \cdot x_1^2 x_2^0 + 0 \cdot 1 \cdot x_1^0 x_2^2,$$

where signs are represented by -1, 0, and 1. Note that we are writing 0 coefficients as $0 \cdot 1$ for notational uniformity. Rewriting this in matrix-vector notation, we have

$$
\begin{bmatrix} \hat{f}_1 \\ \hat{f}_2 \\ \hat{f}_3 \end{bmatrix} = \left(\begin{bmatrix} -1 & 1 & -1 & 0 & 1 \\ 1 & 1 & 1 & -1 & 0 \\ 1 & 1 & -1 & -1 & 0 \end{bmatrix} \circ \begin{bmatrix} 1 & 4 & 2 & 1 & 1 \\ 6 & 1 & 7 & 3 & 1 \\ 4 & 1 & 2 & 5 & 1 \end{bmatrix} \right) \begin{bmatrix} x_1^5 x_2^0 \\ x_1^2 x_2 \\ x_1^2 x_2^0 \\ x_1^0 x_2^3 \\ x_1^0 x_2^2 \end{bmatrix},
$$

where $\circ$ is the component-wise Hadamard product. Pushing this even further, we finally obtain

$$
\begin{bmatrix} \hat{f}_1 \\ \hat{f}_2 \\ \hat{f}_3 \end{bmatrix} = \left(\begin{bmatrix} -1 & 1 & -1 & 0 & 1 \\ 1 & 1 & 1 & -1 & 0 \\ 1 & 1 & -1 & -1 & 0 \end{bmatrix} \circ \begin{bmatrix} 1 & 4 & 2 & 1 & 1 \\ 6 & 1 & 7 & 3 & 1 \\ 4 & 1 & 2 & 5 & 1 \end{bmatrix} \right) \begin{bmatrix} x_1 \\ x_2 \end{bmatrix}^{\begin{bmatrix} 5 & 0 \\ 2 & 1 \\ 2 & 0 \\ 0 & 3 \\ 0 & 2 \end{bmatrix}}.
$$

This concludes our example.

In general, a system $\hat{f} \in \mathbb{R}[x_1, \ldots, x_d]^u$ of multivariate polynomials over the reals will be written compactly as

$$\hat{f} = (s \circ \hat{c})x^e,$$

where

$$\hat{f} = \begin{bmatrix} \hat{f}_1 \\ \vdots \\ \hat{f}_u \end{bmatrix}, \quad s = \begin{bmatrix} s_{11} & \cdots & s_{1v} \\ \vdots & & \vdots \\ s_{u1} & \cdots & s_{uv} \end{bmatrix}, \quad \hat{c} = \begin{bmatrix} \hat{c}_{11} & \cdots & \hat{c}_{1v} \\ \vdots & & \vdots \\ \hat{c}_{u1} & \cdots & \hat{c}_{uv} \end{bmatrix},$$

$$x = \begin{bmatrix} x_1 \\ \vdots \\ x_d \end{bmatrix}, \quad e = \begin{bmatrix} e_1 \\ \vdots \\ e_v \end{bmatrix} = \begin{bmatrix} e_{11} & \cdots & e_{1d} \\ \vdots & & \vdots \\ e_{v1} & \cdots & e_{vd} \end{bmatrix}.$$

We call $s \in \{-1, 0, 1\}^{u \times v}$ the *sign matrix*, $\hat{c} \in \mathbb{R}_+^{u \times v}$ the *coefficient matrix*, and $e \in \mathbb{N}^{v \times d}$ the *exponent matrix* of $\hat{f}$. The rows of the exponent matrix are named $e_1, \ldots, e_v$.

3   Main Result

Definition 2 (Signed Parametric Systems). A *signed parametric system* is given by

$$f = (s \circ c)x^e,$$

where the sign matrix $s \in \{-1, 0, 1\}^{u \times v}$ and the exponent matrix $e \in \mathbb{N}^{v \times d}$ are specified but the coefficient matrix c is unspecified in the sense that it is left parametric. Formally, c is a $u \times v$-matrix of pairwise different indeterminates.

When names of parameters and indeterminates are not important, signed parametric systems are uniquely determined by the sign matrix s and the exponent matrix e.

Example 3. The following is a signed parametric system derived from the system in Example 1:

$$\begin{bmatrix} f_1 \\ f_2 \\ f_3 \end{bmatrix} = \left(\begin{bmatrix} -1 & 1 & -1 & 0 & 1 \\ 1 & 1 & 1 & -1 & 0 \\ 1 & 1 & -1 & -1 & 0 \end{bmatrix} \circ \begin{bmatrix} c_{11} & c_{12} & c_{13} & c_{14} & c_{15} \\ c_{21} & c_{22} & c_{23} & c_{24} & c_{25} \\ c_{31} & c_{32} & c_{33} & c_{34} & c_{35} \end{bmatrix} \right) \begin{bmatrix} x_1 \\ x_2 \end{bmatrix}^{\begin{bmatrix} 5 & 0 \\ 2 & 1 \\ 2 & 0 \\ 0 & 3 \\ 0 & 2 \end{bmatrix}}.$$

This corresponds to

$$f_1 = -c_{11}x_1^5 + c_{12}x_1^2 x_2 - c_{13}x_1^2 + c_{15}x_2^2$$
$$f_2 = c_{21}x_1^5 + c_{22}x_1^2 x_2 + c_{23}x_1^2 - c_{24}x_2^3$$
$$f_3 = c_{31}x_1^5 + c_{32}x_1^2 x_2 - c_{33}x_1^2 - c_{34}x_2^3.$$

Definition 4 (Parametric Positive Solutions). Consider a signed parametric system $f = (s \circ c)x^e$. A *parametric positive solution* of $f(x) > 0$ is a function $z : \mathbb{R}_+^{u \times v} \to \mathbb{R}_+^d$ that maps each possible specification of the coefficient matrix c to a solution of the corresponding non-parametric system, i.e.,

$$\underset{c>0}{\forall}\, f\big(z(c)\big) > 0.$$

Theorem 5 (Main). *Let* $f = (s \circ c)x^e$ *be a signed parametric system. Let*

$$C(n) := \bigwedge_i \bigwedge_{s_{ik}<0} \bigvee_{s_{ij}>0} (e_j - e_k)n \geq 1.$$

Then the following are equivalent:

(i) $f(x) > 0$ *has a parametric positive solution.*
(ii) $C(n)$ *has a solution* $n \in \mathbb{R}^d$.
(iii) $C(n)$ *has a solution* $n \in \mathbb{Z}^d$.

In the positive case, the following function z *is a parametric positive solution of* $f(x) > 0$:

$$z(c) = t^n, \quad \text{where} \quad t = 1 + \sum_{\substack{s_{ij}>0 \\ s_{ik}<0}} \frac{c_{ik}}{c_{ij}}.$$

In fact, we even have $\underset{c>0}{\forall}\,\underset{r \geq t}{\forall}\, f(r^n) > 0.$

Proof. We first show that (i) implies (ii):

$$(i) \iff \underset{c>0}{\forall}\,\underset{x>0}{\exists}\, (s \circ c)x^e > 0$$

$$\iff \underset{c>0}{\forall}\,\underset{x>0}{\exists}\, \bigwedge_i \sum_{s_{ij}>0} c_{ij}x^{e_j} > \sum_{s_{ik}<0} c_{ik}x^{e_k}$$

$$\implies \underset{x>0}{\exists}\, \bigwedge_i \sum_{s_{ij}>0} x^{e_j} > \sum_{s_{ik}<0} 2vx^{e_k}, \quad \text{by instantiating } c$$

$$\implies \underset{x>0}{\exists}\, \bigwedge_i v \max_{s_{ij}>0} x^{e_j} > \max_{s_{ik}<0} 2vx^{e_k}$$

$$\iff \underset{x>0}{\exists}\, \bigwedge_i \max_{s_{ij}>0} x^{e_j} > \max_{s_{ik}<0} 2x^{e_k}$$

$$\iff \underset{x>0}{\exists}\, \bigwedge_i \bigwedge_{s_{ik}<0} \bigvee_{s_{ij}>0} x^{e_j} > 2x^{e_k}$$

$$\iff \underset{x>0}{\exists}\, \bigwedge_i \bigwedge_{s_{ik}<0} \bigvee_{s_{ij}>0} x^{e_j - e_k} > 2$$

$$\iff \underset{x>0}{\exists}\, \bigwedge_i \bigwedge_{s_{ik}<0} \bigvee_{s_{ij}>0} (e_j - e_k)\log_2 x > 1$$

$$\iff \underset{n \in \mathbb{R}^d}{\exists}\, \bigwedge_i \bigwedge_{s_{ik}<0} \bigvee_{s_{ij}>0} (e_j - e_k)n > 1, \quad \text{using } \log_2 : \mathbb{R}_+ \leftrightarrow \mathbb{R}$$

$$\implies (ii).$$

Assume now that (ii) holds. The existence of solutions $n \in \mathbb{R}^d$ and $n \in \mathbb{Q}^d$ of $C(n)$ coincides due to the Linear Tarski Principle: Ordered fields admit quantifier elimination for linear formulas, and therefore $\mathbb{Q}$ is an elementary substructure of $\mathbb{R}$ with respect to linear sentences [29]. Given a solution $n \in \mathbb{Q}^d$, we can use the principal denominator $\delta \geq 1$ of all coordinates of n to obtain a solution $\delta n \in \mathbb{Z}^d$. Hence (iii) holds.

We finally show that (iii) implies (i):

$$(\text{i}) \iff \bigvee_{c>0} \exists_{x>0} (s \circ c) x^e > 0$$

$$\iff \bigvee_{c>0} \exists_{x>0} \bigwedge_i \sum_{s_{ij}>0} c_{ij} x^{e_j} > \sum_{s_{ik}<0} c_{ik} x^{e_k}$$

$$\impliedby \bigvee_{c>0} \exists_{x>0} \bigwedge_i \max_{s_{ij}>0} c_{ij} x^{e_j} > \left(\sum_{s_{ik'}<0} c_{ik'} \right) \max_{s_{ik}<0} x^{e_k}$$

$$\iff \bigvee_{c>0} \exists_{x>0} \bigwedge_i \bigwedge_{s_{ik}<0} \bigvee_{s_{ij}>0} c_{ij} x^{e_j} > \left(\sum_{s_{ik'}<0} c_{ik'} \right) x^{e_k}$$

$$\iff \bigvee_{c>0} \exists_{x>0} \bigwedge_i \bigwedge_{s_{ik}<0} \bigvee_{s_{ij}>0} x^{e_j-e_k} > \sum_{s_{ik'}<0} \frac{c_{ik'}}{c_{ij}}$$

$$\impliedby \bigvee_{c>0} \exists_{x>0} \bigwedge_i \bigwedge_{s_{ik}<0} \bigvee_{s_{ij}>0} x^{e_j-e_k} \geq t, \quad \text{where } t = 1 + \sum_{\substack{s_{i'j'}>0 \\ s_{i'k'}<0}} \frac{c_{i'k'}}{c_{i'j'}}$$

$$\iff \bigvee_{c>0} \exists_{x>0} \bigwedge_i \bigwedge_{s_{ik}<0} \bigvee_{s_{ij}>0} (e_j - e_k) \log_t x \geq 1$$

$$\iff \exists_n \bigwedge_i \bigwedge_{s_{ik}<0} \bigvee_{s_{ij}>0} (e_j - e_k) n \geq 1, \quad \text{using } \log_t : \mathbb{R}_+ \leftrightarrow \mathbb{R}$$

$$\impliedby \exists_{n \in \mathbb{Z}^d} \bigwedge_i \bigwedge_{s_{ik}<0} \bigvee_{s_{ij}>0} (e_j - e_k) n \geq 1$$

$$\iff (\text{iii}).$$

In the proof of the implication from (iii) to (i) we have applied $\log_t$ so that $n = \log_t x$ and, accordingly, $x = t^n$, where t is as stated in the theorem. Notice that any larger choice $r \geq t$ would work there as well. $\square$

Example 6. Consider $\begin{bmatrix} f_1 \\ f_2 \end{bmatrix}$ with f_1, f_2 taken from Example 3:

$$f_1 = -c_{11} x_1^5 + c_{12} x_1^2 x_2 - c_{13} x_1^2 + c_{15} x_2^2$$
$$f_2 = c_{21} x_1^5 + c_{22} x_1^2 x_2 + c_{23} x_1^2 - c_{24} x_2^3.$$

This gives us

$$s = \begin{bmatrix} -1 & 1 & -1 & 0 & 1 \\ 1 & 1 & 1 & -1 & 0 \end{bmatrix} \quad \text{and} \quad e = \begin{bmatrix} 5 & 0 \\ 2 & 1 \\ 2 & 0 \\ 0 & 3 \\ 0 & 2 \end{bmatrix}.$$

```
(set-option :produce-models true)
(set-logic QF_LRA)
(declare-const n1 Real)
(declare-const n2 Real)
(assert (or (>= (+ (* (- 3) n1) (* 1 n2)) 1)
            (>= (+ (* (- 5) n1) (* 2 n2)) 1)))
(assert (or (>= (+ (* 0 n1) (* 1 n2)) 1)
            (>= (+ (* (- 2) n1) (* 2 n2)) 1)))
(assert (or (>= (+ (* 5  n1) (* (- 3) n2)) 1)
            (>= (+ (* 2  n1) (* (- 2) n2)) 1)
            (>= (+ (* 2 n1) (* (- 3) n2)) 1)))
(check-sat)
(get-model)
```

Fig. 1. An SMT-LIB input file for Example 6

We obtain $C(n)$ as follows:

$$(([2\ 1] - [5\ 0])n \geq 1 \vee ([0\ 2] - [5\ 0])n \geq 1) \wedge$$
$$(([2\ 1] - [2\ 0])n \geq 1 \vee ([0\ 2] - [2\ 0])n \geq 1) \wedge$$
$$(([5\ 0] - [0\ 3])n \geq 1 \vee ([2\ 1] - [0\ 3])n \geq 1 \vee ([2\ 0] - [0\ 3])n \geq 1),$$

which simplifies to

$$([-3\ 1]n \geq 1 \vee [-5\ 2]n \geq 1) \wedge ([0\ 1]n \geq 1 \vee [-2\ 2]n \geq 1) \wedge$$
$$([5\ -3]n \geq 1 \vee [2\ -2]n \geq 1 \vee [2\ -3]n \geq 1).$$

This straightforwardly yields the input file shown in Fig. 1. It uses the standardized SMT-LIB language [4] so that it can be directly processed by highly optimized SMT solvers like CVC4 [3], MathSat [12], SMT-RAT [15], Yices [19], or Z3 [16]. All these tools certify satisfiability and give a possible solution for n, which is called a *model* in the SMT world:

$$n = \begin{bmatrix} -\frac{5}{2} \\ -2 \end{bmatrix}.$$

Hence $(s \circ c)x^e > 0$ has a parametric positive solution, e.g.,

$$z(c) = \begin{bmatrix} t^{-\frac{5}{2}} \\ t^{-2} \end{bmatrix}, \quad \text{where} \quad t = 1 + \frac{c_{11}}{c_{12}} + \frac{c_{11}}{c_{15}} + \frac{c_{13}}{c_{12}} + \frac{c_{13}}{c_{15}} + \frac{c_{24}}{c_{21}} + \frac{c_{24}}{c_{22}} + \frac{c_{24}}{c_{23}}.$$

With this solution, in particular the non-parametric subsystem $\begin{bmatrix} \hat{f}_1 \\ \hat{f}_2 \end{bmatrix}$ of Example 1 is feasible. If $\hat{c}$ denotes the coefficient matrix there, then we can compute $t = \frac{719}{28}$ and

$$z(\hat{c}) = \begin{bmatrix} \frac{784\sqrt{20132}}{371694959} \\ \frac{1568}{516961} \end{bmatrix} \approx \begin{bmatrix} 0.0003 \\ 0.0015 \end{bmatrix}.$$

Fig. 2 illustrates the situation.

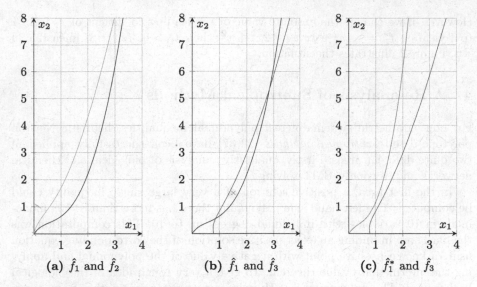

(a) $\hat{f}_1$ and $\hat{f}_2$ **(b)** $\hat{f}_1$ and $\hat{f}_3$ **(c)** $\hat{f}_1^*$ and $\hat{f}_3$

Fig. 2. Implicit plots of varieties of polynomials from Examples 6 and 7. **(a)** Both polynomials are positive in the region containing $(0.0003, 0.0015)$. Since this point is an instance of a parametric positive solution, there will be a suitable point under all modifications of absolute values of coefficients of the polynomials. **(b)** Both polynomials are positive in the region containing $(1.5, 1.5)$. **(c)** After modifying the absolute value of the leading coefficient of $\hat{f}_1$ the polynomials are not simultaneously positive in the first quadrant anymore.

Example 7. We slightly modify Example 6 and consider the subsystem $\begin{bmatrix} f_1 \\ f_3 \end{bmatrix}$ of Example 3:

$$f_1 = -c_{11}x_1^5 + c_{12}x_1^2 x_2 - c_{13}x_1^2 + c_{15}x_2^2$$
$$f_3 = c_{31}x_1^5 + c_{32}x_1^2 x_2 - c_{33}x_1^2 - c_{34}x_2^3.$$

That is

$$s = \begin{bmatrix} -1 & 1 & -1 & 0 & 1 \\ 1 & 1 & -1 & -1 & 0 \end{bmatrix}, \quad e = \begin{bmatrix} 5 & 0 \\ 2 & 1 \\ 2 & 0 \\ 0 & 3 \\ 0 & 2 \end{bmatrix}.$$

Computing $C(n)$ and generating SMT-LIB input analogously to Example 6, SMT solvers will return "unsat," which means that $C(n)$ does not have a solution $n \in \mathbb{R}^2$. Hence $(s \circ c)x^e > 0$, i.e. $f_1 > 0$, $f_3 > 0$, does not have a parametric positive solution.

Nevertheless, with the concrete instantiations $\hat{f}_1$, $\hat{f}_3$ from Example 1 the corresponding system $\hat{f}_1 > 0$, $\hat{f}_3 > 0$ of inequalities is feasible in $\mathbb{R}_+^2$. One possible solution is

$$\begin{bmatrix} \frac{3}{2} \\ \frac{3}{2} \end{bmatrix}.$$

However, if we change the absolute value of the leading coefficient of $\hat{f}_1$ from 1 to 4 yielding $\hat{f}_1^* = -4x_1^5 + 4x_1^2 x_2 - 2x_1^2 + x_2^2$, then $\hat{f}_1^* > 0$, $\hat{f}_3 > 0$ is infeasible in $\mathbb{R}_+^2$. Figure 2 illustrates the situation.

4   A Re-analysis of Subtropical Methods

For non-parametric systems of real polynomial inequalities, heuristic Newton polytope-based *subtropical methods* [22,37] have been successfully applied in two quite different areas: Firstly, qualitative analysis of biological and chemical networks and, secondly, SMT solving.

In the first area, a positive solution of a very large single inequality could be computed. The left hand side polynomial there has more than $8 \cdot 10^5$ monomials in 10 variables with individual degrees up to 10. This computation was the hard step in finding an exact positive solution of the corresponding equation using a known positive point with negative value of the polynomial and applying the intermediate value theorem. To give a very rough idea of the biological background: The polynomial is a Hurwitz determinant originating from a system of ordinary differential equations modeling mitogen-activated protein kinase (MAPK) in the metabolism of a frog. Positive zeros of the Hurwitz determinant point at Hopf bifurcations, which are in turn indicators for possible oscillation of the corresponding reaction network. For further details see [21].

In the second area, a subtropical approach for systems of several polynomial inequalities has been integrated with the SMT solver veriT [6]. That incomplete combination could solve a surprisingly large percentage of SMT benchmarks very fast and thus establishes an interesting heuristic preprocessing step for SMT solving over quantifier-free nonlinear arithmetic (QF_NRA). For detailed statistics see [22].

The goal of this section is to make precise the connections between subtropical methods and our main result here, to use these connections to improve the subtropical methods, and to precisely characterize their incompleteness.

4.1   Subtropical Real Root Finding

In [37] we have studied an incomplete method for heuristically finding a positive solution for a single multivariate polynomial inequality with fixed integer coefficients:

$$[\hat{f}_1] = (s \circ \hat{c})x^e \quad \text{where} \quad s \in \{-1,0,1\}^{1 \times v}, \quad \hat{c} \in \mathbb{Z}_+^{1 \times v}, \quad e \in \mathbb{N}^{v \times d}.$$

The method considers the positive and the negative support, which in terms of our notions is given by

$$S^+ = \{\, e_j \mid s_{1j} > 0 \,\}, \quad S^- = \{\, e_k \mid s_{1k} < 0 \,\}.$$

Then [37, Lemma 4] essentially states that $f_1(x) > 0$ has a positive solution if

$$C' := \bigvee_{e_j \in S^+} \exists_{n \in \mathbb{R}^d} \exists_{\gamma \in \mathbb{R}} \left([-e_j\ 1] \begin{bmatrix} n \\ \gamma \end{bmatrix} \leq -1 \wedge \bigwedge_{\substack{e_k \in S^+ \cup S^- \\ e_k \neq e_j}} [e_k\ -1] \begin{bmatrix} n \\ \gamma \end{bmatrix} \leq -1 \right).$$

Unfortunately, in [37, Lemma 4] vectors $e_l = [0 \cdots 0]$ corresponding to absolute summands are treated specially. We have noted already in [22, p. 192] that an inspection of the proof shows that this is not necessary. Therefore we discuss here a slightly improved and simpler version without that special treatment, which has been explicitly stated as [22, Lemma 2].

The proof of the loop invariant (I_1) in [37, Theorem 5(ii)] shows that the positive support need not be considered in the conjunction:

$$C' \iff \bigvee_{e_j \in S^+} \exists_{n \in \mathbb{R}^d} \exists_{\gamma \in \mathbb{R}} \left([-e_j\ 1] \begin{bmatrix} n \\ \gamma \end{bmatrix} \leq 1 \wedge \bigwedge_{e_k \in S^-} [e_k\ -1] \begin{bmatrix} n \\ \gamma \end{bmatrix} \leq -1 \right).$$

Starting with Fourier–Motzkin elimination [34, Sect. 12.2] of γ, we obtain

$$
\begin{aligned}
C' &\iff \bigvee_{e_j \in S^+} \exists_{n \in \mathbb{R}^d} \bigwedge_{e_k \in S^-} (e_k - e_j)n \leq -2 \\
&\iff \bigvee_{e_j \in S^+} \exists_{n \in \mathbb{R}^d} \bigwedge_{e_k \in S^-} (e_j - e_k)n \geq 1 \\
&\iff \exists_{n \in \mathbb{R}^d} \bigvee_{e_j \in S^+} \bigwedge_{e_k \in S^-} (e_j - e_k)n \geq 1 \\
&\iff \exists_{n \in \mathbb{R}^d} \max_{e_j \in S^+} (e_j n) \geq \max_{e_k \in S^-} (e_k n + 1) \\
&\iff \exists_{n \in \mathbb{R}^d} \bigwedge_{e_k \in S^-} \bigvee_{e_j \in S^+} (e_j - e_k)n \geq 1 \\
&\iff \exists_{n \in \mathbb{R}^d} C(n),
\end{aligned}
$$

with $C(n)$ as in Theorem 5.

Corollary 8. *Let $\hat{f} \in \mathbb{Z}[x_1, \ldots, x_d]$, say, $\hat{f} = (s \circ \hat{c})x^e$, where $s \in \{-1, 0, 1\}^{1 \times v}$, $\hat{c} \in \mathbb{Z}_+^{1 \times v}$, $e \in \mathbb{N}^{v \times d}$. Let $f = (s \circ c)x^e$, where c is a $1 \times v$-matrix of pairwise different indeterminates. Then the following are equivalent:*

(i) *The algorithm* find-positive *[37, Algorithm 1] does not fail, and thus finds a rational solution of $\hat{f} > 0$ with positive coordinates.*
(ii) *There is a row e_j of e with $s_{1j} > 0$ such that the following LP problem has a solution $n \in \mathbb{Q}^d$:*

$$\bigwedge_{s_{1k} < 0} (e_j - e_k)n \geq 1.$$

(iii) $f > 0$ has a parametric positive solution.

In the positive case, $\hat{f}(r^n) > 0$ for all $r \geq 1 + v \sum\limits_{s_{1k}<0} \hat{c}_{1k}$.

Proof. The equivalence between (i), (ii), and (iii) has been derived above.

According to Theorem 5, a solution for $\hat{f} > 0$ can be obtained by plugging $\hat{c}$ into the parametric positive solution $z(c) = t^n$ for f. Since we have positive integer coefficients, we can bound t from above as follows.

$$t = 1 + \sum_{\substack{s_{1j}>0 \\ s_{1k}<0}} \frac{\hat{c}_{1k}}{\hat{c}_{1j}} \leq 1 + \sum_{\substack{s_{1j}>0 \\ s_{1k}<0}} \frac{\hat{c}_{1k}}{1} \leq 1 + v \sum_{s_{1k}<0} \hat{c}_{1k}. \qquad \square$$

In simple words the equivalence between (i) and (iii) in the corollary states the following: The incomplete heuristic [37, Algorithm 1] succeeds *if and only if* not only the inequality for the input polynomial has a solution as required, but also the inequality for all polynomials with the same monomials and signs of coefficients as the input polynomial.

We have added (ii) to the corollary, because we consider this form optimal for algorithmic purposes. Our special case of one single inequality allows to transform the conjunctive normal form provided by Theorem 5 into an equivalent disjunctive normal form without increasing size. This way, a decision procedure can use finitely many LP solving steps [34] instead of employing more general methods like SMT solving [32].

Finally notice that the brute force search for a suitable t in find-positive [37, Algorithm 1, l.10–12] is not necessary anymore. Our corollary computes a suitable number from the coefficients.

4.2   Subtropical Satisfiability Checking

Subsequent work [22] takes an entirely geometric approach to generalize the work in [37] from one polynomial inequality to finitely many such inequalities. Consider a system with fixed integer coefficients in our notation:

$$\hat{f} = \begin{bmatrix} \hat{f}_1 \\ \vdots \\ \hat{f}_u \end{bmatrix} = (s \circ \hat{c})x^e, \quad \text{where} \quad s \in \{-1,0,1\}^{u \times v}, \quad \hat{c} \in \mathbb{Z}_+^{u \times v}, \quad e \in \mathbb{N}^{v \times d}.$$

Then [22, Theorem 12] derives essentially the following sufficient condition for the existence of a positive solution of $\hat{f} > 0$:

$$C'' := \exists_{n \in \mathbb{R}^d} \exists_{\gamma_1 \in \mathbb{R}} \cdots \exists_{\gamma_u \in \mathbb{R}} \bigwedge_{i=1}^{u} \left(\left(\bigvee_{s_{ij}>0} e_j n + \gamma_i > 0 \right) \wedge \bigwedge_{s_{ik}<0} e_k n + \gamma_i < 0 \right).$$

After an equivalence transformation, we can once more apply Fourier–Motzkin elimination [34, Sect. 12.2]:

$$C'' \iff \underset{n \in \mathbb{R}^d}{\exists} \bigwedge_{i=1}^{u} \bigvee_{s_{ij}>0} \underset{\gamma_i \in \mathbb{R}}{\exists} \left(e_j n + \gamma_i > 0 \wedge \bigwedge_{s_{ik}<0} e_k n + \gamma_i < 0 \right)$$

$$\iff \underset{n \in \mathbb{R}^d}{\exists} \bigwedge_{i=1}^{u} \bigvee_{s_{ij}>0} \bigwedge_{s_{ik}<0} (e_j - e_k)n > 0$$

$$\iff \underset{n \in \mathbb{R}^d}{\exists} \bigwedge_{i=1}^{u} \max_{s_{ij}>0} e_j n > \max_{s_{ik}<0} e_k n$$

$$\iff \underset{n \in \mathbb{R}^d}{\exists} \bigwedge_{i=1}^{u} \bigwedge_{s_{ik}<0} \bigvee_{s_{ij}>0} (e_j - e_k)n > 0$$

$$\iff \underset{n \in \mathbb{R}^d}{\exists} \bigwedge_{i=1}^{u} \bigwedge_{s_{ik}<0} \bigvee_{s_{ij}>0} (e_j - e_k)n \geq 1$$

$$\iff \underset{n \in \mathbb{R}^d}{\exists} C(n),$$

with $C(n)$ as in Theorem 5.

Corollary 9. Let $\hat{f} \in \mathbb{Z}[x_1, \ldots, x_d]^u$, say, $\hat{f} = (s \circ \hat{c})x^e$, where $s \in \{-1, 0, 1\}^{u \times v}$, $\hat{c} \in \mathbb{Z}_+^{u \times v}$, $e \in \mathbb{N}^{v \times d}$. Let $f = (s \circ c)x^e$, where c is a $u \times v$-matrix of pairwise different indeterminates. Then the following are equivalent:

   (i) The incomplete subtropical satisfiability checking method for several inequalities over QF_NRA (quantifier-free nonlinear real arithmetic) introduced in [22] succeeds on $\hat{f} > 0$.
   (ii) The following SMT problem with unknowns n is satisfiable over QF_LRA (quantifier-free linear real arithmetic):

$$\bigwedge_{i=1}^{u} \bigwedge_{s_{ik}<0} \bigvee_{s_{ij}>0} (e_j - e_k)n \geq 1.$$

(iii) $f > 0$ has a parametric positive solution.

In the positive case, $\hat{f}(r^n) > 0$ for all $r \geq 1 + v \sum_{s_{ik}<0} \hat{c}_{ik}$.

Proof. The equivalence between (i), (ii), and (iii) has been derived above. About the solution r see the proof of Corollary 8. □

The equivalence between (i) and (iii) in the corollary states the following: The procedure in [22] yields "sat" in contrast to "unknown" *if and only if* not only the input system is satisfiable, but that system with all real choices of coefficients with the same signs as in the input system. While there are no formal algorithms in [22], the work has been implemented within a combination of the veriT solver [6] with the library STROPSAT [22]. Our characterization applies in particular to the completeness of this software.

We have added (ii) to the corollary, because we consider this form optimal for algorithmic purposes. Like the original input C'' used in [22] this is a conjunctive normal form, which is ideal for DPLL-based SMT solvers [32]. Recall that u is the number of inequalities in the input, and d is the number of variables. Let ι and κ be the numbers of positive and negative coefficients, respectively. Then compared to [22] we have reduced $d + u$ variables to d variables, and we have reduced $u\kappa$ clauses with ι atoms each plus u unit clauses to some different $u\kappa$ clauses with ι atoms each but without any additional unit clauses.

With the :produce-models option the SMT-LIB standard [4] supports the computation of a suitable n in (ii), from which one can compute r^n using the bound at the end of the corollary. The work in [22] does not address the computation of solutions. It only mentions that sufficiently large r will work, which implicitly suggests a brute-force search like the one in [37, Algorithm 1, l.10–12].

Acknowledgments. This work has been supported by the European Union's Horizon 2020 research and innovation program under grant agreement No. H2020-FETOPEN-2015-CSA 712689 SC-SQUARE and by the bilateral project ANR-17-CE40-0036 and DFG-391322026 SYMBIONT. The second author would like to thank Georg Regensburger for his hospitality and an interesting week of inspiring discussions around the topic, and Dima Grigoriev for getting him started on the subject.

References

1. Ábrahám, E.: SC^2: satisfiability checking meets symbolic computation. In: Kohlhase, M., Johansson, M., Miller, B., de Moura, L., Tompa, F. (eds.) CICM 2016. LNCS (LNAI), vol. 9791, pp. 28–43. Springer, Cham (2016). https://doi.org/10.1007/978-3-319-42547-4_3
2. Arnon, D.S.: Algorithms for the geometry of semi-algebraic sets. Ph.D. thesis. Technical report 436, Computer Science Department, University of Wisconsin-Madison (1981)
3. Barrett, C., et al.: CVC4. In: Gopalakrishnan, G., Qadeer, S. (eds.) CAV 2011. LNCS, vol. 6806, pp. 171–177. Springer, Heidelberg (2011). https://doi.org/10.1007/978-3-642-22110-1_14
4. Barrett, C., Fontaine, P., Tinelli, C.: The SMT-LIB standard: version 2.6. Technical report, Department of Computer Science, The University of Iowa (2017)
5. Basu, S., Pollack, R., Roy, M.F.: On the combinatorial and algebraic complexity of quantifier elimination. JACM **43**(6), 1002–1045 (1996). https://doi.org/10.1145/235809.235813
6. Bouton, T., Caminha B. de Oliveira, D., Déharbe, D., Fontaine, P.: veriT: an open, trustable and efficient SMT-solver. In: Schmidt, R.A. (ed.) CADE 2009. LNCS (LNAI), vol. 5663, pp. 151–156. Springer, Heidelberg (2009). https://doi.org/10.1007/978-3-642-02959-2_12
7. Bradford, R.: A case study on the parametric occurrence of multiple steady states. In: Burr, M. (ed.) Proceedings of the ISSAC 2017, pp. 45–52. ACM, New York (2017). https://doi.org/10.1145/3087604.3087622

8. Brown, C.W.: Improved projection for CAD's of $\mathbb{R}^3$. In: Traverso, C. (ed.) Proceedings of the ISSAC 2000, pp. 48–53. ACM, New York (2000). https://doi.org/10.1145/345542.345575

9. Brown, C.W.: QEPCAD B: a program for computing with semi-algebraic sets using CADs. ACM SIGSAM Bull. **37**(4), 97–108 (2003). https://doi.org/10.1145/968708.968710

10. Canny, J.: Some algebraic and geometric computations in PSPACE. In: Simon, J. (ed.) Proceedings of the STOC 1988, pp. 460–467. ACM, New York (1988). https://doi.org/10.1145/62212.62257

11. Chen, C., Davenport, J.H., May, J.P., Moreno Maza, M., Xia, B., Xiao, R.: Triangular decomposition of semi-algebraic systems. J. Symb. Comput. **49**, 3–26 (2013). https://doi.org/10.1016/j.jsc.2011.12.014

12. Cimatti, A., Griggio, A., Schaafsma, B.J., Sebastiani, R.: The MathSAT5 SMT solver. In: Piterman, N., Smolka, S.A. (eds.) TACAS 2013. LNCS, vol. 7795, pp. 93–107. Springer, Heidelberg (2013). https://doi.org/10.1007/978-3-642-36742-7_7

13. Collins, G.E., Hong, H.: Partial cylindrical algebraic decomposition for quantifier elimination. J. Symb. Comput. **12**(3), 299–328 (1991). https://doi.org/10.1016/S0747-7171(08)80152-6

14. Collins, G.E.: Quantifier elimination for real closed fields by cylindrical algebraic decompostion. In: Brakhage, H. (ed.) GI-Fachtagung 1975. LNCS, vol. 33, pp. 134–183. Springer, Heidelberg (1975). https://doi.org/10.1007/3-540-07407-4_17

15. Corzilius, F., Kremer, G., Junges, S., Schupp, S., Ábrahám, E.: SMT-RAT: an open source C++ toolbox for strategic and parallel SMT solving. In: Heule, M., Weaver, S. (eds.) SAT 2015. LNCS, vol. 9340, pp. 360–368. Springer, Cham (2015). https://doi.org/10.1007/978-3-319-24318-4_26

16. de Moura, L., Bjørner, N.: Z3: an efficient SMT solver. In: Ramakrishnan, C.R., Rehof, J. (eds.) TACAS 2008. LNCS, vol. 4963, pp. 337–340. Springer, Heidelberg (2008). https://doi.org/10.1007/978-3-540-78800-3_24

17. Dolzmann, A., Seidl, A., Sturm, T.: Efficient projection orders for CAD. In: Gutierrez, J. (ed.) Proceedings of the ISSAC 2004, pp. 111–118. ACM, New York (2004). https://doi.org/10.1145/1005285.1005303

18. Dolzmann, A., Sturm, T.: REDLOG: computer algebra meets computer logic. ACM SIGSAM Bull. **31**(2), 2–9 (1997). https://doi.org/10.1145/261320.261324

19. Dutertre, B.: Yices 2.2. In: Biere, A., Bloem, R. (eds.) CAV 2014. LNCS, vol. 8559, pp. 737–744. Springer, Cham (2014). https://doi.org/10.1007/978-3-319-08867-9_49

20. England, M., Errami, H., Grigoriev, D., Radulescu, O., Sturm, T., Weber, A.: Symbolic versus numerical computation and visualization of parameter regions for multistationarity of biological networks. In: Gerdt, V.P., Koepf, W., Seiler, W.M., Vorozhtsov, E.V. (eds.) CASC 2017. LNCS, vol. 10490, pp. 93–108. Springer, Cham (2017). https://doi.org/10.1007/978-3-319-66320-3_8

21. Errami, H., Eiswirth, M., Grigoriev, D., Seiler, W.M., Sturm, T., Weber, A.: Detection of Hopf bifurcations in chemical reaction networks using convex coordinates. J. Comput. Phys. **291**, 279–302 (2015). https://doi.org/10.1016/j.jcp.2015.02.050

22. Fontaine, P., Ogawa, M., Sturm, T., Vu, X.T.: Subtropical satisfiability. In: Dixon, C., Finger, M. (eds.) FroCoS 2017. LNCS (LNAI), vol. 10483, pp. 189–206. Springer, Cham (2017). https://doi.org/10.1007/978-3-319-66167-4_11

23. González-Vega, L.: A combinatorial algorithm solving some quantifier elimination problems. In: Caviness, B.F., Johnson, J.R. (eds.) Quantifier Elimination and Cylindrical Algebraic Decomposition. Texts and Monographs in Symbolic Computation, pp. 365–375. Springer, Vienna (1998). https://doi.org/10.1007/978-3-7091-9459-1_19

24. Grigoriev, D., Vorobjov, N.: Solving systems of polynomial inequalities in subexponential time. J. Symb. Comput. **5**(1–2), 37–64 (1988). https://doi.org/10.1016/S0747-7171(88)80005-1

25. Hong, H.: An improvement of the projection operator in cylindrical algebraic decomposition. In: Watanabe, S., Nagata, M. (eds.) Proceedings of the ISSAC 1990, pp. 261–264. ACM, New York (1990). https://doi.org/10.1145/96877.96943

26. Hong, H.: Improvements in CAD-based quantifier elimination. Ph.D. thesis, The Ohio State University (1990)

27. Hong, H., Din, M.S.E.: Variant quantifier elimination. J. Symb. Comput. **47**(7), 883–901 (2012). https://doi.org/10.1016/j.jsc.2011.05.014

28. Košta, M.: New concepts for real quantifier elimination by virtual substitution. Doctoral dissertation, Saarland University, Germany (2016). https://doi.org/10.22028/D291-26679

29. Loos, R., Weispfenning, V.: Applying linear quantifier elimination. Comput. J. **36**(5), 450–462 (1993). https://doi.org/10.1093/comjnl/36.5.450

30. McCallum, S.: An improved projection operator for cylindrical algebraic decomposition. Ph.D. thesis, University of Wisconsin-Madison (1984)

31. Müller, S., Feliu, E., Regensburger, G., Conradi, C., Shiu, A., Dickenstein, A.: Sign conditions for injectivity of generalized polynomial maps with applications to chemical reaction networks and real algebraic geometry. Found. Comput. Math. **16**(1), 66–97 (2016). https://doi.org/10.1007/s10208-014-9239-3

32. Nieuwenhuis, R., Oliveras, A., Tinelli, C.: Solving SAT and SAT modulo theories: from an abstract Davis–Putnam–Logemann–Loveland procedure to DPLL(T). JACM **53**(6), 937–977 (2006). https://doi.org/10.1145/1217856.1217859

33. Renegar, J.: On the computational complexity and geometry of the first-order theory of the reals. Part II: the general decision problem. Preliminaries for quantifier elimination. J. Symb. Comput. **13**(3), 301–328 (1992). https://doi.org/10.1016/S0747-7171(10)80004-5

34. Schrijver, A.: Theory of Linear and Integer Programming. Wiley, Chichester (1986)

35. Strzebonski, A.: Cylindrical algebraic decomposition using validated numerics. J. Symb. Comput. **41**(9), 1021–1038 (2006). https://doi.org/10.1016/j.jsc.2006.06.004

36. Sturm, T.: Real quantifier elimination in geometry. Doctoral dissertation, University of Passau, Germany (1999)

37. Sturm, T.: Subtropical real root finding. In: Yokoyama, K., Linton, S., Robertz, D. (eds.) Proceedings of the ISSAC 2015, pp. 347–354. ACM, New York (2015). https://doi.org/10.1145/2755996.2756677

38. Tarski, A.: The Completeness of Elementary Algebra and Geometry. Institute Blaise Pascal, Paris (1930). Reprinted by CNRS 1967

39. Viro, O.: Dequantization of real algebraic geometry on logarithmic paper. CoRR arXiv:math/0005163 (2000)

40. Weber, A., Sturm, T., Abdel-Rahman, E.O.: Algorithmic global criteria for excluding oscillations. Bull. Math. Biol. **73**(4), 899–916 (2011). https://doi.org/10.1007/s11538-010-9618-0

41. Weispfenning, V.: A new approach to quantifier elimination for real algebra. In: Caviness, B.F., Johnson, J.R. (eds.) Quantifier Elimination and Cylindrical Algebraic Decomposition. Texts and Monographs in Symbolic Computation, pp. 376–392. Springer, Vienna (1998). https://doi.org/10.1007/978-3-7091-9459-1_20
42. Wolfram, S.: The Mathematica Book, 5th edn. Cambridge University Press, Cambridge (2003)
43. Yanami, H., Anai, H.: SyNRAC: a Maple toolbox for solving real algebraic constraints. In: Dolzmann, A., Seidl, A., Sturm, T. (eds.) Proceedings of the A3L 2005, pp. 275–279. BoD, Norderstedt (2005)

Qualitative Analysis of a Dynamical System with Irrational First Integrals

Valentin Irtegov and Tatiana Titorenko[✉]

Institute for System Dynamics and Control Theory SB RAS,
134, Lermontov str., Irkutsk 664033, Russia
{irteg,titor}@icc.ru

Abstract. We conduct qualitative analysis for a completely integrable system of differential equations with irrational first integrals. These equations originate from gas dynamics and describe adiabatical motions of a compressible gas cloud with homogeneous deformation. We study the mechanical analog of this gas dynamical system – the rotational motion of a spheroidal rigid body around a fixed point in a potential force field described by an irrational function. Within our study, equilibria, pendulum oscillations and invariant manifolds, which these solutions belong to, have been found. The sufficient conditions of their stability in Lyapunov's sense have been derived and compared with the necessary ones. The analysis has been performed with the aid of computer algebra tools which proved to be essential. The computer algebra system "Mathematica" was employed.

1 Introduction

Many different natural phenomena and processes can be described mathematically by the same equations. Such a mathematical analogy allows one to apply the methods developed for studying and an interpretation of phenomena and processes of one type to phenomena and processes of other type. Let us consider, e.g., the equations of adiabatical motions of an ideal gas in the form [9]:

$$\operatorname{div} \boldsymbol{v} = -\frac{1}{(\gamma - 1)} \frac{d}{dt} \ln T$$

$$\partial_t \boldsymbol{v} = \boldsymbol{v} \wedge \operatorname{rot} \boldsymbol{v} + T \boldsymbol{\nabla} S - \boldsymbol{\nabla}\left(\frac{v^2}{2} + \frac{\gamma T}{\gamma - 1}\right) \tag{1}$$

$$\partial_t S + \boldsymbol{v} \cdot \boldsymbol{\nabla} S = 0,$$

where $\boldsymbol{v}$ is the vector of velocity of the gas, T is the gas temperature, γ is the adiabatical index, and S is the entropy.

As was shown [6,16], in a Lagrangian formalism, when S is a quadratic function of Lagragian coordinates, and $\boldsymbol{v}$ depends on these linearly, partial differential equations (1) are reduced to ordinary ones and describe the motions of an ellipsoidal cloud of a compressible gas expanding freely in vacuum. The mechanical

© Springer Nature Switzerland AG 2018
V. P. Gerdt et al. (Eds.): CASC 2018, LNCS 11077, pp. 254–271, 2018.
https://doi.org/10.1007/978-3-319-99639-4_18

interpretation of these equations was given [6]. They are identical with the equations of motion of a point mass in nine-dimensional Euclidean space. This gas dynamical model was studied in a series of works, e.g., [1,15]. The present paper is based on the results [7,9].

In [7], under some assumptions, such as the gas is monatomic with the adiabatic index $\gamma = 5/3$, and there is neither rotation nor vorticity of the gas cloud, the above gas dynamical model was reduced to three second order ordinary differential equations. It was shown that they are equivalent to the equations of motion of a point mass on the unit 2-sphere, and an additional integral of 3rd degree in momenta has been derived for them. More general case was considered in [9] when the gas ellipsoid rotates around one of its principal axes. Then the equations of motion possess an additional first integral of 6th degree in momenta.

The study of the gas dynamical model proposed [16], [6] is ongoing to the present time towards generalizations of the found integrable cases [8]. A topological analysis of the integrable cases with the additional first integrals of 3rd and 6th degree has been done in [4]. In this work, the mechanical analog for the gas cloud – the motion of a point mass on the 2-sphere – was investigated. According to [3], the dynamics of a point mass on the 2-sphere is equivalent to the motion of a spheroidal rigid body in a potential force field at zeroth level of area integral. Thus, one can use this mechanical analog to study the gas dynamical model and to apply the methods developed for the analysis of dynamical systems of such type.

In the present work, the latter mechanical model is used for the qualitative analysis of the gas dynamical system. We analyze the differential equations of the spheroidal body in the above-mentioned integrable cases and obtain new results for both the gas system and the mechanical one. As is well-known, the problem of the qualitative analysis of differential equations is to find special solutions (equilibria, periodic motions, etc.) of these equations and to study their stability and bifurcations. Based on computer algebra methods, the computer analysis of the above problems can be performed in analytical form. The latter enables us to investigate the properties of the solutions under continuous (smooth) variation of their parameters. The research technique based on computer algebra methods as applied to the qualitative analysis of differential equations with first integrals is presented in the paper. The symbolic analysis is performed using built-in procedures of the computer algebra system "Mathematica" (CAS) and the "Mathematica" software package [2]. The procedures are used to solve computational problems arising in the study and to manipulate mathematical expressions. The package is employed to investigate the stability of the special solutions.

For finding the special solutions, the Routh–Lyapunov method [13] and its generalizations [12] are applied. By these methods, the qualitative analysis of differential equations with polynomial first integrals is reduced to algebraic problems solved efficiently by CAS. The first integrals in the problem under consideration are irrational; that is a special feature of the given problem. To avoid the use of fractional exponents and fractions (that is usually difficult in CAS), we transform irrational expressions to polynomial ones by introducing new variables.

In addition to the above methods for finding the special solutions, the chains of differential consequences [10] are applied. This technique mainly uses symbolic differentiation of expressions and is well suited for both algebraic expressions and irrational ones.

The paper is organized as follows. In Sect. 2, we analyze the equations of motion of the body when these possess the additional cubic integral in momenta. The special solutions of the equations are found and their stability is investigated. In Sect. 3, the same problems are solved for the equations of motion of the body when these have the additional integral of 6th degree in momenta. Finally, we discuss the obtained results and give a conclusion in Sect. 4.

2   The Integrable Case with the Additional Cubic Integral

2.1   Formulation of the Problem

Euler–Poisson's differential equations describing the motion of a spheroidal rigid body around a fixed point in a force field with the potential $2V = 3a\,(\gamma_1^2 + \gamma_2^2 + \gamma_3^2)(\gamma_1\gamma_2\gamma_3)^{-2/3}$ can be written as [5]

$$\dot{M}_1 = -[a\,(\gamma_2^2 - \gamma_3^2)\,(\gamma_1^2 + \gamma_2^2 + \gamma_3^2)]\,(\gamma_2\gamma_3)^{-5/3}\,\gamma_1^{-2/3},\, \dot{\gamma}_1 = \gamma_2 M_3 - \gamma_3 M_2,$$
$$\dot{M}_2 = \ [a\,(\gamma_1^2 - \gamma_3^2)\,(\gamma_1^2 + \gamma_2^2 + \gamma_3^2)]\,(\gamma_1\gamma_3)^{-5/3}\,\gamma_2^{-2/3},\, \dot{\gamma}_2 = \gamma_3 M_1 - \gamma_1 M_3, \quad (2)$$
$$\dot{M}_3 = -[a\,(\gamma_1^2 - \gamma_2^2)\,(\gamma_1^2 + \gamma_2^2 + \gamma_3^2)]\,(\gamma_1\gamma_2)^{-5/3}\,\gamma_3^{-2/3},\, \dot{\gamma}_3 = \gamma_1 M_2 - \gamma_2 M_1,$$

where M_i are the components of the kinetic momentum vector, γ_i are the direction cosines of "the vertical", a is some constant.

The above equations under the corresponding interpretation of the variables describe an expansion of the gas ellipsoid (without rotation) in vacuum. In this case, M_i are the impulses, $\gamma_i = A_i/\sqrt{\sum A_i^2}$, where A_i are the lengths of principal axes of the ellipsoid.

Equation (2) admit the following first integrals:

$$2H = M_1^2 + M_2^2 + M_3^2 + 3a\,(\gamma_1^2 + \gamma_2^2 + \gamma_3^2)\,(\gamma_1\gamma_2\gamma_3)^{-2/3} = 2h,$$
$$V_1 = \gamma_1^2 + \gamma_2^2 + \gamma_3^2 = 1, \ V_2 = M_1\gamma_1 + M_2\gamma_2 + M_3\gamma_3 = 0, \quad (3)$$
$$V_3 = M_1 M_2 M_3 - 3a\,(\gamma_1\gamma_2\gamma_3)^{1/3}(M_1\gamma_1^{-1} + M_2\gamma_2^{-1} + M_3\gamma_3^{-1}) = c_1.$$

Here V_3 is the additional integral derived in [7]. It is cubic with respect to M_1, M_2, M_3. This integral exists when the constant of the integral V_2 is equal to zero.

We can use the integrals having fixed constants for eliminating a part of the variables from differential equations (2) and the rest of the integrals to reduce the dimension of the problem. Let us eliminate the variable M_1 from Eq. (2) with the aid of $V_2 = 0$. They become:

$$\dot{M}_2 = a\,(\gamma_1^2 - \gamma_3^2)\,(\gamma_1^2 + \gamma_2^2 + \gamma_3^2)\,(\gamma_1\gamma_3)^{-5/3}\,\gamma_2^{-2/3},$$
$$\dot{M}_3 = -a\,(\gamma_1^2 - \gamma_2^2)\,(\gamma_1^2 + \gamma_2^2 + \gamma_3^2)\,(\gamma_1\gamma_2)^{-5/3}\,\gamma_3^{-2/3},$$
$$\dot{\gamma}_1 = \gamma_2 M_3 - \gamma_3 M_2,\, \dot{\gamma}_2 = -[M_2\gamma_2 + M_3(\gamma_1^2 + \gamma_3)]\,\gamma_3\gamma_1^{-1}, \quad (4)$$
$$\dot{\gamma}_3 = [M_2(\gamma_1^2 + \gamma_2) + M_3\gamma_3]\,\gamma_2\gamma_1^{-1}.$$

The first integrals of the above equations are:

$$2\tilde{H} = (M_2\gamma_2 + M_3\gamma_3)^2\,\gamma_1^{-2} + M_2^2 + M_3^2 + 3a\,(\gamma_1^2 + \gamma_2^2 + \gamma_3^2)\,(\gamma_1\gamma_2\gamma_3)^{-2/3} = 2\tilde{h},$$
$$V_1 = \gamma_1^2 + \gamma_2^2 + \gamma_3^2 = 1, \tilde{V}_3 = -M_2M_3\,(M_2\gamma_2 + M_3\gamma_3)\,\gamma_1^{-1}$$
$$-3a\,(\gamma_1\gamma_2\gamma_3)^{1/3}[M_2\gamma_2^{-1} + M_3\gamma_3^{-1} - (M_2\gamma_2 + M_3\gamma_3)\,\gamma_1^{-2}] = \tilde{c}_1. \tag{5}$$

Further, we conduct the qualitative analysis of Eq. (4). In the general case, this problem is to find special solutions (equilibria, periodic motions) and to investigate their qualitative properties. In the case of conservative systems, the variety of the special solutions is expanded through stationary sets. By these sets, we mean sets of any finite dimension on which the problem's first integrals (or their combinations) assume a stationary value. Zero-dimensional sets having this property are known as stationary solutions, while we shall call positive dimensional sets the stationary invariant manifolds (IMs).

Our goal is to find the stationary solutions and the IMs of Eq. (4) and to investigate their stability.

2.2   Finding Invariant Manifolds

According to the Routh–Lyapunov method, the stationary solutions and the IMs of the differential equations under consideration can be obtained by solving the conditional extremum problem for the first integrals of these equations. For this purpose, a linear or nonlinear combination of the first integrals (a family of the first integrals) is constructed and the necessary extremum conditions for this family with respect to the phase variables are written. Thus, in the case of algebraic first integrals, the problem of finding stationary solutions and IMs is reduced to solving a system of algebraic equations.

Following the technique chosen, we take the complete linear combination of the first integrals of the problem:

$$2K = 2\lambda_0\tilde{H} - \lambda_1V_1 - 2\lambda_3\tilde{V}_3, \tag{6}$$

where $\lambda_0, \lambda_1, \lambda_3$ are the parameters of the family of the integrals K, and write the necessary conditions for the integral K to have an extremum with respect to the phase variables:

$$\partial K/\partial M_2 = 0, \ \partial K/\partial M_3 = 0, \ \partial K/\partial\gamma_i = 0 \ (i = 1, 2, 3). \tag{7}$$

The solutions of system (7), when it is degenerate (its Jacobian is identically equal to zero), allow one to define the IMs (or their families) for differential equations (4) which correspond to the family of the first integrals K.

System (7) is that of five irrational equations with the parameters $a, \lambda_0, \lambda_1, \lambda_3$. We should transform these equations to polynomial ones to use computer algebra methods, e.g., Gröbner basis method, for finding their solutions. For this purpose, we introduce the new variables:

$$M_2 = M_2, M_3 = M_3, x_1 = \gamma_1, x_2 = \gamma_2^{1/3}\gamma_1^{-1/3}, x_3 = \gamma_3^{1/3}\gamma_1^{-1/3}. \tag{8}$$

In the above variables, the equations of motion (4) and first integrals (5) take the form

$$\dot{M}_2 = -a(x_3^6 - 1)(x_2^6 + x_3^6 + 1)\, x_2^{-2} x_3^{-5}, \quad 3\dot{x}_2 = -M_3\,(x_2^6 + x_3^6 + 1)\, x_2^{-2},$$
$$\dot{M}_3 = a(x_2^6 - 1)(x_2^6 + x_3^6 + 1)\, x_2^{-5} x_3^{-2}, \quad 3\dot{x}_3 = M_2\,(x_2^6 + x_3^6 + 1)\, x_3^{-2}, \qquad (9)$$
$$\dot{x}_1 = x_1\,(M_3 x_2^3 - M_2 x_3^3),$$

$$2\hat{H} = M_2^2 + M_3^2 + (M_2 x_2^3 + M_3 x_3^3)^2 + 3a\,(x_2^6 + x_3^6 + 1)\, x_2^{-2} x_3^{-2} = 2\hat{h},$$
$$\hat{V}_1 = x_1^2\,(x_2^6 + x_3^6 + 1) = 1,$$
$$\hat{V}_3 = -M_2 M_3\,(M_2 x_2^3 + M_3 x_3^3) - 3a x_2 x_3\,[M_2\,(x_2^{-3} - x_2^3) + M_3\,(x_3^{-3} - x_3^3)] = \hat{c}_1,$$

and the conditions for stationarity of the integral K can be written as:

$$[\lambda_0 x_2^2\,(M_2 + x_2^3\,(M_2 x_2^3 + M_3 x_3^3)) + \lambda_3\,(M_3 x_2^2\,(2 M_2 x_2^3 + M_3 x_3^3)$$
$$-3a x_3\,(x_2^6 - 1))]\, x_2^{-2} = 0,$$
$$[\lambda_0 x_3^2\,(M_3 + x_3^3\,(M_2 x_2^3 + M_3 x_3^3)) + \lambda_3\,(M_2 x_3^2\,(M_2 x_2^3 + 2 M_3 x_3^3)$$
$$-3a x_2\,(x_3^6 - 1))]\, x_3^{-2} = 0, \quad \lambda_1 x_1\,(x_2^6 + x_3^6 + 1) = 0, \qquad (10)$$
$$[\lambda_0\,(M_2\, x_2^5 x_3^2\,(M_2 x_2^3 + M_3 x_3^3) + a\,(2 x_2^6 - x_3^6 - 1)) - \lambda_2 x_1^2 x_2^8 x_3^2$$
$$+\lambda_3\,[M_2^2 M_3 x_2^5 x_3^2 - a\,(2 M_2 x_3^3\,(2 x_2^6 + 1) + M_3 x_2^3\,(x_3^6 - 1))]]\, x_2^{-3} x_3^{-2} = 0,$$
$$[\lambda_0\,(a\,(x_2^6 - 2 x_3^6 + 1) - M_3 x_2^2 x_3^5\,(M_2 x_2^3 + M_3 x_3^3)) + \lambda_2 x_1^2 x_2^2 x_3^8$$
$$-\lambda_3\,[M_2 M_3^2 x_2^2 x_3^5 - a\,(M_2 x_3^3\,(x_2^6 - 1) + 2 M_3 x_2^3\,(2 x_3^6 + 1))]]\, x_2^{-2} x_3^{-3} = 0.$$

First, we find the IMs of maximal codimension for Eq. (9). As the first integrals of the problem define IMs and families of IMs of codimension 1, we start with the IMs of codimension 2. As said before, the IMs can be derived as the solutions of system (10) when it is degenerate. To this end, we compute a lexicographical basis for the polynomials in square brackets (10) with respect to a part of the phase variables and the parameters, e.g., $\lambda_0, \lambda_1, M_2, M_3$ (the polynomials have least degrees with respect to these variables). Here the number of the phase variables determines the codimension of the desired IM. This technique enables us to obtain both the IMs and the conditions under which the stationary equations become degenerate (see., e.g., [11]).

The "Mathematica" program *GroebnerBasis* is applied to compute the basis:

```
GroebnerBasis[ polys, {lambda0, lambda2, M2, M3},
  CoefficientDomain -> RationalFunctions]
```

Here *polys* is the list of the polynomials in square brackets (10). All computations are performed on a computer with processor Intel Core 7i (3.6 GHz) and 32 GB RAM. The program has returned the basis in 21 s. So, we have the following system:

$$\sigma_0 M_3^8 + \sigma_2 M_3^6 + \sigma_4 M_3^4 + \sigma_6 M_3^2 + \sigma_8 = 0,$$
$$\sigma M_2 + \sigma_1 M_3^7 + \sigma_3 M_3^5 + \sigma_5 M_3^3 + \sigma_7 M_3 = 0, \qquad (11)$$
$$\lambda_1 = 0, \quad f(M_3, x_2, x_3, \lambda_0, \lambda_3, a) = 0, \qquad (12)$$

where σ_j $(j = 0, \ldots, 8)$, σ are the polynomials of a, x_2, x_3 (their full form is given in the Appendix), f is a linear function of λ_0.

It is easy to verify by IM definition that Eq. (11) determine the IM of codimension 2 of differential equations (9): the derivative of (11) calculated by virtue of Eq. (9) vanishes on the given expressions.

The first of expressions (11) $(\lambda_1 = 0)$ is the condition of degeneration of system (10). The latter expression $(f = 0)$ allows one to derive the first integral for the equations of vector field on IM (11).

By this technique, one can also find an IM of codimension 3. First, under the condition $\lambda_1 = 0$, we compute a Gröbner basis with respect to elimination monomial order for the polynomials in square brackets (10):

```
gb = GroebnerBasis[ polys, {x3}, {M2, M3}, CoefficientDomain ->
RationalFunctions, MonomialOrder -> EliminationOrder]
```

Then, we construct a lexicographical basis:

```
GroebnerBasis[ gb, { M2, M3, x3},
    CoefficientDomain -> RationalFunctions]
```

As a result, we have:

$$\lambda_0^8 x_3^{12} - 2\lambda_0^2 \rho_1 u x_3^6 - 12a_1 \lambda_3^2 \rho_2 x_2^4 x_3^4 + \lambda_0^2 \left(16a_1^3 \lambda_3^6 x_2^6 + \lambda_0^6 v^2\right) = 0,$$

$$2\lambda_0^3 \lambda_3 x_2 [(\lambda_0^{12} + 64a_1^6 \lambda_3^{12}) x_2^6 + 8a_1^3 \lambda_0^6 \lambda_3^6 (x_2^{12} + 1)] M_3 + \lambda_0^4 \left(16a_1^3 \lambda_3^6 \rho_2 + \lambda_0^{12}\right) u x_2^4$$
$$-4a_1^2 \lambda_3^4 [16a_1^3 \lambda_3^6 (\lambda_0^6 v^2 + 12a_1^3 \lambda_3^6 x_2^6) - \lambda_0^{12} (v^2 - x_2^6)] x_3^2 - 2a_1 \lambda_0^2 \lambda_3^2 [\lambda_0^{12} - 32a_1^3$$
$$\times \lambda_3^6 \rho_2] u x_2^2 x_3^4 - \lambda_0^{10} \rho_1 x_2^4 x_3^6 - 2a_1 \lambda_0^8 \lambda_3^2 x_3^8 [2a_1 \lambda_0^4 \lambda_3^2 u - \rho_1 x_2^2 x_3^4] = 0,$$

$$[2\lambda_0 \lambda_3 v (\lambda_0^{12} v^2 + 8a_1^3 \lambda_3^6 x_2^6 (16a_1^3 \lambda_3^6 x_2^6 + \lambda_0^6 (x_2^{12} - 2v + 1)))] M_2 - 2a_1 \lambda_3^2 [8a_1^3 \lambda_3^6 v$$
$$-\lambda_0^6 (v - 2)] \rho_3 x_2^4 x_3 - \lambda_0^2 [\lambda_0^{12} (u + 2) v^2 - 64a_1^6 \lambda_3^{12} (2v + 3v^2) x_2^6$$
$$+8a_1^3 \lambda_0^6 \lambda_3^6 (5v^2 x_2^6 - 2u)] x_3^3 + 4a_1^2 \lambda_0^4 \lambda_3^4 x_2^2 [16a_1^3 \lambda_3^6 ((u + 1) v^2 - 3) \qquad (13)$$
$$+\lambda_0^6 (4 - 3u^2 - v^3 + 16x_2^6)] x_3^5 - \lambda_0^6 x_3^7 [2a_1 \lambda_3^2 x_2^4 \rho_3 - \lambda_0^2 \rho_3 x_3^2$$
$$-4a_1^2 \lambda_0^4 \lambda_3^4 (v^2 - 2) x_2^2 x_3^4] = 0,$$

where $u = x_2^6 + 1$, $v = x_2^6 - 1$, $a_1 = a/3$, $\rho_1 = \lambda_0^6 - 8a_1^3 \lambda_3^6$, $\rho_2 = \lambda_0^6 - 4a_1^3 \lambda_3^6$, $\rho_3 = \lambda_0^6 v^2 + 16a_1^3 \lambda_3^6 x_2^6$. The total time to compute the basis is 8 s.

Likewise as above, it is easy to verify by IM definition that Eq. (13) define the family of IMs of codimension 3 for differential equations (9). Here λ_0, λ_3 are the parameters of the family. In the terms of the paper, it is the family of stationary IMs, since the integral $\hat{K} = \lambda_0 \hat{H} - \lambda_3 \hat{V}_3$ assumes a stationary value on the elements of this family.

One can show that the elements of IMs family (13) are the submanifolds of IM (11). Let us find their intersection. To this end, we compute a lexicographical basis with respect to the variables M_2, M_3, x_3 for the polynomials of the system composed of Eqs. (11), (13). The resulting equations are the family of IMs (13). So, the original assumption is true.

With (8), we can return to the initial variables $M_2, M_3, \gamma_1, \gamma_2, \gamma_3$ in Eqs. (11), (13). In the initial variables, these equations define, respectively, the IM of codimension 2 and the family of IMs of codimension 3 for differential equations (4) that can be verified by IM definition.

Other IMs of codimension 2 for the equations of motion (4) have been obtained by the chains of differential consequences of the kind [10]:

$$W_0' = \varphi_1(x) \, W_1(x), \; W_1' = \varphi_2(x) \, W_2(x), \ldots, \; W_{k-1}' = \varphi_k(x) \, W_k(x), \ldots \quad (14)$$

Here $x = (M_2, M_3, \gamma_1, \gamma_2, \gamma_3)$, and $W_j(x)$ $(j = 0, \ldots)$, $\varphi_m(x)$ $(m = 1, \ldots)$ are some smooth functions of x, W_j' $(j = 1, \ldots)$ are their derivatives by virtue of differential equations (4).

We call the chain of differential consequences (14) cyclical one if for some k:

$$W_k' = \sum_{i=0}^{k} \bar{\varphi}_i(x) \, W_i(x), \quad (15)$$

where $\bar{\varphi}_i(x)$ are the smooth functions.

Statement 1. If system (4) admits cyclical chain (15) then it has the IM defined by the equations $W_0(x) = W_1(x) = \ldots = W_k(x) = 0$. The proof is obvious.

In the given approach, computer algebra tools play an auxiliary role. They give us a possibility to make computational experiments, e.g., for finding the functions W_i that would be most "suitable" to generate the chain. The "Mathematica" program *PolynomialReduce* is used to test criterion (15).

Let be $W_0 = M_2 + M_3$. On differentiating this expression by virtue of Eq. (4) we obtain $W_1 = \gamma_2 - \gamma_3$. The subsequent differentiation of W_1 shows that differential equations (4) admit the following cyclical chain:

$$W_0' = [a \, (\gamma_1^2 + \gamma_2 \gamma_3) \, (\gamma_1^2 + \gamma_2^2 + \gamma_3^2) \, (\gamma_1 \gamma_2 \gamma_3)^{-5/3}] \, W_1,$$
$$W_1' = -[(\gamma_1^2 + \gamma_2^2 + \gamma_2 \gamma_3) \, \gamma_1^{-1}] \, W_0 + [M_3 \, (\gamma_2 + \gamma_3) \, \gamma_1^{-1}] \, W_1.$$

According to Statement 1, the expressions

$$M_2 + M_3 = 0, \; \gamma_2 - \gamma_3 = 0 \quad (16)$$

determine the IM of codimension 2 of differential equations (4).

The vector field on IM (16) is given by

$$\dot{M}_3 = -a \, (\gamma_1^2 - \gamma_3^2) \, (\gamma_1^2 + 2\gamma_3^2) \, \gamma_1^{-5/3} \gamma_2^{-7/3}, \; \dot{\gamma}_1 = 2M_3 \gamma_3, \; \dot{\gamma}_3 = -M_3 \gamma_1. \quad (17)$$

In the same way, the IM defined by the equations

$$M_2 - M_3 = 0, \; \gamma_2 + \gamma_3 = 0 \quad (18)$$

has been derived.

The vector field on this IM is described by

$$\dot{M}_3 = a \, (-\gamma_3)^{1/3} (\gamma_3^2 - \gamma_1^2) \, (\gamma_1^2 + 2\gamma_3^2) \, \gamma_1^{-5/3} \gamma_3^{-8/3},$$
$$\dot{\gamma}_1 = -2M_3 \gamma_3, \; \dot{\gamma}_3 = M_3 \gamma_1. \quad (19)$$

Note that IMs (16), (18) are stationary. The integral $\Omega = \tilde{V}_3^2$ takes a stationary value on them.

All found IMs for differential equations (4) can be "lifted up" into the phase space of system (2). For this purpose, it is sufficient to add expression $V_2 = 0$ (3) to the equations of these IMs. In particular, equations IMs (16), (18) take the form

$$M_2 + M_3 = 0, \quad \gamma_2 - \gamma_3 = 0, \; M_1\gamma_1 = 0 \tag{20}$$

and $M_2 - M_3 = 0, \; \gamma_2 + \gamma_3 = 0, \; M_1\gamma_1 = 0$, respectively,

From the physical viewpoint, in the case of the spheroidal body, the above equations together with (17), (19) define pendulum-like oscillations of the body. From the formulation of the problem it follows that IM (20) is related to the problem of the expanding gas cloud only. Equation (20) together with (17) describe the periodical changes of the cloud sizes.

2.3   Finding Stationary Solutions

As mentioned before, stationary solutions are usually found by the Routh–Lyapunov method from the conditions for stationarity of a family of problem's first integrals. In the case of polynomial first integrals, this approach leads to solving a system of polynomial equations. When the first integrals are not polynomial or the polynomials have high degrees, the technique applied in [11] is more suitable. The given technique is used in the present work.

Equate the right-hand sides of differential equations (4) to zero and add relation $V_1 = 1$ (5) to them:

$$\begin{aligned}
&a\left(\gamma_1^2 - \gamma_3^2\right)\left(\gamma_1^2 + \gamma_2^2 + \gamma_3^2\right)(\gamma_1\gamma_3)^{-5/3}\,\gamma_2^{-2/3} = 0, \gamma_2 M_3 - \gamma_3 M_2 = 0, \\
&-a\left(\gamma_1^2 - \gamma_2^2\right)\left(\gamma_1^2 + \gamma_2^2 + \gamma_3^2\right)(\gamma_1\gamma_2)^{-5/3}\,\gamma_3^{-2/3} = 0, \gamma_1^2 + \gamma_2^2 + \gamma_3^2 - 1 = 0, \\
&-\gamma_3\gamma_1^{-1}\left[M_2\gamma_2 + M_3(\gamma_1^2 + \gamma_3)\right] = 0, \\
&\gamma_2\gamma_1^{-1}\left[M_2(\gamma_1^2 + \gamma_2) + M_3\gamma_3\right] = 0.
\end{aligned} \tag{21}$$

Next, construct a lexicographical Gröbner basis with respect to $M_2, M_3, \gamma_1, \gamma_2, \gamma_3$ for the polynomials of the subsystem

$$\begin{aligned}
&\gamma_1^2 - \gamma_3^2 = 0, \; M_2\gamma_2 + M_3(\gamma_1^2 + \gamma_3) = 0, \; M_2(\gamma_1^2 + \gamma_2) + M_3\gamma_3 = 0, \\
&\gamma_1^2 - \gamma_2^2 = 0, \; \gamma_2 M_3 - \gamma_3 M_2 = 0, \; \gamma_1^2 + \gamma_2^2 + \gamma_3^2 - 1 = 0
\end{aligned}$$

of system (21). As a result, we have:

$$3\gamma_3^2 - 1 = 0, \; 1 - 3\gamma_2^2 = 0, \; 1 - 3\gamma_1^2 = 0, M_2 = 0, \; M_3 = 0.$$

The latter system has the following solutions:

$$\begin{aligned}
&M_2 = 0, \, M_3 = 0, \, \gamma_1 = \pm 3^{-1/2}, \, \gamma_2 = \gamma_3 = 3^{-1/2}, \\
&M_2 = 0, \, M_3 = 0, \, \gamma_1 = \pm 3^{-1/2}, \, \gamma_2 = \gamma_3 = -3^{-1/2}.
\end{aligned} \tag{22}$$

$$M_2 = 0, \ M_3 = 0, \ \gamma_1 = \pm 3^{-1/2}, \ \gamma_2 = -3^{-1/2}, \ \gamma_3 = 3^{-1/2},$$
$$M_2 = 0, \ M_3 = 0, \ \gamma_1 = \pm 3^{-1/2}, \ \gamma_2 = 3^{-1/2}, \ \gamma_3 = -3^{-1/2}. \tag{23}$$

On substituting these solutions into Eq. (4) they are satisfied.

Now, let us derive the family of the integrals which takes a stationary value on solutions (22), (23). When these solutions are substituted into Eq. (7), we find that the equations are satisfied under $\lambda_1 = 0$.

On substituting $\lambda_1 = 0$ into (6), we have:

$$\tilde{K} = \lambda_0 \tilde{H} - \lambda_3 \tilde{V}_3. \tag{24}$$

Thus, the family of the integrals $\tilde{K}$ assumes a stationary value on solutions (22), (23). Each integral belonging to this family also takes a stationary value on the above solutions. It is verified by direct calculation. In particular, the integral $\tilde{V}_3$ is identically equal to zero on all solutions (22), (23).

In the same way as the IMs in Subsect. 2.2, the stationary solutions can be "lifted up" into the phase space of system (2). From the physical viewpoint, in the original phase space, these solutions correspond to the equilibria of the spheroidal body, and only one of these solutions is related to the problem of the expanding gas cloud: $M_1 = M_2 = M_3 = 0, \ \gamma_1 = \gamma_2 = \gamma_3 = 3^{-1/2}$. It was also found in [4]. This solution corresponds to the cloud of the spherical shape without changing sizes.

One can show that stationary solutions (22), (23) belong to IM (11). To this end, we substitute these solutions into the equations of the IM (they must be written in the initial variables $M_2, M_3, \gamma_1, \gamma_2, \gamma_3$). The equations turn into identities. Thus, solutions (22), (23) belong to IM (11).

In the same way, we reveal that solutions (22) and (23) belong to IM (16) and IM (18), respectively. Hence, IM (11) and IM (16) have the common points (i.e., the points of intersection of these IMs) defined by relations (22). Analogously, relations (23) define the points of intersection of IM (11) and IM (18).

2.4   On Stability of Stationary Solutions

The integrals and their families, which take a stationary value on solutions (22), (23), are used to investigate the stability of these solutions by the Routh–Lyapunov method. The problem is to verify the sign-definiteness conditions for the 2nd variation of the family of integrals which is obtained in the neighborhood of the solution under study. These conditions are analyzed on the linear manifold defined by the variations of the "conditional" integrals.

Let us investigate the stability of one of solutions (22), e.g.,

$$M_2 = M_3 = 0, \ \gamma_1 = \gamma_2 = \gamma_3 = 3^{-1/2}, \tag{25}$$

which is related to the problem of the expanding gas cloud.

We use the family of integrals $\tilde{K}$ (24). In the deviations $y_1 = \gamma_1 - 3^{-1/2}$, $y_2 = \gamma_2 - 3^{-1/2}$, $y_3 = \gamma_3 - 3^{-1/2}$, $y_4 = M_2$, $y_5 = M_3$ on the linear manifold $\delta V_1 = 2(y_1 + y_2 + y_3)/\sqrt{3} = 0$, the 2nd variation of $\tilde{K}$ in the neighborhood of solution (25) can be written as:

$$\delta^2 \tilde{K} = \lambda_0 \left[18a \left(y_1^2 + y_1 y_2 + y_2^2 \right) + y_4^2 + y_4 y_5 + y_5^2 \right] + 6\sqrt{3} a \lambda_3 \left[y_1 \left(y_4 + 2y_5 \right) \right.$$
$$\left. + y_2 \left(y_5 - y_4 \right) \right]. \tag{26}$$

The conditions for the quadratic form $\delta^2 \tilde{K}$ to be positive definite in the form of Sylvester's inequalities are given by $a\lambda_0 > 0$, $a^2\lambda_0^2 > 0$, $a^2\lambda_0 (\lambda_0^2 - 6a\lambda_3^2) > 0$, $a^2(\lambda_0^2 - 6a\lambda_3^2)^2 > 0$.

These inequalities are consistent under the following constraints on a, λ_0, λ_3:

$$a > 0, \ \lambda_3 > 0, \ \lambda_0 > \sqrt{6}\sqrt{a}\,\lambda_3. \tag{27}$$

Inequalities (27) are split up into 2 groups. The first ($a > 0$) is the sufficient condition for the stability of solution (25), and the rest of the inequalities separates some subfamily from the family of integrals $\tilde{K}$ (24), the elements of which give us a possibility to derive this condition.

Let us show that the sufficient condition of stability is also necessary. To this end, we use Lyapunov's linear stability theorem [14].

In the case studied, the equations of first approximation, in the deviations y_i $(i = 1, \ldots, 5)$, are:

$$\sqrt{3}\,\dot{y}_1 = y_5 - y_4, \ \sqrt{3}\,\dot{y}_2 = -(y_4 + 2y_5), \ \sqrt{3}\,\dot{y}_3 = 2y_4 + y_5,$$
$$\dot{y}_4 = 6\sqrt{3}a\,(y_1 - y_3), \ \dot{y}_5 = 6\sqrt{3}a\,(y_2 - y_1).$$

The characteristic equation $\lambda\,(\lambda^2 + 18a)^2 = 0$ of the above system has only zero and pure imaginary roots when $a > 0$. On comparing the latter inequality with (27), we conclude that the condition $a > 0$ is necessary and sufficient for the stability of solution (25). For the rest of the stationary solutions, we have obtained similar results.

Now, we investigate the stability of IM (16), which solution (25) belongs to.

For the equations of perturbed motion, in the deviations $y_1 = M_2 + M_3$, $y_2 = \gamma_2 - \gamma_3$, on the linear manifold $\delta V_1 = 2\gamma_3 y_2 = 0$, the 2nd variation of the integral $\Omega = \tilde{V}_3^2$ is:

$$\delta^2 \Omega = [3a\,(\gamma_3^2 - \gamma_1^2) + \gamma_1^{2/3}\gamma_3^{4/3}M_3^2]^2 \, \gamma_1^{-10/3}\gamma_3^{-2/3} \, y_1^2. \tag{28}$$

On IM (16), the integral $\tilde{H}$ assumes the form:

$$\bar{H} = [M_3^2 + 3a\,(\gamma_1^2 + 2\gamma_3^2)]\,(2\gamma_1^{-2/3}\gamma_3^{-4/3}) = h_1. \tag{29}$$

Eliminate M_3 from (28) with (29):

$$4\delta^2 \Omega = (9a\gamma_1^{4/3} - 2\,h_1\gamma_3^{4/3})^2 \, \gamma_1^{-2}\gamma_3^{-2/3} \, y_1^2.$$

Equate the numerator of the latter expression to zero and eliminate γ_1 from the resulting equation with the integral $V_1 = 1$. As a result, we obtain the following boundary value for γ_2:

$$\gamma_2 = \left(\frac{2h_1}{9a} + 1\right)^{3/4},$$

under which there exist the stable oscillations of the spheroidal body. As to the gas ellipsoid, the latter relation allows one to determine the limit values for the lengths of its principal axes under which the periodical changes of the cloud sizes are stable.

When the stability of stationary solutions and IMs is studied on the base of Lyapunov's linear stability theorems and the 2nd Lyapunov method, we need often to derive the sign-definiteness conditions for a quadratic form as well as the characteristic equation for a system of linear differential equations with constant coefficients. The computer program codes of these procedures are included in the "Mathematica" software package [2]. This package has been developed to do the qualitative analysis of conservative systems on the base of the approach described in the this paper. It is applied as an auxiliary tool at different stages of analysis of the systems. In the above calculations, for the given solution and the given combination of the first integrals, the package has constructed the sign-definiteness conditions for the quadratic form $\delta^2 \tilde{K}$ (26) in the form of Sylvester's inequalities. The subsequent analysis of these inequalities was made by computer algebra tools. In a similar manner, the package is used to investigate the stability on the base of Lyapunov's linear stability theorems.

3   The Integrable Case with the Additional 6th Degree Integral

3.1   Formulation of the Problem

The equations of motion of the spheroidal body in a force field with the potential

$$2V = G\left[3a\left(\gamma_1\gamma_2\gamma_3\right)^{-2/3} + 4c^2\left(\gamma_1^2 + \gamma_2^2\right)\left(\gamma_1^2 - \gamma_2^2\right)^{-2}\right]$$

can be written as:

$$
\begin{aligned}
\dot{M}_1 &= -G\left[a\left(\gamma_2^2 - \gamma_3^2\right)\left(\gamma_2\gamma_3\right)^{-5/3}\gamma_1^{-2/3} + 4c^2\gamma_2\gamma_3\left(3\gamma_1^2 + \gamma_2^2\right)\left(\gamma_1^2 - \gamma_2^2\right)^{-3}\right], \\
\dot{M}_2 &= \ \ G\left[a\left(\gamma_1^2 - \gamma_3^2\right)\left(\gamma_1\gamma_3\right)^{-5/3}\gamma_2^{-2/3} - 4c^2\gamma_1\gamma_3\left(\gamma_1^2 + 3\gamma_2^2\right)\left(\gamma_1^2 - \gamma_2^2\right)^{-3}\right], \\
\dot{M}_3 &= -G\left[a\left(\gamma_1^2 - \gamma_2^2\right)\left(\gamma_1\gamma_2\right)^{-5/3}\gamma_3^{-2/3} - 16c^2\gamma_1\gamma_2\left(\gamma_1^2 + \gamma_2^2\right)\left(\gamma_1^2 - \gamma_2^2\right)^{-3}\right], \\
\dot{\gamma}_1 &= \gamma_2 M_3 - \gamma_3 M_2, \quad \dot{\gamma}_2 = \gamma_3 M_1 - \gamma_1 M_3, \quad \dot{\gamma}_3 = \gamma_1 M_2 - \gamma_2 M_1.
\end{aligned}
\tag{30}
$$

Here the variables M_i, γ_i $(i = 1, 2, 3)$ are interpreted as in Sect. 2, $G = \gamma_1^2 + \gamma_2^2 + \gamma_3^2$.

The first integrals of Eq. (30) are given by

$$
\begin{aligned}
2H &= M_1^2 + M_2^2 + M_3^2 + G\left[3a\left(\gamma_1\gamma_2\gamma_3\right)^{-2/3} + 4c^2(\gamma_1^2 + \gamma_2^2)(\gamma_1^2 - \gamma_2^2)^{-2}\right] = 2h, \\
V_1 &= \gamma_1^2 + \gamma_2^2 + \gamma_3^2 = 1, \quad V_2 = M_1\gamma_1 + M_2\gamma_2 + M_3\gamma_3 = 0, \\
V_3 &= (F_3 + F_c)^2 + 4\Phi\left[\bar{\Phi}\gamma_1^2\gamma_3^{-2} + 3a\right]\left[\bar{\Phi}\gamma_2^2\gamma_3^{-2} + 3a\right] = c_1,
\end{aligned}
\tag{31}
$$

where

$$F_3 = M_1 M_2 M_3 - 3a \left(\gamma_1 \gamma_2 \gamma_3\right)^{1/3} \left(M_1 \gamma_1^{-1} + M_2 \gamma_2^{-1} + M_3 \gamma_3^{-1}\right),$$
$$F_c = 4c^2 M_3 \gamma_1 \gamma_2 \gamma_3^2 \left(\gamma_1^2 - \gamma_2^2\right)^{-2},$$
$$\Phi = 4c^2 \gamma_3^2 (\gamma_1 \gamma_2 \gamma_3)^{2/3} \left(\gamma_1^2 - \gamma_2^2\right)^{-2}, \quad \bar{\Phi} = M_1 M_2 \left(\gamma_1 \gamma_2 \gamma_3\right)^{2/3} \gamma_1^{-1} \gamma_2^{-1} + \Phi - 3a.$$

Here V_3 is the additional 6th degree integral with respect to M_1, M_2, M_3. It has been derived in [9]. This integral exists when the constant of the integral V_2 is equal to zero. Note that the potential energy V in this problem has a singularity when $\gamma_1 = \gamma_2$.

Likewise as in Sect. 2, we shall consider the equations of motion of the body on the manifold $V_2 = 0$. On this manifold, differential equations (30) and first integrals (31) take the form:

$$\dot{M}_1 = -G \left[a \left(\gamma_2^2 - \gamma_3^2\right) \left(\gamma_2 \gamma_3\right)^{-5/3} \gamma_1^{-2/3} + 4c^2 \gamma_2 \gamma_3 \left(3\gamma_1^2 + \gamma_2^2\right) \left(\gamma_1^2 - \gamma_2^2\right)^{-3}\right],$$
$$\dot{M}_2 = G \left[a \left(\gamma_1^2 - \gamma_3^2\right) \left(\gamma_1 \gamma_3\right)^{-5/3} \gamma_2^{-2/3} - 4c^2 \gamma_1 \gamma_3 \left(\gamma_1^2 + 3\gamma_2^2\right) \left(\gamma_1^2 - \gamma_2^2\right)^{-3}\right], \quad (32)$$
$$\dot{\gamma}_1 = -[M_1 \gamma_1 \gamma_2 + M_2 \left(\gamma_2^2 + \gamma_3^2\right)] \gamma_3^{-1}, \quad \dot{\gamma}_2 = [M_1 \left(\gamma_1^2 + \gamma_3^2\right) + M_2 \gamma_1 \gamma_2] \gamma_3^{-1},$$
$$\dot{\gamma}_3 = \gamma_1 M_2 - \gamma_2 M_1.$$

$$2\tilde{H} = M_1^2 + M_2^2 + (M_1 \gamma_1 + M_2 \gamma_2)^2 \gamma_3^{-2} + G \left[3a \left(\gamma_1 \gamma_2 \gamma_3\right)^{-2/3}\right.$$
$$+4c^2(\gamma_1^2 + \gamma_2^2) \left(\gamma_1^2 - \gamma_2^2\right)^{-2}] = 2\tilde{h}, \quad V_1 = \gamma_1^2 + \gamma_2^2 + \gamma_3^2 = 1,$$
$$\tilde{V}_3 = (\tilde{F}_3 + \tilde{F}_c)^2 + 4\Phi \left[\bar{\Phi} \gamma_1^2 \gamma_3^{-2} + 3a\right] [\bar{\Phi} \gamma_2^2 \gamma_3^{-2} + 3a] = \tilde{c}_1, \quad \text{where} \quad (33)$$
$$\tilde{F}_3 = -M_1 M_2 \left(M_1 \gamma_1 + M_2 \gamma_2\right) \gamma_3^{-1} - 3a \left(\gamma_1 \gamma_2 \gamma_3\right)^{1/3} \left(M_1 \gamma_1^{-1} + M_2 \gamma_2^{-1}\right.$$
$$-(M_1 \gamma_1 + M_2 \gamma_2) \gamma_3^{-2}), \quad \tilde{F}_c = -4c^2(M_1 \gamma_1 + M_2 \gamma_2)\gamma_1 \gamma_2 \left(\gamma_1^2 - \gamma_2^2\right)^{-2}.$$

They have been derived from (30), (31) by eliminating the variable M_3 from them with the aid of $V_2 = 0$.

In the present work, we restrict our consideration to the problem of finding the stationary solutions for Eq. (32) and the investigation of their stability.

3.2   Finding Stationary Solutions

We apply the same technique as in Subsect. 2.3 to obtain the stationary solutions of differential equations (32). For this purpose, these equations are written in the variables $M_1 = M_1$, $M_2 = M_2$, $x_1 = \gamma_1$, $x_2 = \gamma_2^{1/3} \gamma_1^{-1/3}$, $x_3 = \gamma_3^{1/3} \gamma_1^{-1/3}$:

$$\dot{M}_1 = -\bar{G} \left[a(x_2^6 - 1)^3 (x_2^6 - x_3^6) - 4c^2 x_2^8 x_3^8 \left(x_2^6 + 3\right)\right] x_2^{-5} x_3^{-5} (x_2^6 - 1)^{-3},$$
$$\dot{M}_2 = \bar{G} \left[(4c^2 x_2^2 x_3^8 \left(3x_2^6 + 1\right) - a(x_2^6 - 1)^3 (x_3^6 - 1)\right] x_2^{-2} x_3^{-5} (x_2^6 - 1)^{-3},$$
$$\dot{x}_1 = -[M_1 x_2^3 + M_2(x_2^6 + x_3^6)] x_1 x_3^{-3}, \quad 3\dot{x}_2 = \bar{G} \left[M_1 + M_2 x_2^3\right] x_2^{-2} x_3^{-3}, \quad (34)$$
$$3\dot{x}_3 = (x_2^6 + x_3^6 + 1) M_2 x_3^{-2},$$

where $\bar{G} = x_2^6 + x_3^6 + 1$.

Next, we equate the right-hand sides of Eq. (34) to zero and consider the following subsystem

$$a(x_2^6 - 1)^3(x_2^6 - x_3^6) - 4c^2 x_2^8 x_3^8 (x_2^6 + 3) = 0,$$
$$(4c^2 x_2^2 x_3^8 (3x_2^6 + 1) - a(x_2^6 - 1)^3(x_3^6 - 1) = 0,$$
$$M_1 x_2^3 + M_2(x_2^6 + x_3^6) = 0, \ M_1 + M_2 x_2^3 = 0, \ M_2 = 0 \qquad (35)$$

of the resulting system.

From the latter three equations (35), it follows that $M_1 = M_2 = 0$. For the polynomials of the rest of the equations, we compute a Gröbner basis with respect to the ordering $x_3 > x_2$. Taking into account the above values for M_1, M_2, we have:

$$a^3(x_2^6 - 1)^{12} (x_2^{12} + 6x_2^6 + 1) - 16384\, c^6 x_2^{30}(x_2^6 + 1)^4 = 0,$$
$$16384 a^2 c^2 x_3^2 - 16384 c^6 x_2^{22}(x_2^6 + 1)^3(31x_2^6 + 32)(33x_2^6 + 32) + a^3 x_2^4(x_2^6 - 1)^4$$
$$\times (1023 x_2^{54} - 1021 x_2^{48} - 21488 x_2^{42} + 86920 x_2^{36} - 136858 x_2^{30} + 71014 x_2^{24}$$
$$+72584 x_2^{18} - 138224 x_2^{12} + 88067 x_2^6 - 22529) = 0,$$
$$M_1 = 0, \ M_2 = 0. \qquad (36)$$

It is easy to verify by IM definition that Eq. (36) define the one-dimensional IM of differential equations (34). The vector field on this IM is described by the equation $\dot{x}_1 = 0$. It has the following solution:

$$x_1 = x_1^0 = \text{const.} \qquad (37)$$

Equation (36) together with (37) and the condition

$$x_1^2 (x_2^6 + x_3^6 + 1) = 1, \qquad (38)$$

which is the integral V_1 in the variables x_1, x_2, x_3, determine the set of fixed points for system (34).

In the initial variables $M_1, M_2, \gamma_1, \gamma_2, \gamma_3$, Eqs. (36) and (36)–(38) determine the one-dimensional IM and the set of fixed points for system (32), respectively. In the same way as in Sect. 2, these solutions can be "lifted up" into the phase space of system (30).

From the physical viewpoint, in the original phase space, the solutions defined by (36)–(38) correspond to the equilibria of the spheroidal body (the gas ellipsoid). From equations (36)–(38) it follows that the number of the equilibria is no more than 336 $\forall\, a \neq 0,\, c \neq 0$. One can also see from these equations that they can have one real positive solution only. Thus, in the problem of the expanding gas cloud, there exists no more than one equilibrium position for each fixed pair of values of the parameters $a \neq 0,\, c \neq 0$. The latter agrees with the result [4]. Further, we find the equilibria under some conditions imposed on the parameters a and c.

System (34) is defined in the domain: $x_1^2 (x_2^6 + x_3^6 + 1) = 1,\, x_i \neq 0$ ($i = 1, 2, 3$), $x_2 \neq 1$. We choose a value of x_2 from this domain, e.g. $x_2 = 1/2^{1/6}$, and then substitute it into the 1st equation of system (36). Whence, one can obtain $a = 192 \, (6/17)^{1/3} c^2$. Under the above values of x_2, a, from the rest of Eqs. (36)–(38), we find x_1, x_3. So, for the given values of x_2, a, system (36)–(38) has the following solutions:

$$M_1 = M_2 = 0, \, x_1 = (34/3)^{1/2} \, 5^{-1}, \, x_2 = 2^{-1/6}, \, x_3 = \pm 2^{1/3} \, (3/17)^{1/6},$$
$$M_1 = M_2 = 0, \, x_1 = -(34/3)^{1/2} \, 5^{-1}, \, x_2 = 2^{-1/6}, \, x_3 = \pm 2^{1/3} \, (3/17)^{1/6}.$$

In the initial variables, the above solutions are:

$$M_1 = M_2 = 0, \, \gamma_1 = (34/3)^{1/2} \, 5^{-1}, \, \gamma_2 = (17/3)^{1/2} \, 5^{-1}, \, \gamma_3 = \pm 2\sqrt{2} \, 5^{-1},$$
$$M_1 = M_2 = 0, \, \gamma_1 = -(34/3)^{1/2} \, 5^{-1}, \, \gamma_2 = -(17/3)^{1/2} \, 5^{-1},$$
$$\gamma_3 = \pm 2\sqrt{2} \, 5^{-1}. \tag{39}$$

On substituting these solutions into differential equations (32) they are satisfied.

From the physical viewpoint, in the original phase space, solutions (39) correspond to the equilibria of the spheroidal body. Only one of these solutions is related to the problem of the expanding gas cloud:

$$M_1 = M_2 = M_3 = 0, \, \gamma_1 = (34/3)^{1/2} \, 5^{-1}, \, \gamma_2 = (17/3)^{1/2} \, 5^{-1}, \, \gamma_3 = 2\sqrt{2} \, 5^{-1}.$$

It corresponds to the gas cloud of ellipsoidal shape. This ellipsoid is prolate along its principal axis Ox.

As in Sect. 2, one can show that the family of integrals

$$\tilde{K} = \lambda_0 \tilde{H} - \lambda_3 \tilde{V}_3 \tag{40}$$

(and each integral of this family) assumes a stationary value on solutions (39). The family of integrals $\tilde{K}$ (40) is used for the investigation of stability of the given solutions.

3.3   On Stability of Stationary Solutions

In order to study the stability of stationary solutions (39), we apply the same approach, methods and computing tools as in Sect. 2.

First, let us investigate the stability of one of solutions (39) which is related to the problem of the expanding gas cloud:

$$M_1 = M_2 = 0, \, \gamma_1 = (34/3)^{1/2} \, 5^{-1}, \, \gamma_2 = (17/3)^{1/2} \, 5^{-1}, \, \gamma_3 = 2\sqrt{2} \, 5^{-1}. \tag{41}$$

In the deviations $y_1 = M_1, \, y_2 = M_2, \, y_3 = \gamma_1 - (34/3)^{1/2} \, 5^{-1}, \, y_4 = \gamma_2 - (17/3)^{1/2} \, 5^{-1}, \, y_5 = \gamma_3 - 2\sqrt{2} \, 5^{-1}$, on the linear manifold $\delta V_1 = 2 \, [\sqrt{51} \, (\sqrt{2} y_3 + y_4) + 6\sqrt{2} y_5]/15 = 0$, the 2nd variation of the family of integrals $\tilde{K}$ in the neighborhood of the solution under study is written as: $\delta^2 \tilde{K} = Q_1 + Q_2$, where

$$83521 \, Q_1 = 15000 \, c^2 \, [204 \, (221\lambda_0 - 161792 \, c^4 \lambda_3) \, y_4^2 + \sqrt{102} \, (3961\lambda_0$$
$$+ 7651328 \, c^4 \lambda_3) \, y_4 y_5 + (19822\lambda_0 - 795295744 \, c^4 \lambda_3) \, y_5^2],$$
$$816 \, Q_2 = (986\lambda_0 - 14450688 \, c^4 \lambda_3) \, y_1^2 + 2\sqrt{2} \, (289\lambda_0 + 10764288 \, c^4 \lambda_3) \, y_1 y_2$$
$$+ (697\lambda_0 - 17842176 \, c^4 \lambda_3) \, y_2^2.$$

The conditions for the family of the quadratic forms Q_1, Q_2 to be positive definite are sufficient for the stability of solution (41). In the form of Sylvester's inequalities, they are:

$$221\lambda_0 - 161792\,c^4\lambda_3 > 0, \quad 289\lambda_0^2 - 22282240\,c^4\lambda_0\lambda_3 + 14495514624\,c^8\lambda_3^2 > 0,$$
$$986\lambda_0 - 14450688\,c^4\lambda_3 > 0. \tag{42}$$

Inequalities (42) are compatible under the following constraints on the parameters λ_0, λ_3, c: $17\lambda_0 > 16384\,(40+\sqrt{1546})\,c^4\lambda_3$. The latter condition separates the subfamily from the family of integrals $\tilde{K}$ (40), the elements of which enable us to derive the sufficient conditions for the stability of solution (41). Comparison of the above sufficient condition with the relation $a = 192\,(6/17)^{1/3}c^2$ gives us the following sufficient condition for the stability of solution (41): $a > 0$.

For solution (41), we have also derived the conditions of its stability on the base of Lyapunov's linear stability theorem. The resulting necessary stability conditions coincide with the sufficient ones.

Similar results have been obtained for the rest of solutions (39).

4   Conclusion

In the given work, ordinary differential equations with irrational first integrals were studied. These equations describe a series of dynamical systems, such as an expansion of the gas ellipsoidal cloud in vacuum, the rotation of the spheroidal body in a potential force field, the motion of a point mass on the spherical surface. We analyzed the equations in the cases when they possess the additional first integrals of 3rd and 6th degree in momenta. The purpose of the study was to find the stationary solutions and IMs of the equations and to investigate their stability. To solve these problems, computer algebra methods and tools were applied. The first integrals in the problem are rather complicated irrational functions. Computer algebra methods were used for transforming irrational equations to polynomial ones and for finding their solutions.

In the problem of the expanding gas cloud, in addition to previously known solutions, new IMs of codimension 2, 3 as well one-dimensional IM have been obtained, and the physical interpretation for some of them has been done. It was established that the previously known solutions belong to these IMs. It was also shown that these solutions are stationary. For the stationary solutions and IMs, the sufficient conditions of their stability on the base of the 2nd Lyapunov method have been derived. The "Mathematica" software package developed by the authors together with their colleagues was used to investigate the stability of the found solutions. It should be noted that in the problem of the rotational motion of the spheroidal body, there exists a greater number of stationary solutions and IMs than in the above problem. Some of them have been found and represented in the paper. The analysis of their stability has also been done.

The obtained results, their consistency with those known before, show that the approach used as well the computing tools are rather efficient for the study of the dynamical systems of the considered type.

Acknowledgments. This work was supported by the Russian Foundation for Basic Research (Project 16-07-00201a) and the Program for the Leading Scientific Schools of the Russian Federation (NSh-8081.2016.9).

A   Appendix

The coefficients of equation (11):

$$\sigma_0 = x_2^8 x_3^8 \left(4x_3^6 - (x_2^{6'} - x_3^6 - 1)^2\right), \quad \sigma_2 = -2ax_2^6 \Big[5\,(x_2^6 + 1)\,x_3^{18} - 2\,(8x_2^6$$

$$+5\,(x_2^6 + 1)^2)\,x_3^{12} + (5\,(x_2^{18} + 1) - 9x_2^6\,(x_2^6 + 1))\,x_3^6 + 2x_2^6\,(x_2^6 - 1)^2\Big],$$

$$\sigma_4 = -36\,a^2 x_2^4 x_3^4 \Big[((x_2^6 + 1)^2 + x_2^6)\,x_3^{12} - 2(x_2^{18} + 6x_2^6\,(x_2^6 + 1) + 1)\,x_3^6$$

$$+(x_2^{24} - 2x_2^6\,(x_2^{12} + x_2^6 + 1) + 1)\Big], \quad \sigma_6 = -54\,a^3 x_2^2 x_3^2 \Big[(x_2^{18} + 7x_2^6\,(x_2^6 + 1)$$

$$+1)\,x_3^{12} - 2\,((x_2^{12} + 1)^2 + 14\,x_2^{12} + 7x_2^6\,(x_2^{12} + 1))\,x_3^6 + (x_2^6 - 1)^2\,(x_2^6 + 1)^3\Big],$$

$$\sigma_8 = -27a^4 \Big[((x_2^6 + 1)^2 + 4x_2^6)\,x_3^6 - (x_2^6 + 1)^3\Big]^2,$$

$$\sigma = -18a^3 x_2^3 x_3 \Big[(x_2^6 - 1)\,[(x_2^6 + 1)^2 + 4x_2^6]^2\,x_3^{30} - [3(x_2^{36} - 1) - 8x_2^6(x_2^{24} - 1)$$

$$-153x_2^{12}(x_2^{12} - 1)]\,x_3^{24} + 2\,[x_2^{42} - 15x_2^6(x_2^{30} - 1) - 3x_2^{12}(x_2^{18} - 1)$$

$$+269x_2^{18}(x_2^6 - 1) - 1]\,x_3^{18} + 2\,[x_2^{48} - 2x_2^6(x_2^{36} - 1) - 82x_2^{12}(x_2^{24} - 1)$$

$$+102x_2^{18}(x_2^{12} - 1) - 1]\,x_3^{12} - (x_2^6 + 1)^4\,[3(x_2^{30} - 1) - 23x_2^6(x_2^{18} - 1)$$

$$+86x_2^{12}(x_2^6 - 1)]\,x_3^6 + (x_2^6 - 1)^3\,(x_2^6 + 1)^7\Big],$$

$$\sigma_1 = x_2^6 x_3^{10} \Big[[5\,(x_2^{12} + 1) + 6x_2^6]\,x_3^{30} - 8\,[2(x_2^{18} + 1) + x_2^6(x_2^6 + 1)]\,x_3^{34}$$

$$+2\,[7(x_2^{24} + 1) - 16x_2^6(x_2^{12} + 1) - 30x_2^{12}]\,x_3^{18} + 4\,[(x_2^{30} + 1) + 12x_2^6(x_2^{18} + 1)$$

$$+3x_2^{12}(x_2^6 + 1)]\,x_3^{12} - (x_2^6 - 1)^2\,[11(x_2^{24} + 1) + 20x_2^6(x_2^{12} + 1) + 2x_2^{12}]x_3^6$$

$$+4(x_2^6 - 1)^4\,(x_2^{18} + 1)\Big],$$

$$\sigma_3 = ax_2^4 x_3^2 \Big[4\,[17x_2^6\,(x_2^6 + 1) + 11(x_2^{18} + 1)]\,x_3^{36} - [322x_2^{12} + 292x_2^6\,(x_2^{12} + 1)$$

$$+139(x_2^{24} + 1)]\,x_3^{30} - 4\,[(277x_2^{12}\,(x_2^6 + 1) + 16x_2^6\,(x_2^{18} + 1) - 29(x_2^{30} + 1))\,x_3^{24}$$

$$+2\,[72x_2^{18} + 281x_2^{12}\,(x_2^{12} + 1) + 300x_2^6(x_2^{24} + 1) + 23(x_2^{36} + 1)]\,x_3^{18}$$

$$-4\,[60x_2^{18}\,(x_2^6 + 1) - 191x_2^{12}(x_2^{18} + 1) + 41x_2^6\,(x_2^{30} + 1) + 26\,(x_2^{42} + 1)]\,x_3^{12}$$

$$+(x_2^6 - 1)^2\,[84x_2^{18} - 117x_2^{12}(x_2^{12} + 1) - 90x_2^6\,(x_2^{24} + 1) + 37\,(x_2^{36} + 1)]\,x_3^6$$

$$+6x_2^6\,(x_2^6 - 1)^4(x_2^{18} + 1)\Big],$$

$$\sigma_5 = 3a^2 x_2^2 \Big[(x_2^6 + 1)^2\,[43(x_2^{12} + 1) + 42x_2^6]\,x_3^{36} - 2\,[188x_2^{12}\,(x_2^6 + 1)$$

$$+257x_2^6\,(x_2^{18} + 1) + 67(x_2^{30} + 1)]\,x_3^{30} - 2\,[632x_2^{18} + 701x_2^{12}\,(x_2^{12} + 1)$$

$$-4x_2^6\left(x_2^{24}+1\right)-53(x_2^{36}+1)\right]x_3^{24}+4\left[272x_2^{18}\left(x_2^6+1\right)+119x_2^{12}\left(x_2^{18}+1\right)\right.$$

$$+203x_2^6\left(x_2^{30}+1\right)+14(x_2^{42}+1)\right]x_3^{18}-[510x_2^{24}+20x_2^{18}\left(x_2^{12}+1\right)$$

$$-876x_2^{12}\left(x_2^{24}+1\right)+236x_2^6(x_2^{36}+1)+109(x_2^{48}+1)\right]x_3^{12}+2\left(x_2^6-1\right)^2$$

$$\times[81x_2^{18}\left(x_2^6+1\right)-9x_2^{12}(x_2^{18}+1)-59x_2^6\left(x_2^{30}+1\right)+19(x_2^{42}+1)\right]x_3^6$$

$$-4x_2^6\left(x_2^{12}-1\right)^4\Big],$$

$$\sigma_7=9a^3x_3^4\left[2\left((x_2^6+1)^2+4x_2^6\right)[7\left(x_2^{12}+1\right)+6x_2^6(x_2^{12}+2)]\,x_3^{30}-[37\left(x_2^{36}+1\right)\right.$$

$$+x_2^{24}(208x_2^6+161)+x_2^{12}(16x_2^6-145)+6(32x_2^6+1)]\,x_3^{24}+4\,[7\left(x_2^{42}+1\right)$$

$$+x_2^6(x_2^{30}-14)-2x_2^{24}(22x_2^6+141)-x_2^{12}(13x_2^6+47)+1]\,x_3^{18}+2\,[9(x_2^{48}+1)$$

$$+2x_2^{36}(62x_2^6+83)+4x_2^{24}(128x_2^6-95)+2x_2^{12}(358x_2^6+1)+2(60x_2^6+1)]\,x_3^{12}$$

$$-2\,(x_2^6+1)\,[16\,(x_2^{48}+1)+9x_2^{36}(4x_2^6+1)+x_2^{24}(x_2^{18}+11)-499x_2^{24}(x_2^6-2)$$

$$-x_2^6(8x_2^{12}-23)-35x_2^{12}(11x_2^6-1)+3]\,x_3^6+(x_2^6-1)^2\,(x_2^6+1)^4\,[11\,(x_2^{24}+1)$$

$$-2x_2^6\,(23x_2^{12}+21)+2\,(40x_2^{12}+1)]\Big].$$

References

1. Anisimov, S.I., Lysikov, I.I.: Expansion of a gas cloud in vacuum. J. Appl. Math. Mech. **5**(34), 882–885 (1970)
2. Banshchikov, A.V., Burlakova, L.A., Irtegov, V.D., Titorenko, T.N.: Software Package for Finding and Stability Analysis of Stationary Sets. Certificate of State Registration of Software Programs. FGU-FIPS. No. 2011615235 (2011). (in Russian)
3. Bogoyavlenskii, O.I.: Integrable cases of the dynamics of a rigid body, and integrable systems on the spheres S^n. Math. USSR-Izvestiya **2**(27), 203–218 (1986)
4. Bolsinov, A.V.: Topology and stability of integrable systems. Russ. Math. Surv. **2**(65), 259–318 (2010)
5. Borisov, A.V., Mamaev, I.S.: Rigid body dynamics. Regul. Chaotic Dyn. (2001). NIC, Izhevsk
6. Dyson, F.J.: Dynamics of a spinning gas cloud. J. Math. Mech. **1**(18), 91–101 (1968)
7. Gaffet, B.: Expanding gas clouds of ellipsoidal shape: new exact solutions. J. Fluid Mech. **325**, 113–144 (1996)
8. Gaffet, B.: Spinning gas clouds: liouville integrable cases. Regul. Chaotic Dyn. **4–5**(14), 506–525 (2009)
9. Gaffet, B.: Spinning gas clouds - without vorticity. J. Phys. A: Math. Gen. **33**, 3929–3946 (2000)
10. Irtegov, V.D.: On chains of differential consequences. In: Abstracts of IX Russian Congress on Theoretical and Applied Mechanics, p. 61. N. Novgorod (2006). (in Russian)
11. Irtegov, V.D., Titorenko, T.N.: On invariant manifolds and their stability in the problem of motion of a rigid body under the influence of two force fields. In: Gerdt, V.P., Koepf, W., Seiler, W.M., Vorozhtsov, E.V. (eds.) CASC 2015. LNCS, vol. 9301, pp. 220–232. Springer, Switzerland (2015). https://doi.org/10.1007/978-3-319-24021-3_17

12. Irtegov, V.D., Titorenko, T.N.: The invariant manifolds of systems with first integrals. J. Appl. Math. Mech. **73**(4), 379–384 (2009)
13. Lyapunov, A.M.: On Permanent Helical Motions of a Rigid Body in Fluid. Collected Works, USSR Academy Sciences, Moscow-Leningrad. 1, 276–319 (1954). (in Russian)
14. Lyapunov, A.M.: The General Problem of the Stability of Motion. Taylor & Francis, London (1992)
15. Nemchinov, I.V.: Expansion of a tri-axial gas ellipsoid in a regular behavior. J. Appl. Math. Mech. **1**(29), 143–150 (1965)
16. Ovsiannikov, L.V.: A new solution for hydrodynamic equations. Dokl. Akad. Nauk SSSR. **111**(1), 47–49 (1956). (in Russian)

Effective Localization Using Double Ideal Quotient and Its Implementation

Yuki Ishihara$^{(\boxtimes)}$ and Kazuhiro Yokoyama

Rikkyo University, Tokyo, Japan
{yishihara,kazuhiro}@rikkyo.ac.jp

Abstract. In this paper, we propose a new method for localization of polynomial ideal, which we call "Local Primary Algorithm". For an ideal I and a prime ideal P, our method computes a P-primary component of I after checking if P is associated with I by using *double ideal quotient* $(I : (I : P))$ and its variants which give us a lot of information about localization of I.

Keywords: Gröbner basis · Primary decomposition · Localization

1 Introduction

In commutative algebra, the operation of "localization by a prime ideal" is well-known as a basic tool. To realize it on computer algebra systems, we propose new effective localization using *double ideal quotient* (DIQ) and its variants for ideals, in a polynomial ring over a field. Here, by the words *localization*, we mean the saturation or the contraction of localized ideals.

It is well-known that the localization of an ideal can be computed through its primary decomposition. In more detail, for an ideal I of a polynomial ring $K[X] = K[x_1, \ldots, x_n]$ over a field K and a multiplicatively closed set S in $K[X]$, once a primary decomposition $\mathcal{Q}$ of I is known, the localization (i.e. the contraction of localized ideal) of I by S can be computed by $IK[X]_S \cap K[X] = \bigcap_{Q \in \mathcal{Q}, Q \cap S = \emptyset} Q$ (see Remark 3). Algorithms of primary decomposition have been much studied, for example, by [2,3,5,8]. However, in practice, as such primary decomposition tends to be very time-consuming, use of primary decomposition is not an efficient way and we need an efficient *direct* method without primary decomposition. Toward a direct method of localization, for a given ideal I and a prime ideal P, first we provide several criteria for checking if a primary ideal Q can be a P-primary component of I, and then present a direct method named *Local Primary Algorithm* (LPA) which computes a P-primary component of I. Our method applies different procedures for two cases; isolated and embedded. Both cases use *double ideal quotient and its variants* as a tool for generating and checking primary components. Of course, if we know all associated primes disjoint from a multiplicatively closed set, we get its localization without computing other primary components.

© Springer Nature Switzerland AG 2018
V. P. Gerdt et al. (Eds.): CASC 2018, LNCS 11077, pp. 272–287, 2018.
https://doi.org/10.1007/978-3-319-99639-4_19

For ideals I and J, we call an ideal $(I : (I : J))$ *double ideal quotient* in the paper. Double ideal quotient appears in [10] to check associated primes or compute equidimensional hull, and in [2], to compute equidimensional radical. We survey other properties of double ideal quotient and find that it and its variants have useful information about localization. For instance, for ideals I, J and a primary decomposition $\mathcal{Q}$ of I, a variant of DIQ $(I : (I : J)^\infty)$ coincides with $\bigcap_{Q \in \mathcal{Q}, J \subset IK[X]_{\sqrt{Q}} \cap K[X]} Q$.

To check the practicality of criteria on LPA, we made an implementation on the computer algebra system Risa/Asir [7] and demonstrate the performance in several examples. To evaluate effectiveness coming from its speciality, we compare timings of it to ones of a general algorithm of primary decomposition in Risa/Asir.

For practical implements we devise several efficient techniques for improving our LPA. (For efficient computation of ideal quotient and saturation, see [4,10]). First, instead of computing the *equidimensional hull* hull$(I+P^m)$, we use hull$(I+P_G^{[m]})$ where $P_G^{[m]} = (f_1^m, \ldots, f_r^m)$ for some generator $G = \{f_1, \ldots, f_r\}$ of P. Second, we use *a maximal independent set* of P for computing hull$(\overline{Q})$ where $\overline{Q}$ is a P-hull-primary ideal. Since a maximal independent set U of P is one of $I + P^m$, we obtain hull$(I + P^m) = (I + P^m)K[X]_{K[U]^\times} \cap K[X]$. Moreover, we also use U at the first step of LPA; use $IK[X]_{K[U]^\times} \cap K[X]$ instead of I. By these efficient techniques, our experiment shows certain practicality of our direct localization method.

2 Mathematical Basis

Throughout this paper, we denote a polynomial ring $K[x_1, \ldots, x_n]$ by $K[X]$, where K is a computable field (e.g. the rational field $\mathbb{Q}$ or a finite field $\mathbb{F}_p$) and we denote the set of variables $\{x_1, \ldots, x_n\}$ by X. We write $(f_1, \ldots, f_t)_{K[X]}$ for the ideal generated by elements $f_1, \ldots, f_t$ in $K[X]$. If the ring is obvious, we simply use $(f_1, \ldots, f_t)$. When we simply say I is an ideal, it means the I is an ideal of $K[X]$. Moreover, we denote the radical of I by $\sqrt{I}$.

2.1 Definition of Primary Decomposition and Localization

Here we give the definition of primary decomposition and that of localization which seem slightly different from *standard* ones. We also give fundamental notions and properties related to localization.

Definition 1. *Let I be an ideal of $K[X]$. A set $\mathcal{Q}$ of primary ideals is called a general primary decomposition of I if $I = \bigcap_{Q \in \mathcal{Q}} Q$. A general primary decomposition $\mathcal{Q}$ is called a primary decomposition of I if the decomposition $I = \bigcap_{Q \in \mathcal{Q}} Q$ is an irredundant decomposition. For a primary decomposition of I, each primary ideal is called a primary component of I. The prime ideal associated with a primary component of I is called a prime divisor of I and among all prime divisors, minimal prime ideals are called isolated prime divisors of I and others are called* embedded prime divisors of I. A primary component of I is called isolated if its prime divisor is isolated and embedded if its prime divisor is embedded. We denote by $\mathrm{Ass}(I)$ and $\mathrm{Ass}_{iso}(I)$ the set of all prime divisors of I and the set of all isolated prime divisors respectively.

Definition 2. Let I be an ideal of $K[X]$ and S a multiplicatively closed set in $K[X]$. We denote the set $\{f \in K[X] \mid fs \in I \text{ for some } s \in S\}$ by $IK[X]_S \cap K[X]$, and call it the localization of I with respect to S. For a multiplicatively closed set $K[X] \setminus P$, where P is a prime ideal, we denote simply by $IK[X]_P \cap K[X]$. We assume a multiplicatively closed set S always does not contain 0.

Remark 3. Given a primary decomposition $\mathcal{Q}$ of an ideal I, the localization of I by S is expressed as $\bigcap_{Q \in \mathcal{Q}, Q \cap S = \emptyset} Q$. Moreover, it is also equal to $(I : (\bigcap_{P \in \mathrm{Ass}(I), P \cap S \neq \emptyset} P)^{\infty})$. Thus if we know all primary components or all associated primes, then we can compute localizations of I for any computable multiplicatively closed sets S. (We are thinking mainly about cases where S is finitely generated or the complement of a prime ideal. In these cases, we can decide efficiently whether Q and S intersect or not). However, this method is not a direct method since it computes unnecessary primary components or associated primes.

Lemma 4. Let I be an ideal and P a prime divisor of I. If S is a multiplicatively closed set with $P \cap S = \emptyset$ and Q is a P-primary ideal, then the following conditions are equivalent.

(A) Q is a primary component of I.
(B) Q is a primary component of $IK[X]_S \cap K[X]$.

Proof. First, (A) implies (B) from Proposition 4.9 in [1] . For primary decompositions $\mathcal{Q}$ of I and $\mathcal{Q}'$ of $IK[X]_S \cap K[X]$ with $Q \in \mathcal{Q}'$, we obtain $\{Q' \in \mathcal{Q} \mid Q' \cap S \neq \emptyset\} \cup \mathcal{Q}'$ is also a primary decomposition of I. Hence, (B) implies (A).

Definition 5 ([1], **Chap. 4**). Let I be an ideal. A subset $\mathcal{P}$ of $\mathrm{Ass}(I)$ is said to be isolated if it satisfies the following condition: for a prime divisor $P' \in \mathrm{Ass}(I)$, if $P' \subset P$ for some $P \in \mathcal{P}$, then $P' \in \mathcal{P}$.

Lemma 6 ([1], **Theorem 4.10**). Let I be an ideal and $\mathcal{P}$ an isolated set contained in $\mathrm{Ass}(I)$. For a multiplicatively closed set $S = K[X] \setminus \bigcup_{P \in \mathcal{P}} P$ and a primary decomposition $\mathcal{Q}$ of I, $IK[X]_S \cap K[X] = \bigcap_{Q \in \mathcal{Q}, \sqrt{Q} \in \mathcal{P}} Q$.

Lemma 7. Let $\mathcal{Q}$ be a primary decomposition of I and $Q \in \mathcal{Q}$. For a multiplicatively closed set S, the following conditions are equivalent.

(A) $IK[X]_S \cap K[X] \subset IK[X]_{\sqrt{Q}} \cap K[X]$.
(B) $Q \cap S = \emptyset$.

Proof. Show (A) implies (B). As $IK[X]_{\sqrt{Q}} \cap K[X] \subset Q$, $IK[X]_S \cap K[X] = \bigcap_{Q' \in \mathcal{Q}, Q' \cap S = \emptyset} Q' \subset Q$. Since $\mathcal{Q}$ is irredundant, $IK[X]_S \cap K[X]$ has $\sqrt{Q}$-primary component. Thus, $Q \cap S = \emptyset$. Now, we show (B) implies (A). Then, $\sqrt{Q} \cap S = \emptyset$ and $Q' \cap S = \emptyset$ for any $Q' \in \mathcal{Q}$ s.t. $Q' \subset \sqrt{Q}$. Thus, $IK[X]_{\sqrt{Q}} \cap K[X] = \bigcap_{Q' \subset \sqrt{Q}} Q'$ implies $IK[X]_S \cap K[X] \subset IK[X]_{\sqrt{Q}} \cap K[X]$.

Next we introduce the notion of pseudo-primary ideal.

Definition 8. *Let Q be an ideal. We say Q is* pseudo-primary *if $\sqrt{Q}$ is a prime ideal. In this case, we also say $\sqrt{Q}$-pseudo-primary.*

Definition 9. *Let I be an ideal and P an isolated prime divisor of I. For $\mathcal{P} = \{P' \in \mathrm{Ass}(I) \mid P$ is the unique isolated prime divisor contained in $P'\}$ and $S = K[X] \setminus \bigcup_{P' \in \mathcal{P}} P'$, we call $\overline{Q} = IK[X]_S \cap K[X]$ the P-pseudo-primary component of I. This definition is consistent with one in [8]. We note that the P-pseudo-primary component is determined uniquely and has the P-isolated primary component of I as component.*

Remark 10. *Every P-pseudo-primary component of I is a P-pseudo-primary ideal. Let $\overline{Q}_P$ be the P-pseudo-primary component of I. Then $I = \bigcap_{P \in \mathrm{Ass}_{iso}(I)} \overline{Q}_P \cap I'$ for some I' s.t. $\mathrm{Ass}_{iso}(I') \cap \mathrm{Ass}_{iso}(I) = \emptyset$. This decomposition is called a pseudo-primary decomposition in [8], where it is computed by separators from given $\mathrm{Ass}_{iso}(I)$. Meanwhile, we introduce another method to compute it by using double ideal quotient in Lemma 32.*

Definition 11. *Let I be an ideal and $\mathcal{Q}$ a primary decomposition of I. We call $\mathrm{hull}(I) = \bigcap_{Q \in \mathcal{Q}, \dim(Q)=\dim(I)} Q$ the equidimensional hull of I. Since every primary component Q satisfying $\dim(Q) = \dim(I)$ is isolated, $\mathrm{hull}(I)$ is determined independently from choice of primary decompositions.*

For a given I, $\mathrm{hull}(I)$ can be computed in several manners. For instance, it can be computed by Ext functors [2] or a regular sequence contained in I [10].

Proposition 12 ([2], Theorem 1.1. [10], Proposition 3.41). *Let I be an ideal and $u \subset I$ be a c-length regular sequence, where c is the codimension of I. Then $\mathrm{hull}(I) = ((u) : ((u) : I)) = \mathrm{ann}_{K[X]}(\mathrm{Ext}^c_{K[X]}(K[X]/I, K[X]))$.*

Definition 13. *Let I be an ideal. We say that I is* hull-primary *if $\mathrm{hull}(I)$ is a primary ideal. For a prime ideal P, we say a hull-primary ideal I is P-hull-primary if $P = \mathrm{hull}(\sqrt{I})$.*

Since a pseudo-primary ideal has the unique isolated component, we obtain the following remark.

Remark 14. *A pseudo-primary ideal is hull-primary.*

By the definition of the P-pseudo-primary component of I, it is easy to prove the following lemma.

Lemma 15. *Let P be an isolated prime divisor of I and $\overline{Q}$ a P-pseudo-primary component of I. Then, $\overline{Q}$ is a P-hull-primary and $\mathrm{hull}(\overline{Q})$ is the isolated P-primary component of I.*

Using Lemma 15 and a variant of *double ideal quotient*, we generate the isolated P-primary component of I in Sect. 5.

Lemma 16. *Let Q be a primary ideal. Let I and J be ideals. If $IJ \subset Q$ and $J \not\subset \sqrt{Q}$, then $I \subset Q$. In particular, if $I \cap J \subset Q$ and $J \not\subset \sqrt{Q}$, then $I \subset Q$.*

Proof. Let $f \in I$ and $g \in J \setminus \sqrt{Q}$. Since Q is $\sqrt{Q}$-primary, $fg \in IJ \subset Q$ and thus $f \in Q$.   □

Lemma 17. *Let I be a P-hull-primary and Q a P-primary ideal. If $I \subset Q$, then* hull$(I) \subset Q$.

Proof. Let $\mathcal{Q}$ be a primary decomposition of I and $J = \bigcap_{Q' \in \mathcal{Q}, Q' \neq \text{hull}(I)} Q'$. Then $I = \text{hull}(I) \cap J \subset Q$ and $J \not\subset P$. Since Q is P-primary, we obtain hull$(I) \subset Q$ by Lemma 16.   □

Finally, we recall the famous Prime Avoidance Lemma.

Lemma 18 ([1], Proposition 1.11). *(i) Let $P_1, \ldots, P_n$ be prime ideals and let I be an ideal contained in $\bigcup_{i=1}^{n} P_i$. Then, $I \subset P_i$ for some i.*
(ii) Let $I_1, \ldots, I_n$ be ideals and let P be a prime ideal containing $\bigcap_{i=1}^{n} I_i$. Then $P \supset I_i$ for some i. If $P = \bigcap_{i=1}^{n} I_i$, then $P = I_i$ for some i.

2.2 Fundamental Properties of Ideal Quotient

We introduce fundamental properties of ideal quotient. The first two can be seen in several papers and books ([1], Lemma 4.4. [4], Lemma 4.1.3. [10], a remark before Proposition 3.56). The last two are direct consequences of the first two.

Lemma 19. *Let I and J be ideals, Q a primary ideal and $\mathcal{Q}$ a primary decomposition of I. Then,*

$$
(Q : J) = \begin{cases} Q, \ \text{if } J \not\subset \sqrt{Q}, \\ K[X], \ \text{if } J \subset Q, \\ \sqrt{Q}\text{-primary ideal properly containing } Q, \ \text{if } J \not\subset Q, J \subset \sqrt{Q}, \end{cases}
$$

$$
(Q : J^{\infty}) = (Q : \sqrt{J}^{\infty}) = \begin{cases} Q, \ \text{if } J \not\subset \sqrt{Q}, \\ K[X], \ \text{if } J \subset \sqrt{Q}, \end{cases}
$$

$$
(I : J) = \bigcap_{Q \in \mathcal{Q}, J \not\subset \sqrt{Q}} Q \cap \bigcap_{Q \in \mathcal{Q}, J \not\subset Q, J \subset \sqrt{Q}} (Q : J),
$$

$$
(I : J^{\infty}) = (I : \sqrt{J}^{\infty}) = \bigcap_{Q \in \mathcal{Q}, J \not\subset \sqrt{Q}} Q.
$$

3 Double Ideal Quotient

Double Ideal Quotient (DIQ) is an ideal of shape $(I : (I : J))$ where I and J are ideals. For an ideal I and its primary decomposition $\mathcal{Q}$, we divide $\mathcal{Q}$ into three parts:

$$
\mathcal{Q}_1(J) = \{Q \in \mathcal{Q} \mid J \not\subset \sqrt{Q}\}, \qquad \mathcal{Q}_2(J) = \{Q \in \mathcal{Q} \mid J \subset Q\},
$$
$$
\mathcal{Q}_3(J) = \{Q \in \mathcal{Q} \mid J \not\subset Q, J \subset \sqrt{Q}\}.
$$

Then, our DIQ is expressed precisely by components of them. The following proposition can be proved directly from Lemma 19. We omit an easy but tedious proof.

Proposition 20. *Let I and J be ideals. Then,*

$$(I : (I : J)) = \bigcap_{Q \in \mathcal{Q}_2(J)} \left(Q : \bigcap_{Q' \in \mathcal{Q}_1(J)} Q' \cap \bigcap_{Q' \in \mathcal{Q}_3(J)} (Q' : J) \right)$$

$$\cap \bigcap_{Q \in \mathcal{Q}_3(J)} \left(Q : \bigcap_{Q' \in \mathcal{Q}_1(J)} Q' \cap \bigcap_{Q' \in \mathcal{Q}_3(J)} (Q' : J) \right),$$

$$\sqrt{(I : (I : J))} = \bigcap_{P \in \mathrm{Ass}(I), J \subset P} P.$$

This proposition can be used to prove the following for prime divisors.

Corollary 21 ([10], Corollary 3.4). *Let I be an ideal and P a prime ideal. Then, P belongs to $\mathrm{Ass}(I)$ if and only if $P \supset (I : (I : P))$.*

Proof. We note $P \supset (I : (I : P))$ if and only if $P \supset \sqrt{(I : (I : P))}$. By Proposition 20, $\sqrt{(I : (I : P))} = \bigcap_{P' \in \mathrm{Ass}(I), P \subset P'} P'$. If $P \in \mathrm{Ass}(I)$, then $\sqrt{(I : (I : P))} = \bigcap_{P' \in \mathrm{Ass}(I), P \subset P'} P' \subset P$. On the other hand, if $P \supset \sqrt{(I : (I : P))}$, then there is $P' \in \mathrm{Ass}(I)$ s.t. $P' \subset P$ and $P' \supset P$. Thus $P = P' \in \mathrm{Ass}(I)$. $\qquad\square$

Replacing ideal quotient with saturation in DIQ, we have the following.

Proposition 22. *Let $\mathcal{Q}$ be a primary decomposition of I. Then,*

$$(I : (I : J)^\infty) = \bigcap_{Q \in \mathcal{Q}, J \subset IK[X]_{\sqrt{Q}} \cap K[X]} Q, \tag{1}$$

$$(I : (I : J^\infty)^\infty) = \bigcap_{Q \in \mathcal{Q}, J \subset \sqrt{IK[X]_{\sqrt{Q}} \cap K[X]}} Q, \tag{2}$$

$$(I : (I : J^\infty)) = \bigcap_{Q \in \mathcal{Q}_2(J)} (Q : \bigcap_{Q' \in \mathcal{Q}_1(J)} Q') \cap \bigcap_{Q \in \mathcal{Q}_3(J)} (Q : \bigcap_{Q' \in \mathcal{Q}_1(J)} Q'). \tag{3}$$

We call them the first saturated quotient, the second saturated quotient, and the third saturated quotient, respectively.

Proof. Here, we give an outline of the proof. The formula (1) can be proved by combining the equation

$$(I : (I : J)^\infty) = (I : \sqrt{(I : J)}^\infty) = \bigcap_{Q \in \mathcal{Q}, \bigcap_{Q' \in \mathcal{Q}_1(J)} \sqrt{Q'} \cap \bigcap_{Q' \in \mathcal{Q}_3(J)} \sqrt{Q'} \not\subset \sqrt{Q}} Q$$

by Lemma 19 and the following equivalence

(1-a) $J \subset IK[X]_{\sqrt{Q}} \cap K[X]$.
(1-b) $\bigcap_{Q' \in \mathcal{Q}_1(J)} \sqrt{Q'} \cap \bigcap_{Q' \in \mathcal{Q}_3(J)} \sqrt{Q'} \not\subset \sqrt{Q}$.

for each $Q \in \mathcal{Q}$. The second formula (2) can be proved by combining the equation $(I : (I : J^\infty)^\infty) = (I : (I : J^m)^\infty) = \bigcap_{Q \in \mathcal{Q}, J^m \subset IK[X]_{\sqrt{Q}} \cap K[X]} Q$ for a sufficiently large m from the first formula (1), and the following equivalence

(2-a) $J^m \subset IK[X]_{\sqrt{Q}} \cap K[X]$ for a sufficiently large m.

(2-b) $J \subset \sqrt{IK[X]_{\sqrt{Q}} \cap K[X]}$.

for each $Q \in \mathcal{Q}$. The third formula (3) can be proved directly from Lemma 19. Now, we explain some details. We show (1-a) implies (1-b). If

$$\bigcap_{Q' \in \mathcal{Q}_1(J)} \sqrt{Q'} \cap \bigcap_{Q' \in \mathcal{Q}_3(J)} \sqrt{Q'} \subset \sqrt{Q},$$

then by Lemma 18, $\sqrt{Q'} \subset \sqrt{Q}$ for some $Q' \in \mathcal{Q}_1(J) \cup \mathcal{Q}_3(J)$. Since $Q' \subset \sqrt{Q'} \subset \sqrt{Q}$, we obtain $IK[X]_{\sqrt{Q}} \cap K[X] = \bigcap_{Q'' \in \mathcal{Q}, Q'' \subset \sqrt{Q}} Q'' \subset Q'$. However, since $Q' \in \mathcal{Q}_1(J) \cup \mathcal{Q}_3(J)$, we obtain $J \not\subset Q'$ and this contradicts $J \subset IK[X]_{\sqrt{Q}} \cap K[X] \subset Q'$.

Show (1-b) implies (1-a). Let $Q' \in \mathcal{Q}$ contained $\sqrt{Q}$. Since $\bigcap_{Q'' \in \mathcal{Q}_1(J)} \sqrt{Q''} \cap \bigcap_{Q'' \in \mathcal{Q}_3(J)} \sqrt{Q''} \not\subset \sqrt{Q}$, we obtain $Q' \notin \mathcal{Q}_1(J) \cup \mathcal{Q}_3(J)$ and $Q' \in \mathcal{Q}_2(J)$. Hence, $J \subset Q'$ and $J \subset \bigcap_{Q' \subset \sqrt{Q}} Q' = IK[X]_{\sqrt{Q}} \cap K[X]$.

Trivially, (2-a) implies (2-b) since $J \subset \sqrt{J^m} \subset \sqrt{IK[X]_{\sqrt{Q}} \cap K[X]}$. Show (2-b) implies (2-a). For $Q \in \mathcal{Q}_2(J) \cup \mathcal{Q}_3(J)$, let $m_Q = \min\{m \mid J^m \subset Q\}$ and $m = \max\{m_Q \mid Q \in \mathcal{Q}_2(J) \cup \mathcal{Q}_3(J)\}$. Then, $(I : J^\infty) = (I : J^m)$. Since $IK[X]_{\sqrt{Q}} \cap K[X] = \bigcap_{Q' \in \mathcal{Q}, Q' \subset \sqrt{Q}} Q'$, we obtain $Q' \in \mathcal{Q}_2(J) \cup \mathcal{Q}_3(J)$ for any $Q' \in \mathcal{Q}$ contained in $\sqrt{Q}$. Thus, we obtain $J^m \subset IK[X]_{\sqrt{Q}} \cap K[X]$. □

Using the first saturated quotient, we devise criteria for primary component in Sect. 4. The second saturated quotient can be used to isolated prime divisor check and generate an isolated primary component in Sect. 5. The third saturated quotient gives another prime divisor criterion (Criterion 5 in Sect. 4) other than Corollary 19 by the following proposition.

Proposition 23. *Let I and J be ideals. Then $\sqrt{(I : (I : J^\infty))} = \bigcap_{P \in \mathrm{Ass}(I), J \subset P} P$.*

Proof. Let $\mathcal{Q}$ be a primary decomposition of I. By Proposition 22 (3),

$$\sqrt{(I : (I : J^\infty))} = \bigcap_{Q \in \mathcal{Q}_2(J)} \sqrt{(Q : \bigcap_{Q' \in \mathcal{Q}_1(J)} Q')} \cap \bigcap_{Q \in \mathcal{Q}_3(J)} \sqrt{(Q : \bigcap_{Q' \in \mathcal{Q}_1(J)} Q')}.$$

Since $\mathcal{Q}$ is minimal, we obtain $Q \not\supset \bigcap_{Q' \in \mathcal{Q}_1(J)} Q'$ for any $Q \in \mathcal{Q}_2(J)$ and $Q \not\supset \bigcap_{Q' \in \mathcal{Q}_1(J)} Q'$ for any $Q \in \mathcal{Q}_3(J)$. Thus, by Lemma 19,

$$\sqrt{(I : (I : J^\infty))} = \bigcap_{Q \in \mathcal{Q}_2(J)} \sqrt{(Q : \bigcap_{Q' \in \mathcal{Q}_1(J)} Q')} \cap \bigcap_{Q \in \mathcal{Q}_3(J)} \sqrt{(Q : \bigcap_{Q' \in \mathcal{Q}_1(J)} Q')}$$

$$= \bigcap_{Q \in \mathcal{Q}_2(J)} \sqrt{Q} \cap \bigcap_{Q \in \mathcal{Q}_3(J)} \sqrt{Q} = \bigcap_{P \in \mathrm{Ass}(I), J \subset P} P.$$

□

4   Criteria for Primary Component and Prime Divisor

In this section, we present several criteria for primary component which check if a P-primary ideal Q is a primary component of I or not without computing primary decomposition of I based on the first saturated quotient. We first propose a general criterion applicable to any primary ideal. Later, we propose some specialized criteria aiming for isolated primary components and maximal ones. Finally, we add criteria for prime divisors.

4.1   General Primary Component Criterion

We use the first saturated quotient to check if a given primary ideal is a component or not. We introduce a key notion *saturated quotient invariant*.

Definition 24. *Let I and J be ideals. We say that J is saturated quotient invariant of I if $(I : (I : J)^\infty) = J$.*

Any localization is saturated quotient invariant. Conversely, any proper saturated quotient invariant ideal is some localization of I.

Lemma 25. *Let I be an ideal and J a proper ideal of $K[X]$. Then, the following conditions are equivalent.*

(A) $J = IK[X]_S \cap K[X]$ for some multiplicatively closed set S.
(B) J is saturated quotient invariant of I.

Proof. Let $\mathcal{Q}$ be a primary decomposition. Show (A) implies (B). From Proposition 22 (1),

$$(I : (I : IK[X]_S \cap A)^\infty) = \bigcap_{Q \in \mathcal{Q}, IK[X]_S \cap K[X] \subset IK[X]_{\sqrt{Q}} \cap K[X]} Q. \qquad (4)$$

By Lemma 7, $IK[X]_S \cap K[X] \subset IK[X]_{\sqrt{Q}} \cap K[X]$ if and only if $Q \cap S = \emptyset$. Thus,

$$\bigcap_{Q \in \mathcal{Q}, IK[X]_S \cap K[X] \subset IK[X]_{\sqrt{Q}} \cap K[X]} Q = \bigcap_{Q \in \mathcal{Q}, Q \cap S = \emptyset} Q, \qquad (5)$$

Combining (4), (5) and $IK[X]_S \cap K[X] = \bigcap_{Q \in \mathcal{Q}, Q \cap S = \emptyset} Q$ by Remark 3, we obtain $(I : (I : IK[X]_S \cap A)^\infty) = IK[X]_S \cap K[X]$.

Next, show (B) implies (A). From Proposition 22 (1),

$$(I : (I : J)^\infty) = \bigcap_{J \subset IK[X]_{\sqrt{Q}} \cap K[X]} Q = J. \qquad (6)$$

Let $\mathcal{P} = \{\sqrt{Q} \mid Q \in \mathcal{Q}, J \subset IK[X]_{\sqrt{Q}} \cap K[X]\}$. We may assume $\mathcal{P} \neq \emptyset$, otherwise $\mathcal{P} = \emptyset$ and $J = K[X]$. Then $\mathcal{P}$ is *isolated* since if $P' \in \mathrm{Ass}(I)$ and $P' \subset P$ for some $P \in \mathcal{P}$, then $J \subset IK[X]_P \cap K[X] \subset IK[X]_{P'} \cap K[X]$ and $P' \in \mathcal{P}$. Let $S = K[X] \setminus \bigcup_{P \in \mathcal{P}} P$. By Lemma 6, $IK[X]_S \cap K[X] = \bigcap_{Q \in \mathcal{Q}, \sqrt{Q} \in \mathcal{P}} Q = \bigcap_{J \subset IK[X]_{\sqrt{Q}} \cap K[X]} Q$. By (3), we obtain $IK[X]_S \cap K[X] = J$.   □

Based on Lemma 25, we have the following criterion for primary component.

Theorem 26 (Criterion 1). *Let I be an ideal and P a prime divisor of I. For a P-primary ideal Q, if $Q \not\supset (I : P^\infty)$, then the following conditions are equivalent.*

(A) Q is a P-primary component for some primary decomposition of I.
(B) $(I : P^\infty) \cap Q$ is saturated quotient invariant of I.

Proof. Show (A) implies (B). Let $\mathcal{Q}$ be a primary decomposition. Let $\mathcal{P} = \{P' \in \mathrm{Ass}(I) \mid P \not\subset P' \text{ or } P' = P\}$ and $S = K[X] \setminus \bigcup_{P' \in \mathcal{P}} P'$. Then S is a multiplicatively closed set and $(I : P^\infty) \cap Q \subset IK[X]_S \cap K[X]$ since $(I : P^\infty) \cap Q = \bigcap_{Q' \in \mathcal{Q}, P \not\subset \sqrt{Q'}} Q' \cap Q$. For each $Q' \in \mathcal{Q}$ with $Q' \cap S = \emptyset$, there is $P' \in \mathcal{P}$ such that $\sqrt{Q'} \subset P'$, i.e. $\sqrt{Q'} \in \mathcal{P}$. Thus, $(I : P^\infty) \cap Q \supset IK[X]_S \cap K[X]$ and $(I : P^\infty) \cap Q = IK[X]_S \cap K[X]$. By Lemma 25, $IK[X]_S \cap K[X]$ is saturated quotient invariant of I.

Show (B) implies (A). By Lemma 25, there is a multiplicatively closed set S such that $(I : P^\infty) \cap Q = IK[X]_S \cap K[X]$. Let $\mathcal{Q}$ be a primary decomposition of I. We know $IK[X]_S \cap K[X] = \bigcap_{Q' \in \mathcal{Q}, Q' \cap S = \emptyset} Q'$. By the assumption, $Q \not\supset (I : P^\infty)$ and thus $(I : P^\infty) \cap Q$ has a P-primary component. Then neither $\bigcap_{Q' \in \mathcal{Q}, Q' \cap S \neq \emptyset} Q'$ nor $(I : P^\infty)$ has a P-primary component. Hence,

$$I = (I : P^\infty) \cap Q \cap \bigcap_{Q' \in \mathcal{Q}, Q' \cap S \neq \emptyset} Q' = \bigcap_{Q' \in \mathcal{Q}, P \not\subset \sqrt{Q'}} Q' \cap Q \cap \bigcap_{Q' \in \mathcal{Q}, Q' \cap S \neq \emptyset} Q'$$

is a primary decomposition and Q is its P-primary component. $\qquad\square$

4.2   Other Criteria for Primary Component

Next, we propose criteria for primary components having special properties which can be applied for particular prime divisors. These criteria may be computed more easily than the general one.

Criterion for Isolated Primary Component: If Q is a primary ideal whose radical is an isolated divisor P of an ideal I, then we don't need to compute $(I : P^\infty)$ since the P-primary component of I is the localization of I by P.

Theorem 27 (Criterion 2). *Let I be an ideal and P an isolated prime divisor of I. For a P-primary ideal Q, the following conditions are equivalent.*

(A) Q is the isolated P-primary component of I.
(B) $(I : (I : Q)^\infty) = Q$.

Proof. Show (A) implies (B). Let $S = K[X] \setminus P$. By Lemma 25, $Q = IK[X]_S \cap K[X]$ is saturated quotient invariant of I and thus $(I : (I : Q)^\infty) = Q$. Next, we show (B) implies (A). By Lemma 25, there is a multiplicatively closed set S s.t. $IK[X]_S \cap K[X] = Q$. Since Q is primary, $IK[X]_S \cap K[X]$ is the isolated P-primary component. $\qquad\square$

Criterion for Maximal Primary Component: Each isolated prime divisor is minimal in $\text{Ass}(I)$. On the contrary, we consider "maximal prime divisor" and propose the following criterion for it.

Definition 28. *Let P be a prime divisor of I. We say P is* maximal *if there is no prime divisor P' of I containing P properly.*

Theorem 29 (Criterion 3). *Let I be an ideal and P a maximal prime divisor of I. For P-primary ideal Q, the following conditions are equivalent.*

(A) Q is a P-primary component of I.
(B) $(I : P^\infty) \cap Q = I$.

Proof. Show (A) implies (B). Let $\mathcal{Q}$ be a primary decomposition of I with $Q \in \mathcal{Q}$. Since P is maximal in $\text{Ass}(I)$, $(I : P^\infty) = \bigcap_{Q' \in \mathcal{Q}, \sqrt{Q'} \not\supseteq P} Q' = \bigcap_{Q' \in \mathcal{Q}, Q' \neq Q} Q'$. Thus, $(I : P^\infty) \cap Q = \bigcap_{Q' \in \mathcal{Q}, Q' \neq Q} Q' \cap Q = I$. Next, we show (B) implies (A). Let $\mathcal{Q}'$ be a primary decomposition of $(I : P^\infty)$. Since $\mathcal{Q}'$ does not have P-primary component, $\mathcal{Q}' \cup \{Q\}$ is a primary decomposition of I. $\qquad\square$

Criterion for Another General Primary Component: The general case can be reduced to maximal case via localization by maximal independent set (See [4] the definition of maximal independent and its computation). Letting $S = K[U]^\times = K[U] \setminus \{0\}$, we obtain the following as a special case of Lemma 4.

Theorem 30 (Criterion 4). *Let I be an ideal and P a prime divisor of I. If U is a maximal independent set of P in X and Q is a P-primary ideal, then the following conditions are equivalent.*

(A) Q is a primary component of I.
(B) Q is a primary component of $IK[X]_{K[U]^\times} \cap K[X]$.

4.3   Additional Criterion for Prime Divisor

Here, we add a criterion for prime divisor based on the third saturated quotient.

Theorem 31 (Criterion 5). *Let I be an ideal and P a prime ideal. Then, the following conditions are equivalent.*

(A) $P \in \text{Ass}(I)$.
(B) $P \supset (I : (I : P))$.
(C) $P \supset (I : (I : P^\infty))$.

Proof. By Corollary 21, (A) is equivalent to (B). By Proposition 23, $\sqrt{(I : (I : P))} = \sqrt{(I : (I : P^\infty))} = \bigcap_{P' \in \text{Ass}(I), P \subset P'} P'$. Thus, equivalence between (A) and (C) is proved in a similar way to Corollary 21. $\qquad\square$

Next, we devise criteria for isolated prime divisor based on the second saturated quotient.

Lemma 32. *Let I be an ideal and P an isolated prime divisor of I. If $\overline{Q}$ is the P-pseudo-primary component of I, then $(I : (I : P^\infty)^\infty) = \overline{Q}$.*

Proof. Let $\mathcal{Q}$ be a primary decomposition of I. By Proposition 22 (2),

$$(I : (I : P^\infty)^\infty) = \bigcap\nolimits_{Q \in \mathcal{Q}, P \subset \sqrt{IK[X]_{\sqrt{Q}} \cap K[X]}} Q.$$

Thus it is enough to show that the following statements are equivalent for each $Q \in \mathcal{Q}$.

(1-a) $P \subset \sqrt{IK[X]_{\sqrt{Q}} \cap K[X]}$.

(1-b) P is the unique isolated prime divisor which is contained in $\sqrt{Q}$.

Show (1-a) implies (1-b). As $\sqrt{IK[X]_{\sqrt{Q}} \cap K[X]} \subset \sqrt{Q}$, we know $P \subset \sqrt{Q}$. Then, suppose there is another isolated prime divisor P' contained in $\sqrt{Q}$. We obtain

$$\sqrt{IK[X]_{\sqrt{Q}} \cap K[X]} = \bigcap_{Q' \in \mathcal{Q}, Q' \subset \sqrt{Q}} \sqrt{Q'} \subset P'.$$

However, this implies $P \subset P'$ and contradicts that P' is isolated. It is easy to prove that (1-b) implies (1-a).. $\qquad\square$

Theorem 33 (Criterion 6). *Let I be an ideal and P a prime ideal containing I. Then, the following conditions are equivalent.*

(A) P is an isolated prime divisor of I.
(B) $(I : (I : P^\infty)^\infty) \neq K[X]$.

Proof. Show (A) implies (B). By Lemma 32, $(I : (I : P^\infty)^\infty) = \overline{Q} \neq K[X]$. Show (B) implies (A). By Proposition 22 (2),

$$(I : (I : P^\infty)^\infty) = \bigcap\nolimits_{Q \in \mathcal{Q}, P \subset \sqrt{IK[X]_{\sqrt{Q}} \cap K[X]}} Q \neq K[X]$$

for a primary decomposition $\mathcal{Q}$ of I. Then, there is an isolated prime divisor P' containing P. Since $\sqrt{I} \subset P \subset P'$ and P' is isolated, this implies $P = P'$ is isolated. $\qquad\square$

Since each prime divisor of I contains I, Theorem 33 directly induces the following.

Corollary 34 (Criterion 7). *Let I be an ideal and P a prime divisor of I. Then,*

(i) *P is isolated if $(I : (I : P^\infty)^\infty) \neq K[X]$,*
(ii) *P is embedded if $(I : (I : P^\infty)^\infty) = K[X]$.*

5   Local Primary Algorithm

In this section, we devise Local Primary Algorithm (LPA) which computes P-primary component of I. Our method applies different procedures for two cases; isolated and embedded. Algorithm 1 shows the outline of LPA. Its termination comes from Proposition 35. We remark that, for given prime divisors disjoint from a multiplicatively closed set S, we can compute all primary components disjoint from S by LPA. Then their intersection gives the localization by S.

5.1   Generating Primary Component

First, we introduce several ways to generate primary component through equidimensional hull computation.

Proposition 35 ([2], Sect. 4. [6], Remark 10). *Let I be an ideal and P a prime divisor of I. For any positive integer m, $I + P^m$ is P-hull-primary, and for a sufficiently large integer m, $\mathrm{hull}(I + P^m)$ is a P-primary component appearing in a primary decomposition of I.*

We can use Criteria for Primary Component to check m is large enough or not. If P is an isolated prime divisor, then the component is computed directly by using the second saturated quotient. By Lemmas 15 and 32, we obtain the following theorem.

Theorem 36. *Let I be an ideal and P an isolated prime divisor of I. Then $\mathrm{hull}((I : (I : P^\infty)^\infty))$ is the isolated P-primary component of I.*

Algorithm 1. General Frame of Local Primary Algorithm

Input: I: an ideal, P: a prime ideal
Output: • a P-primary component of I if P is a prime divisor of I
    • "P is not a prime divisor" otherwise
1: **if** P is a prime divisor of I (**Criterion 5**) **then**
2:    **if** P is isolated (**Criteria 6,7**) **then**
3:       $\overline{Q} \leftarrow$ the P-pseudo-primary component of I (**Lemma 32**)
4:       $Q \leftarrow \mathrm{hull}(\overline{Q})$ (**Theorem 36**)
5:       **return** Q is the isolated P primary component
6:    **else**
7:       $m \leftarrow 1$
8:       **while** Q is not primary component of I (**Criteria 1,3,4**) **do**
9:          $\overline{Q} \leftarrow$ a P-hull-primary ideal related to m (**Proposition 35, Lemma 38**)
10:          $Q \leftarrow \mathrm{hull}(\overline{Q})$
11:          $m \leftarrow m + 1$
12:       **end while**
13:       **return** Q is an embedded P-primary component
14:    **end if**
15: **else**
16:    **return** "P is not a prime divisor"
17: **end if**

5.2   Techniques for Improving LPA

We introduce a practical technique for implementing LPA.

5.3   Another Way of Generating Primary Component

Let $G = \{f_1, \ldots, f_r\}$ be a generator of P. Usually we take $\{f_1^{e_1} f_2^{e_2} \cdots f_r^{e_r} \mid e_1 + \cdots + e_r = m\}$ as a generator of P^m for a positive integer m. However, this generator has $\frac{(r+m-1)!}{(r-1)!m!}$ elements and it becomes difficult to compute $\mathrm{hull}(I + P^m)$ when m becomes large. To avoid the explosion of the number of the generator, we can use $P_G^{[m]} = (f_1^m, \ldots, f_r^m)$ instead.

Lemma 37. *Let $\mathcal{Q}$ be a primary decomposition of I and $Q \in \mathcal{Q}$. If $\sqrt{Q}$-hull-primary ideal Q' satisfies $I \subset Q' \subset Q$, then $(\mathcal{Q} \setminus \{Q\}) \cup \{\mathrm{hull}(Q')\}$ is another primary decomposition of I.*

Proof. By Lemma 17, we obtain $I \subsetneq Q' \subset \mathrm{hull}(Q') \subset Q$. Since $I \cap \mathrm{hull}(Q') = I$ and $Q \cap \mathrm{hull}(Q') = \mathrm{hull}(Q')$, we obtain

$$I = I \cap \mathrm{hull}(Q') = \left(\bigcap_{Q'' \in \mathcal{Q}, Q'' \neq Q} Q'' \cap Q \right) \cap \mathrm{hull}(Q') = \bigcap_{Q'' \in \mathcal{Q}, Q'' \neq Q} Q'' \cap \mathrm{hull}(Q').$$

Thus, $(\mathcal{Q} \setminus \{Q\}) \cup \{\mathrm{hull}(Q')\}$ is an irredundant primary decomposition of I. □

Lemma 38. *For any positive integer m, $I + P_G^{[m]}$ is P-hull-primary, and for a sufficiently large m, $\mathrm{hull}(I + P_G^{[m]})$ is a P-primary component appearing in a primary decomposition of I.*

Proof. As $\sqrt{I + P} = \sqrt{I + P_G^{[m]}} = P$, $I + P_G^{[m]}$ is P-hull-primary. By Theorem 35, $\mathrm{hull}(I + P^m)$ is a P-primary component of I for a sufficiently large m. Since $I \subset I + P_G^{[m]} \subset I + P^m \subset \mathrm{hull}(I + P^m)$, $\mathrm{hull}(I + P_G^{[m]})$ is a P-primary component by Lemma 37. □

5.4   Equidimensional Hull Computation with MIS

Next, we devise another computation of $\mathrm{hull}(I + P^m)$ based on *maximal independent set* (MIS) which is much efficient than computations based on Proposition 12. Similarly, by this technique we can replace I with $IK[X]_{K[U]^\times} \cap K[X]$ at the first step of LPA.

Lemma 39. *Let I be a P-hull-primary ideal. For a maximal independent set U of P, $\mathrm{hull}(I) = IK[X]_{K[U]^\times} \cap K[X]$.*

Proof. Let $\mathcal{Q}$ be a primary decomposition of I. Then, $\mathrm{hull}(I)$ is the unique primary component disjoint from $K[U]^\times$. Thus,

$$IK[X]_{K[U]^\times} \cap K[X] = \bigcap_{Q \in \mathcal{Q}, Q \cap K[U]^\times = \emptyset} Q = \mathrm{hull}(I).$$

□

6   Experiments

We made a preliminary implementation on a computer algebra system Risa/Asir [7] and apply it to several examples as naive experiments. Here we show some typical examples. Timings are measured on a PC with Xeon E5-2650 CPU.

First, we see an ideal whose embedded primary components are hard to compute. Let $I_1(n) = (x^2) \cap (x^4, y) \cap (x^3, y^3, (z+1)^n + 1)$. If n is considerably large, it is difficult to compute a full primary decomposition of $I_1(n)$ though the isolated divisor (x) can be detected pretty easily. We apply Local Primary Algorithm (LPA) for this example to compute the isolated primary component for $P_1 = (x)$. We also see another example which is more valuable for mathematics. An ideal $A_{k,m,n}$ is defined in [9] and its primary decomposition has important meanings in Computer Algebra for Statistics. We consider an isolated prime divisor $P_2 = (x_{13}, x_{23}, x_{33}, x_{43})$ of $A_{3,4,5}$ in $\mathbb{Q}[x_{ij} \mid 1 \leq i \leq 4, 1 \leq j \leq 5]$. In Table 1, we can see LPA has certain effectiveness by its speciality comparing a full primary decomposition function noro_pd.syci_dec. From Proposition 12, we also use double ideal quotient to compute equidimensional hull.

Table 1. Local primary algorithm (isolated)

Algorithm	$I_1(100)$	$I_1(200)$	$I_1(300)$	$I_1(400)$	$I_1(500)$	$A_{3,4,5}/P_2$
noro_pd.syci_dec	0.36	15.6	88.3	289	96.0	>3600
LPA	0.02	0.04	0.07	0.11	0.14	14.3

Second, we consider embedded prime divisors; $P_3 = (x_{12}x_{31} - x_{32}x_{11}, x_{42}x_{11} - x_{41}x_{12}, x_{42}x_{31} - x_{41}x_{32}, x_{44}x_{31} - x_{41}x_{34}, x_{44}x_{32} - x_{42}x_{34}, x_{13}, x_{21}, x_{22}, x_{23}, x_{24}, x_{33}, x_{43})$ of $A_{2,4,4}$ in $\mathbb{Q}[x_{ij} \mid 1 \leq i \leq 4, 1 \leq j \leq 4]$ and $P_4 = (x_{16}x_{27} - x_{17}x_{26}, x_{34}x_{13} - x_{33}x_{14}, x_{37}x_{16} - x_{36}x_{17}, x_{36}x_{27} - x_{37}x_{26}, x_{12}, x_{15}, x_{21}, x_{22}, x_{23}, x_{24}, x_{25}, x_{32}, x_{35})$ of $A_{2,3,7}$ in $\mathbb{Q}[x_{ij} \mid 1 \leq i \leq 3, 1 \leq j \leq 7]$. In Table 2, LPA-Pm is an implementation based on Lemma 38 and LPA-MIS is one from Lemma 39 and Criteria 3, 4. Both methods are implemented in LPA-(Pm+MIS). The primitive LPA is not practical since the cost of computing hull$(I + P^m)$ is very high. On the other hand, we can see LPA-(Pm+MIS) has good effectiveness by its speciality comparing a full primary decomposition function noro_pd.syci_dec.

Table 2. Local primary algorithm (embedded) and its improvement

Algorithm	$A_{2,4,4}/P_3$	$A_{2,3,7}/P_4$
noro_pd.syci_dec	3.11	34.8
LPA	>3600	168
LPA-Pm	4.75	29.1
LPA-MIS	0.58	0.38
LPA-(Pm + MIS)	0.15	0.08

7   Conclusion and Future Work

In commutative algebra, the operation of "localization by a prime ideal" is well-known as a basic tool. However, its computation through primary decomposition is very difficult. Thus, we devise a new effective localization *Local Primary Algorithm* (LPA) using Double Ideal Quotient(DIQ) and its variants without computing unnecessary primary components for localization. For our construction of LPA, we devise several criteria for primary component based on DIQ and its variants. We take preliminary benchmarks for some examples to examine certain effectiveness of LPA coming from its speciality. To make our LPA very practical we shall continue to improve it through obtaining timing data for a lot of larger examples.

In future work, we are finding a way to compute "sample points" of prime divisors. For localization it does not need all divisors; it is enough to find $f_P \in P \cap S$ for each prime divisor P with $P \cap S \neq \emptyset$ and we obtain $IK[X]_S \cap K[X] = (I : (\prod_{P \cap S \neq \emptyset} f_P)^{\infty})$. Another work is to apply our primary component criteria to *probabilistic or inexact* methods for primary decomposition, such as numerical ones. Probabilistic or inexact ways have low computational costs, however, they have low accuracy for outputs. Hence, our criterion using double ideal quotient may help to guarantee their outputs. Finally, localization in general setting, that is localization by a prime ideal not necessary associated is interesting work.

Acknowledgment. The authors would like to thank the referees for their helpful comments to improve the presentation of this paper. The authors are also grateful to Masayuki Noro for technical assistance with the computer experiments and coding on Risa/Asir.

References

1. Atiyah, M.F., MacDonald, I.G.: Introduction to Commutative Algebra. Addison-Wesley Series in Mathematics. Avalon Publishing, New York (1994)
2. Eisenbud, D., Huneke, C., Vasconcelos, W.: Direct methods for primary decomposition. Inventi. Math. **110**(1), 207–235 (1992)
3. Gianni, P., Trager, B., Zacharias, G.: Gröbner bases and primary decomposition of polynomial ideals. J. Symb. Comput. **6**(2), 149–167 (1988)
4. Greuel, G.-M., Pfister, G.: A Singular Introduction to Commutative Algebra. Springer, Heidelberg (2002). https://doi.org/10.1007/978-3-662-04963-1
5. Kawazoe, T., Noro, M.: Algorithms for computing a primary ideal decomposition without producing intermediate redundant components. J. Symb. Comput. **46**(10), 1158–1172 (2011)
6. Matzat, B.H., Greuel, G.-M., Hiss, G.: Primary decomposition: algorithms and comparisons. In: Matzat, B.H., Greuel, G.M., Hiss, G. (eds.) Algorithmic Algebra and Number Theory, pp. 187–220. Springer, Heidelberg (1999). https://doi.org/10.1007/978-3-642-59932-3_10
7. The Risa/Asir developing team: Risa/Asir. A computer algebra system. http://www.math.kobe-u.ac.jp/Asir

8. Shimoyama, T., Yokoyama, K.: Localization and primary decomposition of polynomial ideals. J. Symb. Comput. **22**(3), 247–277 (1996)
9. Sturmfels, B.: Solving systems of polynomial equations. In: CBMS Regional Conference Series. American Mathematical Society, no. 97 (2002)
10. Vasconcelos, W.: Computational Methods in Commutative Algebra and Algebraic Geometry. Algorithms and Computation in Mathematics. Springer, Heidelberg (2004)

A Purely Functional Computer Algebra System Embedded in Haskell

Hiromi Ishii[✉]

University of Tsukuba, Tsukuba, Ibaraki 305-8571, Japan
h-ishii@math.tsukuba.ac.jp

Abstract. We demonstrate how methods in *Functional Programming* can be used to implement a computer algebra system. As a proof-of-concept, we present the `computational-algebra` package. It is a computer algebra system implemented as an embedded domain-specific language in *Haskell*, a purely functional programming language. Utilising methods in functional programming and prominent features of Haskell, this library achieves safety, composability, and correctness at the same time. To demonstrate the advantages of our approach, we have implemented advanced Gröbner basis algorithms, such as Faugère's F_4 and F_5, in a composable way.

Keywords: Gröbner basis · Signature-based algorithms
Computational algebra · Functional programming · Haskell
Type system · Formal methods · Property-based testing
Implementation report

1 Introduction

In the last few decades, the area of computational algebra has grown larger. Many algorithms have been proposed, and there have emerged plenty of computer algebra systems. Such systems must achieve *correctness*, *composability* and *safety* so that one can implement and examine new algorithms within them. More specifically, we want to achieve the following goals:

Composability means that users can easily implement algorithms or mathematical objects so that they work seamlessly with existing features.
Safety prevents users and implementors from writing "wrong" code. For example, elements in different rings, e.g. $\mathbb{Q}[x, y, z]$ and $\mathbb{Q}[w, x, y]$, should be treated differently and must not directly be added. Also, it is convenient to have handy ways to convert, inject, or coerce such values.
Correctness of algorithms, with respect to prescribed formal specifications, should be guaranteed with a high assurance.

We apply methods in the area of *functional programming* to achieve these goals. As a proof-of-concept, we present the `computational-algebra` package [12]. It is implemented as an embedded domain-specific language in the

© Springer Nature Switzerland AG 2018
V. P. Gerdt et al. (Eds.): CASC 2018, LNCS 11077, pp. 288–303, 2018.
https://doi.org/10.1007/978-3-319-99639-4_20

Table 1. Symbols in code fragments

Symbol	Code	Symbol	Code	Symbol	Code	Symbol	Code
$\underline{\mathbb{N}}$	Nat	$\mathbb{Z}$	Integer	$\mathbb{Q}$	Rational	$\mathbb{F}_p$	F p
::	::	$=$	==	$\neq$	/=	$\lambda\ \vec{x} \to e$	\x -> e
$\times$	*	$\ltimes$	!*	$\frown$	++	$\ominus$	%-
$\simeq$	:~:	$\sim$	~	$\to$	->	$\leftarrow$	<-
$\implies$	==>	$\Rightarrow$	=>	$:=$	.=	$:\Leftarrow$	.%=
$\subseteq$	'Subset'	$\leq$	<=	$\circ$	.	$\wedge$	.&&.
$\bullet$	.*	$\langle\$\rangle$	<$>	$\langle\circledast\rangle$	<*>	$\forall$	forall

Haskell Language [10]. More precisely, we adopt the *Glasgow Haskell Compiler* (GHC) [7] as our hosting language. We use GHC because: its *type-system* allows us to build a safe and composable interface for computer algebra; *lazy evaluation* enables us to treat infinite objects intuitively; *declarative style* sometimes reduces a burden of writing mathematical programs; *purity* permits a wide range of equational optimisation; and there are plenty of libraries for functional methods, especially *property-based testing*. These methods are not widely adopted in this area; an exception is *DoCon* [23], a pioneering work combining Haskell and computer algebra. Our system is designed with more emphasis on safety and correctness than DoCon, adding more ingredients. Although we use a functional language, some methods in this paper are applicable in imperative languages.

This paper is organised as follows. In Sect. 2, we discuss how the progressive type-system of GHC enables us to build a safe and expressive type-system for a computer algebra. Then, in Sect. 3, we see how the method of *property-based testing* can be applied to verify the correctness of algebraic programs in a lightweight and top-down manner. To demonstrate the practical advantages of Haskell, Sect. 4 gives a brief description of the current implementations of the Hilbert-driven, F_4 and F_5 algorithms. We also take a simple benchmark there. We summarise the paper and discuss related and future works in Sect. 5.

In what follows, we use symbols in Table 1 in code fragments for readability.

2   Type System for Safety and Composability

In this section, we will see how the progressive type-level functionalities of GHC can be exploited to construct a safe, composable and flexible type-system for a computer algebra system. There are several existing works on type-systems for computer algebra, such as in Java and Scala [15,18], and DoCon. However, none of them achieves the same level of safety and composability as our approach, which utilises the power of *dependent types* and *type-level functions*.

2.1   Type Classes to Encode Algebraic Hierarchy

We use *type-classes*, an ad-hoc polymorphism mechanism in Haskell, to encode an algebraic hierarchy. This idea is not particularly new (for example, see

Mechveliani [23] or Jolly [15]), and we build our system on top of the existing `algebra` package [17], which provides a fine-grained abstract algebraic hierarchy.

Code 1. Group structure, coded in the `algebra` package

```
1   class Additive a where
2     (+) :: a → a → a
3   class Additive a ⇒ Monoidal a where
4     zero :: a
5   class Monoidal a ⇒ Group a where
6     negate :: a → a
```

Code 1 illustrates a simplified version of the algebraic hierarchy up to `Group` provided by the `algebra` package. Each statement between `class` or ⇒ and `where`, such as `Additive` a or `Monoidal` a, expresses the constraint for types. For example, Lines 1 and 2 express "a type a is `Additive` if it is endowed with a binary operation +", and Lines 3 and 4 that "a type a is `Monoidal` if it is `Additive` and has a distinguished element called `zero`".

Note that none of these requires the "proof" of algebraic axioms. Hence, one can accidentally write a non-associative `Additive`-instance, or non-distributive `Ring`-instance[1]. This sounds rather "unsafe", and we will see how this could be addressed reasonably in Sect. 3.

2.2 Classes for Polynomials and Dependent Types

Expressing algebraic hierarchy using type-class hierarchy, or class inheritance, is not so new and they are already implemented in DoCon or JAS. However, these systems lack a functionality to distinguish the arity of polynomials or the denominator of a quotient ring. In particular, DoCon uses sample arguments to indicate such parameters, and they cannot be checked at compile-time. To overcome these restrictions, we use *Dependent Types*.

For example, Code 2 presents the simplified definition of the class `IsOrdPoly` for polynomials. We provide an abstract class for polynomials, not just an implementation, to enable users to choose appropriate internal representations fitting their use-cases.

The class definition includes not only functions, but also *associated types*, or *type-level functions*: `Arity`, `MOrder` and `Coeff`. Respectively, they correspond to the number of variables, the monomial ordering and the coefficient ring.

Note that `liftMap` corresponds to the universality of the polynomial ring $R[X_1, \ldots, X_n]$; i.e. the free associative commutative R-algebra over $\{1, \ldots, n\}$.

---

[1] Indeed, one can use *dependent types*, described in the next subsection, to require such proofs. However, this is too heavy for the small outcome, and does not currently work for primitive types.

Code 2. A type-class for polynomials

```
1    class (Module (Coeff poly) poly, Commutative poly, Ring poly,
2             CoeffRing (Coeff poly), IsMonomialOrder (MOrder poly))
3          ⇒ IsOrdPoly poly where
4      type Arity  poly ::  N
5      type MOrder poly ::  Type
6      type Coeff  poly ::  Type
7      liftMap ::  (Module (Scalar (Coeff poly)) alg, Ring alg)
8               ⇒ (N_{<Arity poly} → alg) → poly → alg
9      leadTerm ::  poly → (Coeff poly, OrdMonom (MOrder poly) n)
10     ...
```

Code 3. Examples for polynomial instances

```
1    instance (IsMonomialOrder ord, CoeffRing r)
2             ⇒ IsOrdPoly (OrdPoly r ord n) where
3      type Arity  (OrdPoly r ord n) = n
4      type MOrder (OrdPoly r ord n) = ord
5      type Coeff  (OrdPoly r ord n) = r
6      ...
7
8    f ::  OrdPoly Q Grevlex 3
9    f = let [x,y,z] = vars in x ^ 2 × y + 3 × x + z + 1
10
11   instance (CoeffRing r) ⇒ IsOrdPoly (Unipol r) where
12     type Arity  (OrdPoly r ord n) = 1
13     type MOrder (OrdPoly r ord n) = Lex
14     type Coeff  (OrdPoly r ord n) = r
15     ...
```

In theory, this function suffices to characterise the polynomial ring. However, for the sake of efficiency, we also include some other operations in the definition.

Code 3 shows example instance definitions for the standard multivariate and univariate polynomial ring types. Note that, in Lines 8 and 12, number literal *expressions* 1 and 3 occur in *type* contexts. Types depending on expressions are called *Dependent Types* in type theory. GHC supports them via the *Promoted Data-types* language extension [27] since version 7.4. Our library heavily uses this functionality, and achieves the type-safety preventing users from unintendedly confusing elements from different rings.

2.3 Proofs in Dependent Types and Type-Driven Casting Function

In theory, we can use liftMap to cast between any elements of "compatible" polynomial rings. To reduce the burden to write boilerplate casting functions, our library comes with smart functions, as shown in Code 4. The convPoly function maps a polynomial into one with the same setting but different representation; e.g. OrdPoly Q Lex 1 into Unipol Q. The next injVars function maps an

---

Code 4. Various casting function, with simplified type-signatures

```
1    convPoly :: (Coeff r ~ Coeff r', MOrder r ~ MOrder r',
2                 Arity r ~ Arity r')
3              ⇒ r → r'
4    injVars :: (Arity r ≤ Arity r', Coeff r ~ Coeff r')
5              ⇒ r → r'
6    injVarsOffset :: (n + Arity r ≤ Arity r', Coeff r ~ Coeff r')
7                   ⇒ Sing n → r → r'
```

---

element of $R[X_1, \ldots, X_n]$ into another polynomial ring with the same coefficient ring, but with more number of variables, e.g. $R[X_1, \ldots, X_{n+m}]$, regardless of ordering. For example, it maps Unipol $\mathbb{Q}$ into OrdPoly $\mathbb{Q}$ Grevelx 3. Then, injVarsOffset is a variant of injVars which maps variables with offset; for example,

```
1    injVarsOffset [sn|3|] :: Unipol ℚ → Polynomial ℚ 5
```

maps $\mathbb{Q}[X]$ into $\mathbb{Q}[X_0, \ldots, X_4]$ with $X \mapsto X_3$. Here, [sn|3|] is called a *singleton* for the type-level natural number 3, first introduced by Eisenberg et al. [4]. More precisely, for any *type-level natural* n, there is the unique *expression* sing :: Sing n and we can use it as a tag for type-level arguments.

To work with type-level naturals, we sometimes have to *prove* some constraints. For example, suppose we want to write a variant of injVars mapping variables to *the end of* those of the target polynomial ring, instead of *the beginning*. We might first write it as follows:

```
1    injVarsAtEnd :: (Arity r ≤ Arity r', Coeff r ~ Coeff r')
2                  ⇒ r → r'
3    injVarsAtEnd =
4      let sn = sing :: Sing (Arity r)
5          sm = sing :: Sing (Arity r')
6      in injVarsOffset (sm ⊖ sn) -- Errors!
```

However, GHC cannot see Arity r' - Arity r + Arity r ≤ Arity r'. Although this constraint is rather clear to us, we have to give the compiler its proof. We have developed the type-natural package [14] which includes typical "lemmas". For example, we can use the minusPlus lemma to fix this:

```
1    -- From type-natural:
2    minusPlus :: Sing n → Sing m
3               → IsTrue (m ≤ n) → ((n - m) + m) ≃ n
4
5    injVarsAtEnd :: (Arity r ≤ Arity r', Coeff r ~ Coeff r')
6                  ⇒ r → r'
7    injVarsAtEnd =
8      let sn = sing :: Sing (Arity r)
9          sm = sing :: Sing (Arity r')
10     in withRefl (minusPlus sm sn Witness) $
11        injVarsOffset (sm ⊖ sn)
```

Since giving such a proof each time is rather tedious, we can use type-checker plugins to let the compiler try to prove constraints automatically. In particular, the author developed the `ghc-typelits-presburger` plugin [13] to resolve propositions in Presburger arithmetic at compile time.

Our library also provides the `LabPoly` type, which converts existing polynomial types into "*labelled*" ones. For example, one can write as follows:

```
1  f :: LabPoly (Polynomial Q 3) '["x", "y", "z"]
2  f = 5 × #x ^ 2 × #y ^ 3 - #y × #z + 1
```

This relies on the `DataKinds` and `OverloadedLabels` language extensions of GHC. GHC's type system is strong enough to reject illegal terms and types, such as `#w :: LabPoly (Unipol Q)'["a"]` (w is not listed as a variable) or `LabPoly (Polynomial Q 3) '["x", "y", "x"]` (the variable x occurs twice). Using the type-level information, one can invoke the canonical inclusion maps naturally as follows:

```
1  f :: LabPoly' Q Grevlex '["x", "y", "z"]
2  f = #x × #y × #z + 2 × #y - 3  × #z × #x + 1
3  g :: LabPoly' Q Lex '["w", "z", "y", "u", "x"]
4  g = canonicalMap f
5
6  -- Where:
7  canonicalMap :: (xs ⊆ ys, Wraps xs poly, Wraps ys poly',
8                   IsPolynomial poly, IsPolynomial poly',
9                   Coeff poly ~ Coeff poly')
10                 ⇒ LabPoly poly xs → LabPoly poly' ys
```

2.4 Optimising Casting Functions with Rewriting Rules

Since the casting functions are implemented generically, they sometimes introduce unnecessary overhead. For example, if one uses `injVars` with the *same* source and target types, it should just be the identity function. Fortunately, we can use the type-safe *Rewriting Rule* functionality of GHC to achieve this:

```
1  {-# RULES "injVars/identity" injVars = id #-}
```

Each rewriting rule fires at compile-time, if there is a term matching the left-hand side of the rule and having the same type as the right-hand side.

In Haskell, it suffices just to consider algebraic laws to write down custom rewriting rules. This is due to the *purity* of Haskell. That is, every expression in Haskell is pure, in a sense that they evaluate to the same result when given the same arguments. Note that this does not mean that Haskell cannot treat values with side-effects; indeed, the type-system of Haskell distinguishes pure and impure values at type-level, and one can treat impure operations without violating purity as a whole. The trick behind this situation is to describe side-effects as some kind of abstract instructions, instead of treating impure values directly. Hence, for example, duplicating the same term does not make any difference in its meaning, provided that it is algebraically correct. Such a rewriting rule is used extensively in Haskell. For example, Stream Fusion [3] uses them to eliminate unnecessary intermediate expressions and fuse complicated functions into efficient one-path constructions. Yet, DoCon did not do any optimisation using rewriting rules.

In our library, we also use rewriting rules to remove idempotent applications such as "grading" a monomial ordering twice, e.g:

```
1   {-# RULES "graded/graded" ∀ ord.
2      graded (graded ord) = graded ord #-}
```

2.5   Notes on Applicability in Imperative Languages

The safety we achieved in this section cannot be achieved at compile-time without dependent types and type-level functions. Existing works using type-classes or class inheritance to encode algebraic hierarchy, such as JAS or DoCon, lack this level of safety. In theory, one can achieve the same level of safety even in a statically-typed *imperative* language, if it supports a kind of dependent types. For example, in C++, templates with non-type arguments can be used to simulate dependent types. On the other hand, in Java, Generics do not allow non-type arguments and we need to mimic Peano numerals with classes. In either case, it requires much effort to prove the properties of naturals within them, because they lack dedicated support for type-level naturals or type-checker plugins.

On the other hand, to make use of rewriting rules, we need purity as discussed above.

3   Lightweight Correctness: Property-Based Testing

3.1   Property-Based Testing Introduced

In this section, we will address the correctness issue, in a top-down, or *lightweight* manner. Especially, we apply the method of *property-based testing* [1] to verify the correctness of our implementation. The idea is that one specifies the formal properties that the implemented algorithms and types must satisfy, and checks if they hold by testing them against randomly or exhaustively generated inputs. Although it is not as rigorous as a theorem proving, it still gives a guarantee of the correctness at high assurance, after repeating tests time after time.

Code 5 presents the example specifications for algebraic programs. In Lines 1 through 4, `prop_division` states that the implementation of $\mathbb{Q}$ must satisfy the axioms of division ring. The `prop_passesSTest` function demand the result of `calcGroebnerBasis` to pass the S-test. The tester accepts the specifications above, generates a specified number of inputs (default: 100) and tests against them. If all the inputs satisfy the specifications, it successfully halts; otherwise, it reports counterexamples, which is useful while debugging.

Code 5. Formal Specification of Algebraic Programs

```
1   prop_division :: ℚ → Property
2   prop_division q =
3        q ≠ 0 ⟹ (recip q × q = 1 ∧ q × recip q = 1)
4     ∧ q × 1 = q ∧ 1 × q = q
5
6   prop_passesSTest n =
7     forAll (idealOfArity n) $ λ ideal →
8     let gs = calcGroebnerBasis (toIdeal ideal)
9     in all (isZero ∘ ('modPoly' gs))
10            [sPoly f g | f ← gs, g ← gs, f ≠ g]
```

3.2 Discussion

There are several libraries for property-based testing adopting different strategies to generate inputs. For example, QuickCheck [1] generates inputs randomly, while SmallCheck [26] exhaustively enumerates inputs in the depth-increasing order. Even though there are other implementations of property-based testers in languages other than Haskell [11], it does not seem that it is applied in existing systems, such as Singular [9], JAS or DoCon.

By its *generative* nature, property-based testing has several drawbacks and pitfalls. First, evidently, it cannot assure the validity as rigorously as the *formal theorem proving*, unless the input space is finite. There are several pieces of research that combine formal theorem proving and computational algebra to rigorously certify correctness of implementations (for example, [2,24]). These first formalise the theory of Gröbner basis in the constructive type-theory. Then, they execute them within the host theorem proving language, or extract the program into other languages. However, by its nature, this approach requires everything to be proven formally. It is not so easy a task to prove the correctness of every part of a program, even with help from automatic provers. Even if one manages to finish the proof of the validity of some algorithm, when one wants to optimise it afterwards, then one must prove the "equivalence" or validity of that optimisation. Moreover, it is sometimes the case that the validity, or even termination, of the algorithm remains unknown when it is implemented; e.g. the correctness and termination of Faugerè's F_5 [6] are proven very recently [25]. Furthermore, there is an obvious restriction that we can extract programs only into the languages supported by the theorem prover. We consider these conditions too restrictive, and decided to adopt theorem proving only in trivial arity arithmetic.

Secondly, if the algorithm has a bad time complexity, property-based tests can easily explode. Specifically, since Gröbner bases have double-exponential worst time complexity, randomly generated input can take much time to be processed. One might reduce the burden by combining randomised and enumerative generation strategies carefully, but there is still a possibility that there are small inputs which take much time. To avoid such a circumstance, one can reduce the number of inputs, however it also reduces the assurance of validity.

Finally, they are not so good at treating *existential properties*. Although SmallCheck provides the existential quantifier in its vocabulary, it just tries to find solutions up to a prescribed depth. If solutions are relatively "larger" than their inputs, this results in *false-negative* failures. For example, one can write the following specification that demands each element of the result of `calcGroebnerBasis` to be a member of the original ideal, however it does not work as expected:

```
1  prop_gbInc ideal =
2    let j = calcGroebnerBasis ideal
3    in exists $ λ cs →
4        and (zipWith (λ f gs → f = dot ideal gs) j cs)
```

In the above, `dot i g` denotes the "dot-product". As a workaround, we currently combine inter-process communication with property-based testing. More specifically, we invoke a reliable existing implementation, such as SINGULAR, inside the spec as follows:

```
1  prop_gbInc = forAll arbitrary $ λ i → monadicIO $ do
2    let gs = calcGroebnerBasis i
3    is ← evalSingularIdealWith [] [] $
4        funE "reduce" [
5            idealE gs, funE "groebner" [idealE i]]
6    return $ all isZero is
```

Thus, if the existential property in question is decidable and has an existing reliable implementation, then it might be better to call it inside specifications.

4 Case Study: Implementing the Hilbert-Driven, F_4 and F_5 Algorithms for Calculating Gröbner Bases

In this section, we will focus on three algorithms as case-studies: the Hilbert-driven, F_4 and F_5 algorithms. Firstly, we demonstrate the power of laziness and parallelism by the Hilbert-driven algorithm. Then by the F_4 interface, we illustrate the practical example of composability. Finally, we skim through the simplified version of the main routine of F_5 and see how imperative programming with mutable states can be written purely in Haskell. For our purpose, we will discuss only a fragment of implementations that elucidates the advantages of Haskell, rather than the entire implementation and theoretical details.

4.1   Homogenisation and Hilbert-Driven Basis Conversion

Homogenisation is a powerful tool in Gröbner basis computation. If $I \subseteq k[\mathbf{X}]$ is a non-homogeneous ideal and $\bar{I} \subseteq k[x, \mathbf{X}]$ its homogenisation, then one can get a Gröbner basis for I by unhomogenising the Gröbner basis $\bar{G}$ for $\bar{I}$ w.r.t. a suitably induced monomial ordering. In this way, any Gröbner basis algorithm for homogeneous ideals can be converted into one for non-homogeneous ones.

Code 6. Basic API for homogenisation

```
1   data Homogenised poly
2   instance IsOrdPoly poly ⇒ IsOrdPoly (Homogenised poly) where
3      type Arity  (Homogenised poly) = 1 + Arity poly
4      type MOrder (Homogenised poly) = HomogOrder (MOrder poly)
5      type Coeff  (Homogenised poly) = Coeff poly
6      ...
7   homogenise   :: IsOrdPoly poly ⇒ poly → Homogenised poly
8   unhomogenise :: IsOrdPoly poly ⇒ Homogenised poly → poly
9
10  calcGBViaHomog :: (Field (Coeff poly), IsOrdPoly poly)
11                    ⇒ (∀ r. (Field (Coeff r), IsOrdPoly r)
12                         ⇒ Ideal r → [r])
13                    → Ideal poly → [poly]
14  calcGBViaHomog calc i
15    | all isHomogeneous i = (calc) i
16    | otherwise = map unhomogenise ( calc  (fmap homogenise i))
```

Code 6 is an API for these operations. The type Homogenised poly represents polynomials obtained by homogenising polynomials of type poly. Then
calcGBViaHomog calc i first checks if the input i is homogeneous. If it is so,
then it applies the argument (calc) to its input directly (Line 15); otherwise,
it first homogenises the input, applies calc, and then unhomogenises it to get
the final result (Line 16). Note that, though it uses the same term calc in
both cases, they have different types. In the first case, since it just feeds an
input directly, (calc) has type Ideal poly → [poly]. On the other hand, in
the non-homogeneous case, it is applied *after* homogenisation, hence it is of type
Ideal (Homogenised poly) → [Homogenised poly]. Thus, calcGBViaHomog
takes a *polymorphic function* as its first argument and this is why we have ∀
inside the type of the first argument. Such a nested polymorphic type is called
a *rank n polymorphic type*, and it is supported by GHC's RankNTypes language
extension[2].

For example, one can use the so-called *Hilbert-driven algorithm* as the first
argument to calcGBViaHomog. It first computes a Gröbner basis w.r.t. a lighter
monomial ordering, compute the Hilbert–Poincaré series (HPS) with it and use it
to compute Gröbner basis w.r.t. the heavier ordering. In this procedure, we need
the following operations on HPS: Equality test on HPS's, n^{th} Taylor coefficient
of the given HPS, and the $\mathbb{Z}[X]$-module operation on HPS. Code 7 illustrates
such an interface for HPS. For equality test, we use the numerator hpsNumerator
of the closed form, and an *infinite list* taylor maintains Taylor coefficients. By
the *lazy* nature of Haskell, we can intuitively treat infinite lists and write a
convolution on them. In Line 12, (par) and (seq) specify the *evaluation strategy*.
In brief, expressions x and y in " x `par` y" (resp. (seq)) are evaluated *parallelly*

---

[2] This can be achieved in object-oriented language with subtyping and Generics.

Code 7. Data-type of and operations on Hilbert–Poincaré series

```
1   data HPS n = HPS { taylor :: [ Z ], hpsNumerator :: Unipol Z }
2
3   instance Eq (HPS a) where
4     ( = ) = ( = ) 'on' hpsNumerator
5   instance Additive (HPS n) where
6     HPS cs f + HPS ds g = HPS (zipWith (+) cs ds) (f + g)
7   instance LeftModule (Unipol Z ) (HPS n) where
8     f • HPS cs g = HPS (conv (taylor f ⌢ repeat 0) cs) (f × g)
9
10  conv :: [ Z ] → [ Z ] → [ Z ]
11  conv (x : xs) (y : ys) =
12    let parSum a b c = a 'par' ' b 'par' c 'seq' (a + b + c) in
13    x × y :
14      zipWith3 parSum (map (x×) ys) (map (y×) xs) (0 : conv xs ys)
```

(resp. *sequentially*). Since every expression is pure in Haskell, we can safely take advantage of parallelism, without a possibility of changing results.

4.2   A Composable Implementation of F_4

F_4 is one of the most efficient algorithms for Gröbner basis computation and was introduced by Faugère [5]. Briefly, F_4 reduces more than two polynomials at once, replacing S-polynomial remaindering in the Buchberger Algorithm with the *Gaussian elimination* of the matrices. This means that the efficiency of F_4 reduces to that of Gaussian elimination and the internal representation of matrices. Thus, it is useful if we can easily switch internal representations and elimination algorithms. For this purpose, we provide type-classes for mutable and immutable matrices which admit row operations and a dedicated Gaussian elimination. Code 8 demonstrates the interface for immutable and mutable matrices (`Matrix` and `MMatrix`) and the type signature of our F_4 implementation (`f4`). In Lines 1 and 6, the last type argument `a` of `Matrix` and `MMatrix` corresponds to the type of coefficients. Note that one can give different instance definitions for the same `mat` but different coefficient types `a`. For example, one can implement efficient Gaussian elimination on $\mathbb{F}_p$ for `Matrix` Mat $\mathbb{F}_p$, and then use it in the definition of `Matrix` Mat $\mathbb{Q}$, with the Hensel lifting or Chinese remaindering.

In Line 15, the first argument of `f4` of type `proxy mat` specifies the internal representation `mat` of matrices. In addition, `f4` takes a *selection strategy* as the second argument. Here, the selection strategy is abstracted as a weighting function to some ordered types, and we store intermediate polynomials in a heap and select all the polynomials with the minimum weight at each iteration.

4.3   The F_5 Algorithm

Finally, we present the simplified version of the main routine of Faugère's F_5 [6] (Code 9). Readers may be surprised that the code looks much imperative. This is

---

Code 8. Matrix classes and the F_4 function

```
1  class MMatrix mat a where
2    fromRows :: [Vector a] → ST s (mat s a)
3    scaleRow :: Multiplicative a ⇒ Int → a → mat s a → ST s ()
4    ...
5
6  class MMatrix (Mutable mat) a ⇒ Matrix mat a where
7    type Mutable mat :: ⋆ → ⋆
8    freeze :: Mutable mat s a → ST s (mat a)
9    ...
10   gaussReduction :: Field a ⇒ mat a → mat a
11
12 type Strategy f w = f → f → w
13 f4 :: (Ord w, IsOrdPoly poly, Field (Coeff poly),
14        Matrix mat (Coeff poly))
15    ⇒ proxy mat → Strategy poly w → Ideal poly → [poly]
```

---

made possible by the *ST monad* [19], which encapsulates side-effects introduced by mutable states and prevents them from leaking outside. We use a functional heap to choose the polynomial vectors with the least signature, demonstrating the fusion of functional and imperative styles.

4.4 Benchmarks

We also take a simple benchmark and the result is shown in Table 2 (examples are taken from Giovini et al. [8]). This compares the algorithms implemented in our **computational-algebra** package and Singular. The first four rows correspond to the alrorithms implemented in our library; i.e. the Buchberger algorithm optimised with syzygy and sugar strategy (B), the degree-by-degree algorithm for homogeneous ideals (DbyD), the Hilbert-driven algorithm (Hilb), and F_5. S(gr) and S(sba) stand for the **groebner** and **sba** functions in the Singular computer algebra system 4.0.3. The complete source-code is available on GitHub [12][3]. The benchmark program is compiled with GHC 8.2.2 with flags -O2 -threaded -rtsopts -with-rtsopts=-N, and ran on an Intel Xeon E5-2690 at 2.90 GHz, RAM 128GB, Linux 3.16.0-4 (SMP), using 10 cores in parallel. We used the Gauge framework to report the run-time of our library, and the **rtimer** primitive for Singular. For actual benchmark codes, see http://bit.ly/hbench1 and hbench2. Unfortunately, in our system, F_4 takes much more computing time, hence we did not include the result. The results show that, among the algorithms implemented in our system, F_5 works fine in general, though it takes much time in some specific cases. Nevertheless, there remains much room for improvement to compete with the state-of-the-art implementations such as Singular, although there is one case where our implementation is slightly faster than Singular's **groebner** function.

---

[3] More specifically, we used the implementation in commit 70e6e7b.

Code 9. Main Routine of the F_5 Algorithm

```
1   f5 :: (Field (Coeff pol), IsOrdPoly pol)
2      ⇒ Vector pol → [(Vector pol, pol)]
3   f5 (map monoize → i0) = runST $ do
4     let n = length i0
5     gs  ← newSTRef []
6     ps  ← newSTRef $ H.fromList [ basis n i | i ← [0..n-1]]
7     syzs ← newSTRef
8        [ sVec (i0 ! m) (i0 ! n) | m ← [0..n-1], n ← [0..j-1] ]
9     whileJust_ (H.viewMin <$> readSTRef ps) $
10    λ (Entry sig g, ps') → do
11      ps := ps'
12      (gs0, ss0) ← (,) <$> readSTRef gs <*> readSTRef syzs
13      unless (standardCriterion sig ss0) $ do
14        let (h, ph) = reduceSignature i0 g gs0
15            h' = map (× injectCoeff (recip $ leadingCoeff ph)) h
16        if isZero ph then syzs :⇐ (mkEntry h : )
17          else do
18          let adds = fromList $ mapMaybe (regSVec (ph, h')) gs0
19          ps :⇐ H.union adds
20          gs :⇐ ((monoize ph, mkEntry h') :)
21    map (λ (p, Entry _ a) → (a, p)) <$> readSTRef gs
```

Table 2. Benchmark results (ms)

	I_1 (Lex)	I_1 (Grevlex)	I_2 (Lex)	I_2 (Grevlex)	I_3 (Grevlex)
B	1.820×10^0	1.593×10^1	1.400×10^1	4.129×10^0	6.689×10^2
DbyD	6.364×10^1	9.162×10^2	1.147×10^2	5.647×10^1	4.125×10^2
Hilb	1.644×10^2	2.313×10^2	5.265×10^1	3.414×10^1	9.645×10^3
F_5	1.851×10^0	4.314×10^2	7.129×10^0	2.648×10^0	1.290×10^3
S(gr)	2.300×10^0	8.493×10^{-1}	2.651×10^0	8.210×10^{-1}	9.511×10^{-1}
S(sba)	2.279×10^{-1}	8.711×10^{-1}	2.343×10^{-1}	7.958×10^{-1}	1.541×10^{-1}

$I_1 := \langle 35y^4 - 30xy^2 - 210y^2z + 3x^2 + 30xz - 105z^2 + 140yt - 21u,$
$\quad 5xy^3 - 140y^3z - 3x^2y + 45xyz - 420yz^2 + 210y^2t - 25xt + 70zt + 126yu \rangle$
$I_2 := \langle w + x + y + z, wx + xy + yz + zw, wxy + xyz + yzw + zwx, wxyz - 1 \rangle$
$I_3 := \langle x^{31} - x^6 - x - y, x^8 - z, x^{10} - t \rangle$

5   Conclusions

In this paper, we have demonstrated how we can adopt the methods developed in the area of functional programming to build a computer algebra system. Some of these methods are also applicable in imperative languages.

In Sect. 2, we presented a type-system strong enough to detect algebraic errors at compile-time. For example, our system can distinguish number of variables of polynomial rings at type-level thanks to dependent types. It also enables us to automatically generate casting functions and we saw how their overhead can be reduced using rewriting rules. As for type-systems for a computer algebra system, there are several existing works {18,23]. However, these systems are not safe enough for discriminating variable arity at type-level and don't make use of rewriting rules.

In Sect. 3, we successfully applied the method of *property-based testing* for verification of the implementation, which is lightweight compared to the existing theorem-prover based approach [2,24]. Although property-based testing is not as rigorous as theorem proving, it is lightweight and can be applied to algorithms not yet proven to be valid or terminate and available also for imperative languages.

We have seen that, in Sect. 4, other features of Haskell, such as higher-order polymorphism, parallelism and laziness, can also be easily applied to computer algebra by actual examples. Even though they are shown as fragments of code, we expect them to be convincing.

Since some of the methods in this paper, such as dependent types or property-based testing, are not limited to the functional paradigm, it might be interesting to investigate their applicability in the imperative settings.

From the viewpoint of efficiency, there is much to be done. For example, efficiency of our current F_4 implementation is far inferior to that of the naïve Buchberger algorithm, and other algorithms are far much slower than state-of-the-art implementations such as Singular. To optimise implementations, we can make more use of Rewriting Rules and efficient data structures. Also, the parallelism must undoubtedly play an important role. Fortunately, there are plenty of the parallel computation functionalities in Haskell, such as Regular Parallel Arrays [16] and `parallel` package [22], and another book by Marlow [21] on general topics in parallelism in Haskell. Also, there is an existing work by Lobachev et al. [20] on parallel symbolic computation in Eden, a dialect of Haskell with parallelism support. Although Eden is retired, the methods introduced there might be helpful.

Acknowledgments. The author would like to thank my supervisor, Prof. Akira Terui, for discussions, and to anonymous reviewers for helpful comments. This research is supported by Grant-in-Aid for JSPS Research Fellow Number 17J00479, and partially by Grants-in-Aid for Scientific Research 16K05035.

References

1. Claessen, K., Hughes, J.: QuickCheck: a lightweight tool for random testing of Haskell programs. In: Proceedings of the Fifth ACM SIGPLAN International Conference on Functional Programming, ICFP 2000, pp. 268–279. ACM, New York (2000). https://doi.org/10.1145/351240.351266
2. Coquand, T., Persson, H.: Gröbner bases in type theory. In: Altenkirch, T., Reus, B., Naraschewski, W. (eds.) TYPES 1998. LNCS, vol. 1657, pp. 33–46. Springer, Heidelberg (1999). https://doi.org/10.1007/3-540-48167-2_3

3. Coutts, D., Leshchinskiy, R., Stewart, D.: Stream fusion. from lists to streams to nothing at all. In: Proceedings of the 12th ACM SIGPLAN International Conference on Functional Programming, ICFP 2007 (2007)

4. Eisenberg, R.A., Weirich, S.: Dependently typed programming with singletons. ACM SIGPLAN Not. **47**(12), 117–130 (2012). Haskell 2012

5. Faugére, J.-C.: A new efficient algorithm for computing Gröbner bases (F_4). J. Pure Appl. Algebra **139**(1), 61–88 (1999)

6. Faugére, J.-C.: A new efficient algorithm for computing Gröbner bases without reduction to zero (F_5). In: Proceedings of the 2002 International Symposium on Symbolic and Algebraic Computation, pp. 75–83. ACM, Lille (2002)

7. GHC Team: The Glasgow Haskell Compiler (2018). https://www.haskell.org/ghc/. Accessed 2018

8. Giovini, A., Mora, T., Niesi, G., Robbiano, L., Traverso, C.: "One sugar cube, please" or selection strategies in the Buchberger algorithm. In: Proceedings of the 1991 International Symposium on Symbolic and Algebraic Computation, ISSAC 1991, pp. 5–4. ACM (1991)

9. Greuel, G.-M., Pfister, G.: A Singular Introduction to Commutative Algebra, 2nd edn. Springer, Heidelberg (2007). https://doi.org/10.1007/978-3-662-04963-1

10. Haskell Committee: The Haskell Programming Language. http://haskell.org/

11. Hypothesis: Most testing is ine ective - Hypothesis (2018). https://hypothesis. works. Accessed 06 May 2018

12. Ishii, H.: The computational-algebra package (2018). https://konn.github.io/computational-algebra

13. Ishii, H.: The ghc-typelits-presburger package (2017). http://hackage.haskell.org/package/ghc-typelits-presburger

14. Ishii, H.: The type-natural package (2013). http://hackage.haskell.org/package/type-natural

15. Jolly, R.: Categories as type classes in the scala algebra system. In: Gerdt, V.P., Koepf, W., Mayr, E.W., Vorozhtsov, E.V. (eds.) CASC 2013. LNCS, vol. 8136, pp. 209–218. Springer, Cham (2013). https://doi.org/10.1007/978-3-319-02297-0_18

16. Keller, G., Chakravarty, M.M., Leshchinskiy, R., Peyton Jones, S., Lippmeier, B.: Regular, shape-polymorphic, parallel arrays in Haskell. In: Proceedings of the 15th ACM SIGPLAN International Conference on Functional Programming, ICFP 2010, pp. 261–272. ACM, Baltimore (2010)

17. Kmett, E.A.: The algebra package (2011). http://hackage.haskell.org/package/algebra. Accessed 2018

18. Kredel, H., Jolly, R.: Generic, type-safe and object oriented computer algebra software. In: Gerdt, V.P., Koepf, W., Mayr, E.W., Vorozhtsov, E.V. (eds.) CASC 2010. LNCS, vol. 6244, pp. 162–177. Springer, Heidelberg (2010). https://doi.org/10.1007/978-3-642-15274-0_14

19. Launchbury, J., Peyton Jones, S.L.: Lazy functional state threads. In: Proceedings of the ACM SIGPLAN 1994 Conference on Programming Language Design and Implementation, PLDI 1994, pp. 24–35. ACM, Orlando (1994)

20. Lobachev, O., Loogen, R.: Implementing data parallel rational multiple-residue arithmetic in Eden. In: Gerdt, V.P., Koepf, W., Mayr, E.W., Vorozhtsov, E.V. (eds.) CASC 2010. LNCS, vol. 6244, pp. 178–193. Springer, Heidelberg (2010). https://doi.org/10.1007/978-3-642-15274-0_15

21. Marlow, S.: Parallel and Concurrent Programming in Haskell: Techniques for Multicore and Multithreaded Programming. O'Reilly Media, Sebastopol (2013)

22. Marlow, S., Maier, P., Loidl, H.-W., Aswad, M.K., Trinder, P.: Seq no more: better strategies for parallel Haskell. In: Proceedings of the Third ACM Haskell Symposium on Haskell, Haskell 2010, pp. 91–102. ACM, Baltimore (2010). https://doi.org/10.1145/1863523.1863535
23. Mechveliani, S.D.: Computer algebra with Haskell: applying functional-categorial-"lazy" programming. In: Proceedings of International Workshop CAAP, pp. 203–211 (2001)
24. Mechveliani, S.D.: DoCon-A a Provable Algebraic Domain Constructor (2018). http://www.botik.ru/pub/local/Mechveliani/docon-A/2.02/manual.pdf. Accessed 06 May 2018
25. Pan, S., Hu, Y., Wang, B.: The termination of the F5 algorithm revisited. In: Proceedings of the 38th International Symposium on Symbolic and Algebraic Computation, ISSAC 2013, pp. 291–298. ACM, Boston (2013)
26. Runciman, C., Naylor, M., Lindblad, F.: SmallCheck and Lazy SmallCheck: automatic exhaustive testing for small values. In: Proceedings of the First ACM SIGPLAN Symposium on Haskell, Haskell 2008, pp. 37–48. ACM, Victoria (2008). https://doi.org/10.1145/1411286.1411292
27. Yorgey, B.A., Weirich, S., Cretin, J., Peyton Jones, S., Vytiniotis, D., Magalhães, J.P.: Giving Haskell a promotion. In: Proceedings of the 8th ACM SIGPLAN Workshop on Types in Language Design and Implementation, TLDI 2012, pp. 53–66. ACM, Philadelphia (2012)

Splitting Permutation Representations of Finite Groups by Polynomial Algebra Methods

Vladimir V. Kornyak$^{(\boxtimes)}$

Laboratory of Information Technologies, Joint Institute for Nuclear Research,
141980 Dubna, Russia
vkornyak@gmail.com

Abstract. An algorithm for splitting permutation representations of a finite group over fields of characteristic zero into irreducible components is described. The algorithm is based on the fact that the components of the invariant inner product in invariant subspaces are operators of projection into these subspaces. An important part of the algorithm is the solution of systems of quadratic equations. A preliminary implementation of the algorithm splits representations up to dimensions of hundreds of thousands. Examples of computations are given in the appendix.

1 Introduction

One of the central problems of group theory and its applications in physics is the decomposition of linear representations of groups into irreducible components. In general, the problem of splitting a module over an associative algebra into irreducible submodules is quite nontrivial. An overview of the algorithmic aspects of this problem can be found in Chap. 7 of [1]. For vector spaces over finite fields, the most efficient is the Las Vegas algorithm [1] called *MeatAxe* [2]. This algorithm played an important role in solving the problem of classifying finite simple groups. However, the approach used in the *MeatAxe* is ineffective in characteristic zero, whereas quantum-mechanical problems are formulated just in Hilbert spaces over zero characteristic fields. Our algorithm deals with representations over such fields, and its implementation copes with dimensions up to hundreds of thousands, which is not less than the dimensions achievable for the *MeatAxe*. The algorithm requires knowledge of the centralizer ring of the considered group representation. In the general case, the calculation of the centralizer ring is a problem of linear algebra, namely, solving matrix equations of the form $AX = XA$. For permutation representations, there is an efficient way to compute the centralizer ring, which reduces to constructing the set of orbitals. In addition, permutation representations are fundamental in the sense that any linear representation of a finite group is a subrepresentation of some permutation

---

[1] A *Las Vegas algorithm* is a randomized algorithm, each iteration of which either produces the correct result, or reports a failure. An algorithm of this type always gives the correct answer, but the run time is indeterminate.

© Springer Nature Switzerland AG 2018
V. P. Gerdt et al. (Eds.): CASC 2018, LNCS 11077, pp. 304–318, 2018.
https://doi.org/10.1007/978-3-319-99639-4_21

representation, and we use this fact in some quantum mechanical considerations [3,4]. Thus, we consider only permutation representations here.

2   Mathematical Preliminaries

Let G (or $\mathsf{G}(\Omega)$) be a *transitive* permutation group on a set $\Omega \cong \{1, \ldots, \mathsf{N}\}$. We will denote the action of $g \in \mathsf{G}$ on $i \in \Omega$ by i^g. A representation of G in an N-dimensional vector space over a field $\mathcal{F}$ by the matrices $\mathrm{P}(g)$ with the entries $\mathrm{P}(g)_{ij} = \delta_{i^g j}$, where δ_{ij} is the Kronecker delta, is called a *permutation representation*. We assume that the permutation representation space is a Hilbert space $\mathcal{H}_\mathsf{N}$. We will assume that the base field $\mathcal{F}$ is a constructive *splitting field* of the group G. In particular, such a field can be a subfield of the mth cyclotomic field, where m is a suitable divisor of the exponent of G. Such a constructive field $\mathcal{F}$, being an abelian extension of the field $\mathbb{Q}$, is a dense subfield of $\mathbb{R}$ or $\mathbb{C}$.

An orbit of G on the Cartesian square $\Omega \times \Omega$ is called an *orbital* [5]. The number of orbitals, called the *rank* of $\mathsf{G}(\Omega)$, will be denoted by R. An orbital Δ is called *self-paired*, if $(i,j) \in \Delta \Rightarrow (j,i) \in \Delta$, i.e., $\Delta = \Delta^\mathrm{T}$. Among the orbitals of a transitive group, there is one *diagonal* orbital, $\Delta_1 = \{(i,i) \mid i \in \Omega\}$, which will always be fixed as the first element in the list of orbitals: $\{\Delta_1, \ldots, \Delta_\mathrm{R}\}$. For transitive action of G, there is a natural one-to-one correspondence between the orbitals of G and the orbits of a point stabilizer G_i:

$$\Delta \longleftrightarrow \Sigma_i = \{j \in \Omega \mid (i,j) \in \Delta\}.$$

The G_i-orbits are called *suborbits* and their cardinalities will be called the *suborbit lengths*. Note that $|\Delta| = \mathsf{N}\,|\Sigma_i|$.

The invariance condition for a bilinear form A in the Hilbert space $\mathcal{H}_\mathsf{N}$ can be written as the system of equations $A = \mathrm{P}(g)\,A\mathrm{P}(g^{-1})$, $g \in \mathsf{G}$. It is easy to verify that in terms of the entries these equations have the form $(A)_{ij} = (A)_{i^g j^g}$. Thus, the basis of all invariant bilinear forms is in one-to-one correspondence with the set of orbitals: with each orbital $\Delta_r \in \{\Delta_1, \ldots, \Delta_\mathrm{R}\}$ is associated a *basis matrix* $\mathcal{A}_r$ of the size $\mathsf{N} \times \mathsf{N}$ with the entries

$$(\mathcal{A}_r)_{ij} = \begin{cases} 1, & \text{if } (i,j) \in \Delta_r, \\ 0, & \text{if } (i,j) \notin \Delta_r. \end{cases}$$

It is clear that the matrix of a self-paired orbital is symmetric. For the diagonal orbital, we have $\mathcal{A}_1 = \mathbb{1}_\mathsf{N}$. The matrices

$$\mathcal{A}_1, \mathcal{A}_2, \ldots, \mathcal{A}_\mathrm{R} \tag{1}$$

form a basis of the *centralizer ring* (or *centralizer algebra*) of the representation P. The multiplication table for this basis has the form

$$\mathcal{A}_p \mathcal{A}_q = \sum_{r=1}^{\mathrm{R}} C_{pq}^r \mathcal{A}_r, \tag{2}$$

where C_{pq}^r are non-negative integers. The commutativity of the centralizer ring indicates that the permutation representation P is *multiplicity-free*.

3   Algorithm

Let T be a transformation (we can assume that T is a unitary matrix) that splits the permutation representation P into M irreducible components:

$$T^{-1}\mathrm{P}(g)\,T = 1 \oplus \mathsf{U}_{d_2}(g) \oplus \cdots \oplus \mathsf{U}_{d_m}(g) \oplus \cdots \oplus \mathsf{U}_{d_M}(g)\,,$$

where U_{d_m} is a d_m-dimensional irreducible subrepresentation, $\oplus$ denotes the direct sum of matrices, i.e., $A \oplus B = \mathrm{diag}(A, B)$.

The identity matrix $\mathbb{1}_\mathsf{N}$ is the *standard inner product* in any orthonormal basis. In the splitting basis, we have the following decomposition of the standard inner product

$$\mathbb{1}_\mathsf{N} = \mathbb{1}_{d_1 = 1} \oplus \cdots \oplus \mathbb{1}_{d_m} \oplus \cdots \oplus \mathbb{1}_{d_M}\,.$$

The *inverse image* of this decomposition in the original permutation basis has the form

$$\mathbb{1}_\mathsf{N} = \mathcal{B}_1 + \cdots + \mathcal{B}_m + \cdots + \mathcal{B}_M, \tag{3}$$

where $\mathcal{B}_m$ is defined by the relation

$$T^{-1}\mathcal{B}_m T = \mathbb{0}_{1+d_2+\cdots+d_{m-1}} \oplus \mathbb{1}_{d_m} \oplus \mathbb{0}_{d_{m+1}+\cdots+d_M} \equiv \mathcal{D}_m. \tag{4}$$

The set $\mathcal{B}_1, \ldots, \mathcal{B}_M$ contains complete information about irreducible decomposition of the representation P. In particular, the transformation matrix can be obtained from the linear system $\mathcal{B}_1 T - T\mathcal{D}_1 = \cdots = \mathcal{B}_M T - T\mathcal{D}_M = \mathbb{0}_\mathsf{N}$.

The main idea of the algorithm is based on the fact that $\mathcal{B}_m$'s form a complete set of *orthogonal projectors*, i.e., in addition to the *completeness* (3), we have the *idempotency*

$$\mathcal{B}_m^2 = \mathcal{B}_m \tag{5}$$

and the mutual *orthogonality*

$$\mathcal{B}_m \mathcal{B}_{m'} = \mathbb{0}_\mathsf{N} \text{ if } m \neq m'. \tag{6}$$

It follows from (4) that

$$\mathrm{tr}\,\mathcal{B}_m = d_m. \tag{7}$$

We see that all $\mathcal{B}_m$'s can be obtained as solutions of the equation

$$X^2 - X = \mathbb{0}_\mathsf{N} \tag{8}$$

for the generic invariant form

$$X = x_1 \mathcal{A}_1 + \cdots + x_\mathsf{R} \mathcal{A}_\mathsf{R}.$$

Using the multiplication table (2), we can write the left-hand side of (8) as a set of R polynomials (we will call them *idempotency polynomials*)

$$E(x_1, \ldots, x_\mathsf{R}) = \{E_1(x_1, \ldots, x_\mathsf{R}), \ldots, E_\mathsf{R}(x_1, \ldots, x_\mathsf{R})\} \tag{9}$$

and Eq. (8) can be written symbolically as

$$E(x_1, \ldots, x_R) = 0. \tag{10}$$

Each polynomial in (9) has the structure $E_r(x_1, \ldots, x_R) = Q_r(x_1, \ldots, x_R) - x_r$, where $Q_r(x_1, \ldots, x_R)$ is a homogeneous quadratic polynomial in the indeterminates $x_1, \ldots, x_R$.

In the basis (1), the projector $\mathcal{B}_m$ can be represented as

$$\mathcal{B}_m = b_{m,1}\mathcal{A}_1 + b_{m,2}\mathcal{A}_2 + \cdots + b_{m,R}\mathcal{A}_R,$$

where the vector $B_m = [b_{m,1}, \ldots, b_{m,R}]$ is a solution of Eq. (10).

Since only $\mathcal{A}_1$ has nonzero diagonal elements, we have

$$\operatorname{tr} \mathcal{B}_m = b_{m,1}\mathsf{N}.$$

Combining this with (7) we can fix the coefficient $b_{m,1}$:

$$b_{m,1} = d_m/\mathsf{N}.$$

Thus, the only relevant values of x_1 in (10) are d/N for some d's from the interval $[1, \ldots, \mathsf{N} - 1]$. Any relevant natural number d is either an irreducible dimension or a sum of such dimensions. Using the orthogonality condition (6) for the irreducible projectors, we can exclude the consideration of dimensions that are sums of irreducible ones. The generic orthogonality condition can be written as

$$BX = 0, \tag{11}$$

where $B = b_1\mathcal{A}_1 + \cdots + b_R\mathcal{A}_R$. Equation (11) is a system of linear equations for the indeterminates $x_1, \ldots, x_R$ with the parameters $b_1, \ldots, b_R$. Again, using the multiplication table (2), we can write the left-hand side of (11) as a system of R bilinear forms, which we denote by

$$O(b_1, \ldots, b_R; x_1, \ldots, x_R) \tag{12}$$

and call *orthogonality polynomials*.

The core part of the algorithm is a loop on dimensions that starts with $d = 1$ and ends when the sum of irreducible dimensions becomes equal to N.

The current d is processed as follows.

- We solve [2] the system of equations $E(d/\mathsf{N}, x_2, \ldots, x_R) = 0$.
- If the system is incompatible, then go to the next d.
- If $E(d/\mathsf{N}, x_2, \ldots, x_R)$ describes a zero-dimensional ideal, then we have k (including $k = 1$) different d-dimensional irreducible subrepresentations.

---

[2] The solution is always algorithmically realizable, since the problem involves only polynomial equations with abelian Galois groups.

- If the polynomial ideal has dimension $h > 0$, then we encounter an irreducible component with a multiplicity $k > 1$. The corresponding component of the centralizer algebra has the form $A \otimes \mathbb{1}_d$, where A is an arbitrary $k \times k$ matrix, and $\otimes$ denotes the Kronecker product. The idempotency condition $(A \otimes \mathbb{1}_d)^2 = A \otimes \mathbb{1}_d$ implies $A^2 - A = 0$. The complete family of solutions of this equation [3] is a manifold of dimension $\lfloor k^2/2 \rfloor = h$. In this case, we select, by a somewhat arbitrary procedure, k convenient mutually orthogonal representatives in the family of equivalent subrepresentations.
- In any case, if at the moment we have a solution $\mathcal{B}_m$, we append $\mathcal{B}_m$ to the list of irreducible projectors, and exclude from the further consideration the corresponding invariant subspace by adding the linear orthogonality polynomials $\mathcal{B}_m X$ to the polynomial system:

$$E(x_1, x_2, \ldots, x_R) \leftarrow E(x_1, x_2, \ldots, x_R) \cup \{\mathcal{B}_m X\}.$$

- After processing all $\mathcal{B}_m$'s of dimension d, go to the next d.

4 Implementation

Our approach involves some widely used methods of polynomial computer algebra. Therefore, it is reasonable, at least for the preliminary experience, to take advantage of computer algebra systems with developed tools for working with polynomials.

The complete algorithm is implemented by two procedures, the pseudocodes of which are given below.

1. The procedure *PreparePolynomialData* is a program written in **C**. The input data for this program is a set of permutations of Ω that generates the group $\mathsf{G}(\Omega)$. The program computes the basis of the centralizer ring and its multiplication table, constructs the idempotency and orthogonality polynomials, and generates the code of the procedure *SplitRepresentation* that processes the polynomial data. The main parameter that determines the run time for *PreparePolynomialData* is the dimension of the representation. The example in Sect. A.3 shows that the PC implementation copes with a dimension of about one hundred thousand in a time of about one hour.
2. The procedure *SplitRepresentation* implements the above described loop on dimensions that splits the representation of the group into irreducible components. It is generated by the **C** program *PreparePolynomialData*. Currently, the code is generated in the **Maple 2017.3** language, and the polynomial equations are processed by the **Maple** implementation of the Gröbner bases algorithms. The run time for *SplitRepresentation* depends mainly on the rank of the representation. Problems of rank R = 17 take about 8 hours on a PC.

---

[3] It is well known that any solution of the matrix equation $A^2 = A$ can be represented as $A = Q^{-1}(\mathbb{1}_r \oplus \mathbb{0}_{k-r})Q$, where Q is an arbitrary invertible $k \times k$ matrix and $r \in [0, k]$.

Input: $S = \{s_1, \ldots, s_K\}$ // set of permutations of Ω that generates group G
Output: $E(x_1, \ldots, x_R)$, $O(b_1, \ldots, b_R; x_1, \ldots, x_R)$, *SplitRepresentation*
 1: compute basis of centralizer ring $\mathcal{A}_1, \ldots, \mathcal{A}_R$
 2: compute multiplication table $\mathcal{A}_p \mathcal{A}_q = \sum\limits_{r=1}^{R} C_{pq}^r \mathcal{A}_r$
 3: construct idempotency polynomials $E(x_1, \ldots, x_R)$
 4: construct orthogonality polynomials $O(b_1, \ldots, b_R; x_1, \ldots, x_R)$
 5: construct code *SplitRepresentation* for processing polynomial data
 6: **return** *SplitRepresentation* $(E(x_1, \ldots, x_R), O(b_1, \ldots, b_R; x_1, \ldots, x_R))$

Algorithm 1: *PreparePolynomialData*

Input: $E(x_1, \ldots, x_R)$, $O(b_1, \ldots, b_R; x_1, \ldots, x_R)$
Output: $IrreducibleProjectors = [(1, \mathcal{B}_1), \ldots, (d_m, \mathcal{B}_m) \ldots, (d_M, \mathcal{B}_M)]$
 1: $IrreducibleProjectors \leftarrow \left[\left(1, \frac{1}{N}[1, \ldots, 1]\right)\right]$ // trivial subrepresentation
 2: $E(x_1, \ldots, x_R) \leftarrow E(x_1, \ldots, x_R) \cup O(1, \ldots, 1; x_1, \ldots, x_R)$
 3: $Sdim \leftarrow 1$ // sum of dimensions, global variable
 4: $D \leftarrow 0$ // current dimension, global variable
 5: **while** $Sdim < \mathsf{N}$ **do**
 6:      $D \leftarrow NextRelevantDimension(\mathbf{D})$
 7:      $all\_solutions \leftarrow SolveAlgebraicSystem(E(D/N, x_2, \ldots, x_R))$
 8:      **if** $all\_solutions \neq \varnothing$ **then**
 9:          $h \leftarrow NumberOfFreeParameters(all\_solutions)$
10:         **if** $h = 0$ **then**
11:             **for** $solution \in all\_solutions$ **do**
12:                 $UseSingleSolution(solution)$
13:         **else**
14:             **repeat**
15:                 $solution \leftarrow PickBestSolution(all\_solutions)$
16:                 $UseSingleSolution(solution)$
17:                 $all\_solutions \leftarrow SolveAlgebraicSystem(E(D/N, x_2, \ldots, x_R))$
18:             **until** $all\_solutions = \varnothing$
19: **return** $IrreducibleProjectors$

Algorithm 2: *SplitRepresentation*

Input: $solution = [\beta_1, \ldots, \beta_R]$
 1: $E(x_1, \ldots, x_R) \leftarrow E(x_1, \ldots, x_R) \cup O(\beta_1, \ldots, \beta_R; x_1, \ldots, x_R)$
 2: $IrreducibleProjectors \leftarrow [IrreducibleProjectors, (D, solution)]$
 3: $Sdim \leftarrow Sdim + D$

Algorithm 3: *UseSingleSolution*

Comments on the procedure *SplitRepresentation*:

- The procedure *NextRelevantDimension* can be implemented in different ways, depending on the available information about the group and the representation:
  - The simplest implementation is "$D \leftarrow D + 1$".
  - The implementation "**repeat** $D \leftarrow D + 1$ **until** $D \mid \mathrm{Ord}(G)$" is about 25% faster than the simplest one. In fact, the size of the group is always known.
  - Knowledge of the character decomposition provides the most efficient loop on dimensions. Sometimes this information is available. Actually, computing the character decomposition is much easier than computing the decomposition of the representation.
- The procedures *SolveAlgebraicSystem* and *NumberOfFreeParameters* involve the polynomial algebra functions available in the computer algebra system used. At present, we use the **Maple** implementation of Gröbner basis techniques.
- The *PickBestSolution* procedure is applied in the case of nontrivial multiplicity of the irreducible component. It selects a particular solution in the parametric set of solutions. Currently, the choice of solutions with zero values of parameters is used. Such an oversimplified approach sometimes leads to "ugly roots" that go beyond the "natural" splitting field. This can be illustrated by the example of a 29155-dimensional representation of the Held group whose decomposition into irreducible components is given in Sect. A.2. The decomposition contains a 1275-dimensional irreducible component of multiplicity two. Representatives of this component obtained by the simple version of *PickBestSolution* contain irrationality $i\sqrt{231}$ (see $\mathcal{B}_{1275}^{(1)}$ and $\mathcal{B}_{1275}^{(2)}$ expressions), which belongs to the quadratic field $\mathbb{Q}(\sqrt{-231})$, while the representation in question splits over the "much smaller" field $\mathbb{Q}(\sqrt{-7})$. Therefore, the *PickBestSolution* procedure requires improvement using strategies that lead to minimal extensions of the field $\mathbb{Q}$.

4.1   Comparison with the Magma Implementation of the *MeatAxe*

The **Magma** database contains a 3906-dimensional permutation representation of the exceptional group of Lie type $G_2(5)$. The decomposition into irreducible components of this representation over the field $\mathrm{GF}(2)$ is given in [6] as an illustration of the possibilities of the *MeatAxe*.

The application of our algorithm to this problem shows that in the characteristic zero, the considered representation is split over the field $\mathbb{Q}$. The calculation produces the following data:

Rank: 4. Suborbit lengths: $1, 30, 750, 3125$.

$$\underline{3906} \cong 1 \oplus 930 \oplus 1085 \oplus 1890$$

$$\mathcal{B}_1 = \frac{1}{3906} \sum_{k=1}^{4} \mathcal{A}_k$$

$$\mathcal{B}_{930} = \frac{5}{21} \left(\mathcal{A}_1 + \frac{3}{10} \mathcal{A}_2 + \frac{1}{50} \mathcal{A}_3 - \frac{1}{125} \mathcal{A}_4 \right)$$

$$\mathcal{B}_{1085} = \frac{5}{18} \left(\mathcal{A}_1 - \frac{1}{5} \mathcal{A}_2 + \frac{1}{25} \mathcal{A}_3 - \frac{1}{125} \mathcal{A}_4 \right)$$

$$\mathcal{B}_{1890} = \frac{15}{31} \left(\mathcal{A}_1 - \frac{1}{30} \mathcal{A}_2 - \frac{1}{30} \mathcal{A}_3 + \frac{1}{125} \mathcal{A}_4 \right)$$

Time **C**: 0.5 s. Time **Maple**: 0.8 s.

Magma failed to split the 3906-dimensional representation over the field $\mathbb{Q}$ due to memory exhaustion after long computation, but we can simulate to some extent the case of characteristic zero, using a field of a characteristic that does not divide $\mathrm{Ord}(G_2(5))$. The smallest such field is $\mathrm{GF}(11)$.

Below is the session of the corresponding **Magma V2.21-1** computation on a computer with two Intel Xeon E5410 2.33 GHz CPUs (time is given in seconds).

```
> load "g25";
Loading "/opt/magma.21-1/libs/pergps/g25"
The Lie group G( 2, 5 ) represented as a permutation
group of degree 3906.
Order: 5 859 000 000 = 2^6 * 3^3 * 5^6 * 7 * 31.
Group: G
> time Constituents(PermutationModule(G,GF(11)));
[
    GModule of dimension 1 over GF(11),
    GModule of dimension 930 over GF(11),
    GModule of dimension 1085 over GF(11),
    GModule of dimension 1890 over GF(11)
]
Time: 282.060
```

5   Conclusion

The algorithm described here is based on the use of methods of polynomial algebra, which are considered algorithmically difficult. However, our approach leads to a small number (in typical cases) of low-degree polynomials. Recall that the idempotency system (9) is a set of R square polynomials. Calculations of Gröbner bases in **Maple** on PC are limited in practice to R = 17. Among the 886 permutation representations available in the ATLAS [7], 761 (i.e., 86%) have ranks R ≤ 17. As can be seen in Appendix A, even a straightforward implementation of the approach can cope with rather large tasks. The data presented in the appendix shows that the most restrictive parameter for the **Maple** part of the implementation is the rank of representations, i.e., the number of polynomial indeterminates. A possible way to improve performance is to try to develop specialized algorithms that take into account the very special type of polynomial equations that arise in the problem instead of the universal Gröbner basis methods.

Acknowledgments. I am grateful to Yu.A. Blinkov, V.P. Gerdt and R.A. Wilson for fruitful discussions and valuable advice.

A   Examples of Computations

- Generators of representations are taken from the section "Sporadic groups" of the ATLAS [7].
- For a group G
  - M(G) denotes the *Schur multiplier*, the 2nd homology group $H_2(G, \mathbb{Z})$,
  - Out(G) denotes the *outer automorphism group* of G,
  - $n.G$ denotes a *covering group* of G, a *central extension* of G by C_n.
- The results presented below assume the following ordering for the centralizer ring basis matrices

$$\mathcal{A}_1 = \mathbb{1}_N, \ \underbrace{\mathcal{A}_2, \dots, \mathcal{A}_k,}_{\text{symmetric matrices}} \ \underbrace{\mathcal{A}_{k+1}, \mathcal{A}_{k+2} = \mathcal{A}_{k+1}^T, \dots, \mathcal{A}_{R-1}, \mathcal{A}_R = \mathcal{A}_{R-1}^T}_{\text{asymmetric matrices}}.$$

The matrices within the first sublist are ordered by the rule: $A < B$ if $i_A < i_B$, where $i_X = \min(i \mid (X)_{i1} = 1)$. The same rule is applied to the first elements of the pairs of asymmetric matrices.
- Representations are denoted by their dimensions in bold (possibly with some signs added to distinguish different representations of the same dimension). Permutation representations are underlined. Multiple subrepresentations are underbraced in the decompositions.
- We omit the irreducible projectors related to the trivial subrepresentation: these projectors have the standard form $\mathcal{B}_1 = \frac{1}{N} \sum_{k=1}^{R} \mathcal{A}_k$.
- All timing data refer to a PC with 3.30 GHz Intel Core i3 2120 CPU.

A.1   Higman–Sims Group HS

Main properties: $\text{Ord}(HS) = 44352000 = 2^9 \cdot 3^2 \cdot 5^3 \cdot 7 \cdot 11$.
$M(HS) = C_2$. $\text{Out}(HS) = C_2$.

11200-dimensional Representation of 2.HS
Rank: 16. Suborbit lengths: $1^2, 110, 132^2, 165^2, 660^2, 792^2, 990, 1320^2, 1980^2$.

$$\underline{\mathbf{11200}} \cong \mathbf{1} \oplus \mathbf{22} \oplus \mathbf{56} \oplus \mathbf{77} \oplus \mathbf{154} \oplus \mathbf{175} \oplus \mathbf{176} \oplus \overline{\mathbf{176}} \oplus \mathbf{616} \oplus \overline{\mathbf{616}}$$
$$\oplus \mathbf{770} \oplus \mathbf{825} \oplus \mathbf{1056} \oplus \mathbf{1980} \oplus \overline{\mathbf{1980}} \oplus \mathbf{2520}$$

$$\mathcal{B}_{22} = \frac{11}{5600} \left(\mathcal{A}_1 + \frac{13}{33}\mathcal{A}_2 - \frac{7}{33}\mathcal{A}_3 + \frac{1}{11}\mathcal{A}_4 + \frac{1}{11}\mathcal{A}_5 + \frac{13}{33}\mathcal{A}_6 + \frac{1}{11}\mathcal{A}_7 - \frac{7}{33}\mathcal{A}_8 \right.$$

$$+ \frac{13}{33}\mathcal{A}_9 + \mathcal{A}_{10} - \frac{7}{33}\mathcal{A}_{11} - \frac{7}{33}\mathcal{A}_{12} + \frac{1}{11}\mathcal{A}_{13} + \frac{1}{11}\mathcal{A}_{14} - \frac{17}{33}\mathcal{A}_{15}$$

$$\left. - \frac{17}{33}\mathcal{A}_{16} \right)$$

$$\mathcal{B}_{56} = \frac{1}{200} \left(\mathcal{A}_1 + \frac{1}{4}\mathcal{A}_3 + \frac{1}{4}\mathcal{A}_4 - \frac{1}{4}\mathcal{A}_5 + \frac{1}{4}\mathcal{A}_6 - \frac{1}{4}\mathcal{A}_8 - \frac{1}{4}\mathcal{A}_9 - \mathcal{A}_{10} \right)$$

$$\mathcal{B}_{77} = \frac{11}{1600} \left(\mathcal{A}_1 + \frac{1}{11}\mathcal{A}_2 + \frac{17}{132}\mathcal{A}_3 - \frac{23}{132}\mathcal{A}_4 - \frac{23}{132}\mathcal{A}_5 + \frac{37}{132}\mathcal{A}_6 - \frac{4}{11}\mathcal{A}_7 \right.$$

$$+ \frac{17}{132}\mathcal{A}_8 + \frac{37}{132}\mathcal{A}_9 + \mathcal{A}_{10} - \frac{2}{33}\mathcal{A}_{11} - \frac{2}{33}\mathcal{A}_{12} + \frac{1}{66}\mathcal{A}_{13} + \frac{1}{66}\mathcal{A}_{14}$$

$$\left. + \frac{8}{33}\mathcal{A}_{15} + \frac{8}{33}\mathcal{A}_{16} \right)$$

$$\mathcal{B}_{154} = \frac{11}{800} \left(\mathcal{A}_1 + \frac{3}{55}\mathcal{A}_2 + \frac{7}{55}\mathcal{A}_3 + \frac{1}{11}\mathcal{A}_4 + \frac{1}{11}\mathcal{A}_5 - \frac{1}{11}\mathcal{A}_6 - \frac{19}{55}\mathcal{A}_7 + \frac{7}{55}\mathcal{A}_8 \right.$$

$$- \frac{1}{11}\mathcal{A}_9 + \mathcal{A}_{10} - \frac{1}{55}\mathcal{A}_{11} - \frac{1}{55}\mathcal{A}_{12} - \frac{3}{55}\mathcal{A}_{13} - \frac{3}{55}\mathcal{A}_{14} - \frac{7}{55}\mathcal{A}_{15}$$

$$\left. - \frac{7}{55}\mathcal{A}_{16} \right)$$

$$\mathcal{B}_{175} = \frac{1}{64} \left(\mathcal{A}_1 + \frac{7}{55}\mathcal{A}_2 - \frac{1}{15}\mathcal{A}_3 + \frac{1}{33}\mathcal{A}_4 + \frac{1}{33}\mathcal{A}_5 + \frac{1}{33}\mathcal{A}_6 + \frac{7}{55}\mathcal{A}_7 - \frac{1}{15}\mathcal{A}_8 \right.$$

$$+ \frac{1}{33}\mathcal{A}_9 + \mathcal{A}_{10} + \frac{1}{33}\mathcal{A}_{11} + \frac{1}{33}\mathcal{A}_{12} - \frac{1}{15}\mathcal{A}_{13} - \frac{1}{15}\mathcal{A}_{14} + \frac{37}{165}\mathcal{A}_{15}$$

$$\left. + \frac{37}{165}\mathcal{A}_{16} \right)$$

$$\mathcal{B}_{176} = \frac{11}{700} \left(\mathcal{A}_1 + \frac{2}{33}\mathcal{A}_3 - \frac{1}{11}\mathcal{A}_4 + \frac{1}{11}\mathcal{A}_5 + \frac{7}{33}\mathcal{A}_6 - \frac{2}{33}\mathcal{A}_8 - \frac{7}{33}\mathcal{A}_9 - \mathcal{A}_{10} \right.$$

$$\left. + i\frac{1}{33}\mathcal{A}_{11} - i\frac{1}{33}\mathcal{A}_{12} + i\frac{2}{33}\mathcal{A}_{13} - i\frac{2}{33}\mathcal{A}_{14} + i\frac{7}{33}\mathcal{A}_{15} - i\frac{7}{33}\mathcal{A}_{16} \right)$$

$$\mathcal{B}_{616} = \frac{11}{200} \left(\mathcal{A}_1 - \frac{7}{132}\mathcal{A}_3 + \frac{1}{44}\mathcal{A}_4 - \frac{1}{44}\mathcal{A}_5 + \frac{13}{132}\mathcal{A}_6 + \frac{7}{132}\mathcal{A}_8 - \frac{13}{132}\mathcal{A}_9 \right.$$

$$- \mathcal{A}_{10} - i\frac{1}{66}\mathcal{A}_{11} + i\frac{1}{66}\mathcal{A}_{12} - i\frac{1}{33}\mathcal{A}_{13} + i\frac{1}{33}\mathcal{A}_{14} + i\frac{4}{33}\mathcal{A}_{15}$$

$$\left. - i\frac{4}{33}\mathcal{A}_{16} \right)$$

$$\mathcal{B}_{770} = \frac{11}{160}\left(\mathcal{A}_1 - \frac{1}{165}\mathcal{A}_2 - \frac{1}{60}\mathcal{A}_3 - \frac{1}{44}\mathcal{A}_4 - \frac{1}{44}\mathcal{A}_5 + \frac{13}{132}\mathcal{A}_6 - \frac{4}{55}\mathcal{A}_7\right.$$
$$-\frac{1}{60}\mathcal{A}_8 + \frac{13}{132}\mathcal{A}_9 + \mathcal{A}_{10} + \frac{7}{165}\mathcal{A}_{11} + \frac{7}{165}\mathcal{A}_{12} - \frac{1}{110}\mathcal{A}_{13}$$
$$\left.-\frac{1}{110}\mathcal{A}_{14} - \frac{16}{165}\mathcal{A}_{15} - \frac{16}{165}\mathcal{A}_{16}\right)$$

$$\mathcal{B}_{825} = \frac{33}{448}\left(\mathcal{A}_1 + \frac{13}{495}\mathcal{A}_2 + \frac{7}{220}\mathcal{A}_3 - \frac{13}{396}\mathcal{A}_4 - \frac{13}{396}\mathcal{A}_5 - \frac{1}{12}\mathcal{A}_6 + \frac{12}{55}\mathcal{A}_7\right.$$
$$+\frac{7}{220}\mathcal{A}_8 - \frac{1}{12}\mathcal{A}_9 + \mathcal{A}_{10} - \frac{1}{990}\mathcal{A}_{13} - \frac{1}{990}\mathcal{A}_{14} - \frac{8}{165}\mathcal{A}_{15}$$
$$\left.-\frac{8}{165}\mathcal{A}_{16}\right)$$

$$\mathcal{B}_{1056} = \frac{33}{350}\left(\mathcal{A}_1 - \frac{23}{495}\mathcal{A}_2 + \frac{3}{220}\mathcal{A}_3 + \frac{1}{36}\mathcal{A}_4 + \frac{1}{36}\mathcal{A}_5 + \frac{13}{132}\mathcal{A}_6 + \frac{6}{55}\mathcal{A}_7\right.$$
$$+\frac{3}{220}\mathcal{A}_8 + \frac{13}{132}\mathcal{A}_9 + \mathcal{A}_{10} - \frac{1}{55}\mathcal{A}_{11} - \frac{1}{55}\mathcal{A}_{12} - \frac{2}{495}\mathcal{A}_{13}$$
$$\left.-\frac{2}{495}\mathcal{A}_{14} + \frac{4}{165}\mathcal{A}_{15} + \frac{4}{165}\mathcal{A}_{16}\right)$$

$$\mathcal{B}_{1980} = \frac{99}{560}\left(\mathcal{A}_1 + \frac{1}{132}\mathcal{A}_3 - \frac{1}{396}\mathcal{A}_4 + \frac{1}{396}\mathcal{A}_5 - \frac{7}{132}\mathcal{A}_6 - \frac{1}{132}\mathcal{A}_8 + \frac{7}{132}\mathcal{A}_9\right.$$
$$\left.-\mathcal{A}_{10} - \mathbf{i}\frac{1}{33}\mathcal{A}_{11} + \mathbf{i}\frac{1}{33}\mathcal{A}_{12} + \mathbf{i}\frac{1}{99}\mathcal{A}_{13} - \mathbf{i}\frac{1}{99}\mathcal{A}_{14}\right)$$

$$\mathcal{B}_{2520} = \frac{9}{40}\left(\mathcal{A}_1 - \frac{1}{165}\mathcal{A}_2 - \frac{1}{60}\mathcal{A}_3 + \frac{1}{396}\mathcal{A}_4 + \frac{1}{396}\mathcal{A}_5 - \frac{7}{132}\mathcal{A}_6 - \frac{4}{55}\mathcal{A}_7\right.$$
$$-\frac{1}{60}\mathcal{A}_8 - \frac{7}{132}\mathcal{A}_9 + \mathcal{A}_{10} - \frac{1}{330}\mathcal{A}_{11} - \frac{1}{330}\mathcal{A}_{12} + \frac{1}{90}\mathcal{A}_{13}$$
$$\left.+\frac{1}{90}\mathcal{A}_{14} + \frac{4}{165}\mathcal{A}_{15} + \frac{4}{165}\mathcal{A}_{16}\right)$$

Time **C**: 8 s. Time **Maple**: 1 h 39 min 6 s.

A.2   Held Group He

Main properties: Ord$(He) = 4030387200 = 2^{10} \cdot 3^3 \cdot 5^2 \cdot 7^3 \cdot 17$.
$M(He) = 1$. Out$(He) = \mathsf{C}_2$.

29155-dimensional Representation of He

Rank: 12. Suborbit lengths: $1, 90, 120, 384, 960^2, 1440, 2160, 2880^2, 5760,$ 11520.

$$\underline{29155} \cong 1 \oplus 51 \oplus \overline{51} \oplus 680 \oplus \underbrace{(1275 \oplus 1275)} \oplus 1920 \oplus 4352 \oplus 7650$$
$$\oplus\, 11900$$

$$\mathcal{B}_{51} = \frac{3}{1715}\left\{ \mathcal{A}_1 + \frac{5}{12}\mathcal{A}_2 - \frac{1}{48}\mathcal{A}_3 + \frac{1}{8}\mathcal{A}_4 + \frac{1}{8}\mathcal{A}_5 + \frac{13}{48}\mathcal{A}_6 - \frac{1}{6}\mathcal{A}_7 - \frac{1}{6}\mathcal{A}_8 \right.$$

$$- \frac{1}{32}\left(3 - i\frac{7\sqrt{7}}{3} \right)\mathcal{A}_9 - \frac{1}{32}\left(3 + i\frac{7\sqrt{7}}{3} \right)\mathcal{A}_{10}$$

$$\left. + \frac{1}{96}\left(5 + 7i\sqrt{7} \right)\mathcal{A}_{11} + \frac{1}{96}\left(5 - 7i\sqrt{7} \right)\mathcal{A}_{12} \right\}$$

$$\mathcal{B}_{680} = \frac{8}{343}\left(\mathcal{A}_1 + \frac{3}{10}\mathcal{A}_2 - \frac{1}{48}\mathcal{A}_3 - \frac{23}{1440}\mathcal{A}_4 - \frac{1}{20}\mathcal{A}_5 + \frac{1}{8}\mathcal{A}_6 + \frac{1}{120}\mathcal{A}_7 \right.$$

$$\left. + \frac{13}{90}\mathcal{A}_8 + \frac{1}{36}\mathcal{A}_9 + \frac{1}{36}\mathcal{A}_{10} + \frac{1}{15}\mathcal{A}_{11} + \frac{1}{15}\mathcal{A}_{12} \right)$$

$$\mathcal{B}_{1275}^{(1)} = \frac{15}{343}\left\{ \mathcal{A}_1 + \frac{1}{4280}\left(\frac{331}{3} - 7i\sqrt{231} \right)\mathcal{A}_2 - \frac{1}{25680}\left(13 - i\frac{7\sqrt{231}}{3} \right)\mathcal{A}_3 \right.$$

$$- \frac{1}{25680}\left(\frac{1381}{3} + 7i\sqrt{231} \right)\mathcal{A}_4 + \frac{1}{25680}\left(2101 + 7i\sqrt{231} \right)\mathcal{A}_5$$

$$- \frac{1}{1712}\left(13 - i\frac{7\sqrt{231}}{3} \right)\mathcal{A}_6 + \frac{1}{2568}\left(\frac{109}{3} - i\frac{7\sqrt{231}}{5} \right)\mathcal{A}_7$$

$$+ \frac{1}{4815}\left(1571 - i\frac{7\sqrt{231}}{2} \right)\mathcal{A}_8 - \frac{1}{38520}\left(467 - 7i\sqrt{231} \right)\mathcal{A}_9$$

$$\left. - \frac{1}{38520}\left(467 - 7i\sqrt{231} \right)\mathcal{A}_{10} \right\}$$

$$\mathcal{B}_{1275}^{(2)} = \frac{15}{343}\left\{ \mathcal{A}_1 + \frac{1}{4280}\left(\frac{1381}{3} + 7i\sqrt{231} \right)\mathcal{A}_2 + \frac{1}{25680}\left(227 - i\frac{7\sqrt{231}}{3} \right)\mathcal{A}_3 \right.$$

$$- \frac{1}{25680}\left(\frac{331}{3} - 7i\sqrt{231} \right)\mathcal{A}_4 - \frac{1}{25680}\left(389 + 7i\sqrt{231} \right)\mathcal{A}_5$$

$$+ \frac{1}{1712}\left(227 - i\frac{7\sqrt{231}}{3} \right)\mathcal{A}_6 + \frac{1}{2568}\left(\frac{319}{3} + i\frac{7\sqrt{231}}{5} \right)\mathcal{A}_7$$

$$- \frac{1}{4815}\left(394 - i\frac{7\sqrt{231}}{2} \right)\mathcal{A}_8 - \frac{7}{38520}\left(\frac{157}{2} + i\sqrt{231} \right)\mathcal{A}_9$$

$$\left. - \frac{7}{38520}\left(\frac{157}{2} + i\sqrt{231} \right)\mathcal{A}_{10} - \frac{1}{16}\mathcal{A}_{11} - \frac{1}{16}\mathcal{A}_{12} \right\}$$

$$\mathcal{B}_{1920} = \frac{384}{5831}\left(\mathcal{A}_1 + \frac{1}{120}\mathcal{A}_2 - \frac{7}{384}\mathcal{A}_3 + \frac{1}{120}\mathcal{A}_4 + \frac{7}{160}\mathcal{A}_5 - \frac{7}{384}\mathcal{A}_6 + \frac{1}{120}\mathcal{A}_7 \right.$$

$$\left. - \frac{2}{15}\mathcal{A}_8 + \frac{5}{192}\mathcal{A}_9 + \frac{5}{192}\mathcal{A}_{10} - \frac{13}{480}\mathcal{A}_{11} - \frac{13}{480}\mathcal{A}_{12} \right)$$

$$\mathcal{B}_{4352} = \frac{256}{1715}\left(\mathcal{A}_1 + \frac{1}{8}\mathcal{A}_2 + \frac{7}{768}\mathcal{A}_3 - \frac{5}{576}\mathcal{A}_4 - \frac{7}{128}\mathcal{A}_6 - \frac{1}{48}\mathcal{A}_7\right.$$
$$\left. - \frac{1}{18}\mathcal{A}_8 + \frac{1}{576}\mathcal{A}_9 + \frac{1}{576}\mathcal{A}_{10} - \frac{1}{192}\mathcal{A}_{11} - \frac{1}{192}\mathcal{A}_{12}\right)$$
$$\mathcal{B}_{7650} = \frac{90}{343}\left(\mathcal{A}_1 - \frac{1}{20}\mathcal{A}_2 + \frac{1}{120}\mathcal{A}_4 - \frac{7}{360}\mathcal{A}_5 - \frac{1}{90}\mathcal{A}_7 + \frac{1}{10}\mathcal{A}_8 + \frac{1}{240}\mathcal{A}_9\right.$$
$$\left. + \frac{1}{240}\mathcal{A}_{10} - \frac{1}{80}\mathcal{A}_{11} - \frac{1}{80}\mathcal{A}_{12}\right)$$
$$\mathcal{B}_{11900} = \frac{20}{49}\left(\mathcal{A}_1 - \frac{1}{20}\mathcal{A}_2 - \frac{1}{720}\mathcal{A}_4 + \frac{1}{120}\mathcal{A}_7 - \frac{1}{18}\mathcal{A}_8 - \frac{1}{180}\mathcal{A}_9 - \frac{1}{180}\mathcal{A}_{10}\right.$$
$$\left. + \frac{1}{60}\mathcal{A}_{11} + \frac{1}{60}\mathcal{A}_{12}\right)$$

Time **C**: 47 s. Time **Maple**: 15 s.

A.3   Suzuki Group *Suz*

Main properties: Ord(Suz) = 448345497600 = $2^{13} \cdot 3^7 \cdot 5^2 \cdot 7 \cdot 11 \cdot 13$.
M(Suz) = C$_6$. Out(Suz) = C$_2$.

65520-dimensional Representation of 2.Suz
Rank: 10. Suborbit lengths: $1^2, 891^2, 2816^2, 3960, 12672, 20736^2$.

$$\underline{65520} \cong 1 \oplus 143 \oplus 364_\alpha \oplus 364_\beta \oplus \overline{364_\beta} \oplus 5940 \oplus 12012 \oplus 14300$$
$$\oplus 16016 \oplus \overline{16016}$$

$$\mathcal{B}_{143} = \frac{11}{5040}\left(\mathcal{A}_1 + \mathcal{A}_2 + \frac{2}{11}\mathcal{A}_3 - \frac{1}{11}\mathcal{A}_4 + \frac{2}{11}\mathcal{A}_5 - \frac{1}{11}\mathcal{A}_6 + \frac{3}{11}\mathcal{A}_9 + \frac{3}{11}\mathcal{A}_{10}\right)$$
$$\mathcal{B}_{364_\alpha} = \frac{1}{180}\left(\mathcal{A}_1 + \mathcal{A}_2 + \frac{1}{16}\mathcal{A}_3 + \frac{1}{6}\mathcal{A}_4 + \frac{1}{16}\mathcal{A}_5 - \frac{1}{24}\mathcal{A}_6 - \frac{1}{144}\mathcal{A}_7 - \frac{1}{144}\mathcal{A}_8\right.$$
$$\left. - \frac{1}{9}\mathcal{A}_9 - \frac{1}{9}\mathcal{A}_{10}\right)$$
$$\mathcal{B}_{364_\beta} = \frac{1}{180}\left(\mathcal{A}_1 - \mathcal{A}_2 - \frac{1}{8}\mathcal{A}_3 + \frac{1}{8}\mathcal{A}_5 + i\frac{\sqrt{3}}{72}\mathcal{A}_7 - i\frac{\sqrt{3}}{72}\mathcal{A}_8\right.$$
$$\left. + i\frac{\sqrt{3}}{9}\mathcal{A}_9 - i\frac{\sqrt{3}}{9}\mathcal{A}_{10}\right)$$
$$\mathcal{B}_{5940} = \frac{33}{364}\left(\mathcal{A}_1 + \mathcal{A}_2 + \frac{1}{352}\mathcal{A}_3 + \frac{1}{66}\mathcal{A}_4 + \frac{1}{352}\mathcal{A}_5 + \frac{1}{66}\mathcal{A}_6 - \frac{7}{864}\mathcal{A}_7\right.$$
$$\left. - \frac{7}{864}\mathcal{A}_8 + \frac{1}{27}\mathcal{A}_9 + \frac{1}{27}\mathcal{A}_{10}\right)$$

$$\mathcal{B}_{12012} = \frac{11}{60}\left(\mathcal{A}_1 + \mathcal{A}_2 + \frac{1}{88}\mathcal{A}_3 - \frac{1}{66}\mathcal{A}_4 + \frac{1}{88}\mathcal{A}_5 + \frac{1}{264}\mathcal{A}_6 - \frac{1}{33}\mathcal{A}_9 - \frac{1}{33}\mathcal{A}_{10}\right)$$

$$\mathcal{B}_{14300} = \frac{55}{252}\left(\mathcal{A}_1 + \mathcal{A}_2 - \frac{5}{352}\mathcal{A}_3 + \frac{1}{330}\mathcal{A}_4 - \frac{5}{352}\mathcal{A}_5 - \frac{1}{132}\mathcal{A}_6 + \frac{1}{288}\mathcal{A}_7\right.$$
$$\left. + \frac{1}{288}\mathcal{A}_8 + \frac{1}{99}\mathcal{A}_9 + \frac{1}{99}\mathcal{A}_{10}\right)$$

$$\mathcal{B}_{16016} = \frac{11}{45}\left(\mathcal{A}_1 - \mathcal{A}_2 + \frac{1}{352}\mathcal{A}_3 - \frac{1}{352}\mathcal{A}_5 - i\frac{\sqrt{3}}{288}\mathcal{A}_7 + i\frac{\sqrt{3}}{288}\mathcal{A}_8\right.$$
$$\left. + i\frac{\sqrt{3}}{99}\mathcal{A}_9 - i\frac{\sqrt{3}}{99}\mathcal{A}_{10}\right)$$

Time **C**: 6 min 3 s. Time **Maple**: 10 s.

98280-dimensional Representation of 3.*Suz*

Rank: 14. Suborbit lengths: $1^3, 891^3, 2816^3, 5940, 19008, 20736^3$.

$$\underline{\mathbf{98280}} \cong 1 \oplus 78 \oplus \overline{78} \oplus 143 \oplus 364 \oplus 1365 \oplus \overline{1365} \oplus 4290 \oplus \overline{4290}$$
$$\oplus\, 5940 \oplus 12012 \oplus 14300 \oplus 27027 \oplus \overline{27027}$$

$$\mathcal{B}_{78} = \frac{1}{1260}\left(\mathcal{A}_1 - \frac{1}{12}\mathcal{A}_2 - \frac{1}{3}\mathcal{A}_4 + \frac{1}{4}\mathcal{A}_6 - \frac{r}{12}\mathcal{A}_7 - \frac{r^2}{12}\mathcal{A}_8\right.$$
$$\left. + \frac{r}{4}\mathcal{A}_9 + \frac{r^2}{4}\mathcal{A}_{10} - \frac{r^2}{3}\mathcal{A}_{11} - \frac{r}{3}\mathcal{A}_{12} + r\mathcal{A}_{13} + r^2\mathcal{A}_{14}\right)$$

$$\mathcal{B}_{143} = \frac{11}{7560}\left(\mathcal{A}_1 - \frac{1}{11}\mathcal{A}_3 + \frac{3}{11}\mathcal{A}_4 - \frac{1}{11}\mathcal{A}_5 + \frac{2}{11}\mathcal{A}_6 + \frac{2}{11}\mathcal{A}_9\right.$$
$$\left. + \frac{2}{11}\mathcal{A}_{10} + \frac{3}{11}\mathcal{A}_{11} + \frac{3}{11}\mathcal{A}_{12} + \mathcal{A}_{13} + \mathcal{A}_{14}\right)$$

$$\mathcal{B}_{364} = \frac{1}{270}\left(\mathcal{A}_1 - \frac{1}{144}\mathcal{A}_2 + \frac{1}{6}\mathcal{A}_3 - \frac{1}{9}\mathcal{A}_4 - \frac{1}{24}\mathcal{A}_5 + \frac{1}{16}\mathcal{A}_6 - \frac{1}{144}\mathcal{A}_7\right.$$
$$\left. - \frac{1}{144}\mathcal{A}_8 + \frac{1}{16}\mathcal{A}_9 + \frac{1}{16}\mathcal{A}_{10} - \frac{1}{9}\mathcal{A}_{11} - \frac{1}{9}\mathcal{A}_{12} + \mathcal{A}_{13} + \mathcal{A}_{14}\right)$$

$$\mathcal{B}_{1365} = \frac{1}{72}\left(\mathcal{A}_1 + \frac{1}{144}\mathcal{A}_2 + \frac{1}{9}\mathcal{A}_4 + \frac{1}{16}\mathcal{A}_6 + \frac{r}{144}\mathcal{A}_7 + \frac{r^2}{144}\mathcal{A}_8\right.$$
$$\left. + \frac{r}{16}\mathcal{A}_9 + \frac{r^2}{16}\mathcal{A}_{10} + \frac{r^2}{9}\mathcal{A}_{11} + \frac{r}{9}\mathcal{A}_{12} + r\mathcal{A}_{13} + r^2\mathcal{A}_{14}\right)$$

$$\mathcal{B}_{4290} = \frac{11}{252}\left(\mathcal{A}_1 + \frac{1}{72}\mathcal{A}_2 - \frac{5}{99}\mathcal{A}_4 + \frac{1}{88}\mathcal{A}_6 + \frac{r}{72}\mathcal{A}_7 + \frac{r^2}{72}\mathcal{A}_8\right.$$
$$\left. + \frac{r}{88}\mathcal{A}_9 + \frac{r^2}{88}\mathcal{A}_{10} - \frac{5r^2}{99}\mathcal{A}_{11} - \frac{5r}{99}\mathcal{A}_{12} + r\mathcal{A}_{13} + r^2\mathcal{A}_{14}\right)$$

$$\mathcal{B}_{5940} = \frac{11}{182}\left(\mathcal{A}_1 - \frac{7}{864}\mathcal{A}_2 + \frac{1}{66}\mathcal{A}_3 + \frac{1}{27}\mathcal{A}_4 + \frac{1}{66}\mathcal{A}_5 + \frac{1}{352}\mathcal{A}_6 - \frac{7}{864}\mathcal{A}_7 \right.$$

$$\left. - \frac{7}{864}\mathcal{A}_8 + \frac{1}{352}\mathcal{A}_9 + \frac{1}{352}\mathcal{A}_{10} + \frac{1}{27}\mathcal{A}_{11} + \frac{1}{27}\mathcal{A}_{12} + \mathcal{A}_{13} + \mathcal{A}_{14} \right)$$

$$\mathcal{B}_{12012} = \frac{11}{90}\left(\mathcal{A}_1 - \frac{1}{66}\mathcal{A}_3 - \frac{1}{33}\mathcal{A}_4 + \frac{1}{264}\mathcal{A}_5 + \frac{1}{88}\mathcal{A}_6 + \frac{1}{88}\mathcal{A}_9 \right.$$

$$\left. + \frac{1}{88}\mathcal{A}_{10} - \frac{1}{33}\mathcal{A}_{11} - \frac{1}{33}\mathcal{A}_{12} + \mathcal{A}_{13} + \mathcal{A}_{14} \right)$$

$$\mathcal{B}_{14300} = \frac{55}{378}\left(\mathcal{A}_1 + \frac{1}{288}\mathcal{A}_2 + \frac{1}{330}\mathcal{A}_3 + \frac{1}{99}\mathcal{A}_4 - \frac{1}{132}\mathcal{A}_5 - \frac{5}{352}\mathcal{A}_6 + \frac{1}{288}\mathcal{A}_7 \right.$$

$$\left. + \frac{1}{288}\mathcal{A}_8 - \frac{5}{352}\mathcal{A}_9 - \frac{5}{352}\mathcal{A}_{10} + \frac{1}{99}\mathcal{A}_{11} + \frac{1}{99}\mathcal{A}_{12} + \mathcal{A}_{13} + \mathcal{A}_{14} \right)$$

$$\mathcal{B}_{27027} = \frac{11}{40}\left(\mathcal{A}_1 - \frac{1}{432}\mathcal{A}_2 + \frac{1}{297}\mathcal{A}_4 - \frac{1}{176}\mathcal{A}_6 - \frac{r}{432}\mathcal{A}_7 - \frac{r^2}{432}\mathcal{A}_8 \right.$$

$$\left. - \frac{r}{176}\mathcal{A}_9 - \frac{r^2}{176}\mathcal{A}_{10} + \frac{r^2}{297}\mathcal{A}_{11} + \frac{r}{297}\mathcal{A}_{12} + r\mathcal{A}_{13} + r^2\mathcal{A}_{14} \right)$$

$r = \exp(2\pi i/3) = -\frac{1}{2} + i\frac{\sqrt{3}}{2}$ is the basic primitive 3rd root of unity.
Time **C**: 57 min 58 s. Time **Maple**: 7 min 41 s.

References

1. Holt, D.F., Eick, B., O'Brien, E.A.: Handbook of Computational Group Theory. Chapman & Hall/CRC, Boca Raton (2005)
2. Parker, R.: The computer calculation of modular characters (the Meat-Axe). In: Atkinson, M.D. (ed.) Computational Group Theory, pp. 267–274. Academic Press, London (1984)
3. Kornyak, V.V.: Quantum models based on finite groups. J. Phys.: Conf. Ser. **965**, 012023 (2018). http://stacks.iop.org/1742-6596/965/i=1/a=012023
4. Kornyak, V.V.: Modeling quantum behavior in the framework of permutation groups. EPJ Web Conf. **173**, 01007 (2018). https://doi.org/10.1051/epjconf/201817301007
5. Cameron, P.J.: Permutation Groups. Cambridge University Press, Cambridge (1999)
6. Bosma, W., Cannon, J., Playoust, C., Steel, A.: Solving Problems with Magma. University of Sydney. http://magma.maths.usyd.edu.au/magma/pdf/examples.pdf
7. Wilson, R., et al.: Atlas of finite group representations. http://brauer.maths.qmul.ac.uk/Atlas/v3

Factoring Multivariate Polynomials
with Many Factors and Huge Coefficients

Michael Monagan$^{(\boxtimes)}$ and Baris Tuncer

Department of Mathematics, Simon Fraser University,
Burnaby, BC V5A 1S6, Canada
{mmonagan,ytuncer}@sfu.ca

Abstract. The standard approach to factor a multivariate polynomial in $\mathbb{Z}[x_1, x_2, \ldots, x_n]$ is to factor a univariate image in $\mathbb{Z}[x_1]$ then recover the multivariate factors from their images using a process known as multivariate Hensel lifting. For the case when the factors are expected to be sparse, at CASC 2016, we introduced a new approach which uses sparse polynomial interpolation to solve the multivariate polynomial diophantine equations that arise inside Hensel lifting.

In this work we extend our previous work to the case when the number of factors to be computed is more than 2. Secondly, for the case where the integer coefficients of the factors are large we develop an efficient p-adic method. We will argue that the probabilistic sparse interpolation method introduced by us provides new options to speed up the factorization for these two cases. Finally we present some experimental data comparing our new methods with previous methods.

Keywords: Polynomial factorization
Sparse polynomial interpolation · Multivariate Hensel lifting
Polynomial diophantine equations

1 Introduction

Suppose we seek to factor a multivariate polynomial $a \in R = \mathbb{Z}[x_1, \ldots, x_n]$. Today many modern computer algebra systems, such as Maple, Magma and Singular, use Wang's incremental design of multivariate Hensel lifting (MHL) to factor multivariate polynomials over integers. MHL was developed by Yun [15] and improved by Wang [13,14].

To factor $a(x_1, \ldots, x_n)$ the first step is to choose a main variable, say x_1, then compute the content of a in x_1 and remove it from a. If $a = \sum_{i=0}^{d} a_i(x_2, \ldots, x_n) x_1^i$, the content of a is $\gcd(a_0, a_1, \ldots, a_d)$, a polynomial in one fewer variables which is factored recursively. Let us assume this has been done.

The second step identifies any repeated factors in a by doing a *square-free factorization*. See Chap. 8 of [2]. In this step one obtains the factorization $a = b_1 b_2^2 b_3^3 \cdots b_k^k$ such that each factor b_i has no repeated factors and $\gcd(b_i, b_j) = 1$.

© Springer Nature Switzerland AG 2018
V. P. Gerdt et al. (Eds.): CASC 2018, LNCS 11077, pp. 319–334, 2018.
https://doi.org/10.1007/978-3-319-99639-4_22

Let us assume this has also been done. So let $a = f_1 f_2 \dots f_r$ be the irreducible factorization of a over $\mathbb{Z}$. Also, let $\#f$ denote the number of terms of a polynomial f and $\mathrm{Supp}(f)$ denote the support f, i.e., the set of monomials in f.

MHL chooses an evaluation point $\alpha = (\alpha_2, \alpha_3, \dots, \alpha_n) \in \mathbb{Z}^{n-1}$ where the α_i's are preferably small and contain many zeros. Then $a(x_1, \alpha)$ is factored over $\mathbb{Z}$. The evaluation point α must satisfy

 (i) $L(\alpha) \neq 0$ where L is the leading coefficient of a in x_1,
 (ii) $a(x_1, \alpha)$ must have no repeated factors in x_1 and
 (iii) $f_i(x_1, \alpha)$ must be irreducible over $\mathbb{Q}$.

If any condition is not satisfied the algorithm must restart with a new evaluation point. Conditions (i) and (ii) may be imposed in advance of the next step. One way to ensure that condition (iii) is true with high probability is to pick a second evaluation point $\beta = (\beta_2, \dots, \beta_n) \in \mathbb{Z}^{n-1}$, factor $a(x_1, \beta)$ over $\mathbb{Z}$ and check that the two factorizations have the same degree pattern before proceeding.

For simplicity let us assume a is monic and suppose we have obtained the monic factors $f_i(x_1, \alpha)$ in $\mathbb{Z}[x_1]$. Next the algorithm picks a prime p which is big enough to cover the coefficients of a and the factors f_i of a.

The input to MHL is $a, \alpha, f_i(x_1, \alpha)$ and p such that $a(x_1, \alpha) = \prod_{i=1}^{r} f_i(x_1, \alpha)$ where $\gcd(f_i(x_1, \alpha), f_j(x_1, \alpha)) = 1$ in $\mathbb{Z}_p[x_1]$ for $i \neq j$. If the gcd condition is not satisfied, the algorithm chooses a new prime p until it is.

There are two main subroutines in the design of MHL. For details see Chap. 6 of [2]. The first one is the leading coefficient correction algorithm (LCC). The most well-known is the Wang's heuristic LCC [14] which works well in practice and is the one Maple currently uses. There are other approaches by Kaltofen [6] and most recently by Lee [9]. In our implementation we use Wang's LCC.

In a typical application of Wang's LCC, one first factors the leading coefficient of a, a polynomial in $\mathbb{Z}[x_2, \dots, x_n]$, by a recursive call and then one applies LCC before the jth step of MHL. Then the total cost of the factorization is given by the cost of LCC + the cost of factoring $a(x_1, \alpha)$ over $\mathbb{Z}$ + the cost of MHL. One can easily construct examples where LCC or factoring $a(x_1, \alpha)$ dominates the cost. However this is not typical. Usually MHL dominates the cost.

The second main subroutine solves a multivariate polynomial diophantine problem (MDP). In MHL, for each j with $2 \leq j \leq n$, Wang's design of MHL must solve many instances of the MDP in $\mathbb{Z}_p[x_1, \dots, x_{j-1}]$. Wang's method for solving an MDP (see Algorithm 2) is recursive. Although Wang's method performs significantly better than the previous algorithm that he developed with Rothschild in [14], it does not explicitly take sparsity into account. During computation, the ideal-adic representation of factors is dense when the evaluation points $\alpha_2, \dots, \alpha_n$ are non-zero. In practice, conditions (i) and (iii) of LCC may force many non-zero α_j's. This makes Wang's approach exponential in n.

Zippel's sparse interpolation [18] was the first probabilistic method aimed at taking sparsity into account. Based on sparse interpolation and multivariate Newton's iteration, Zippel then introduced a sparse Hensel lifting (ZSHL) algorithm in [17,19], which uses a MHL organization different from Wang's.

Another approach for sparse Hensel lifting for the sparse case was proposed by Kaltofen (KSHL) in [6]. Kaltofen's method is also based on Wang's incremental design of MHL but it uses a LCC different from Wang's LCC and offers a distinct solution to the multivariate diophantine problem (MDP) that appears in Wang's design of MHL.

At CASC 2016 the authors proposed a new practical sparse Hensel Lifting algorithm (MTSHL) [11]. It is also based on Wang's incremental design of MHL and LCC but offers a solution to the MDP different from those of Zippel and Kaltofen. To solve the MDP problem appearing in MHL, MTSHL exploits the fact that at each step of MHL, the solutions to MDP's, which are just Taylor polynomial coefficients, are structurally related. At the jth step of MHL we are recovering x_j in the factors. Let f be one such factor in $\mathbb{Z}_p[x_1, x_2, \ldots, x_j]$ and let $f = \sum_{k=0}^{l} f_k(x_j - \alpha_j)^k$ be its Taylor representation. At this point we know only f_0. But $\operatorname{Supp}(f_k) \subseteq \operatorname{Supp}(f_{k-1})$ with high probability if α_j is chosen randomly from $[0, p-1]$ and p is sufficiently large. MTSHL is built on this key observation.

In this paper we consider the case where a has $r > 2$ factors and secondly the case where the factors have large integer coefficients. When $r > 2$, the MDP problem is called a multiterm MDP problem and an approach to the solution to this problem is described in [2]. It reduces the multiterm MDP problem to $r - 1$ two term MDP problems. Our previous implementation of MTSHL described in [11] also used this approach.

In Sect. 2 we define the MDP problem in the context of MHL. See Algorithms 1 and 2. In Sect. 3 we discuss main ideas for the solution to the MDP used by MTSHL and present it as Algorithm 3 to make our explanation precise. We call Algorithm 3 MTSHL-d (d stands for direct), since it differs from our previous version of MTSHL (Algorithm 4 in [11]) in how it solves MDP problems when $r > 2$. For $r = 2$ it is the same as Algorithm 4 in [11].

In Sect. 4 we discuss the case $r > 2$. We argue that the probabilistic sparse interpolation method used in the design of MTSHL allows us to reduce the time spent solving multiterm MDP's by up to a factor of $r - 1$. Because our proposal also reduces the multiplication cost in the previous approach described in [2], the observed speedup is sometimes greater than $r - 1$.

In Sect. 5, we study the case where the integer coefficients of the factors are large. The current approach (see [2]) chooses a prime p and $l > 0$ such that p^l bounds any coefficients in the factors f_i of a. We show that the sparse MDP solver developed in [11] renders an improved option. Suppose one factor $f \in \mathbb{Z}[x_1, \ldots, x_n]$ has a p-adic representation $f = \sum_{k=0}^{l} f_k p^k$. We show that in this case also $\operatorname{Supp}(f_k) \subseteq \operatorname{Supp}(f_{k-1})$ with high probability if p is chosen randomly. Therefore we propose first to factor a in $\mathbb{Z}_p[x_1, \ldots, x_n]$ by doing all arithmetic mod p where p is a machine prime (e.g. 63 bits on a 64 bit computer), i.e. run the entire Hensel lifting modulo a machine prime. Then lift the solution to $\mathbb{Z}_{p^l}[x_1, \ldots, x_n]$ by computing f_k, again by solving each MDP appearing in the lifting process using the sparse interpolation developed in the design of MTSHL. Using this approach most of the computation is modulo p a machine prime.

In Sect. 6 we present some timing data to compare our new approaches with previous approaches and end with some concluding remarks.

In the paper we assume the input polynomial a is monic in x_1 so as not to complicate the presentation with LCC. We note that what we explain remains true for the non-monic case with slight modifications. Our implementation uses Wang's LCC for the non-monic case.

2   The Multivariate Diophantine Problem (MDP)

The Multivariate Diophantine Problem (MDP) arises naturally as a subproblem of the incremental design of MHL developed by Wang. For completeness we provide the jth step of MHL as Algorithm 1 for the monic case and Wang's solution to the MDP as Algorithm 2.

Algorithm 1. j^{th} step of Multivariate Hensel Lifting for $j > 1$.

Input : $\alpha_j \in \mathbb{Z}_p,\, a_j \in \mathbb{Z}_p[x_1,\dots,x_j],\, f_{j-1,1},\dots,f_{j-1,r} \in \mathbb{Z}_p[x_1,\dots,x_{j-1}]$ where $a_j, f_{j-1,i}$ are monic in x_1 and $a_j(x_j = \alpha_j) = \prod_{i=1}^{r} f_{j-1,i}$.

Output : $f_{j,1},\dots,f_{j,r} \in \mathbb{Z}_p[x_1,\dots,x_j]$ such that $f_{j,i}(x_j = \alpha_j) = f_{j-1,i}$ and $a_j = \prod_{i=1}^{r} f_{j,i}$ or FAIL.

1: **for** i from 1 to r **do**
2:     $\sigma_{0,i} \leftarrow f_{j-1,i};\; f_{j,i} \leftarrow \sigma_{0,i};\; b_{j,i} \leftarrow \prod_{k=1,k\neq i}^{r} f_{j-1,k}.$
3: **end for**
4: $error \leftarrow a_j - \prod_{i=1}^{r} f_{j,i}.$
5: **for** k from 1 while $error \neq 0$ and $\sum_{i=1}^{r} \deg_{x_j} f_{j,i} < \deg_{x_j} a_j$ **do**
6:     $c_k \leftarrow$ Taylor coefficient of $(x_j - \alpha_j)^k$ of $error$ at $x_j = \alpha_j$
7:     **if** $c_k \neq 0$ **then**
8:         Solve MDP$_{j,k}$: $\sigma_{k,1}b_{j,1} + \dots + \sigma_{k,r}b_{j,r} = c_k$ for $\sigma_{k,i} \in \mathbb{Z}_p[x_1,\dots,x_{j-1}]$.
9:         **for** i from 1 to r **do**
10:             $f_{j,i} \leftarrow f_{j,i} + \sigma_{k,i} \times (x_j - \alpha_j)^k$
11:         **end for**
12:         $error \leftarrow a_j - \prod_{i=1}^{r} f_{j,i}.$
13:     **end if**
14: **end for**
15: **if** $error = 0$ **then** return $f_{j,1},\dots,f_{j,r}$ **else** return FAIL **end if**

The MDP appears at line 8 of Algorithm 1. Consider the case where the number of factors r to be computed is 2, i.e., $r = 2$. We discuss the case $r > 2$ in Sect. 4.

Let $u, w, c \in \mathbb{Z}_p[x_1,\dots,x_j]$ with u and w monic with respect to the variable x_1 and let $I_j = \langle x_2 - \alpha_2,\dots,x_j - \alpha_j \rangle$ be an ideal of $\mathbb{Z}_p[x_1,\dots,x_j]$ with $\alpha_i \in \mathbb{Z}$. The MDP is to find multivariate polynomials $\sigma, \tau \in \mathbb{Z}_p[x_1,\dots,x_j]$ that satisfy

$$\sigma u + \tau w = c \mod I_j^{d_j+1} \tag{1}$$

with $\deg_{x_1}(\sigma) < \deg_{x_1}(w)$ where d_j is the maximal degree of σ and τ with respect to the variables $x_2, \ldots, x_j$ and it is given that

$$\mathrm{GCD}\,(u \bmod I_j, w \bmod I_j) = 1 \text{ in } \mathbb{Z}_p[x_1].$$

It can be shown that the solution (σ, τ) exists and is unique and independent of the choice of the ideal I_j. For $j = 1$ the MDP is in $\mathbb{Z}_p[x_1]$ and can be solved with the extended Euclidean algorithm (see Chap. 2 of [2]).

To solve the MDP for $j > 1$, Wang uses the same approach as for Hensel Lifting, that is, an ideal-adic lifting approach. See Algorithm 2.

Algorithm 2. WMDS (Wang's multivariate diophantine solver)

Input A point $\alpha_j \in \mathbb{Z}_p$, polynomials c, $f_{j,k} \in \mathbb{Z}_p[x_1, \ldots, x_j]$ for $k = 1, \ldots, r$ and an ideal $I = \langle x_2 - \alpha_2, \ldots, x_n - \alpha_n \rangle$ with $n \geq j$ where $\gcd(f_{j,k} \bmod I, f_{j,l} \bmod I) = 1$ in $\mathbb{Z}_p[x_1]$ for $k \neq l$ and a degree bound d_j satisfying $d_j \geq \max(\deg_{x_j} \sigma_k)$ for $2 \leq i \leq n$. (One may use $d_j = \deg_{x_j} a$)

Output $(\sigma_1, \ldots, \sigma_r) \in \mathbb{Z}_p[x_1, \ldots, x_j]$ satisfying $\sum_{k=1}^r \sigma_k b_k = c \in \mathbb{Z}_p[x_1, \ldots, x_j]$ where $b_k = \prod_{i \neq k}^r f_{j,i}$ and $\deg_{x_1} \sigma_k < \deg_{x_1} f_{j,k}$ or FAIL if no such solution exists.

1: $b_k \leftarrow \prod_{i \neq k}^r f_{j,i}$ for $k = 1, \ldots, r$
2: **if** $j = 1$ **then** use extended Euclidean algorithm (see Ch 2 of [2] for $r = 2$ and Section 4 for $r > 2$) **end if**
3: $(\sigma_{1,0}, \ldots, \sigma_{r,0}) \leftarrow$ WMDS$(f_{j,k}(x_j = \alpha_j), c(x_j = \alpha_j), I)$
4: **if** WMDS output FAIL **then** return FAIL **end if**
5: $\sigma_k \leftarrow \sigma_{k,0}$ for $k = 1, \ldots, r$; $error \leftarrow c - \sum_{k=1}^r \sigma_k b_k$
6: **for** $i = 1, 2, \ldots, d_j$ while $error \neq 0$ **do**
7: $\quad c_i \leftarrow$ Taylor coeff$(error, (x_j - \alpha_j)^i)$
8: $\quad$ **if** $c_i \neq 0$ **then**
9: $\quad\quad (s_1, \ldots, s_r) \leftarrow$ WMDS(σ_k, c_i, I)
10: $\quad\quad$ **if** WMDS output FAIL **then** return FAIL **end if**
11: $\quad\quad \sigma_k \leftarrow \sigma_k + s_k \times (x_j - \alpha_j)^i$ for $k = 1, \ldots, r$.
12: $\quad\quad error \leftarrow error - \sum_{k=1}^r \sigma_k b_k$
13: $\quad$ **end if**
14: **end for**
15: **if** $error = 0$ **then** return $(\sigma_1, \ldots, \sigma_r)$ **else** return FAIL **end if**

In general, if $\alpha_j \neq 0$ the Taylor series expansion of σ and τ about $x_j = \alpha_j$ is dense in x_j so the $c_i \neq 0$. Then the number of calls to the Euclidean algorithm of Wang's solution to MDP is exponential in n. It is this exponential behaviour that the design of MTSHL eliminates. On the other hand, if MHL can choose some α_j to be 0, for example, if the input polynomial $a(x_1, \ldots, x_n)$ is monic in x_1 then this exponential behaviour may not occur for sparse f and g.

3   MTSHL's Solution to the MDP via Sparse Interpolation

We consider whether we can interpolate $x_2, \ldots, x_j$ in σ and τ in (1) using sparse interpolation methods. If $\beta \in \mathbb{Z}_p$ with $\beta \neq \alpha_j$, then

$$\sigma(x_j = \beta)u(x_j = \beta) + \tau(x_j = \beta)w(x_j = \beta) = c(x_j = \beta) \bmod I_{j-1}^{d_{j-1}+1}.$$

For $K_j = \langle x_2 - \alpha_2, \ldots, x_{j-1} - \alpha_{j-1}, x_j - \beta \rangle$ and $G_j = \mathrm{GCD}(u \bmod K_j, w \bmod K_j)$, we obtain a unique solution $\sigma(x_j = \beta)$ iff $G_j = 1$. However $G_j \neq 1$ is possible. Let $R = \mathrm{res}_{x_1}(u, w)$ be the Sylvester resultant of u and w taken in x_1. Since u, w are monic in x_1 one has[1]

$$G_j \neq 1 \iff \mathrm{res}_{x_1}(u \bmod K_j, w \bmod K_j) = 0 \iff R(\alpha_2, \ldots, \alpha_{j-1}, \beta) = 0.$$

Let $r = R(\alpha_2, \ldots, \alpha_{j-1}, x_j) \in \mathbb{Z}_p[x_j]$ so that $R(\alpha_2, \ldots, \alpha_{j-1}, \beta) = r(\beta)$. Also $\deg(R) \leq \deg(u)\deg(w)$ [1]. Now if β is chosen at random from $\mathbb{Z}_p$ and $\beta \neq \alpha_j$ then

$$\Pr[G_j \neq 1] = \Pr[r(\beta) = 0] \leq \frac{\deg(r, x_j)}{p-1} \leq \frac{\deg(u)\deg(w)}{p-1}.$$

This bound for $\Pr[G_j \neq 1]$ is a worst case bound. In [10] we show that the average probability for $\Pr[G_j \neq 1] = 1/(p-1)$. Thus if p is large, the probability that $G_j = 1$ is high. Interpolation is thus an option to solve the MDP.

As can be seen from line 10 of Algorithm 1, the solutions to the MDP are the Taylor coefficients of the factors to be computed at the jth step. As such, if $\sigma_{0,i}$ is sparse then the $\sigma_{k,i}$ are also sparse. In line 5 of Algorithm 1, as k increases, on average, the number of terms of the $\sigma_{k,i}$ decrease even for dense cases. That is, on average $\#\sigma_{k,i} < \#\sigma_{k-1,i}$. A natural idea then is to use sparse interpolation techniques to solve the MDP. However, the sparse technique proposed by Zippel [16] is also iterative; it recovers x_2 then x_3 etc. To make one more step in this direction consider the following Lemma whose proof can be found in [11].

Lemma 1. *Let $f \in \mathbb{Z}_p[x_1, \ldots, x_n]$ and let α be a randomly chosen element in $\mathbb{Z}_p$ and $f = \sum_{i=0}^{d_n} b_i(x_1, \ldots, x_{n-1})(x_n - \alpha)^i$ where $d_n = \deg_{x_n} f$. Then*

$$\Pr[\mathrm{Supp}(b_{j+1}) \not\subseteq \mathrm{Supp}(b_j)] \leq |\mathrm{Supp}(b_{j+1})| \frac{d_n - j}{p - d_n + j + 1} \text{ for } 0 \leq j < d_n.$$

Lemma 1 says that for the sparse case, if p is big enough then the probability of $\mathrm{Supp}(b_{j+1}) \subseteq \mathrm{Supp}(b_j)$ is high. This observation suggests, during MHL we use $\sigma_{k-1,i}$ as a form of the solution of $\sigma_{k,i}$. That is, the solutions to the MDP's are related. During MHL, these problems shouldn't be treated independently as previous approaches do. In light of the key role this assumption plays at

---

[1] This argument also works for the non-monic case if the leading coefficients of u and w w.r.t. x_1 do not vanish at $(\alpha_2, \ldots, \alpha_n)$ modulo p, conditions which we note are imposed by Wang's LCC.

each MHL step $j > 1$, for each factor f_i, we call this assumption $\mathrm{Supp}(\sigma_{k,i}) \subseteq \mathrm{Supp}(\sigma_{k-1,i})$ for all $k > 0$ the **strong SHL assumption**.

Algorithms 3 and 4 below show how this assumption can be combined with the sparse interpolation idea of Zippel [16] to reduce the solution to the MDP problem to solving linear systems over $\mathbb{Z}_p$. To see how MTSHL works on a concrete example for $r = 2$ and how MTSHL decreases the evaluation cost that sparse interpolation brings see [11].

We present the jth step of the new version of MTSHL in Algorithm 4 below and call it as MTSHL-d, as a shortcut for MTSHL direct. For $r = 2$ MTSHL-d is equivalent to MTSHL described in [11]. In the following section we discuss the case $r > 2$ and make it clear why we call it MTSHL direct.

4   The Multiterm Diophantine Problem

Let the input polynomial $a(x_1, \ldots, x_n)$ be square-free with total degree d and irreducible factorization of a be

$$a = f_1 \cdots f_r \in \mathbb{Z}[x_1, \ldots, x_n].$$

We consider the case $r > 2$. We start with the unique factorization of $a_1(x_1) = a(x_1, \alpha) = u_1(x_1) \cdots u_r(x_1) \in \mathbb{Z}[x_1]$. By Hilbert's irreducibility theorem [7] most probably $u_i(x_1) = f_i(x_1, \alpha)$. Next we choose a prime p which is big enough to cover the coefficients occurring in each f_i and then pass to mod p

$$a(x_1, \alpha) = u_1(x_1) \cdots u_r(x_1) \in \mathbb{Z}_p[x_1].$$

We need $\gcd(u_i, u_j) = 1 \in \mathbb{Z}_p[x_1]$ for all $1 \leq i < j \leq r$. Otherwise we choose a different prime and repeat the process.

Suppose $f_i = \sum_{k=0} \sigma_{i,k}(x_j - \alpha_j)^k$. So $\sigma_{i,k}$ is the kth Taylor coefficient of the ith factor to be computed in the jth step of MHL. (See line 10 of Algorithm 1.) During the jth step of MHL, for each iteration $k > 0$, the algorithm computes $\sigma_{k,i}$, by solving the multiterm Diophantine problem (multi-MDP), which is a natural generalization of the MDP defined in Sect. 2 and denoted as $\mathrm{MDP}_{j,k}$ in line 8 of Algorithm 1. It has the form

$$\mathrm{MDP}_{j,k} : \ \sigma_{k,1} b_1 + \cdots + \sigma_{k,r} b_r = c_k,$$

where $b_k = \prod_{i=1, i \neq k}^{r} f_{j-1,i}(x_1, \ldots, x_{j-1})$. So, given b_k and c_k in $\mathbb{Z}_p[x_1, \ldots, x_{j-1}]$, the goal is to find $\sigma_{k,i}$ for each i.

The current approach to solve a multiterm MDP is to reduce it into $r - 1$ two term MDP's. We describe the idea with an example. Let $r = 4$ and to save some space let $u_i = f_{j-1,i}$. Then

$$\begin{aligned}
c_k &= \sigma_{k,1} b_1 + \sigma_{k,2} b_2 + \sigma_{k,3} b_3 + \sigma_{k,4} b_4 \\
&= \sigma_{k,1} u_2 u_3 u_4 + \sigma_{k,2} u_1 u_3 u_4 + \sigma_{k,3} u_1 u_2 u_4 + \sigma_{k,4} u_1 u_2 u_3 \\
&= \sigma_{k,1} u_2 u_3 u_4 + u_1(\sigma_{k,2} u_3 u_4 + u_2(\sigma_{k,3} u_4 + \sigma_{k,4} u_3)).
\end{aligned}$$

Algorithm 3. SparseInt: solve an MDP using a sparse interpolation

Input: Polynomials $f_i, \sigma_i, c \in \mathbb{Z}_p[x_1, x_2, \ldots, x_{j-1}]$ for $i = 1, \ldots, r$. f_i are monic in x_1 and p a prime.

Output: Update σ_i so that they form a solution to the multi-MDP $\sigma_1 b_1 + \cdots + \sigma_r b_r = c$ in $\mathbb{Z}_p[x_1, x_2, \ldots, x_{j-1}]$ where $b_i = \prod_{k=1, k \neq i}^r f_i$ or FAIL.

1: **for** i from 1 to r **do**
2:     $\sigma_i \leftarrow \sum_{l,k} c_{ilk}(x_3, \ldots, x_{j-1}) x_1^l x_2^k$ where $c_{ilk} = \sum_{w=1}^{s_{ilk}} c_{ilkw} M_{ilkw}$ with c_{ilkw} are unknown coefficients to be solved for and $x_1^l x_2^k M_{ilkw}$ are the monomials in Supp(σ_i).
3: **end for**
4: Let $t = \max_{i=1}^r \{\max s_{ilk} = \max \# c_{ilkw}\}$
5: Pick $(\beta_3, \ldots \beta_{j-1}) \in (\mathbb{Z}_p \backslash \{0\})^{j-3}$ at random.
6: **for** s from 1 to t **do** (*Precomputation.(see [11]*))
7:     Let $Y_s = (x_3 = \beta_3^s, \ldots, x_{j-1} = \beta_{j-1}^s)$.
8:     Evaluate $c(x_1, x_2, Y_s)$ and $f_i(x_1, x_2, Y_s)$ for $1 \leq i \leq r$.
9: **end for**
10: **for** i from 1 to r **do**
11:     Compute $b_i(x_1, x_2, Y_i) = \prod_{k=1, k \neq i}^r f_i(x_1, x_2, Y_i)$ in $\mathbb{Z}_p[x_1, x_2]$.b
12: **end for**
13: **for** i from 1 to r **do**
14:     Compute monomial evaluation sets for σ_i

$$\{S_{ilk} = \{m_{ilkw} = M_{ilkw}(\beta_3, \ldots, \beta_{j-1}) : 1 \leq w \leq s_{ilk}\} \text{ for each } l, k\}.$$

15:     If $|S_{ikl}| \neq s_{ikl}$ for some ikl try a different choice for $(\beta_3, \ldots, \beta_{j-1})$.
16:     If this fails, **return** FAIL. (*p is not big enough*)
17:     Let $t_i = \max_{l,k} s_{ilk}$
18:     **for** s from 1 to t_i **do** (*Compute the bivariate images of σ_i*)
19:         Solve $\tilde{\sigma}_1(x_1, x_2) f_1(x_1, x_2, Y_i) + \cdots + \tilde{\sigma}_r(x_1, x_2) f_r(x_1, x_2, Y_i) = c(x_1, x_2, Y_i)$ in $\mathbb{Z}_p[x_1, x_2]$ for $\tilde{\sigma}_i(x_1, x_2)$ using multi-BDP (see section 4).
20:         **if** multi-BDP returns FAIL **then return** FAIL **end if**
        (multi-BDP fails if it choses $\gamma \in \mathbb{Z}_p$ with $\gcd(f_i(x_1, \gamma, Y_i), f_j(x_1, \gamma, Y_i)) \neq 1$ for some $i \neq j$).
21:     **end for**
22:     **for** each l, k **do**
23:         Construct and solve the $s_{ilk} \times s_{ilk}$ linear system

$$\left\{ \sum_{w=1}^{s_{ilk}} c_{ilkw} m_{ilkw}^n = \text{coefficient of } x_1^l x_2^k \text{ in } \tilde{\sigma}_i(x_1, x_2) \text{ for } 1 \leq n \leq s_{ilk} \right\}$$

        for the coefficients c_{ilkw} of $c_{ilk}(x_3, \ldots, x_{j-1})$. Because it is a Vandermonde system in m_{iklw} which are distinct by Step 15 it has a unique solution.
24:     **end for**
25:     Substitute the solutions for c_{ilkw} into σ_i
26: **end for**
27: Verify probabilistically whether $\sum_{i=1}^r \sigma_i b_i = c$:
        Pick $\beta = (\beta_1, \ldots \beta_{j-1}) \in \mathbb{Z}_p^{j-1}$ at random.
        if $\sum_{i=1}^r \sigma_i(\beta) b_i(\beta) \neq c(\beta)$ **then return** FAIL **end if**
28: **return** $\sigma_1, \ldots, \sigma_r$

Algorithm 4. j^{th} step of MTSHL-d for $j > 1$.

Input : $\alpha_j \in \mathbb{Z}_p$, $a_j \in \mathbb{Z}_p[x_1,\ldots,x_j]$, $f_{j-1,1},\ldots,f_{j-1,r} \in \mathbb{Z}_p[x_1,\ldots,x_{j-1}]$ where $a_j, f_{j-1,i}$ are monic in x_1 and $a_j(x_j = \alpha_j) = \prod_{i=1}^{r} f_{j-1,i}$.
Output : $f_{j,1},\ldots,f_{j,r} \in \mathbb{Z}_p[x_1,\ldots,x_j]$ such that $f_{j,i}(x_j = \alpha_j) = f_{j-1,i}$ and $a_j = \prod_{i=1}^{r} f_{j,i}$ or FAIL.

1: **for** i from 1 to r **do** $f_{j,i} \leftarrow f_{j-1,i}$, $\sigma_{0,i} \leftarrow f_{j-1,i}$ **end do**
2: $error \leftarrow a_j - \prod_{i=1}^{r} f_{j,i}$
3: **for** $k = 1, 2, 3, \ldots$ **while** $error \neq 0$ and $\sum_{i=1}^{r} \deg(f_{j,i}, x_j) < \deg(a_j, x_j)$ **do**
4: $\quad$ $c_k \leftarrow$ Taylor coefficient of $(x_j - \alpha_j)^k$ of $error$ at $x_j = \alpha_j$
5: $\quad$ **if** $c_k \neq 0$ **then**
6: $\qquad$ Solve the MDP$_{j,k}$ (see line 8 of Alg. 1) without computing $b_{j,i}$ as follows:
7: $\qquad$ **for** i from 1 to r **do** $\sigma_{k,i} \leftarrow \sigma_{k-1,i}$ **end do**   (Strong SHL assumption.)
8: $\qquad$ $(\sigma_{k,1},\ldots,\sigma_{k,r}) \leftarrow$ **SparseInt**$(f_{j-1,i}, c_k, \sigma_{k,i}, i = 1,\ldots,r)$ (see Alg. 3)
9: $\qquad$ **if** $(\sigma_{k,1},\ldots,\sigma_{k,r})$=FAIL **then** restart MTSHL-d with a new α **end if**
10: $\qquad$ **for** i from 1 to r **do** $f_{j,i} \leftarrow f_{j,i} + \sigma_{k,i} \times (x_j - \alpha_j)^k$ **end do**
11: $\qquad$ $error \leftarrow a_j - \prod_{i=1}^{r} f_{j,i}$
12: $\quad$ **end if**
13: **end for**
14: **if** $error = 0$ **then** return $f_{j,1},\ldots,f_{j,r}$ **else** return FAIL **end if**

We first solve the MDP $\sigma_{k,1}u_2u_3u_4 + u_1w_1 = c_k$ for $\sigma_{k,1}$ and w_1. Then we solve $\sigma_{k,2}u_3u_4 + u_2w_2 = w_1$ for $\sigma_{k,2}$ and w_2. Finally we solve $\sigma_{k,3}u_4 + \sigma_{k,4}u_3 = w_2$ to compute $\sigma_{k,3}$ and $\sigma_{k,4}$. Let us call this approach as the iterative approach to solve the multiterm MDP.

Note that Wang's approach to solve the MDP is recursive. So when $r > 2$, the iterative approach to solve multiterm MDP makes Wang's design highly recursive. Also, if the polynomials u_i have many terms then the b_i's will be large and expensive to compute. If we use the probabilistic sparse MDP solver of MTSHL as described in [11] for each of these MDP's, then we will first compute the b_i's and then evaluate b_i's at random points. But evaluation is one of the most costly operations in sparse interpolation and this cost increases as the size of the polynomial to be evaluated increases.

However, the probabilistic non-recursive sparse interpolation idea used to solve the MDP's in MHL renders another simple and efficient option. One can invoke the sparse MDP solver to compute the $\sigma_{k,i}$'s simultaneously without reducing MDP$_{j,k}$ to $r - 1$ two term MDP's in the following way.

According to Lemma 1, if α_j is random and p is big, then for each factor $f_{j,i}$, with probability $\geq 1 - |\text{Supp}(\sigma_{k,i})|\frac{d_i - i}{p - d_i + j + 1}$ one has $\text{Supp}(\sigma_{k,i}) \subseteq \text{Supp}(\sigma_{k-1,i})$ for $k = 1,..,d_i$ where $\sigma_{0,i}$ is defined as $\sigma_{0,i} := f_{j-1,i}$ and $d_i = \deg_{x_j}(f_{j,i})$. Therefore to solve MDP$_{j,k}$ we use $\text{Supp}(\sigma_{k-1,i})$ as a skeleton of the solution of $\sigma_{k,i}$. That is, if $\sigma_{k-1,i} = \sum_{l,k} m_{ilk}M_{ilk}$ for $m_{ilk} \in \mathbb{Z}_p - \{0\}$ with distinct monomials in $M_{ilk} \in \mathbb{Z}_p[x_1,\ldots,x_{j-1}]$, then we construct $\bar{\sigma}_{k,i} = \sum_{l,k} c_{ilk}M_{ilk}$ as a solution form (skeleton) of $\sigma_{k,i}$, where c_{ilk} are to be computed.

At the kth iteration suppose that we need t_i evaluations to recover the coefficients c_{ilk} (see line 17 of Algorithm 3). Let $\beta = (\beta_2, \ldots \beta_{j-1})$ where $\beta_i \in \mathbb{Z}_p - \{0\}$ be a random evaluation point. Consider the t_i consecutive univariate multiterm MDP's

$$\tilde{\sigma}_{k,1} b_1(x_1, \beta^s) + \cdots + \tilde{\sigma}_{k,r} b_r(x_1, \beta^s) = c_i(x_1, \beta^s) \text{ for } 1 \le s \le t_i, \qquad (2)$$

where the $\tilde{\sigma}_{k,i}$ are to be computed. By uniqueness of the solutions to the multiterm MDP, with average probability $\binom{r}{2} \frac{1}{p}$ one has $\tilde{\sigma}_{k,i} = \sigma_{k,i}(x_1, \beta^s)$.

Equation 2 can be solved efficiently for $\tilde{\sigma}_{k,i}$ using the iterative approach in the univariate domain $\mathbb{Z}_p[x_1]$. Next the univariate images $\bar{\sigma}_{k,i}(x_1, \beta^s)$ of $\bar{\sigma}_{k,i}$ are used to compute the coefficient c_{ilk} of $\bar{\sigma}_{k,i}$ by solving Vandermonde systems which are constructed by equating the coefficients of $\sigma_{k,i}(x_1, \beta^j)$ and $\tilde{\sigma}_{k,i}$ (see line 23 of Algorithm 3). Again, if the strong SHL assumption is true, then by following Zippel's analysis in [16], one can show that with probability $\ge 1 - \frac{(\#f_i)^2}{2(p-1)}$, we have a unique solution to Vandermonde systems.

At this stage we have candidate solutions $\bar{\sigma}_{k,i}$ for the actual solutions $\sigma_{k,i}$ of $\text{MDP}_{j,k}$. Because our assumption $\text{Supp}(\sigma_{k,i}) \subseteq \text{Supp}(\sigma_{k-1,i})$ may be false, we need to verify if $\bar{\sigma}_{k,i} = \sigma_{k,i}$. We do this using a random evaluation in line 27 of Algorithm 3.

What does this approach bring us? First, MTSHL-d essentially follows MTSHL but eliminates an iteration at the cost of an increase in the probability of failure. However this probability is negligible if p is big enough. In our implementation we used a 31 bit prime and MTSHL-d never failed. Since it is an iteration on r, we expect MTSHL-d to solve multi-MDP's faster than MTSHL by a factor of $\mathcal{O}(r)$. This is verified by the experimental data in Table 1 of Sect. 6.

Second, $b_k(x_1, \beta^s) = \prod_{i=1, i \ne k}^{r} f_i(x_1, \beta^s)$, so we don't need to compute $b_k \in \mathbb{Z}_p[x_1, \ldots, x_{j-1}]$. All we need to do is to compute and multiply their univariate images $f_i(x_1, \beta^s)$ of f_i to obtain $b_k(x_1, \beta^j)$.

Finally in MTSHL-d, like MTSHL, we may evaluate down to $\mathbb{Z}[x_1, x_2]$ instead of $\mathbb{Z}[x_1]$ to decrease the number of evaluations t_i needed and the size of the Vandermonde systems (Line 17 in Algorithm 3). To do this MTSHL-d uses multi-Bivariate Diophant Solver (multi-BDP). We implemented Multi-BDP in C. It solves the bivariate multi-MDP by the iterative approach and uses evaluation and interpolation on x_2 to reduce to the univariate case.

5 The Case Modulo p^l with $l > 1$

When the integer coefficients of a or the factors of a to be computed are huge the current strategy implemented by most of the computer algebra platforms, including Maple, Singular [9] and Magma [12], is the following. For details see [2]. First we pick a prime p and a natural number $l > 0$ such that the ring $\mathbb{Z}_{p^l}$ can be identified with the ring $\mathbb{Z}$. That is, we find a bound B such that the integer coefficients of the polynomial a to be factored and its irreducible factors are bounded by B. One way to choose such an upper bound B is given by [4]. Then

---

Algorithm 5. LiftTheFactors for $r = 2$ (optimized)

Input : $a \in \mathbb{Z}[x_1, \ldots, x_n]$, $f_0, g_0 \in \mathbb{Z}_p[x_1, \ldots, x_n]$ where a, f_0, g_0 are monic in x_1
and $a = f_0 g_0$ in $\mathbb{Z}_p[x_1, \ldots, x_n]$. Also an integer bound $l > 0$ (For example, [Lemma 14, [4]]).
Output : $f, g \in \mathbb{Z}[x_1, \ldots, x_j]$ such that $a = fg \in \mathbb{Z}[x_1, \ldots, x_n]$ or FAIL
 1: $(f, g) \leftarrow (\text{mods}(u_0, p), \text{mods}(w_0, p))$. (*# use symmetric range*)
 2: $modulus \leftarrow 1$.
 3: $error \leftarrow (a - fg)/p$, $(\sigma_f, \sigma_g) \leftarrow (f, g)$
 4: **for** i from 1 to l **while** $error \neq 0$ **do**
 5:     $modulus \leftarrow modulus \times p$, $c \leftarrow error \mod p$
 6:     *# Solve the MDP $\sigma u_0 + \tau w_0 = c$ for σ and τ in $\mathbb{Z}_p[x_1, \ldots, x_n]$:*
 7:     $(\sigma, \tau) \leftarrow \textbf{SparseInt}(f, g, \sigma_f, \sigma_g, c)$ (Algorithm 3)
 8:     **if SparseInt output FAIL then return FAIL end if**
 9:     $(\sigma, \tau) \leftarrow (\text{mods}(\sigma, p), \text{mods}(\tau, p))$. (*# use symmetric range*)
10:     $(\sigma_f, \sigma_g) \leftarrow (\sigma, \tau)$, $error \leftarrow (error - (f\tau + g\sigma) + \sigma\tau \times modulus)/p$
11:     $(f, g) \leftarrow (f + \sigma \times modulus, g + \tau \times modulus)$.
12: **end for**
13: **if** $error \neq 0$ **then return** FAIL **else return** (f, g) **end if**

---

we choose l such that $p^l > 2B$. Next the MDP solution in $\mathbb{Z}_p[x_1]$ is lifted to
the solution in $\mathbb{Z}_{p^l}[x_1]$. The second step is to lift the solution from $\mathbb{Z}_{p^l}[x_1]$ to
$\mathbb{Z}_{p^l}[x_1, \ldots, x_n]$. Note that in the second step all arithmetic is in $\mathbb{Z}_{p^l}$ with $p^l > 2B$.
In this section we question whether this strategy is the best approach for the
case $l > 1$.

Suppose for example that the coefficients of the factors are bounded by p^{10}.
Before the factorization we don't have this information. Since most likely the
coefficient bound $B > p^{20}$, this means that throughout MHL all integer arithmetic is modulo p^{20} which is expensive.

MTSHL's sparse multivariate diophantine solver allows us to propose an
approach that eliminates most of the multi-precision arithmetic and allows us
to lift up to the size of the actual coefficients in the factors, thus avoiding B.

- First choose a random $(m + 1)$-bit machine prime p, i.e. $p \in [2^m < p < 2^{m+1}]$
  and compute the factorization of a by lifting the factorization in $\mathbb{Z}_p[x_1]$ to in
  $\mathbb{Z}_p[x_1, \ldots, x_n]$ with MTSHL-d. Most of this work is mod p.
- Next compute a lifting bound B. One may use Lemma 14 of [4] for this
  purpose. Now pick the smallest l such that $p^l > 2B$.
- Then as a second stage do a p-adic lift of the factorization from $\mathbb{Z}_p[x_1, \ldots, x_n]$
  stopping when f and g are recovered or we exceed p^l. The p-adic lift is presented as Algorithm 5. It reduces to solving MDPs in $\mathbb{Z}_p[x_1, \ldots, x_n]$.

To make the following explanation easier we assume $r = 2$ and suppose that
$a = uw$ where $a, u, w \in \mathbb{Z}[x_1, \ldots, x_n]$ and u, w are unknown to us. As a first
step we choose an evaluation ideal $I = \langle x_2 - \alpha_2, \ldots, x_n - \alpha_n \rangle$ with randomly
chosen α_i from $[0, p - 1]$ such that conditions (i) and (ii) for MHL are satisfied
with $l = 1$. Then there is a factorization $a = u^{(n)} w^{(n)} \in \mathbb{Z}_p[x_1, \ldots, x_n]$. This
factorization is computed using MTSHL-d.

Now suppose that u (similarly w) has the form

$$u = \sum_{j=1}^{t} c_j M_j(x_1, \ldots, x_n) = \sum_{j=1}^{t} \sum_{i=0}^{l-1} s_{ji} p^i M_j(x_1, \ldots, x_n),$$

where the M_j are distinct monomials and $0 \neq c_j \in \mathbb{Z}$ with $c_j = \sum_{i=0}^{l-1} s_{ji} p^j$ where $-p^l/2 < s_{ji} < p^l/2$. Then we have

$$u = \sum_{i=0}^{l-1} \left(\sum_{j=1}^{t} s_{ji} M_j(x_1, \ldots, x_n) \right) p^i = \sum_{i=0}^{l-1} u_i p^i.$$

It follows that

$$\frac{u - \sum_{i=0}^{k-1} u_i p^i}{p^k} = \sum_{j=1}^{t} \left(\sum_{i=k}^{l-1} s_{ji} p^{i-k} \right) M_j(x_1, \ldots, x_n).$$

Also, we have $u_0 = u \mod p \neq 0$ since in the first stage u is lifted from u_0. Now we make a key observation: If p is chosen at random such that $2^m < p < 2^{m+1}$, the probability that $p \mid c_i$ is $\Pr[p \mid c_i] = \frac{\#\text{distinct } (m+1)\text{bit prime divisors of } c_i}{\#m \text{ bit primes}}$. Let $\pi(s)$ be the number of primes $\leq s$. Since there are at most $\lfloor \log_{2^m}(c_i) \rfloor$ many $(m+1)$-bit primes dividing c_i we have

$$\Pr[p \mid c_i] \leq \frac{\lfloor \log_{2^m}(c_i) \rfloor}{\pi(2^{m+1}) - \pi(2^m)} \leq \frac{l}{\pi(2^{m+1}) - \pi(2^m)}$$

This probability is very small because according to the prime number theorem $\pi(s) \sim s/\log(s)$ and hence $\pi(2^{m+1}) - \pi(2^m) \sim \frac{2^m}{m \log(2)}$.

It has been shown in [8] that the exact number of 31-bit primes ($m = 30$) is 50697537. Therefore in our implementation the support of u_0 will contain all monomials M_i and $\text{Supp}\{u_j\} \subseteq \text{Supp}\{u_0\}$ with probability $> 1 - \frac{tl}{5 \cdot 10^7}$.

We make one more key observation and claim that $\text{Supp}\{u_j\} \subseteq \text{Supp}\{u_{j-1}\}$ for $1 \leq j \leq l$ with high probability: We have

$$u_j = s_{0j} M_0 + s_{1j} M_1 + \cdots + s_{kj} M_t,$$
$$u_{j+1} = s_{0,j+1} M_0 + s_{1,j+1} M_1 + \cdots + s_{k,j+1} M_t.$$

For a given $j > 0$, if $s_{i,j+1} \neq 0$, but $s_{ij} = 0$ then $M_i \in \text{Supp}(u_{j+1})$ but $M_i \notin \text{Supp}(u_j)$. We consider $\Pr[s_{ij} = 0 \mid s_{i,j+1} \neq 0]$. If A is the event that $s_{ij} = 0$ and B is the event that $s_{i,j+1} = 0$ then

$$\Pr[A \mid B^c] = \frac{\Pr[A] - \Pr[B] \Pr[A \mid B]}{\Pr[B^c]} \leq \frac{\Pr[A]}{\Pr[B^c]}.$$

It follows that

$$\frac{\Pr[A]}{\Pr[B^c]} \leq \frac{l/(\pi(2^{m+1}) - \pi(2^m))}{1 - l/(\pi(2^{m+1}) - \pi(2^m))} = \frac{l}{(\pi(2^{m+1}) - \pi(2^m)) - l}.$$

Hence,

$$\Pr[\mathrm{Supp}\{u_j\} \subseteq \mathrm{Supp}\{u_{j-1}\} \,|\, 1 \leq j \leq l] > 1 - \frac{t\,l}{((\pi(2^{m+1}) - \pi(2^m)) - l}.$$

As an example for $m = 30, l = 5, t = 500$, this probability is >0.99993.

Hardy and Ramanujan [5] proved that for almost all integers, the number of distinct primes dividing a number s is $\omega(s) \approx \log\log(s)$. This theorem was generalized by Erdős-Kac which shows that $\omega(s)$ is essentially normally distributed [3]. By this approximation note that

$$\frac{\Pr[A]}{\Pr[B^c]} \leq \frac{\log\log(s_{ij})/(\pi(2^{m+1}) - \pi(2^m))}{1 - \log\log(s_{i,j+1})/(\pi(2^{m+1}) - \pi(2^m))} = \frac{\log(l\log p)}{(\pi(2^{m+1}) - \pi(2^m)) - \log(l\log p)}.$$

Hence the probability that $\mathrm{Supp}\{u_j\} \subseteq \mathrm{Supp}\{u_{j-1}\}$ is $\gtrsim 1 - t\frac{m\log(lm)}{2^m - m\log(lm)}$. As an example for $m = 30, l = 5, t = 500$, this probability is >0.99995.

What does this mean in the context of multivariate factorization over mod $\mathbb{Z}_{p^l}$ for $l > 1$? It means that the solutions to the multivariate diophantine problems occurring in the lifting process will, with high probability, be a subset of the monomials of the solutions of the previous step and these solutions can be computed simply by solving Vandermonde systems by using a machine prime p and hence by an efficient arithmetic using a sparse MDP solver as described in Algorithm 3.

We sum up the observations made in this section in Theorem 1 below.

Theorem 1. *Let p be a randomly chosen m-bit prime, i.e. $p \in [2^m < p < 2^{m+1}]$. With the notation introduced in this section*

$$\Pr(\mathrm{Supp}\{u_j\} \subseteq \mathrm{Supp}\{u_{j-1}\} \text{ for all } 1 \leq j \leq l) > 1 - \frac{t\,l}{((\pi(2^{m+1}) - \pi(2^m)) - l}.$$

This probability can be approximated by

$$\Pr[\mathrm{Supp}\{u_j\} \subseteq \mathrm{Supp}\{u_{j-1}\} \text{ for all } 1 \leq j \leq l] \gtrsim 1 - \frac{t\,m\log(lm)}{2^m - m\log(lm)}.$$

6 Timing Data

In this section we give some experimental data to verify the effectiveness of the methods described in Sects. 4 and 5. In the tables that follow all timings are in CPU seconds and were obtained on an Intel Core i5–4670 CPU running at 3.40 GHz with 16 GB of RAM. For all Maple timings, we set `kernelopts(numcpus=1);` to restrict Maple to use only one core as otherwise it will do polynomial multiplications and divisions in parallel.

6.1   Iterative vs Direct

In this section, we give some data in Table 1 to compare MTSHL-d with the current approach, i.e. implementing MTSHL so that it solves multi-MDP's using iterative approach as explained in Sect. 4. We include also timings for Wang's algorithm which also uses the iterative approach.

We generated r random polynomials in n variables of total degree d with T terms and coefficients from $[1, 99]$ using Maple's `randpoly` command thus `x1^(d+1)+randpoly([x1,\ldots,xn],degree=d,terms=T,coeffs=rand(1..99))` and multiplied them. Then we factored these polynomials using (i) Wang's algorithm, (ii) MTSHL and (iii) MTSHL-d (our new method explained in Sect. 4). All implementations are in Maple. $tX(tY)$ means that the algorithm factored the polynomial in tX CPU seconds and spent tY CPU seconds solving multiterm MDPs. OOM stands for out of memory. As can be seen from the data, MTSHL is significantly faster than Wang's algorithm and the MDP time in MTSHL-d is less than the MDP time in MTSHL by a factor of $r - 1$ or more.

Table 1. Timings for Wang, MTSHL vs MTSHL-d with $r > 2$.

$r/n/d/T$	Wang (MDP)	MTSHL (MDP)	MTSHL-d (MDP)
3/9/10/30	18.94 (16.00)	2.26 (0.60)	1.36 (0.30)
4/9/15/30	OOM	104.72 (23.23)	90.04 (6.55)
3/9/10/50	251.20 (240.77)	8.87 (2.28)	4.99 (0.71)
3/9/15/100	2302.69 (2235.2)	122.36 (28.58)	99.28 (8.17)
3/11/15/100	OOM	272.78 (42.74)	208.35 (11.51)
3/11/10/100	515.98 (424.76)	189.07 (23.90)	146.80 (6.25)
3/11/20/100	OOM	316.12 (66.7)	256.79 (19.22)

6.2   The p^L Case

In this section, we give some data in Table 2 to compare the current approach, i.e. implementing MTSHL so that it computes a bound l_B and factors staying in modulo $\mathbb{Z}_{p^{l_B}}$ arithmetic, with the p-adic lifting at the last step approach, i.e. the -staying in $\mathbb{Z}_p$ arithmetic approach-, as explained in this Sect. 5.

We generated 2 random polynomials in n variables of total degree d with T with coefficients in $[0, p^l)$ for $p = 2^{31} - 1$. Then we multiplied the two factors over $\mathbb{Z}$ and then factored the product with MTSHL. Since MTSHL does not know what the actual value of l is, it needs to compute the coefficient bound l_B (using Lemma 14 of [4]) and stays in the $\mathbb{Z}_{p^{l_B}}$ arithmetic. It factored the polynomial in $tX(tY)$ seconds where tY denotes the time spent on solving MDP's. Then we factored the polynomial with MTSHL-d which uses p-adic lifting to recover the integer coefficients as explained in Sect. 5. The timings in column MTSHL-d (MDP) (Lift) are the total time, the time spent in MDP and the time spent doing l lifts. The data in Table 2 shows that doing a p-adic lift is much faster than the previous approach.

Table 2. Timings for MTSHL vs MTSHL-d for large integer coefficients.

$n/d/T_{f_i}$	t_{f_i}	l	l_B	MTSHL (MDP)	MTSHL-d (MDP) (Lift)		
5/10/300	0.07	2	5	5.866 (5.101)	0.438	(0.132)	(0.241)
5/10/500	0.11	2	5	9.265 (7.937)	1.194	(0.186)	(0.480)
5/10/1000	0.23	2	5	14.448 (12.826)	2.202	(0.264)	(1.332)
5/10/300	0.07	4	9	6.923 (6.104)	1.067	(0.156)	(0.553)
5/10/500	0.11	4	9	10.971 (9.737)	1.854	(0.219)	(1.231)
5/10/1000	0.23	4	9	16.943 (15.183)	3.552	(0.350)	(2.632)
5/10/300	0.07	8	17	8.638 (7.596)	2.553	(0.201)	(2.076)
5/10/500	0.11	8	17	13.118 (11.686)	3.101	(0.280)	(2.396)
5/10/1000	0.23	8	17	19.031 (17.225)	4.905	(0.459)	(4.032)

7 Conclusion

We have shown that when the number of factors to be computed ≥ 2 and for
the case where the coefficients of the factors are huge, sparse interpolation techniques can be used to speed up multivariate polynomial factorization. The second
author has integrated our code into Maple under a MITACS internship with Dr.
Jürgen Gerhard of Maplesoft. The new code will become the default factorization algorithm used by Maple's `factor` command for multivariate polynomials
with integer coefficients. The old code will still be accessible as an option.

References

1. Cox, D., Little, J., O'Shea, D.: Ideals, Varieties and Algorithms, 3rd edn. Springer, New York (2007). https://doi.org/10.1007/978-0-387-35651-8
2. Geddes, K.O., Czapor, S.R., Labahn, G.: Algorithms for Computer Algebra. Kluwer, Boston (1992)
3. Erdős, P., Kac, M.: The Gaussian law of errors in the theory of additive number theoretic functions. Am. J. Math. **62**, 738–742 (1940)
4. Gelfond, A.O.: Transcendental and Algebraic Numbers. GITTL, Moscow (1952). English translation by Leo F. Boron, Dover, New York (1960)
5. Hardy, G.H., Ramanujan, S.: The normal number of prime factors of a number n. Q. J. Math. **48**, 76–92 (1917)
6. Kaltofen, E.: Sparse hensel lifting. In: Caviness, B.F. (ed.) EUROCAL 1985. LNCS, vol. 204, pp. 4–17. Springer, Heidelberg (1985). https://doi.org/10.1007/3-540-15984-3_230
7. Lang, S.: Diophantine Geometry. Wiley, Hoboken (1962)
8. Law, M.: Computing characteristic polynomials of matrices of structured polynomials, Masters thesis (2017)
9. Lee, M.M.: Factorization of multivariate polynomials. Ph.D. thesis (2013)
10. Monagan, M., Tuncer, B.: Some results on counting roots of polynomials and the Sylvester resultant. In: Proceedings of FPSAC 2016, pp. 887–898. DMTCS (2016)

11. Monagan, M., Tuncer, B.: Using sparse interpolation in hensel lifting. In: Gerdt, V.P., Koepf, W., Seiler, W.M., Vorozhtsov, E.V. (eds.) CASC 2016. LNCS, vol. 9890, pp. 381–400. Springer, Cham (2016). https://doi.org/10.1007/978-3-319-45641-6_25
12. Steel, A.: Private communication
13. Wang, P.S.: An improved multivariate polynomial factoring algorithm. Math. Comput. **32**, 1215–1231 (1978)
14. Wang, P.S., Rothschild, L.P.: Factoring multivariate polynomials over the integers. Math. Comput. **29**, 935–950 (1975)
15. Yun, D.Y.Y.: The Hensel lemma in algebraic manipulation. Ph.D. thesis (1974)
16. Zippel, R.: Probabilistic algorithms for sparse polynomials. In: Ng, E.W. (ed.) Symbolic and Algebraic Computation. LNCS, vol. 72, pp. 216–226. Springer, Heidelberg (1979). https://doi.org/10.1007/3-540-09519-5_73
17. Zippel, R.E.: Newton's iteration and the sparse Hensel algorithm. In: Proceedings of SYMSAC 1981, pp. 68–72. ACM (1981)
18. Zippel, R.E.: Interpolating polynomials from their values. J. Symb. Comput. **9**(3), 375–403 (1990)
19. Zippel, R.E.: Effective Polynomial Computation. Kluwer, Boston (1993)

Beyond the First Class of Analytic Complexity

T. M. Sadykov[✉]

Plekhanov Russian University, Stremyanny 36, Moscow 125993, Russia
Sadykov.TM@rea.ru

Abstract. We investigate the notion of analytic complexity of a bivariate holomorphic function by means of computer algebra tools. An estimate from below on the number of terms in the differential polynomials defining classes of analytic complexity is established. We provide an algorithm which allows one to explicitly compute the differential membership criteria for certain families of bivariate analytic functions in the second complexity class. The presented algorithm is implemented in the computer algebra system Singular 4-1-1.

Keywords: Analytic complexity · Differential polynomial
Differentially algebraic function

1 Introduction

The notion of analytic complexity of a bivariate holomorphic function stems from Hilbert's 13th problem on the possibility to represent the algebraic function implicitly defined by the reduced septic equation with three parameters through compositions of functions in at most two variables. For continuous functions, the positive answer is given in a much more general setup by the celebrated Kolmogorov–Arnold theorem [1].

Theorem 1. *(See* [1].*) Any continuous function defined on a compact subset of $\mathbb{R}^n$ can be represented as a finite superposition of univariate continuous functions and a single bivariate function $s(x,y)$ which can be chosen to be the addition: $s(x,y) = x + y$.*

Such a representation is only possible due to the vastness of the space of all continuous functions of real variables defined on a compact set. In fact, the construction in the proof of the Kolmogorov–Arnold theorem uses continuous functions that are not analytic in any open set. In the analytic category, the problem of representing a holomorphic function as a finite superposition of holomorphic functions in fewer variables turns out to be much more subtle. It leads to the concept of classes of analytic complexity defined inductively as finite superpositions of univariate functions and a fixed bivariate analytic function.

© Springer Nature Switzerland AG 2018
V. P. Gerdt et al. (Eds.): CASC 2018, LNCS 11077, pp. 335–344, 2018.
https://doi.org/10.1007/978-3-319-99639-4_23

Apart from trivial examples, computing or estimating the analytic complexity of a bivariate holomorphic function is a difficult task which requires full use of elimination theory and heavily relies on computer algebra tools.

In the present paper, we investigate the notion of analytic complexity of a bivariate holomorphic function by means of computer algebra tools. An estimate from below on the number of terms in the differential polynomials defining classes of analytic complexity is established. We provide an algorithm which allows one to explicitly compute the differential membership criteria for certain families of bivariate analytic functions in the second complexity class. The presented algorithms are implemented in computer algebra system Singular 4-1-1. All examples in the paper have been computed on Intel Core i5-4440 CPU clocked at 3.10 GHz with 16 Gb RAM under MS Windows 7 Ultimate SP1.

The author is thankful to V. Beloshapka for the numerous fruitful discussions on the analytic complexity of holomorphic functions and related topics.

2   Analytic Complexity of Bivariate Functions

Throughout the paper we denote by (x, y) the coordinates in the two-dimensional complex space $\mathbb{C}^2$. We denote by $\mathcal{O}(U)$ the space of functions that are holomorphic in the domain $U \subset \mathbb{C}^2$. A (multi-valued) analytic function will be identified with its germ unless explicitly stated otherwise. The next definition is central to the paper.

Definition 1. (See [2].) The class of functions of analytic complexity zero Cl_0 is defined to comprise the functions that depend on at most one of the variables. A function $F(x, y)$ is said to belong to the class Cl_n of functions with analytic complexity $n > 0$ if and only if the following two conditions are satisfied:

(1)   There exists a point $(x_0, y_0) \in \mathbb{C}^2$ and a germ $\mathfrak{F}(x, y) \in \mathcal{O}(U(x_0, y_0))$ of this function holomorphic at (x_0, y_0) such that

$$\mathfrak{F}(x, y) = c(a(x, y) + b(x, y))$$

for some germs of holomorphic functions $a, b \in Cl_{n-1}$ and $c \in Cl_0$;
(2)   No relation of this form exists for $a, b \in Cl_k$ with $k < n - 1$.

If there is no such representation for any finite n then the function F is said to be of infinite analytic complexity.

Thus, a function of two complex variables is said to have analytic complexity zero if and only if it only depends on one of the variables or is identically constant. A function belongs to the first class of analytic complexity if it admits a representation of the form $c(a(x) + b(y))$ for certain univariate analytic functions a, b, c in some open subset of $\mathbb{C}^2$.

Typically, the analytic complexity of a bivariate holomorphic function is rather difficult to estimate and even more difficult to compute exactly. The inductive definition of analytic complexity leads to a wealth of counterintuitive examples. For instance, both the generic linear function $\alpha x + \beta y$ and the product $x \cdot y$ of the variables clearly belong to the first class of analytic complexity. The sum of two functions in Cl_1 is usually a function in the second complexity class. However, for any $\alpha, \beta, \gamma \in \mathbb{C}^*$ the polynomial $\alpha x + \beta y + \gamma xy$ is still a function in Cl_1 since

$$\alpha x + \beta y + \gamma xy = -\frac{\alpha\beta}{\gamma} + \frac{\alpha\beta}{\gamma} + \alpha x + \beta y + \gamma xy =$$

$$-\frac{\alpha\beta}{\gamma} + \left(\frac{\beta}{\sqrt{\gamma}} + \sqrt{\gamma}\,x\right)\left(\frac{\alpha}{\sqrt{\gamma}} + \sqrt{\gamma}\,y\right).$$

A differential monomial with the unknown function $F(x,y)$ is the product of integer powers of F and its partial derivatives, i.e., an expression of the form $F^{p_{00}} F_x^{p_{10}} F_y^{p_{01}} F_{xx}^{p_{20}} F_{xy}^{p_{11}} F_{yy}^{p_{02}} \cdots$. (the product is finite). By a differential polynomial over a field $\mathbb{K}$ with the unknown function $F(x,y)$, we will mean a finite linear combination of differential monomials with coefficients in $\mathbb{K}$.

The next result due to V.K. Beloshapka shows that classes of analytic complexity for bivariate holomorphic functions admit membership criteria defined by differential polynomials with integer coefficients.

Theorem 2. *(See [2].) The set of bivariate analytic functions whose analytic complexity does not exceed n coincides with the set of holomorphic solutions to a finite number of differential polynomials $\triangle_n$ with integer coefficients, i.e.,*

$$Cl_n = \{F(x,y) : \triangle_n(F(x,y)) \equiv 0\}.$$

Due to the conservation principle, the analytic continuation of a solution to a system of partial differential equations whose coefficients are entire analytic functions along any path also satisfies the same system of equations. Thus, it suffices to apply a differential membership criterion for a class of analytic complexity to any germ of the holomorphic function in question.

Theorem 2 implies that any function of finite analytic complexity is differentially algebraic, i.e., satisfies a (typically nonlinear) partial differential equation with constant coefficients. Thus, any differentially transcendental function (e.g. the polylogarithm $\mathrm{Li}_x(y) := \sum\limits_{n=1}^{\infty} \frac{y^n}{n^x}$) necessarily has infinite analytic complexity. In fact, the set of holomorphic functions of finite analytic complexity is a set of first category in the space $\mathcal{O}(U)$ for any domain $U \subseteq \mathbb{C}^2$.

Unfortunately, explicit differential membership criteria for the classes of analytic complexity or other families of (bivariate) analytic functions are in general very difficult to compute. The only known elements of the family $\triangle_n$ are $\triangle_0(F) = F_x F_y$ and

$$\triangle_1(F) = F_x^2 F_{xy} F_{yy} - F_x^2 F_{xyy} F_y - F_{xx} F_{xy} F_y^2 + F_x F_{xxy} F_y^2,$$

which have been found in [2]. The latter can be computed as the numerator of the expression $\frac{\log(F_x/F_y)}{\partial x \partial y}$ which clearly vanishes for any $F \in Cl_1$, i.e., for $F(x,y) = c(a(x) + b(y))$.

Differential membership criteria $\triangle_n$ for the classes of analytic complexity are so difficult to compute because for $n \geq 2$, they are themselves incredibly complex. Examples in Sect. 5 suggest that explicit computation of $\triangle_2$ is probably beyond the capacity of modern computer algebra tools or requires a completely new insight into the issue. An important tractable subset of the second analytic complexity class, the so-called $Cl_{3/2}$, has been considered in great detail in [6]. In the next section, we provide a rough estimate from below for the number of differential monomials in the differential polynomial $\triangle_n(F(x,y))$ and describe an algorithm which is later used to compute defining differential polynomials for certain families of functions in the second class of analytic complexity. All of the found differential polynomials are particular cases of the membership criterion for the second class of analytic complexity which appears to be out of reach for today's computer algebra systems.

3 Estimating the Number of Terms in the Differential Membership Criteria for Complexity Classes

The structure of nonlinear differential equations with both constant and variable coefficients that define families of analytic functions depending on arbitrary univariate functions was since long ago the focus of intensive research of numerous authors (see [2,3,10] and the references therein). The complexity of such differential equations typically grows very quickly with the number of univariate functions that encode the family in question. Although these equations usually enjoy a rich differential-algebraic structure, one of the most important ways of estimating their complexity is by counting the number of differential monomials in their irreducible factors. The following theorem provides a rough estimate from below for the number of such monomials in the differential polynomials defining classes of analytic complexity.

Theorem 3. *The number of differential monomials in the differential membership criterion for the n th class of analytic complexity is greater than $(2^{n-1}+1)!$*

Proof. We prove the estimate by considering a family of functions for which the defining differential polynomial is known. Namely, let

$$S_k = \{F(x,y) : F(x,y) = \sum_{j=1}^{k} a_j(x)b_j(y)\} \tag{1}$$

be the family of bivariate analytic functions which can (locally) be represented as the scalar product of univariate vector-valued functions $(a_1(x),\ldots,a_k(x))$ and $(b_1(y),\ldots,b_k(y))$. Induction shows that for generic univariate analytic functions $a_j(x)$ and $b_j(y)$, the analytic complexity of $F(x,y) \in S_{2^p-1}$ equals p. Indeed, by definition, the analytic complexity of $a_1(x)b_1(y)$ equals 1 while adding together two generic elements in S_k results in the unit increment of the analytic complexity.

For the sake of brevity, we use the notation $F_{x^k y^\ell} = \frac{\partial^{k+\ell} F}{\partial x^k \partial y^\ell}$. It has been announced by C. Stéphanos and later proved in [8] (see also [9]) that the family of functions $\mathcal{S}_n$ is the set of all solutions to the partial differential equation

$$\begin{vmatrix} F & F_x & \cdots & F_{x^n} \\ F_y & F_{xy} & \cdots & F_{x^n y} \\ \vdots & \vdots & \ddots & \vdots \\ F_{y^n} & F_{xy^n} & \cdots & F_{x^n y^n} \end{vmatrix} = 0. \tag{2}$$

By the construction of the family $\mathcal{S}_k$, the number of differential monomials in the differential membership criterion $\triangle_p$ for the nth class of analytic complexity cannot be smaller than the number of monomials in the differential polynomial defining $\mathcal{S}_{2^{p-1}}$. The left-hand side of (2) is a differential polynomial with $(n+1)!$ differential monomials which concludes the proof.

Intensive computer experiments suggest that the determinant in the left-hand side of (2) is irreducible as long as the function $F(x, y)$ is sufficiently general. Yet, no proof of this fact appears to be present in the literature.

Examples in Sect. 5 show that Theorem 3 gives a rather weak estimate on the number of terms in the differential membership criterion $\triangle_n(F(x, y))$. In the next section, we discuss a symbolic computational approach towards the structure of this differential polynomial.

4 Algorithmic Computation of Differential Membership Criteria

Efficient symbolic computation of partial differential relations defining families of bi- and multivariate analytic functions is the focus of intensive research by numerous authors. It is central to the fundamental monograph [10]. Most of the bivariate analytic functions considered in [10] have finite analytic complexity. An attempt to derive differential polynomials for compositions of analytic functions is described in [7]. Such polynomials for certain families of functions can also be computed by means of characteristic sets theory (see [4] and the references therein).

Any family of bivariate analytic functions of finite analytic complexity is typically annihilated by an infinite hierarchy of differential relations which is heavily dependent on the field of allowed coefficients (see [6] and the example in Sect. 5.2 below). Even in the case when the ideal of relations is principal, it is in general not possible to minimize the differential order and the algebraic degree of the generator simultaneously. For these reasons, efficient computation of an annihilating differential polynomial for a given family of bivariate analytic functions requires a thorough analysis of the structure of the generic element in the family. When the family in question involves a function which satisfies a linear partial differential equation, it is often beneficial to find a differential polynomial whose coefficients only depend on this function.

Algorithm 1. Algorithm for computing an annihilating differential polynomial for a family of bivariate analytic functions

---

Require: List of complex variables *vars_list*; list of univariate functions *fcns_list*; list of the numeric parameters of the equation defining the family of bivariate analytic functions *p_list*; equation *eqn* defining a generic element in the family as a function of the elements in *vars_list, fcns_list,* and *p_list*.

Ensure: List of differential monomials with integer coefficients and the unknown function depending on the variables in *vars_list* whose sum gives the defining relation for the family of functions under study.

1: **procedure** DIFFPOLY(*vars_list, fcns_list, p_list, eqn*)
2:    *J_list* := empty list
3:    *FJ_list* := empty list
4:    *D_list* := empty list
5:    *dp_poly* := 1
6:    *d* := 1   ▷ The order of the jet space where the differential polynomial is to be found
7:    **repeat**
8:      **for** $k = 0 : d : 1$ **do**
9:        **for** $j = 0 : k : 1$ **do**
10:           Add $\frac{\partial^k (F(x,y) - eqn)}{\partial x^j \partial y^{k-j}}$ to *J_list*   ▷ Forming the jet space of order d
11:        **end for**
12:        Add *J_list* to *FJ_list*
13:      **end for**
14:      **for** $k = 0 : d : 1$ **do**
15:        **for** $j = 1 : \text{Length}(fcns\_list) : 1$ **do**
16:          $u(x,y) = fcns\_list[j]$
17:           Add $\frac{\partial^k u}{\partial x^k}$ and $\frac{\partial^k u}{\partial x^k}$ to *D_list*   ▷ Forming the list of differential variables to be eliminated
18:        **end for**
19:      **end for**
20:        *dp_poly* = elimination ideal obtained by eliminating the elements of *D_list* out of the relations *FJ_list*
21:      *d* := *d* + 1
22:    **until** *dp_poly* ≠ 0
23:    **return** *dp_poly*
24: **end procedure**

The next algorithm was used to compute differential membership criteria for a number of families in the second class of analytic complexity.

The key component and the bottleneck of the algorithm is of course the elimination of differential variables out of a differential ideal. It makes an extensive use of both built-in and custom-designed methods of elimination and cannot be consistently described in a short research paper. For each family of functions in the below examples, a particular version of the elimination procedure taking into account the key differential-algebraic properties of the generic representative of the family has been used.

5   Computing Differential Membership Criteria for Families in the Second Class of Analytic Complexity

We now employ Algorithm 1 to produce differential polynomials with integer coefficients for certain families of bivariate analytic functions whose generic elements belong to the second class of analytic complexity. The generic element of this class is a function that admits a local representation of the form

$$f(c(a(x) + b(y)) + w(u(x) + v(y)))$$

for univariate analytic functions $a, \ldots, w$ such that the above composition is well defined and analytic in some domain in $\mathbb{C}^2$. The below examples are obtained by specifying some of these univariate functions in a certain concrete way.

5.1   A Differential Polynomial for the Family of Functions $F(x, y) = b(a(x) + y) + c(x + y)$

Let $a(\cdot), b(\cdot), c(\cdot)$ be arbitrary univariate analytic functions such that the composition $F(x, y) = b(a(x) + y) + c(x + y)$ is well defined for (x, y) in some domain in the complex space $\mathbb{C}^2$. Using Algorithm 1 we compute the following differential polynomial with integer coefficients which vanishes on any function in this family:

$$
\begin{aligned}
&F_y F_{yyy}^2 F_{xy} - F_y F_{yy} F_{yyyy} F_{xy} - F_{yyy}^2 F_x F_{xy} + F_{yy} F_{yyyy} F_x F_{xy} + F_y F_{yyyy} F_{xy}^2 - \\
&F_{yyyy} F_x F_{xy}^2 - F_y F_{yy} F_{yyy} F_{xyy} + F_y^2 F_{yyyy} F_{xyy} + F_{yy} F_{yyy} F_x F_{xyy} - \\
&2 F_y F_{yyyy} F_x F_{xyy} + F_{yyyy} F_x^2 F_{xyy} - F_y F_{yyy} F_{xy} F_{xyy} + F_{yyy} F_x F_{xy} F_{xyy} + \\
&F_y F_{yy} F_{xyy}^2 - F_{yy} F_x F_{xyy}^2 + F_y F_{yy}^2 F_{xyyy} - F_y^2 F_{yyy} F_{xyyy} - F_{yy}^2 F_x F_{xyyy} + \\
&2 F_y F_{yyy} F_x F_{xyyy} - F_{yyy} F_x^2 F_{xyyy} - F_y F_{yy} F_{xy} F_{xyyy} + F_{yy} F_x F_{xy} F_{xyyy} - \\
&F_y F_{yyy}^2 F_{xx} + F_y F_{yy} F_{yyyy} F_{xx} + F_{yyy}^2 F_x F_{xx} - F_{yy} F_{yyyy} F_x F_{xx} - \\
&F_y F_{yyyy} F_{xy} F_{xx} + F_{yyyy} F_x F_{xy} F_{xx} + 2 F_y F_{yyy} F_{xyy} F_{xx} - 2 F_{yyy} F_x F_{xyy} F_{xx} - \\
&F_y F_{xyy}^2 F_{xx} + F_x F_{xyy}^2 F_{xx} - F_y F_{yy} F_{xyyy} F_{xx} + F_{yy} F_x F_{xyyy} F_{xx} + \\
&F_y F_{xy} F_{xyyy} F_{xx} - F_x F_{xy} F_{xyyy} F_{xx} + F_y F_{yy} F_{yyy} F_{xxy} - F_y^2 F_{yyyy} F_{xxy} - \\
&F_{yy} F_{yyy} F_x F_{xxy} + 2 F_y F_{yyyy} F_x F_{xxy} - F_{yyyy} F_x^2 F_{xxy} - F_y F_{yyy} F_{xy} F_{xxy} + \\
&F_{yyy} F_x F_{xy} F_{xxy} - F_y F_{yy} F_{xyy} F_{xxy} + F_{yy} F_x F_{xyy} F_{xxy} + F_y F_{xy} F_{xyy} F_{xxy} - \\
&F_x F_{xy} F_{xyy} F_{xxy} + F_y^2 F_{xyyy} F_{xxy} - 2 F_y F_x F_{xyyy} F_{xxy} + F_x^2 F_{xyyy} F_{xxy} - \\
&F_y F_{yy}^2 F_{xxyy} + F_y^2 F_{yyy} F_{xxyy} + F_{yy}^2 F_x F_{xxyy} - 2 F_y F_{yyy} F_x F_{xxyy} + \\
&F_{yyy} F_x^2 F_{xxyy} + 2 F_y F_{yy} F_{xy} F_{xxyy} - 2 F_{yy} F_x F_{xy} F_{xxyy} - F_y F_{xy}^2 F_{xxyy} + \\
&F_x F_{xy}^2 F_{xxyy} - F_y^2 F_{xyy} F_{xxyy} + 2 F_y F_x F_{xyy} F_{xxyy} - F_x^2 F_{xyy} F_{xxyy}.
\end{aligned}
$$

An alternative way of computing this differential polynomial can be based on the main result of [6]. For families of polynomial instances of special functions of hypergeometric type [5], the differential membership criteria computed by means of Algorithm 1 get greatly simplified.

Since the above differential polynomial has differential order 4, the general theory of partial differential equations suggests that its general solution depends on four univariate analytic functions. Thus, the initial family of functions $\{b(a(x) + y) + c(x + y)\}$ cannot exhaust the whole solution space of the obtained differential polynomial and additional relations must be computed. Similar arguments apply to the examples in the subsections that follow.

5.2   A Differential Polynomial for the Family of Functions $F(x, y) = c(a(e^x + y) + b(x + y))$

The family of bivariate analytic functions comprising functions of the form $c(a(d(x) + y) + b(x + y))$ is one step closer to the generic element of the second class of analytic complexity than the previous example. Unfortunately, numerous computer experiments suggest that computation of the annihilating differential polynomial with integer coefficients for this family of functions is probably out of reach for the present day's computer algebra systems. However, it turns out to be possible to treat a subfamily of this class of functions corresponding to $d(x) = e^x$ since all of the derivatives of this function coincide which brings symbolic elimination within manageable range. Using Algorithm 1 we compute the following defining polynomial for this family with the coefficients in the ring $\mathbb{Z}[e^x]$:

$$
\begin{aligned}
&e^{4x}(-2F_y F_{yy}^2 F_x + F_y^2 F_{yyy} F_x + F_{yy}^2 F_x^2 - F_y F_{yyy} F_x^2 + 2F_y^2 F_{yy} F_{xy} + \\
&2F_y F_{yy} F_x F_{xy} - F_{yy} F_x^2 F_{xy} - 2F_y^2 F_{xy}^2 - F_y^3 F_{xyy} + F_y F_x^2 F_{xyy} - \\
&F_y^2 F_{yy} F_{xx} + F_y^2 F_{xy} F_{xx} + F_y^3 F_{xxy} - F_y^2 F_x F_{xxy}) + \\
&e^{3x}(F_y^4 - 2F_y^3 F_x + 4F_y F_{yy}^2 F_x - 2F_y^2 F_{yyy} F_x + F_y^2 F_x^2 + F_y F_{yy} F_x^2 - \\
&F_{yy}^2 F_x^2 + F_y F_{yyy} F_x^2 - F_{yy} F_x^3 + F_{yyy} F_x^3 - 4F_y^2 F_{yy} F_{xy} - 2F_y^2 F_x F_{xy} - \\
&2F_y F_{yy} F_x F_{xy} + 2F_y F_x^2 F_{xy} - 2F_{yy} F_x^2 F_{xy} + 2F_y^2 F_{xy}^2 + 2F_x^2 F_{xy}^2 + \\
&2F_y^3 F_{xyy} - F_y F_x^2 F_{xyy} - F_x^3 F_{xyy} + F_y^3 F_{xx} + F_y^2 F_{yy} F_{xx} - F_y^2 F_x F_{xx} + \\
&F_{yy} F_x^2 F_{xx} + 2F_y^2 F_{xy} F_{xx} - 2F_y F_x F_{xy} F_{xx} - F_y^2 F_{xx}^2 - F_y^3 F_{xxy} + \\
&F_y^2 F_x F_{xxy} - F_y^3 F_{xxx} + F_y^2 F_x F_{xxx}) + \\
&e^{2x}(-F_y^2 F_{yy} F_x - 2F_y F_{yy}^2 F_x + F_y^2 F_{yyy} F_x - F_{yy}^2 F_x^2 + F_y F_{yyy} F_x^2 + \\
&F_{yy} F_x^3 - 2F_{yyy} F_x^3 + F_y^3 F_{xy} + 2F_y^2 F_{yy} F_{xy} + 3F_y^2 F_x F_{xy} - \\
&2F_y F_{yy} F_x F_{xy} - 5F_y F_x^2 F_{xy} + 6F_{yy} F_x^2 F_{xy} + F_x^3 F_{xy} + 2F_y^2 F_{xy}^2 - \\
&2F_x^2 F_{xy}^2 - F_y^3 F_{xyy} + F_x^3 F_{xyy} - 3F_y^3 F_{xx} + F_y^2 F_{yy} F_{xx} + 4F_y^2 F_x F_{xx} - \\
&F_y F_x^2 F_{xx} - F_{yy} F_x^2 F_{xx} - 6F_y^2 F_{xy} F_{xx} + 2F_y F_x F_{xy} F_{xx} - 2F_x^2 F_{xy} F_{xx} + \\
&F_y^2 F_{xx}^2 + 2F_y F_x F_{xx}^2 - F_y^3 F_{xxy} + F_x^3 F_{xxy} + 2F_y^3 F_{xxx} - F_y^2 F_x F_{xxx} - F_y F_x^2 F_{xxx}) + \\
&e^x(-F_y^3 F_x + F_y^2 F_{yy} F_x + 2F_y^2 F_x^2 - F_y F_{yy} F_x^2 + F_{yy}^2 F_x^2 - F_y F_{yyy} F_x^2 - \\
&F_y F_x^3 + F_{yyy} F_x^3 - F_y^3 F_{xy} - F_y^2 F_x F_{xy} + 2F_y F_{yy} F_x F_{xy} + 3F_y F_x^2 F_{xy} - \\
&2F_{yy} F_x^2 F_{xy} - F_x^3 F_{xy} - 2F_y^2 F_{xy}^2 - 2F_x^2 F_{xy}^2 - F_y F_x^2 F_{xyy} + F_x^3 F_{xyy} + \\
&2F_y^3 F_{xx} - F_y^2 F_{yy} F_{xx} - 3F_y^2 F_x F_{xx} + F_y F_x^2 F_{xx} - F_{yy} F_x^2 F_{xx} + 2F_y^2 F_{xy} F_{xx} + \\
&2F_y F_x F_{xy} F_{xx} + 4F_x^2 F_{xy} F_{xx} + F_y^2 F_{xx}^2 - 4F_y F_x F_{xx}^2 + F_y^3 F_{xxy} + F_y^2 F_x F_{xxy} - \\
&2F_x^3 F_{xxy} - F_y^3 F_{xxx} - F_y^2 F_x F_{xxx} + 2F_y F_x^2 F_{xxx}) - \\[4pt]
&F_{yy} F_x^2 F_{xy} + 2F_x^2 F_{xy}^2 + F_y F_x^2 F_{xyy} - F_x^3 F_{xyy} + F_{yy} F_x^2 F_{xx} + F_x^2 F_{xy} F_{xx} - \\
&2F_y F_x F_{xy} F_{xx} - 2F_x^2 F_{xy} F_{xx} - F_y^2 F_{xx}^2 + 2F_y F_x F_{xx}^2 - F_y^2 F_x F_{xxx} + F_x^3 F_{xxy} + \\
&F_y^2 F_x F_{xxx} - F_y F_x^2 F_{xxx}.
\end{aligned}
$$

Treating e^x as a new independent variable, we differentiate the above differential polynomial with respect to y and eliminate e^x out of the obtained ideal. The result is given by

$$
\begin{aligned}
&-18F_y^{18}F_{yy}^7F_x^6F_{xy} + 84F_y^{17}F_{yy}^8F_x^6F_{xy} - 120F_y^{16}F_{yy}^9F_x^6F_{xy} + \\
&48F_y^{15}F_{yy}^{10}F_x^6F_{xy} + 24F_y^{19}F_{yy}^5F_{yyy}F_x^6F_{xy} - 144F_y^{18}F_{yy}^6F_{yyy}F_x^6F_{xy} + \\
&240F_y^{17}F_{yy}^7F_{yyy}F_x^6F_{xy} - 96F_y^{16}F_{yy}^8F_{yyy}F_x^6F_{xy} - 6F_y^{20}F_{yy}^3F_{yyy}^2F_x^6F_{xy} + \\
&72F_y^{19}F_{yy}^4F_{yyy}^2F_x^6F_{xy} - 162F_y^{18}F_{yy}^5F_{yyy}^2F_x^6F_{xy} + \\
&2002F_y^{20}F_x^5F_{xx}^2F_{xxy}F_{xxxy}^4 - 3003F_y^{19}F_x^6F_{xx}^2F_{xxy}F_{xxxy}^4 + \\
&3432F_y^{18}F_x^7F_{xx}^2F_{xxy}F_{xxxy}^4 - 3003F_y^{17}F_x^8F_{xx}^2F_{xxy}F_{xxxy}^4 + \\
&2002F_y^{16}F_x^9F_{xx}^2F_{xxy}F_{xxxy}^4 - 1001F_y^{15}F_x^{10}F_{xx}^2F_{xxy}F_{xxxy}^4 + \\
&364F_y^{14}F_x^{11}F_{xx}^2F_{xxy}F_{xxxy}^4 - 91F_y^{13}F_x^{12}F_{xx}^2F_{xxy}F_{xxxy}^4 + \\
&14F_y^{12}F_x^{13}F_{xx}^2F_{xxy}F_{xxxy}^4 - F_y^{11}F_x^{14}F_{xx}^2F_{xxy}F_{xxxy}^4 + 2\,731\,601 \text{ other terms.}
\end{aligned}
$$

The complexity of this differential polynomial suggests that the defining relations for the second class of analytic complexity are far beyond the capacity of today's computer algebra systems.

5.3   A Differential Polynomial for the Family of Functions $F(x,y) = b(a(x) + e^{\alpha y}) + c(x)$

We now consider a family of bivariate analytic functions whose elements depend on a complex parameter apart from arbitrary univariate functions: $\{F(x,y) = b(a(x) + e^{\alpha y}) + c(x),\ \alpha \in \mathbb{C}^*\}$. Using Algorithm 1, we compute the following defining differential polynomial with integer coefficients for this family:

$$
\begin{aligned}
&F_{yyy}^2F_{xy}^6 - 4F_{yy}F_{yyy}F_{xy}^5F_{xyy} - F_{yy}^2F_{xy}^4F_{xyy}^2 + 4F_yF_{yyy}F_{xy}^4F_{xyy}^2 + \\
&2F_yF_{yy}F_{xy}^3F_{xyy}^3 - F_y^2F_{xy}^2F_{xyy}^4 + 4F_{yy}^2F_{xy}^5F_{xyyy} - 2F_yF_{yyy}F_{xy}^5F_{xyyy} - \\
&4F_yF_{yy}F_{xy}^4F_{xyy}F_{xyyy} + F_y^2F_{xy}^4F_{xyyy}^2 + 4F_{yy}^3F_{xy}^3F_{xyy}F_{xxy} - \\
&4F_yF_{yy}F_{yyy}F_{xy}^3F_{xyy}F_{xxy} - 2F_yF_{yy}^2F_{xy}^2F_{xyy}^2F_{xxy} + 3F_y^2F_{yyy}F_{xy}^2F_{xyy}^2F_{xxy} - \\
&F_y^2F_{yy}F_{xy}F_{xyy}^3F_{xxy} - F_y^3F_{xyy}^4F_{xxy} + F_y^3F_{xy}F_{xyy}^2F_{xyyy}F_{xxy} - F_y^2F_{yy}^2F_{xy}^2F_{xyy}^2F_{xxy} + \\
&F_y^3F_{yyy}F_{xy}^2F_{xyy}^2F_{xxy} - 4F_{yy}^3F_{xy}^4F_{xxyy} + 4F_yF_{yy}F_{yyy}F_{xy}^4F_{xxyy} + \\
&2F_yF_{yy}^2F_{xy}^3F_{xyy}F_{xxyy} - 3F_y^2F_{yyy}F_{xy}^3F_{xyy}F_{xxyy} + F_y^2F_{yy}F_{xy}^2F_{xyy}^2F_{xxyy} + \\
&F_y^3F_{xy}F_{xyy}^3F_{xxyy} - F_y^3F_{xy}^2F_{xyy}F_{xyyy}F_{xxyy} + 2F_y^2F_{yy}F_{xy}F_{xyy}F_{xxy}F_{xxyy} - \\
&2F_y^3F_{yyy}F_{xy}F_{xyy}F_{xxy}F_{xxyy} - F_y^2F_{yy}^2F_{xy}^2F_{xxyy}^2 + F_y^3F_{yyy}F_{xy}^2F_{xxyy}^2.
\end{aligned}
$$

We emphasize that the above polynomial annihilates any function in the family under study and does not depend on the choice of the parameter $\alpha \in \mathbb{C}^*$.

Acknowledgments. This research has been performed in the framework of the basic part of the scientific research state task in the field of scientific activity of the Ministry of Education and Science of the Russian Federation, project No. 2.9577.2017/8.9.

References

1. Arnold, V.I.: On the representation of continuous functions of three variables by superpositions of continuous functions of two variables. Mat. Sb. **48**(1), 3–74 (1959)
2. Beloshapka, V.K.: Analytic complexity of functions of two variables. Russ. J. Math. Phys. **14**(3), 243–249 (2007)
3. Beloshapka, V.K.: Algebraic functions of complexity one, a Weierstrass theorem, and three arithmetic operations. Russ. J. Math. Phys. **23**(3), 343–347 (2016)
4. Boulier, F., Lemaire, F., Maza, M.: Computing differential characteristic sets by change of ordering. J. Symb. Comput. **45**(1), 124–149 (2010)
5. Dickenstein, A., Sadykov, T.M.: Algebraicity of solutions to the Mellin system and its monodromy. Dokl. Math. **75**(1), 80–82 (2007)
6. Krasikov, V.A., Sadykov, T.M.: On the analytic complexity of discriminants. Proc. Steklov Inst. Math. **279**, 78–92 (2012)
7. Mansfield, E.L.: Differential Gröbner bases. Ph.D. thesis, University of Sydney (1991)
8. Neuman, F.: Factorizations of matrices and functions of two variables. Chechoslovak Math. J. **32**(4), 582–588 (1982)
9. Neuman, F.: Finite sums of products of functions in single variables. Linear Algebra Appl. **134**, 153–164 (1990)
10. Robertz, Daniel: Formal Algorithmic Elimination for PDEs. LNM, vol. 2121. Springer, Cham (2014). https://doi.org/10.1007/978-3-319-11445-3

A Theory and an Algorithm
for Computing Sparse Multivariate
Polynomial Remainder Sequence

Tateaki Sasaki[✉]

University of Tsukuba, Tsukuba-shi, Ibaraki 305-8571, Japan
sasaki@math.tsukuba.ac.jp

Abstract. This paper presents an algorithm for computing the polynomial remainder sequence (PRS) and corresponding cofactor sequences of sparse multivariate polynomials over a number field $\mathbb{K}$. Most conventional algorithms for computing PRSs are based on the pseudo remainder (Prem), and the celebrated subresultant theory for the PRS has been constructed on the Prem. The Prem is uneconomical for computing PRSs of sparse polynomials. Hence, in this paper, the concept of sparse pseudo remainder (spsPrem) is defined. No subresultant-like theory has been developed so far for the PRS based on spsPrem. Therefore, we develop a matrix theory for spsPrem-based PRSs. The computational formula for PRS, regardless of whether it is based on Prem or spsPrem, causes a considerable intermediate expression growth. Hence, we next propose a technique to suppress the expression growth largely. The technique utilizes the power-series arithmetic but no Hensel lifting. Simple experiments show that our technique suppresses the intermediate expression growth fairly well, if the sub-variable ordering is set suitably.

Keywords: Multivariate polynomial remainder sequence
Cofactor sequence · Sparse multivariate polynomials
Pseudo remainder · Sparse pseudo remainder · Subresultant
Hearn's trial-division algorithm

1 Introduction

The multivariate *polynomial remainder sequence (PRS)* is now scarcely studied. However, some researchers are becoming interested in PRSs of sparse multivariate polynomials. The first reason of revival of the study is due to applications. Let G and H be relatively prime multivariate polynomials in $\mathbb{K}[x, u]$, where $(u) = (u_1, \ldots, u_\ell)$. Currently, we can compute the lowest-order element of the elimination ideal $\langle G, H \rangle \cap \mathbb{K}[u]$ through the last element of $\mathrm{PRS}(G, H)$ and its cofactors [12]. The second reason is that conventional PRS algorithms are based

Work supported by Japan Society for Promotion of Science KAKENHI Grant number 15K00005.

© Springer Nature Switzerland AG 2018
V. P. Gerdt et al. (Eds.): CASC 2018, LNCS 11077, pp. 345–360, 2018.
https://doi.org/10.1007/978-3-319-99639-4_24

on the *pseudo remainder (Prem)*, but the Prem is not suited for sparse polynomials, and researchers are now investigating another remainder which is more reasonable than the Prem. We call the new Prem suited for sparse polynomials *sparse Prem (spsPrem)*. Then, we need a new theory for computing PRSs based on the spsPrem. We call the spsPrem-based PRS *sparse PRS (spsPRS)*.

Starting from G and H, we can generate a PRS $(P_1 = G, P_2 = H, \ldots, P_i, P_{i+1}, \ldots)$ w.r.t. x, by the formula $P_{i+1} = \mathrm{rem}(\alpha_i P_{i-1}, P_i)/\beta_i$, where $\alpha_i, \beta_i \in \mathbb{K}[\boldsymbol{u}]$. The α_i makes the remainder in $\mathbb{K}[x, \boldsymbol{u}]$ and β_i makes P_{i+1} simple by removing a common factor contained in the coefficients, hence we have $P_{i+1} \in \mathbb{K}[x, \boldsymbol{u}]$. Conventionally, the α_i is set as $\alpha_i = \mathrm{lc}(P_i)^{\delta_i}$, with $\delta_i = \deg(P_{i-1}) - \deg(P_i) + 1$, where $\mathrm{lc}(P)$ and $\deg(P)$ denote the leading coefficient and the degree of P, respectively, w.r.t. x. The remainder with this choice of α_i is the $\mathrm{Prem}(P_{i-1}, P_i)$. We note that the subresultant theory for the PRS is critically dependent on the Prem. For the subresultant theory, see [1–5,7].

Now, consider that the given polynomials G and H are sparse w.r.t. x. Then, $\mathrm{Prem}(G, H)$ is uneconomical. For example, if $(G(x, \boldsymbol{u}), H(x, \boldsymbol{u})) = (\widetilde{G}(x^l, \boldsymbol{u}), \widetilde{H}(x^l, \boldsymbol{u}))$, then the α in $\mathrm{Prem}(\widetilde{G}(x, \boldsymbol{u}), \widetilde{H}(x, \boldsymbol{u}))$ is $\alpha = \mathrm{lc}(\widetilde{H})^{\deg(\widetilde{G}) - \deg(\widetilde{H}) + 1}$. Obviously, the leading term of G can be eliminated by H with the same α, while the multiplier in $\mathrm{Prem}(G, H)$ is $\mathrm{lc}(H)^{l \deg(\widetilde{G}) - l \deg(\widetilde{H}) + 1}$. Hence, it is natural to introduce the spsPrem in which the multiplier α is made as small as possible. We give a procedure of spsPrem in Sect. 2. The concept of spsPrem is not new; Loos has defined the same concept in [9]. The problem for spsPrem is that we have no subresultant-like theory for the PRS based on spsPrem, so we cannot determine β_i in actual computation. In fact, in [9], Loos used only β_i determined by the subresultant theory.

Therefore, the first aim of this paper is to develop a subresultant-like theory for spsPrem-based PRSs. Now, we have already subresultant-like theories [10, 11]. Hence, following such theories, we develop a theory for spsPRSs in Sect. 3. Currently, the theory is not complete for determining a theoretical formula for β_i, but it is sufficient for determining β_i by Hearn's trial-division algorithm [8].

The PRS computation causes a considerable intermediate expression growth, regardless of whether Prem or spsPrem is used. The computational formula given above for P_{i+1} is executed by two steps: $P'_{i+1} := \mathrm{rem}(\alpha_i P_{i-1}, P_i) \Rightarrow P_{i+1} := P'_{i+1}/\beta_i$. The expression size of P'_{i+1} is often very large compared with P_{i+1}. If the PRS is "normal", i.e., $\deg(P_{i-1}) - \deg(P_i) = 1$, then Collins' algorithm sets $\alpha_i = \mathrm{lc}(P_i)^2$ and $\beta_i = \alpha_{i-1}$. For abnormal PRSs, Brown-Traub's algorithm is available in which the intermediate expression growth is much larger in general. Enhancing the Prem-based PRS algorithms have been challenged by several authors; see Ducos' paper [6] and references in it. Probably, Ducos' algorithm is currently most efficient. However, even in his algorithm, the division is necessary for computing P_{i+1}.

In Sect. 4, we present a new simple algorithm which suppresses the expression growth largely. The idea is to use power-series multiplication and division. We see that the division P'_{i+1}/β_i is exact, hence only a part of dividend is enough to compute the quotient. Therefore, we cut off unnecessary part of P'_{i+1} by introducing the power-series variable for sub-variables. Speeding-up the polynomial operations by using the power-series arithmetic is not new but done in [13].

2 Sparse Pseudo Remainder (spsPrem)

Let $\mathbb{K}$ be a field of numbers. In this paper, by $F(x, \boldsymbol{u})$, we denote a polynomial in $\mathbb{K}[x, \boldsymbol{u}]$, where x and $(\boldsymbol{u}) = (u_1, \ldots, u_\ell)$ are the main variable and the sub-variables, respectively; we usually treat the case of $\ell \geq 2$. Let $F(x, \boldsymbol{u})$ be expressed as $F(x, \boldsymbol{u}) = f_d(\boldsymbol{u}) x^d + f_{d-1}(\boldsymbol{u}) x^{d-1} + \cdots + f_0(\boldsymbol{u})$. By $\deg(F), \mathrm{lc}(F)$, and $\mathrm{ltm}(F)$, we denote the degree, the leading coefficient, and the leading term, respectively, of F w.r.t. x: $\deg(F) = d$, $\mathrm{lc}(F) = f_d(\boldsymbol{u})$, $\mathrm{ltm}(F) = f_d(\boldsymbol{u}) x^d$. By $\mathrm{rest}(F)$ and $\mathrm{Rest}(F, i)$, with $i \in \{1, 2, \ldots\}$, we denote the rest terms of F and the i-th rest terms of F, respectively: $\mathrm{rest}(F) = F - \mathrm{ltm}(F)$, $\mathrm{Rest}(F, i) = f_{d-i} x^{d-i} + f_{d-i-1} x^{d-i-1} + \cdots$; if F is sparse w.r.t. x, we skip the 0-coefficient terms. By $\gcd(G, H, \ldots)$ we denote the greatest common divisor (GCD) of $G, H, \ldots$. By $\mathrm{cont}(F)$ we denote the content of F w.r.t. x; $\mathrm{cont}(F) = \gcd(f_d(\boldsymbol{u}), \ldots, f_0(\boldsymbol{u}))$. By $\mathrm{rem}(G, H)$, we denote the remainder of G divided by H w.r.t. x. If $\mathrm{rem}(G, H) = 0$ then we say that H divides G and express this as $H \mid G$.

Although the procedure of spsPrem has been given in [12], we describe it below to make the paper self-contained. The cofactors A_{i+1} and B_{i+1}, satisfying $A_{i+1} G + B_{i+1} H = P_{i+1}$, play a crucial role in many cases. So, we show the procedure of spsPrem for computing $(P_{i+1}, A_{i+1}, B_{i+1})$ below, where $(P_1, A_1, B_1) = (G, 1, 0)$ and $(P_2, A_2, B_2) = (H, 0, 1)$.

> Procedure $\mathtt{spsPrem}((P_{i-1}, A_{i-1}, B_{i-1}), (P_i, A_i, B_i)) ==$
> (1) $c_j := \mathrm{lc}(P_j)$, $d_j := \deg(P_j)$ $(j = i-1, i)$;
> (2) **while** $\delta := d_{i-1} - d_i \geq 0$ **do**
> (3) $\qquad (P_{i-1}, A_{i-1}, B_{i-1}) :=$
> $\qquad\qquad c_i (P_{i-1}, A_{i-1}, B_{i-1}) - c_{i-1} x^\delta (P_i, A_i, B_i);$
> (4) $\qquad c_{i-1} := \mathrm{lc}(P_{i-1})$; $d_{i-1} := \deg(P_{i-1})$; **enddo**;
> (5) **return** $(P_{i+1}, A_{i+1}, B_{i+1}) := (P_{i-1}, A_{i-1}, B_{i-1})$.

By repeating spsPrem, we can generate spsPrem-based PRS which we call *sparse polynomial remainder sequence (spsPRS)*.

Just the same as the conventional PRS computed by using Prem, the spsPRS will be such that the coefficients of each remainder P_{i+1} $(i \geq 3)$ will contain a big common factor, let it be β_i, and we will compute P_{i+1} by removing β_i. So, we redefine the output of spsPrem to be $(P'_{i+1}, A'_{i+1}, B'_{i+1})$, and redefine $(P_{i+1}, A_{i+1}, B_{i+1})$ to be as follows.

$$\begin{cases} (P'_{i+1}, A'_{i+1}, B'_{i+1}) = \mathtt{spsPrem}((P_{i-1}, A_{i-1}, B_{i-1}), (P_i, A_i, B_i)), \\ (P_{i+1}, A_{i+1}, B_{i+1}) = (P'_{i+1}, A'_{i+1}, B'_{i+1})/\beta_i, \quad \text{where } \beta_2 = 1. \end{cases} \quad (2.1)$$

We determine β_i $(i \geq 3)$ to be a product of $\mathrm{lc}(P_j)$, where $2 \leq j \leq i-1$. This is the same as in the conventional algorithms. However, we determine β_i very differently from the conventional way, because our subresultant-like theory for the spsPRS computation is not well developed to give a theoretical formula for β_i. Our algorithm is executed in two phases. In the first phase, we determine the form of β_i by computing spsPRS of a simplified system. Then, in the second phase, we compute spsPRS by using the form of β_i determined in the first phase. For details, see Sect. 4.

3   A Matrix Theory for Sparse PRS

Let $\mathcal{M} = (c_{i,j})$, with $1 \leq i \leq m$ and $1 \leq j \leq m+n$, be an $m \times (m+n)$ matrix over $\mathbb{K}[u]$, where we assume that the leading $m-1$ columns of $\mathcal{M}$ are linearly independent. Furthermore, we assume that, for any $j \geq 0$, the $(m+j)$-th column corresponds to $x^{e_{n-j}}$, where $e_n > e_{n-1} > \cdots > e_0$. Following Collins [4], we define the *associate polynomial*, to be expressed as $\mathrm{assP}(\mathcal{M})$, as follows.

$$\mathrm{assP}(\mathcal{M}) \stackrel{\text{def}}{=} \sum_{j=0}^{n} \begin{vmatrix} c_{1,1} & \cdots & c_{1,m-1} & c_{1,m+j} \\ \vdots & \ddots & \vdots & \vdots \\ c_{m,1} & \cdots & c_{m,m-1} & c_{m,m+j} \end{vmatrix} x^{e_{n-j}}. \tag{3.1}$$

3.1   Elimination Matrix and Inverse Elimination

Although the targets of this paper are sparse polynomials, we explain the elimination matrix by dense polynomials $P_{i-2} = c_{i-2}^{(e+2)} x^{e+2} + c_{i-2}^{(e+1)} x^{e+1} + \cdots + c_{i-2}^{(0)}$, $P_{i-1} = c_{i-1}^{(e+1)} x^{e+1} + c_{i-1}^{(e)} x^e + \cdots + c_{i-1}^{(0)}$ and $P_i = c_i^{(e)} x^e + c_i^{(e-1)} x^{e-1} + \cdots + c_i^{(0)}$. Put $P'_i = \mathrm{rem}(c_{i-1}^2 P_{i-2}, P_{i-1})$ and $P'_{i+1} = \mathrm{rem}(c_i^2 P_{i-1}, P_i)$, where $c_{i-1} \stackrel{\text{def}}{=} c_{i-1}^{(e+1)}$ and $c_i \stackrel{\text{def}}{=} c_i^{(e)}$. Then, P'_{i+1} can be expressed as $P'_{i+1} = \mathrm{assP}(\mathcal{M}_{i+1}^{(i)})$, where

$$\mathcal{M}_{i+1}^{(i)} = \begin{pmatrix} c_i^{(e)} & c_i^{(e-1)} & c_i^{(e-2)} & \cdots \\ & c_i^{(e)} & c_i^{(e-1)} & \cdots \\ c_{i-1}^{(e+1)} & c_{i-1}^{(e)} & c_{i-1}^{(e-1)} & \cdots \end{pmatrix}. \tag{3.2}$$

We call the rows of $\mathcal{M}_{i+1}^{(i)}$ *coefficient vectors* or *coef-vectors* in short: the 1st, the 2nd and the 3rd rows are coef-vectors of xP_i, P_i and P_{i-1}, respectively, and the 1st, the 2nd and the 3rd columns correspond to x^{e+1}-, x^e- and x^{e-1}-terms, respectively. By upper-triangularizing the matrix $\mathcal{M}_{i+1}^{(i)}$, the bottom row of the triangularized matrix gives the coef-vector of P'_{i+1}. So, we call such a matrix as $\mathcal{M}_{i+1}^{(i)}$ *elimination matrix*.

Now, we will express P'_{i+1} by the coef-vectors of P_{i-1} and P_{i-2}; we neglect the $\pm$-sign for simplicity below. We add two coef-vectors of $x^2 P_{i-1}$ and xP_{i-1} to the above $\mathcal{M}_{i+1}^{(i)}$; let the matrix obtained be $\mathcal{M}'_{i+1}$.

Since $P_i = c_{i-1}^2 P_{i-2} - (q_{i,1} x + q_{i,0}) P_{i-1}$, with $q_{i,1}, q_{i,0} \in \mathbb{K}[u]$, we can replace two coef-vectors of P_i of $\mathcal{M}'_{i+1}$ by those of P_{i-2}. By this, we can convert $\mathcal{M}'_{i+1}$ to the following matrix $\mathcal{M}_{i+1}^{(i-1)}$.

$$\mathcal{M}_{i+1}^{(i-1)} = \begin{pmatrix} c_{i-2}^{(e+2)} & c_{i-2}^{(e+1)} & c_{i-2}^{(e)} & c_{i-2}^{(e-1)} & c_{i-2}^{(e-2)} & \cdots \\ & c_{i-2}^{(e+2)} & c_{i-2}^{(e+1)} & c_{i-2}^{(e)} & c_{i-2}^{(e-1)} & \cdots \\ c_{i-1}^{(e+1)} & c_{i-1}^{(e)} & c_{i-1}^{(e-1)} & c_{i-1}^{(e-2)} & c_{i-1}^{(e-3)} & \cdots \\ & c_{i-1}^{(e+1)} & c_{i-1}^{(e)} & c_{i-1}^{(e-1)} & c_{i-1}^{(e-2)} & \cdots \\ & & c_{i-1}^{(e+1)} & c_{i-1}^{(e)} & c_{i-1}^{(e-1)} & \cdots \end{pmatrix} \tag{3.3}$$

We call the operation which derives $\mathcal{M}_{i+1}^{(i-1)}$ from $\mathcal{M}_{i+1}^{(i)}$ *inverse elimination*.

We can find a relation between $\mathrm{assP}(\mathcal{M}_{i+1}^{(i)})$ and $\mathrm{assP}(\mathcal{M}_{i+1}^{(i-1)})$ easily, as follows; note that $c_i = c_i^{(e_i,1)}$. Definition of $\mathrm{assP}(\mathcal{M})$ in (3.1) gives $\mathrm{assP}(\mathcal{M}_{i+1}') = (c_{i-1})^2 \mathrm{assP}(\mathcal{M}_{i-1}^{(i)})$. Replacing the coef-vectors of P_i by those of P_{i-2}, we obtain $\mathrm{assP}(\mathcal{M}_{i+1}^{(i-1)}) = (\beta_{i-1}/c_{i-1}^2)^2 \mathrm{assP}(\mathcal{M}_{i+1}')$. Therefore, we find $\mathrm{assP}(\mathcal{M}_{i+1}^{(i)}) = \mathrm{assP}(\mathcal{M}_{i+1}^{(i-1)})(c_{i-1}/\beta_{i-1})^2$.

Similarly, we can express the cofactors A_{i+1} and B_{i+1} by determinants easily. We explain this by dense polynomials, by putting $P_1 = G = g_{e+1}x^{e+1} + g_e x^e + g_{e-1}x^{e-1} + \cdots$ and $P_2 = H = h_e x^e + h_{e-1}x^{e-1} + h_{e-2}x^{e-2} + \cdots$. We can express $P_3' := \mathrm{Prem}(P_1, P_2)$ and its cofactors A_3' and B_3' as follows.

$$P_3' = \mathrm{assP}\left(\begin{pmatrix} h_e & h_{e-1} & h_{e-2} & \cdots \\ & h_e & h_{e-1} & \cdots \\ g_{e+1} & g_e & g_{e-1} & \cdots \end{pmatrix} \right) = \begin{vmatrix} h_e & h_{e-1} & \mathrm{Rest}(x^1 P_2, 2) \\ & h_e & \mathrm{Rest}(x^0 P_2, 1) \\ g_{e+1} & g_e & \mathrm{Rest}(x^0 P_1, 2) \end{vmatrix},$$
$$A_3' = \begin{vmatrix} h_e & h_{e-1} & 0 \\ & h_e & 0 \\ g_{e+1} & g_e & x^0 \end{vmatrix}, \quad B_3' = \begin{vmatrix} h_e & h_{e-1} & x^1 \\ & h_e & x^0 \\ g_{e+1} & g_e & 0 \end{vmatrix}. \tag{3.4}$$

In fact, the above determinants for A_3' and B_3' give $A_3'G + B_3'H = P_3'$. The rightmost column of the determinant for P_3' may be ${}^t(x^1 P_2, x^0 P_2, x^0 P_1)$; the columns for the x^{e+1}- and x^e-terms of ${}^t(x^1 P_2, x^0 P_2, x^0 P_1)$ give no contribution because they are the same as the first and the second columns of the determinant, respectively. It is easy to generalize the above representations to A_i' and B_i'.

3.2 Constructing the Elimination Matrix $\mathcal{M}_{i+1}^{(i-1)}$

The above method is applicable to sparse polynomials too, although the matrices become pretty complicated; see an illustrative example in Subsect. 3.4.

The matrix $\mathcal{M}_{i+1}^{(i-1)}$ is now for sparse polynomials; note that, although the zero-coefficient terms are skipped, we must pad 0-elements in the matrix so that each column corresponds to the same exponent w.r.t. x. Let Q_{i-1} and Q_i be quotients in $\mathrm{spsPrem}(P_{i-2}, P_{i-1})$ and $\mathrm{spsPrem}(P_{i-1}, P_i)$, respectively, and let Q_{i-1} and Q_i consist of μ and ν terms, respectively, as follows.

$$\begin{cases} P_i' := \mathrm{spsPrem}(P_{i-2}, P_{i-1}) = \mathrm{lc}(P_{i-1})^\mu P_{i-2} - Q_{i-1}P_{i-1}, \\ Q_{i-1} = q_{i-1,\mu}x^{\delta_\mu} + \cdots + q_{i-1,1}x^{\delta_1}, \quad \delta_\mu > \cdots > \delta_1 \geq 0. \end{cases} \tag{3.5}$$

$$\begin{cases} P_{i+1}' := \mathrm{spsPrem}(P_{i-1}, P_i) = \mathrm{lc}(P_i)^\nu P_{i-1} - Q_iP_i, \\ Q_i = q_{i,\nu}x^{\delta_\nu'} + \cdots + q_{i,1}x^{\delta_1'}, \quad \delta_\nu' > \cdots > \delta_1' \geq 0. \end{cases} \tag{3.6}$$

We note that μ and ν depend on i. (We may better express μ and ν as μ_{i-1} and μ_i, respectively, which leads to complicated expressions in Q_{i-1} and Q_i above.) We will see later that the exponent-sets of Q_{i-1} and Q_i etc. are quite important. So, we define Qelist as follows.

$$\text{Qelist} := (\ldots, (i-1 : \delta_\mu, \ldots, \delta_1), (i : \delta'_\nu, \ldots, \delta'_1), \ldots). \tag{3.7}$$

In constructing the elimination matrix, x-*support*, i.e., the support w.r.t. x, plays an important role. For polynomial $P = c_n x^{e_n} + c_{n-1} x^{e_{n-1}} + \cdots + c_0 x^{e_0}$, where $e_n > e_{n-1} > \cdots > e_0$, the x-support is defined to be $\text{supp}_x(P) \overset{\text{def}}{=} \{x^{e_n}, x^{e_{n-1}}, \ldots, x^{e_0}\}$. For quotients Q_{i-1} and Q_i in (3.5) and (3.6), we have $\text{supp}_x(Q_{i-1}) = \{x^{\delta_\mu}, \ldots, x^{\delta_1}\}$ and $\text{supp}_x(Q_i) = \{x^{\delta'_\nu}, \ldots, x^{\delta'_1}\}$. We define $\mathcal{S}$ to be the x-support for all the polynomials appearing in $\mathcal{M}_{i+1}^{(i-1)}$, as follows.

$$\mathcal{S} = \left(\cup_{j=1}^{\nu} \text{supp}_x(x^{\delta'_j} P_{i-2}) \right) \cup \text{supp}_x(P_{i-1})$$
$$\cup \left(\cup_{k=1}^{\mu} \cup_{l=0}^{\nu} \text{supp}_x(x^{\delta_k + \delta'_l} P_{i-1}) \right). \tag{3.8}$$

The $\mathcal{M}_{i+1}^{(i)}$ consists of ν coef-vectors of $x^{\delta'_\nu} P_i, \ldots, x^{\delta'_1} P_i$ and one coef-vector of P_{i-1}. We can construct the $\mathcal{M}_{i+1}^{(i-1)}$ directly from P_{i-2} and P_{i-1}, as follows.

Rule-1. Let coef-vectors of $x^{\delta'_\nu} P_{i-2}, \ldots, x^{\delta'_1} P_{i-2}$ be upper ν rows of $\mathcal{M}_{i+1}^{(i-1)}$.

Rule-2. For each $j \in \{1, \ldots, \nu\}$, generate μ coef-vectors of $x^{\delta_\mu + \delta'_j} P_{i-1}, \ldots, x^{\delta_1 + \delta'_j} P_{i-1}$. Thus, we have $\mu \times \nu + 1$ coef-vectors of P_{i-1}; the last one is the coef-vector of P_{i-1}. Among these coef-vectors, let only mutually different ones be the lower rows of $\mathcal{M}_{i+1}^{(i-1)}$.

Rule-3. Arrange the elements of $\mathcal{S}$ in (3.8) in high-to-low degree order, and let each element of $\mathcal{S}$ correspond to only one column of $\mathcal{M}_{i+1}^{(i-1)}$.

Rule-4. Let $\hat{\mu}$ be the number of lower rows of $\mathcal{M}_{i+1}^{(i-1)}$. Check the $\hat{\mu}$ lower rows from the top: if the $(\nu+j)$-th row is such that $\text{lc}(P_{i-1})$ is not the $(\nu+j, j)$-element, hence the element is 0, then delete the j-th column from $\mathcal{M}_{i+1}^{(i-1)}$.

Remark 1. *The* **Rule-4** *is for the case that the leading-term elimination eliminates some lower terms, too, but it is messy to check this case in the runtime. As we will mention in Sect. 4, we will compute a PRS of a simplified system to know the sets of exponents of x, of Q_{i-1} and Q_i. Once we know the exponent-sets, constructing the elimination matrices is easy.*

Thus, we obtain the following matrix as $\mathcal{M}_{i+1}^{(i-1)}$.

$$
\mathcal{M}_{i+1}^{(i-1)} = \left. \left(\begin{array}{c} \text{coefficient vector of } x^{\delta'_\nu} P_{i-2} \\ \ddots \quad \ddots \quad \ddots \quad \ddots \\ \text{coefficient vector of } x^{\delta'_1} P_{i-2} \\ \text{coefficient vector of } x^{\delta_\mu + \delta'_\nu} P_{i-1} \\ \text{coefficient vector of } x^{\delta_\mu - 1 + \delta'_\nu} P_{i-1} \\ \ddots \quad \ddots \quad \ddots \quad \ddots \\ \text{coefficient vector of } x^{\delta_1 + \delta'_1} P_{i-1} \\ \text{coefficient vector of } P_{i-1} \end{array} \right) \right\} \begin{array}{l} \\ \\ \end{array} \qquad (3.9)
$$

$\left. \right\} \nu$ rows

$\left. \right\} \widehat{\mu}$ rows

Theorem 1. *Let the quotients Q_{i-1} in $P'_i := \text{spsPrem}(P_{i-2}, P_{i-1})$ and Q_i in $P'_{i+1} := \text{spsPrem}(P_{i-1}, P_i)$ consist of μ and ν nonzero terms, respectively, as in (3.5) and (3.6). Then, the elimination matrix $\mathcal{M}_{i+1}^{(i-1)}$ which expresses P'_{i+1} in terms of coef-vectors of P_{i-2} and P_{i-1} is given by that in (3.9) uniquely up to the exchange of rows.*

Proof. It is enough to show that the **Rule-1–Rule-4** specify the elimination matrix $\mathcal{M}_{i+1}^{(i-1)}$ uniquely. First, $\mathcal{M}_{i+1}^{(i-1)}$ must contain ν rows of P_{i-2}, hence the **Rule-1** specifies the upper rows uniquely. Then, for each upper row, μ rows of P_{i-1} are necessary, but duplicated rows are unnecessary. Hence, the **Rule-2** specifies the lower rows uniquely. The $\mathcal{S}$ specifies enough columns for $\mathcal{M}_{i+1}^{(i-1)}$. The $\mathcal{M}_{i+1}^{(i-1)}$ contains $\widehat{\mu} + \nu$ rows, and its leading $\widehat{\mu} + \nu - 1$ columns must be linearly independent. The successive leading-term elimination of P_{i-2} by P_{i-1} is nothing but the upper-triangularization of matrix $\mathcal{M}_{i+1}^{(i-1)}$. Hence, if the leading j columns, $1 < j < \widehat{\mu} + \nu$, are linearly dependent then the **Rule-4** detects the dependence and reforms the matrix. $\qquad \square$

3.3 Relation Between $\text{assP}(\mathcal{M}_{i+1}^{(i)})$ and $\text{assP}(\mathcal{M}_{i+1}^{(i-1)})$

We denote $\text{lc}(P_i)$ and $\deg(P_i)$ by c_i and d_i, respectively, as before. The matrix $\mathcal{M}_{i+1}^{(i)}$ for $P'_{i+1} = \text{spsPrem}(P_{i-1}, P_i)$ contains ν coef-vectors of P_i and one coef-vector of P_{i-1}. On the other hand, the matrix $\mathcal{M}_{i+1}^{(i-1)}$ contains ν coef-vectors of P_{i-2} and $\widehat{\mu}$ coef-vectors of P_{i-1}. Considering the reformation of $\text{assP}(\mathcal{M}_{i+1}^{(i)})$ to $\text{assP}(\mathcal{M}'_{i+1})$ in Subsect. 3.1, we see that adding $\widehat{\mu} - 1$ coef-vectors of P_{i-1} to $\mathcal{M}_{i+1}^{(i)}$ is equal to multiply $(c_{i-1})^{\widehat{\mu}-1}$ to the matrix. Since $P_i = [(c_{i-1})^\mu P_{i-2} - Q_{i-1} P_{i-1}]/\beta_{i-1}$, replacement of each coef-vector of P_i in $\mathcal{M}_{i+1}^{(i)}$ by that of P_{i-2} is equal to multiplying $(c_{i-1})^\mu / \beta_{i-1}$ to $\text{assP}(\mathcal{M}_{i+1}^{(i-1)})$. Therefore, we obtain the following theorem (we neglect the $\pm$-sign).

Theorem 2. *We have the following relation for $i \geq 3$.*

$$\mathrm{assP}(\mathcal{M}_{i+1}^{(i)}) = \mathrm{assP}(\mathcal{M}_{i+1}^{(i-1)}) \cdot \frac{(c_{i-1})^{\lambda_i}}{(\beta_{i-1})^{\nu}}, \quad \text{where} \qquad (3.10)$$

$$\nu \leq \widehat{\mu} \leq \mu\nu + 1, \quad \lambda_i \overset{\mathrm{def}}{=} \mu\nu - \widehat{\mu} + 1 \geq 0. \qquad (3.11)$$

Proof. Derivation of equation in (3.10) has been explained above. In the matrix $\mathcal{M}_{i+1}^{(i)}$, at least one coef-vector of P_{i-1} is necessary to eliminate the leading element of each coef-vector of P_{i-2}. Hence, we have $\nu \leq \widehat{\mu}$. In Rule-2 above, $\mu\nu + 1$ coef-vectors of P_{i-1} are generated, and $\widehat{\mu}$ is the number of mutually different ones among them. Hence, we have $\widehat{\mu} \leq \mu\nu + 1$. □

Remark 2. *One may think that the expression in the r.h.s. of (3.10) is a rational function, but it is wrong. Eliminating upper ν rows of $\mathcal{M}_{i+1}^{(i-1)}$ by $\widehat{\mu}$ lower rows, just similarly as the determinant computation, each upper row is converted to a coef-vector of P_i' which can be divided by β_{i-1}. Hence, in determining β_i for P_{i+1}', we may neglect the factor $(\beta_{i-1})^{\nu}$ in (3.10).*

Remark 3. *One may think that the β_i is determined by only the factor $c_{i-1}^{\mu\nu - \widehat{\mu} + 1}$ in (3.10), but it is wrong. If $\mathrm{lc}(P_{i-2})$ is a factor of $\mathrm{lc}(P_{i-1})$ then the $\mathrm{lc}(P_{i-2})$ is contained in β_i, as we will show in the next subsection.*

3.4 An Illustrative Example

We explain the construction of the elimination matrix explicitly by an example, and show that we can determine the β_i once the elimination matrix is constructed. However, the determination of β_i is rather complicated as we have mentioned in Remark 3; we have chosen the example to show this clearly.

The coef-vectors of P_{i-1} added to $\mathcal{M}_{i+1}^{(i)}$ are specified by $\mathrm{supp}_x(Q_{i-1}Q_i)$. Actually, we use the exponent-set of $\mathrm{supp}_x(Q_{i-1}Q_i)$. The exponent-sets of Q_{i-1} and Q_i are $\{\delta_\mu, \ldots, \delta_1\}$ and $\{\delta_\nu', \ldots, \delta_1'\}$, respectively, and the exponent-set of $Q_{i-1}Q_i$ is computed as $\{\delta_\mu, \ldots, \delta_1\} \oplus \{\delta_\nu', \ldots, \delta_1'\} \overset{\mathrm{def}}{=} \{\delta_\mu + \delta_\nu', \ldots, \delta_1 + \delta_\nu', \ldots, \delta_1 + \delta_1'\}$. We neglect the $\pm$-sign of the elimination matrix below.

Example 1. Let P_1 and P_2 be the following polynomials.

$$\begin{cases} P_1 = x^{10} \times (y+z) + x^7 \times (2y-z) + x^5 \times (3y) - x^3 \times (2z) + (2y-3z), \\ P_2 = x^{10} \times (y-z) + x^7 \times (y-3z) - x^5 \times (5z) + x^3 \times (4y) + (3y+5z). \end{cases} \quad (3.12)$$

This example and the following remainder polynomials were given in [12]; we can compute the polynomials by applying procedures spsPrem and reducePrem given in the next section.

$$\begin{cases} P_3 = c_{3,7}x^7 + c_{3,5}x^5 + c_{3,3}x^3 + c_{3,0} & \Longleftarrow \mathrm{rem}(c_2^1 P_1, P_2)/\,1, \\ P_4 = c_{4,6}x^6 + c_{4,5}x^5 + c_{4,4}x^4 + c_{4,3}x^3 + c_{4,1}x + c_{4,0} & \Longleftarrow \mathrm{rem}(c_3^3 P_2, P_3)/c_2, \\ P_5 = c_{5,5}x^5 + c_{5,4}x^4 + c_{5,3}x^3 + c_{5,2}x^2 + c_{5,1}x + c_{5,0} & \Longleftarrow \mathrm{rem}(c_4^2 P_3, P_4)/c_3^2, \quad (3.13) \\ P_6 = c_{6,4}x^4 + c_{6,3}x^3 + c_{6,2}x^2 + c_{6,1}x + c_{6,0} & \Longleftarrow \mathrm{rem}(c_5^2 P_4, P_5)/c_4^2 c_3, \\ P_{i+1}\ (i = 6, 7, 8, 9) \text{ are omitted} & \Longleftarrow \beta_i = \alpha_{i-1} = c_{i-1}^2, \end{cases}$$

where $c_3 \stackrel{\text{def}}{=} \mathrm{lc}(P_3) = y^2 - yz + 4z^2$, $c_4 \stackrel{\text{def}}{=} \mathrm{lc}(P_4) = c_3 \times (13y^4 + 14y^3 z + 42y^2 z^2 + 46yz^3 + 17z^4)$, and $c_j \stackrel{\text{def}}{=} \mathrm{lc}(P_j)$ $(j \geq 5)$ are irreducible. The Qelist defined in (3.7) is Qelist $= ((2:\ 0),\ (3:\ 3\ 1\ 0),\ (4:\ 1\ 0),\ (5:\ 1\ 0),\ \dots)$ and we have $(\mu, \nu) = (1, 3), (3, 2), (2, 2)$ for P_4', P_5', P_6', respectively. Below, we consider P_4', P_5', P_6'.

$$P_4' = \mathrm{assP}\left(\begin{array}{ccccccc} x^{10} & x^8 & x^7 & \cdots & x^3 & x^1 & x^0 \\ \hline c_{3,7} & c_{3,5} & & \cdots & c_{3,0} & & \\ & c_{3,7} & & \cdots & & c_{3,0} & \\ & & c_{3,7} & \cdots & c_{3,3} & & c_{3,0} \\ c_{2,10} & & c_{2,7} & \cdots & c_{2,3} & & c_{2,0} \end{array} \right) \quad \Leftarrow\ \mathrm{rem}(c_3^3 P_2, P_3)\ (3.14)$$

$(\ \{0\} \oplus \{3, 1, 0\} = \{3, 1, 0\}\ \Rightarrow\ $ add coef-vectors of $x^3 P_2, x P_2\ )$

$$= \mathrm{assP}\left(\begin{array}{cccccccc} x^{13} & x^{11} & x^{10} & x^8 & x^7 & x^6 & \cdots & x^0 \\ \hline c_{1,10} & & c_{1,7} & c_{1,5} & & c_{1,3} & \cdots & \\ & c_{1,10} & & c_{1,7} & & c_{1,5} & \cdots & \\ & & c_{1,10} & & c_{1,7} & & \cdots & c_{1,0} \\ c_{2,10} & & c_{2,7} & c_{2,5} & & c_{2,3} & \cdots & \\ & c_{2,10} & & c_{2,7} & & c_{2,5} & \cdots & \\ & & c_{2,10} & & c_{2,7} & & \cdots & c_{2,0} \end{array} \right) \times \frac{(c_2/\beta_2)^3}{(c_2)^2}. \quad (3.15)$$

By the right factor in (3.15) and $\beta_2 = 1$, we can set β_3 to be c_2.

$$P_5' = \mathrm{assP}\left(\begin{array}{ccccc} x^7 & x^6 & x^5 & \cdots & x^1 & x^0 \\ \hline c_{4,6} & c_{4,5} & c_{4,4} & \cdots & c_{4,0} & \\ & c_{4,6} & c_{4,5} & \cdots & & c_{4,0} \\ c_{3,7} & & c_{3,5} & \cdots & & c_{3,0} \end{array} \right) \quad \Leftarrow\ \mathrm{rem}(c_4^2 P_3, P_4) \qquad (3.16)$$

$(\ \{3, 1, 0\} \oplus \{1, 0\} = \{4, 3, 2, 1, 0\}\ \Rightarrow\ $ add 4 coef-vectors of $P_3\ )$

$$= \mathrm{assP}\left(\begin{array}{cccccccc} x^{11} & x^{10} & x^9 & x^8 & x^7 & x^6 & x^5 & \cdots & x^0 \\ \hline c_{2,10} & & & c_{2,7} & & c_{2,5} & & \cdots & \\ & c_{2,10} & & & c_{2,7} & & c_{2,5} & \cdots & c_{2,0} \\ c_{3,7} & & c_{3,5} & & c_{3,3} & & & \cdots & \\ & c_{3,7} & & c_{3,5} & & c_{3,3} & & \cdots & \\ & & c_{3,7} & & c_{3,5} & & c_{3,3} & \cdots & \\ & & & c_{3,7} & & c_{3,5} & & \cdots & \\ & & & & c_{3,7} & & c_{3,5} & \cdots & c_{3,0} \end{array} \right) \times \frac{(c_3^3/\beta_3)^2}{(c_3)^4}. \quad (3.17)$$

The right factor in (3.17) gives us $\beta_4 = c_3^2$; we need not consider $(1/\beta_3)^2$ due to Remark 2. In fact, eliminating the 1st row (resp. the 2nd row) of the matrix in (3.17) by 3rd, 5th and 6th rows (resp. 4th, 6th and 7th rows), we see that the resulting row contains a factor β_3. On the other hand, determination of β_5 is complicated.

$$P_6' = \mathrm{assP}\left(\begin{pmatrix} x^6 & x^5 & x^4 & \cdots & x^1 & x^0 \\ \hline c_{5,5} & c_{5,4} & c_{5,3} & \cdots & c_{5,0} & \\ & c_{5,5} & c_{5,4} & \cdots & c_{5,1} & c_{5,0} \\ c_{4,6} & c_{4,5} & c_{4,4} & \cdots & c_{4,1} & c_{4,0} \end{pmatrix}\right) \quad \Leftarrow \mathrm{rem}(c_5^2 P_4, P_5) \qquad (3.18)$$

$$(\ \{1,0\} \oplus \{1,0\} = \{2,1,0\} \ \Rightarrow \ \text{add coef-vectors of } x^2 P_4, x P_4\ )$$

$$= \mathrm{assP}\left(\begin{pmatrix} x^8 & x^7 & x^6 & x^5 & x^4 & \cdots & x^0 \\ \hline c_{3,7} & & c_{3,5} & c_{3,4} & c_{3,3} & \cdots & \\ & c_{3,7} & & c_{3,5} & c_{3,4} & \cdots & c_{3,0} \\ c_{4,6} & c_{4,5} & c_{4,4} & c_{4,3} & c_{4,2} & \cdots & \\ & c_{4,6} & c_{4,5} & c_{4,4} & c_{4,3} & \cdots & \\ & & c_{4,6} & c_{4,5} & c_{4,4} & \cdots & c_{4,0} \end{pmatrix}\right) \times \frac{(c_4^2/\beta_4)^2}{(c_4)^2}. \qquad (3.19)$$

In this case, the right factor of (3.19) gives c_4^2 and, since $c_{4,6}$ is a multiple of $c_3(= c_{3,4})$, the first column of the matrix in (3.19) gives c_3, hence we can set $\beta_5 = c_4^2 c_3$. $\qquad\Box$

4   An Algorithm for Computing SpsPRS

In this and the next sections, we assume that $G, H \in \mathbb{Z}[x, \boldsymbol{u}]$.

One will be able to find a theoretical formula of β_i if one repeats the inverse elimination until P_{i+1}' is expressed by a matrix $\mathcal{M}_{i+1}^{(2)}$ the rows of which are coef-vectors of $P_1 = G$ and $P_2 = H$. For performing this plan, one must know the quotient sequence $(Q_2, Q_3, \ldots, Q_{k-1})$. However, the sequence is complicated in general if G and H are sparse. So, in the first half of this section, we will show that we can estimate $\beta_2, \ldots, \beta_{k-1}$ by Hearn's trial-division algorithm given below. In the second half of this section, we propose a very simple but efficient algorithm which allows us to suppress the intermediate expression growth caused by the PRS formula in (2.1)

4.1   Hearn's Trial-Division Algorithm

In the computation of Prem-based PRS, β_i is chosen to be $\beta_i = \prod_{j=2}^{i-1}(\mathrm{lc}(P_j))^{n_{ij}}$, where $n_{i,j} \in \mathbb{Z}$; the case of negative $n_{i,j}$ appears only in the "abnormal PRS" in which $\deg(P_{i'-1}) - \deg(P_{i'}) > 1$ for some $i' < i$. In the Prem-based PRS, even if $\mathrm{lc}(P_{i-1})$ is a factor of $\mathrm{lc}(P_i)$ hence P_{i+1}' is obviously a multiple of $\mathrm{lc}(P_{i-1})$, this factor is not included in β_i.

In Hearn's algorithm, we assume that β_i is of the form $\beta_i = \prod_{j=2}^{i-1}(\mathrm{lc}(P_j))^{\nu_{ij}}$, where $\nu_{i,j} \geq 0$. This assumption is verified by Theorem 2 with Remark 2. With this assumption only, we can determine the values of $\nu_{i,i-1}, \ldots, \nu_{i,2}$, by successive trial-divisions of P_{i+1}' by $\mathrm{lc}(P_j)$. It should be emphasized that, if $\mathrm{lc}(P_{i-1})$ is a factor of $\mathrm{lc}(P_i)$ hence P_{i+1}' contains $\mathrm{lc}(P_{i-1})$ as a factor, then Hearn's algorithm removes $\mathrm{lc}(P_{i-1})$ from P_{i+1}'.

Hearn's algorithm uses a list `Alphs` which is a list of (c_j, μ_j), $j = 2, 3, \ldots$, where $c_j = \mathrm{lc}(P_j)$ and μ_j denotes the number of times c_j is multiplied to P_{j-1}: $P_{j+1}' = \mathrm{rem}((c_j)^{\mu_j} P_{j-1}, P_j)$. The algorithm performs the trial-division of P_{i+1}'

by c_j, from $j = i-1$ to $j = 2$ successively; we try the division from bigger to smaller divisors, because c_j may contain $c_{j'}$ $(j' < j)$. If the trial-division by c_j succeeds then we decrease μ_j by 1 and continue the trial-division by c_j so long as $\mu_j > 0$. We do not perform the trial-division by c_j if $\mu_j = 0$; since $(c_j)^{\mu_j}$ is multiplied to P_{j-1}, the c_j can be removed only μ_j times at most.

> Procedure $\texttt{reducePrem}(i+1, (P_{i+1}, A_{i+1}, B_{i+1}), \texttt{Alphs}) ==$
> %% Hearn's trial-division algorithm [8]
> (1) **for** $j = i - 1$ **to** 2 **step** -1 **do**
> (2) $c_j := \text{1-st}(j\text{-th}(\texttt{Alphs}));$ $\mu_j := \text{2-nd}(j\text{-th}(\texttt{Alphs}));$
> (3)   **while** $\mu_j > 0$ **and** c_j divides A_{i+1}, B_{i+1} **do**
> (4) $(P_{i+1}, A_{i+1}, B_{i+1}) := (P_{i+1}, A_{i+1}, B_{i+1})/c_j;$
> $\mu_j := \mu_j - 1;$ **enddo;**
> (5)   **enddo; return** $(P_{i+1}, A_{i+1}, B_{i+1})$.

We check the exact-division of A_{i+1}, B_{i+1} by c_j first in line (3), because if the divisions succeed then P_{i+1} is always divisible by c_j, but the converse is not always true. Hearn's algorithm is practically quite good for sparse polynomials.

4.2   Avoiding Intermediate Expression Growth

As we have mentioned in Sect. 1, our idea is to compute the products and the quotients of multivariate polynomials in formulas in (2.1) by the power-series arithmetic w.r.t. sub-variables; the power-series arithmetic allows us to compute only the lower-power terms of the products which are necessary to obtain the quotients exactly.

In order to execute the above plan, the forms of $\beta_3, \ldots, \beta_{k-1}$ must be known before the computation of spsPRS. We get this information by computing the spsPRS of a simplified system $(\widetilde{G}, \widetilde{H})$. By spsPRS$(\widetilde{G}, \widetilde{H})$, we compute $\texttt{prsHist}$, the history of PRS-computation. The $\texttt{prsHist}$ is a list that the i-th element of which is $(i, (\text{Mul } \mu_i), (\text{Div } \nu_{i,i-1}, \ldots, \nu_{i,2}))$, showing that $\alpha_i = (\text{lc}(P_i))^{\mu_i}$ and $\beta_i = \prod_{j=i-1}^{2}(\text{lc}(P_j))^{\nu_{ij}}$. We propose two choices for specifying $(\widetilde{G}, \widetilde{H})$.

- **Choice-S:** Substitute different **small** prime numbers for the sub-variables of G and H, and compute the PRS$(\widetilde{G}(x), \widetilde{H}(x))$ over $\mathbb{Z}$; currently, each prime p satisfies $|p| \geq 5$.
- **Choice-L:** Substitute different **large** random integers for sub-variables except one, of G, H, and compute the PRS$(\widetilde{G}(x, u_1), \widetilde{H}(x, u_1))$ over p, where p is a large prime number (word-size, say).

The Choice-S is for small systems (a few sub-variables, low degrees and small numerical coefficients), and the Choice-L is for large systems of many sub-variables.

We next explain how the sub-variables are treated as power-series. We introduce a system variable T, with the variable-order $x \succ T \succ u_1, \ldots u_\ell$, and treat T as the power-series variable. We multiply T to sub-variables, according to one of the next two choices.

– **Choice-H:** We multiply T to sub-variable except for the first sub-variable u_1: $(u_2, \ldots, u_\ell) \mapsto (Tu_2, \ldots, Tu_\ell)$.
– **Choice-A:** We multiply T to all sub-variables: $(u_1, \ldots, u_\ell) \mapsto (Tu_1, \ldots, Tu_\ell)$.

Thus, T denotes the "total-degree" of sub-variables being multiplied by T. The Choice-H is suited for the case in which coefficients of P_i (especially P_k) are nearly homogeneous in the sub-variables; in this case the Choice-A is very ineffective for cutting off the higher degree terms. The Choice-A is suited for the case in which coefficients of P_i (especially P_k) consist of terms with total-degrees distributed widely in each sub-variable.

We explain the power-series arithmetic briefly. We assume that the recursive representation is adopted to express polynomials and power-series inside the computer: let $F(x, \boldsymbol{u}) = \sum_{i=0}^{d} f_i(\boldsymbol{u}) x^i$, then F is represented by a list $((d, f_d) \ldots, (i, f_i), \ldots)$, and each coefficient f_i is also represented by a list recursively. Since $x \succ T \succ u_1, u_2, \ldots, u_\ell$, the power-series operations are executed only on the coefficients w.r.t. x. Each coefficient w.r.t. x is a power-series in T, and the leading terms of the coefficient are the terms of the lowest degree w.r.t. T; in the Choice-H, a polynomial in u_1 is the leading terms. In the power-series arithmetic, the "cutoff-degree" T_{cut} must be set, and the products and the quotients are computed only up to T_{cut} w.r.t. T, from low-to-high degrees. Therefore, the above variable-ordering and the recursive representation are very suited for executing the power-series arithmetic efficiently.

We set the cutoff-degree T_{cut} for P_{i+1} as follows.

$$T_{\text{cut}}(P_{i+1}) := \deg_T(P_{i-1}) + \mu_i \times \deg_T(\text{lc}(P_i)) - (\deg_T(\beta_i) - \text{ord}_T(\beta_i)), \quad (4.1)$$

where $\deg_T(P)$ denotes the degree of $P \in \mathbb{Z}[x, T, \boldsymbol{u}]$, w.r.t. T, and $\text{ord}_T(\beta)$ denotes the "order" of $\beta \in \mathbb{Z}[T, \boldsymbol{u}]$, i.e. the lowest power w.r.t. T, of terms of β.

Summarizing the above, our new algorithm which we call spsPcPRS is executed as follows; by "Pc" we mean "Power-series coefficients".

> Procedure spsPcPRS(G, H) ==
> %% use spsPrem for the remainder computation.
> %% use reducePrem for computing prsHist.
> %% for simplicity, we omit A_i and B_i below.
> (1) **construct** a simplified system $(\widetilde{G}, \widetilde{H})$;
> (2) **compute** prsHist, as mentioned above;
> (3) **define** the power-series variable T, and
>   **multiply** to $(\boldsymbol{u})$ as mentioned above;
> (4) **while** $\deg(P_i) > 0$ **do**
>   { **compute** $C_i := T_{\text{cut}}(P_{i+1})$ by (4.1);
>   **compute** P'_{i+1} up to C_i w.r.t. T;
>   **compute** $P_{i+1} := P'_{i+1}/\beta_i$ up to C_i
>   by the power-series division };
> (5) **return** $\{G, H, P_3, \ldots, P_k\}$.

4.3   Simple Experiments and Remarks

We implemented procedure spsPcPRS on our algebra system GAL which was developed mainly in Sasaki's Lab., and made simple experiments on spsPcPRS. We have adopted **Choice-S** for constructing $\{\widetilde{G}, \widetilde{H}\}$ and **Choice-H** for multiplying T to sub-variables. The experiment was done on a computer with Intel(R)-U2300 (1.20 GHz), operated by Linux 3.4.100.

Experiment 1 (Computation of prsHist). Let G and H be as follows.

$$\begin{cases} G := x^7 \times (y+z) + x^5 \times (y-2z) + x^2 \times (2y-z) + (2y-3z), \\ H := x^7 \times (y-z) + x^5 \times (2y+z) + x^2 \times (y-3z) + (3y+5z). \end{cases} \quad (4.2)$$

Substituting 5 and -7 for y and z of G and H, we obtain

$$\widetilde{G} = -2x^7 + 19x^5 + 17x^2 + 31, \qquad \widetilde{H} = 12x^7 + 3x^5 + 26x^2 - 20.$$

Let $\widetilde{P}'_{i+1}$ and $\widetilde{\beta}_i$ be the i-th element of spsPRS$(\widetilde{G}, \widetilde{H})$ and corresponding divisor, respectively. We show $\widetilde{P}'_{i+1}$ and $\widetilde{\beta}_i$ for $i \geq 4$.

$$\widetilde{P}'_5 = -6077916\,x^3 - 15335424\,x^2 + 25899588\,x - 19888128,$$
$$:\ \widetilde{\beta}_4 = 234 \ (= \mathrm{lc}(\widetilde{P}_3)),$$
$$\widetilde{P}'_6 = -411977666259432\,x^2 + \cdots - 408338884048680,$$
$$:\ \widetilde{\beta}_5 = 234 \ (= \mathrm{lc}(\widetilde{P}_3)),$$
$$\widetilde{P}'_7 = -896164609226696558184252 2112\,x + 476806936412081921559562 32192,$$
$$:\ \widetilde{\beta}_6 = (-25974)^2 \times (-59904) \ (= \mathrm{lc}(\widetilde{P}_5))^2\mathrm{lc}(\widetilde{P}_4)),$$
$$\widetilde{P}'_8 = -190277237388214975621248015433997975 9143232,$$
$$:\ \widetilde{\beta}_7 = (-1760588317348)^2 \ (= (\mathrm{lc}(\widetilde{P}_6))^2).$$

We obtained the same prsHist for $(G(x,7,-11), H(x,7,-11))$.   □

Experiment 2 (Computation of P_{i+1} with Power-series). Let G and H be as follows, where X is the main variable (this example was treated in [12]).

$$\begin{cases} G = X^6\,(u+2v+w) + X^4\,(u-2x-z) + X^2\,(v+3y-z) + (v+2w+y), \\ H = X^6\,(v-w+2x) - X^4\,(v+y-2z) + X^2\,(w-2x+y) + (u-v+2z). \end{cases} \quad (4.3)$$

From G and H, we generate polynomial pair $(G_\ell, H_\ell) =:$ **Ex-ℓ**, as follows.

$$\begin{aligned}
&\textbf{Ex-6} : (G_6, H_6) := (G, H), \\
&\textbf{Ex-5} : (G_5, H_5) := \text{replace} \quad (z) \text{ by } (w) \quad \text{in } (G, H), \\
&\textbf{Ex-4} : (G_4, H_4) := \text{replace} \quad (y, z) \text{ by } (v, w) \quad \text{in } (G, H), \\
&\textbf{Ex-3} : (G_3, H_3) := \text{replace } (x, y, z) \text{ by } (u, v, w) \text{ in } (G, H).
\end{aligned}$$

The G and H in (4.3) suggest that the last element of spsPRS(G, H) is nearly homogeneous in the sub-variables, so we employ **Choice-H**. In each **Ex-i**, the PRS is $(G_i, H_i, P_3, P_4, P_5)$.

We compared our new algorithm based on power-series arithmetic with old one based on Hearn's trial-division only, and the result is shown in Table 1, where "#tms(P)" denotes the number of monomials contained in polynomial P, (CPU) and (GC) denote "Central Processing Unit" and "Garbage Collection", respectively. The unit of time is milliseconds.

Table 1. Comparison of new algorithm with old one.

Ex-ℓ	(**old**) trial-division		(**new**) power-series division	
	#tms(P_5', A_5', B_5')	#tms(P_5, A_5, B_5)	#tms(P_5', A_5', B_5')	#tms(β_4)
Ex-3	$(65, 163, 163)$	$(28, 62, 62)$	$(27, 62, 63)$	15
	time(CPU)3.70 + (GC)0.00		time(CPU)2.77 + (GC)0.00	
Ex-4	$(279, 603, 603)$	$(81, 154, 160)$	$(81, 163, 164)$	28
	time(CPU)13.6 + (GC)2.01		time(CPU)6.12 + (GC)0.53	
Ex-5	$(961, 1880, 1880)$	$(201, 329, 312)$	$(206, 353, 330)$	51
	time(CPU)48.3 + (GC)8.13		time(CPU)12.9 + (GC)1.66	
Ex-6	$(2815, 5192, 5192)$	$(445, 665, 671)$	$(455, 728, 705)$	84
	time(CPU)165. + (GC)32.5		time(CPU)28.9 + (GC)4.94	

Table 1 shows a very nice performance of our new algorithm: unnecessary terms are cut off by power-series almost completely. We must say, however, that the data in Table 1 are too nice. Performance of our algorithm depends on the "sub-variable ordering" very much, and the above data were obtained by choosing the best sub-variable ordering.   □

The performance of our new algorithm will be good (resp. bad) if the number of lowest-degree terms w.r.t. T, of β_i (especially β_{k-1}) is small (resp. large). Furthermore, (4.1) tells that $T_{\text{cut}}(P_{i+1})$ becomes larger if $\deg_T(\beta_i) - \text{ord}_T(\beta_i) > 0$. Hence, we test our algorithm based on power-series division, by changing the ordering of sub-variables.

Experiment 3 (Dependence on Sub-variable Ordering). We use the system **Ex-6** given in Experiment 2. Most computation of each PRS is occupied by that of P_5. Hence, we show the the numbers of terms of P_5', A_5', B_5' as well as the total computation times. Note that #tms(P_5', A_5', B_5') = (2815, 5192, 5192) by the old algorithm (Table 2).

The timing data are classified into three classes, Class-(1), Class-(2) and Class-(3), in which leading sub-variables are in $\{v, x\}$, $\{y, z\}$, and $\{u, w\}$, respectively,

Table 2. Efficiency depends subVar-ordering strongly.

Ordering	#tms($P_5', A_5', B_5', \beta_4$)	Comput. time (msec)
$v \succ u \succ w \succ x \succ y \succ z$	(462, 756, 752, 84)	(CPU)32.0 + (GC)5.74
$x \succ u \succ v \succ w \succ y \succ z$	(455, 728, 705,84)	(CPU)28.9 + (GC)4.94
$y \succ u \succ v \succ w \succ x \succ z$	(1144, 1861, 1815, 84)	(CPU)82.9 + (GC)14.6
$z \succ u \succ v \succ w \succ x \succ y$	(1174, 1944, 1955, 84)	(CPU)87.7 + (GC)15.3
$u \succ v \succ w \succ x \succ y \succ z$	(1270, 2150, 2246, 84)	(CPU)102. + (GC)17.7
$w \succ u \succ v \succ x \succ y \succ z$	(1270, 2275, 2278, 84)	(CPU)104. + (GC)18.6

$v : \beta_4 = 4v^4 + T \times (5 \text{ terms}) + T^2 \times (13 \text{ terms}) + T^3 \times (28 \text{ terms}) + T^4 \times (37 \text{ terms})$,

$x : \beta_4 = 16x^4 + T \times (4 \text{ terms}) + T^2 \times (12 \text{ terms}) + T^3 \times (25 \text{ terms}) + T^4 \times (42 \text{ terms})$,

$y : \beta_4 = T^2 \times (6 \text{ terms}) + T^3 \times (26 \text{ terms}) + T^4 \times (52 \text{ terms})$,

$z : \beta_4 = T^2 \times (10 \text{ terms}) + T^3 \times (26 \text{ terms}) + T^4 \times (48 \text{ terms})$,

$u : \beta_4 = T^2 \times (15 \text{ terms}) + T^3 \times (27 \text{ terms}) + T^4 \times (42 \text{ terms})$,

$w : \beta_4 = T^2 \times (15 \text{ terms}) + T^3 \times (27 \text{ terms}) + T^4 \times (42 \text{ terms})$.

We see that our new algorithm shows the best performance when β_{k-1} contains a term of the form u_j^l and the sub-variable u_j is set to be of highest order.   □

Remark 4 (On Setting Sub-variable Ordering Optimally). *Since we know the β_i before the computation of P_{i+1}', one idea of setting the sub-variable ordering is to investigate β_i whether or not it contains a term of single sub-variable or terms of a very small total-degree w.r.t. sub-variables. An optimal ordering may depend on i. Hence, this check may be done for β_{k-1} only.*   □

Finally, we comment on the time complexity of our new algorithm. The complexity analysis of arithmetic operations of sparse multivariate polynomials is not so easy because there are many models of polynomials; see [12] for one of such models and a complexity analysis based on the model. As for the complexity of the spsPrem-based old algorithm, see the analysis given in [12]. In this paper, we show a comparison of the old algorithm and the new one using power-series arithmetic, which is easy.

Let C_{old} and C_{new} be time-complexities of computing (P_k', A_k', B_k') by the old and the new algorithms, respectively, and $\|P_{k,old}'\|$ and $\|P_{k,new}'\|$ be the numbers of terms of P_k's computed by the old and new algorithms, respectively. Since the computation of spsPRS is dominated by that of (P_k', A_k', B_k'), and since complexity of P_{i+1} by formulas in (2.1) is approximated by the size of P_{i+1}', we have the following.

$$C_{new}/C_{old} = O(\|P_{k,new}'\|/\|P_{k,old}'\|). \tag{4.4}$$

References

1. Brown, W.S.: On Euclid's algorithm and the computation of polynomial greatest common divisors. JACM **18**(4), 478–504 (1971)
2. Brown, W.S., Traub, J.F.: On Euclid's algorithm and the theory of subresultants. JACM **18**(4), 505–515 (1971)
3. Brown, W.S.: The subresultant PRS algorithm. ACM TOMS **4**, 237–249 (1978)
4. Collins, G.E.: Polynomial remainder sequences and determinants. Am. Math. Mon. **71**, 708–712 (1966)
5. Collins, G.E.: Subresultants and reduced polynomial remainder sequences. JACM **14**, 128–142 (1967)
6. Ducos, L.: Optimizations of the subresultant algorithm. J. Pure Appl. Algebra **145**, 149–163 (2000)
7. Habicht, W.: Zur inhomogenen Eliminationstheorie. Comm. Math. Helvetici **21**, 79–98 (1948)
8. Hearn, A.C.: Non-modular computation of polynomial GCDS using trial division. In: Ng, E.W. (ed.) Symbolic and Algebraic Computation. LNCS, vol. 72, pp. 227–239. Springer, Heidelberg (1979). https://doi.org/10.1007/3-540-09519-5_74
9. Loos, R.: Generalized polynomial remainder sequence. In: Buchberger, B., Collins, G.E., Loos, R. (eds.) Computer Algebra. Computing Supplementum, vol. 4, pp. 115–137. Springer, Vienna (1982). https://doi.org/10.1007/978-3-7091-3406-1_9
10. Sasaki, T.: A subresultant-like theory for Buchberger's procedure. JJIAM (Jap. J. Indust. Appl. Math.) **31**, 137–164 (2014)
11. Sasaki, T., Furukawa, A.: Theory of multiple polynomial remainder sequence. Publ. RIMS (Kyoto Univ.) **20**, 367–399 (1984)
12. Sasaki, T., Inaba, D.: Simple relation between the lowest-order element of ideal $\langle G, H \rangle$ and the last element of polynomial remainder sequence. In: Proceedings of SYNASC 2017 (Symbolic and Numeric Algorithms for Scientific Computing), IEEE Computer Society (2017, in printing)
13. Sasaki, T., Suzuki, M.: Three new algorithms for multivariate polynomial GCD. J. Symb. Comput. **13**, 395–411 (1992)

A Blackbox Polynomial System Solver on Parallel Shared Memory Computers

Jan Verschelde[✉]

Department of Mathematics, Statistics, and Computer Science,
University of Illinois at Chicago, 851 S. Morgan Street (m/c 249),
Chicago, IL 60607-7045, USA
janv@uic.edu
http://www.math.uic.edu/~jan

Abstract. A numerical irreducible decomposition for a polynomial system provides representations for the irreducible factors of all positive dimensional solution sets of the system, separated from its isolated solutions. Homotopy continuation methods are applied to compute a numerical irreducible decomposition. Load balancing and pipelining are techniques in a parallel implementation on a computer with multicore processors. The application of the parallel algorithms is illustrated on solving the cyclic n-roots problems, in particular for $n = 8, 9$, and 12.

Keywords: Homotopy continuation
Numerical irreducible decomposition · Mathematical software
Multitasking · Pipelining · Polyhedral homotopies
Polynomial system · Shared memory parallel computing

1 Introduction

Almost all computers have multicore processors enabling the simultaneous execution of instructions in an algorithm. The algorithms considered in this paper are applied to solve a polynomial system. Parallel algorithms can often deliver significant speedups on computers with multicore processors.

A blackbox solver implies a fixed selection of algorithms, run with default settings of options and tolerances. The selected methods are homotopy continuation methods to compute a numerical irreducible decomposition of the solution set of a polynomial system. As the solution paths defined by a polynomial homotopy can be tracked independently from each other, there is no communication and no synchronization overhead. Therefore, one may hope that with p threads, the speedup will be close to p.

The number of paths that needs to be tracked to compute a numerical irreducible decomposition can be a multiple of the number of paths defined by a

This material is based upon work supported by the National Science Foundation under Grant No. 1440534.

homotopy to approximate all isolated solutions. Nevertheless, in order to properly distinguish the isolated singular solutions (which occur with multiplicity two or higher) from the solutions on positive dimensional solutions, one needs a representation for the positive dimensional solution sets.

On parallel shared memory computers, the work crew model is applied. In this model, threads are collaborating to complete a queue of jobs. The pointer to the next job in the queue is guarded by a semaphore so only one thread can access the next job and move the pointer to the next job forwards. The design of multithreaded software is described in [17].

The development of the blackbox solver was targeted at the cyclic n-roots systems. Backelin's Lemma [2] states that, if n has a quadratic divisor, then there are infinitely many cyclic n-roots. Interesting values for n are thus 8, 9, and 12, respectively considered in [4,7,16].

Problem Statement. The top down computation of a numerical irreducible decomposition requires first the solving of a system augmented with as many general linear equations as the expected top dimension of the solution set. This first stage is then followed by a cascade of homotopies to compute candidate generic points on lower dimensional solution sets. In the third stage, the output of the cascades is filtered and generic points are classified along their irreducible components. In the application of the work crew model with p threads, the problem is to study if the speedup will converge to p, asymptotically for sufficiently large problems. Another interesting question concerns *quality up*: if we can afford the same computational time as on one thread, then by how much can we improve the quality of the computed results with p threads?

Prior Work. The software used in this paper is PHCpack [20], which provides a numerical irreducible decomposition [18]. For the mixed volume computation, MixedVol [8] and DEMiCs [14] are used. An introduction to the homotopy continuation methods for computing positive dimensional solution sets is described in [19]. The overhead of double double and quad double precision [9] in path trackers can be compensated on multicore workstations by parallel algorithms [21]. The factorization of a pure dimensional solution set on a distributed memory computer with message passing was described in [10].

Related Work. A numerical irreducible decomposition can be computed by a program described in [3], but that program lacks polyhedral homotopies, needed to efficiently solve sparse polynomial systems such as the cyclic n-roots problems. Parallel algorithms for mixed volumes and polyhedral homotopies were presented in [5,6]. The computation of the positive dimensional solutions for the cyclic 12-roots problem was reported first in [16]. A recent parallel implementation of polyhedral homotopies was announced in [13].

Contributions and Organization. The next section proposes the application of pipelining to interleave the computation of mixed cells with the tracking of solution paths to solve a random coefficient system. The production rate of mixed cells relative to the cost of path tracking is related to the pipeline latency. The third section describes the second stage in the solver and examines the speedup for tracking paths defined by sequences of homotopies. In Sect. 4, the speedup of the application of the homotopy membership test is defined. One outcome of this research is free and open software to compute a numerical irreducible decomposition on parallel shared memory computers. Computational experiments with the software are presented in Sect. 5.

2   Solving the Top Dimensional System

There is only one input to the blackbox solver: the expected top dimension of the solution set. This input may be replaced by the number of variables minus one. However, entering an expected top dimension that is too high may lead to a significant computational overhead.

2.1   Random Hyperplanes and Slack Variables

A system is called *square* if it has as many equations as unknowns. A system is *underdetermined* if it has fewer equations than unknowns. An underdetermined system can be turned into a square system by adding as many linear equations with randomly generated complex coefficients as the difference between the number of unknowns and equations. A system is *overdetermined* if there are more equations than unknowns. To turn an overdetermined system into a square one, add repeatedly to every equation in the overdetermined system a random complex constant multiplied by a new slack variable, repeatedly until the total number of variables equals the number of equations.

The top dimensional system is the given polynomial system, augmented with as many linear equations with randomly generated complex coefficients as the expected top dimension. To the augmented system as many slack variables are added as the expected top dimension. The result of adding random linear equations and slack variables is called an *embedded* system. Solutions of the embedded system with zero slack variables are generic points on the top dimensional solution set. Solutions of the embedded system with nonzero slack variables are start solutions in cascades of homotopies to compute generic points on lower dimensional solution sets.

Example 1. (embedding a system) The equations for the cyclic 4-roots problem are

$$\mathbf{f}(\mathbf{x}) = \begin{cases} x_1 + x_2 + x_3 + x_4 = 0 \\ x_1 x_2 + x_2 x_3 + x_3 x_4 + x_4 x_1 = 0 \\ x_1 x_2 x_3 + x_2 x_3 x_4 + x_3 x_4 x_1 + x_4 x_1 x_2 = 0 \\ x_1 x_2 x_3 x_4 - 1 = 0. \end{cases} \tag{1}$$

The expected top dimension equals one. The system is augmented by one linear equation and one slack variable z_1. The embedded system is then the following:

$$E_1(\mathbf{f}(\mathbf{x}), z_1) = \begin{cases} x_1 + x_2 + x_3 + x_4 + \gamma_1 z_1 = 0 \\ x_1 x_2 + x_2 x_3 + x_3 x_4 + x_4 x_1 + \gamma_2 z_1 = 0 \\ x_1 x_2 x_3 + x_2 x_3 x_4 + x_3 x_4 x_1 + x_4 x_1 x_2 + \gamma_3 z_1 = 0 \\ x_1 x_2 x_3 x_4 - 1 + \gamma_4 z_1 = 0 \\ c_0 + c_1 x_1 + c_2 x_2 + c_3 x_3 + c_4 x_4 + z_1 = 0. \end{cases} \quad (2)$$

The constants γ_1, γ_2, γ_3, γ_4 and c_0, c_1, c_2, c_3, c_4 are randomly generated complex numbers.

The system $E_1(\mathbf{f}(\mathbf{x}), z_1) = \mathbf{0}$ has 20 solutions. Four of those 20 solutions have a zero value for the slack variable z_1. Those four solutions satisfy thus the system

$$E_1(\mathbf{f}(\mathbf{x}), 0) = \begin{cases} x_1 + x_2 + x_3 + x_4 = 0 \\ x_1 x_2 + x_2 x_3 + x_3 x_4 + x_4 x_1 = 0 \\ x_1 x_2 x_3 + x_2 x_3 x_4 + x_3 x_4 x_1 + x_4 x_1 x_2 = 0 \\ x_1 x_2 x_3 x_4 - 1 = 0 \\ c_0 + c_1 x_1 + c_2 x_2 + c_3 x_3 + c_4 x_4 = 0. \end{cases} \quad (3)$$

By the random choice of the constants c_0, c_1, c_2, c_3, and c_4, the four solutions are generic points on the one dimensional solution set. Four equals the degree of the one dimensional solution set of the cyclic 4-roots problem.

For systems with sufficiently general coefficients, polyhedral homotopies are generically optimal in the sense that no solution path diverges. Therefore, the default choice to solve the top dimensional system is the computation of a mixed cell configuration and the solving of a random coefficient start system. Tracking the paths to solve the random coefficient start system is a pleasingly parallel computation, which with dynamic load balancing will lead to a close to optimal speedup.

2.2 Pipelined Polyhedral Homotopies

The computation of all mixed cells is harder to run in parallel, but fortunately the mixed volume computation takes in general less time than the tracking of all solution paths and, more importantly, the mixed cells are not obtained all at once at the end, but are produced in sequence, one after the other. As soon as a cell is available, the tracking of as many solution paths as the volume of the cell can start. Figure 1 illustrates a 2-stage pipeline with p threads.

Figure 2 illustrates the application of pipelining to the solving of a random coefficient system where the subdivision of the Newton polytopes has six cells. The six cells are computed by the first thread. The other three threads take the cells and run polyhedral homotopies to compute as many solutions as the volume of the corresponding cell.

Counting the horizontal span of time units in Fig. 2, the total time equals 9 units. In the corresponding sequential process, it takes 24 time units. This particular pipeline with 4 threads gives a speedup of $24/9 \approx 2.67$.

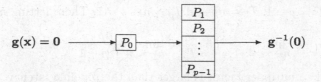

Fig. 1. A 2-stage pipeline with thread P_0 in the first stage to compute the cells to solve the start systems with paths to be tracked in the second stage by $p-1$ threads P_1, P_2, ..., P_{p-1}. The input to the pipeline is a random coefficient system $\mathbf{g}(\mathbf{x}) = \mathbf{0}$ and the output are its solutions in the set $\mathbf{g}^{-1}(\mathbf{0})$.

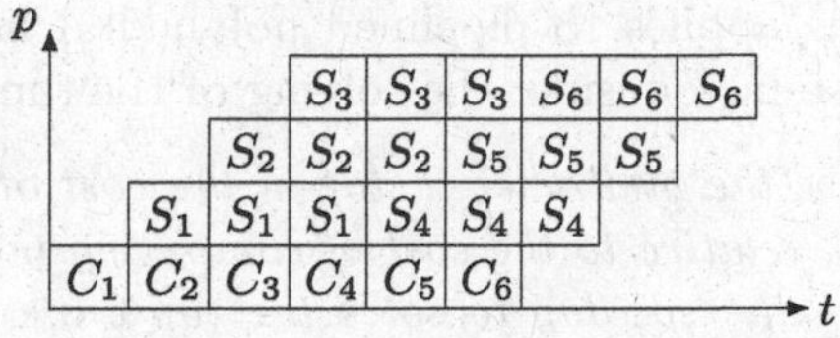

p

			S_3	S_3	S_3	S_6	S_6	S_6
		S_2	S_2	S_2	S_5	S_5	S_5	
	S_1	S_1	S_1	S_4	S_4	S_4		
C_1	C_2	C_3	C_4	C_5	C_6			

t

Fig. 2. A space time diagram for a 2-stage pipeline with one thread to produce 6 cells C_1, C_2, ..., C_6 and 3 threads to solve the corresponding 6 start systems S_1, S_2, ..., S_6. For regularity, it is assumed that solving one start system takes three times as many time units as it takes to produce one cell.

2.3 Speedup

As in Fig. 1, consider a scenario with p threads:

– the first thread produces n cells; and
– the other $p-1$ threads track all paths corresponding to the cells.

Assume that tracking all paths for one cell costs F times the amount of time it takes to produce that one cell. In this scenario, the sequential time T_1, the parallel time T_p, and the speedup S_p are defined by the following formulas:

$$T_1 = n + Fn, \quad T_p = p - 1 + \frac{Fn}{p-1}, \quad S_p = \frac{T_1}{T_p} = \frac{n(1+F)}{p-1+\frac{Fn}{p-1}}. \qquad (4)$$

The term $p-1$ in T_p is *the pipeline latency*, the time it takes to fill up the pipeline with jobs. After this latency, the pipeline works at full speed.

The formula for the speedup S_p in (4) is rather too complicated for direct interpretation. Let us consider a special case. For large problems, the number n of cells is larger than the number p of threads, $n \gg p$. For a fixed number p of threads, let n approach infinity. Then an optimal speedup is achieved, if the pipeline latency $p-1$ equals the multiplier factor F in the tracking of all paths relative to the time to produce one cell. This observation is formalized in the following theorem.

Theorem 1. *If $F = p-1$, then $S_p = p$ for $n \to \infty$.*

Proof. For $F = p - 1$, $T_1 = np$ and $T_p = n + p - 1$. Then, letting $n \to \infty$,

$$\lim_{n \to \infty} S_p = \lim_{n \to \infty} \frac{T_1}{T_p} = \lim_{n \to \infty} \frac{np}{n + p - 1} = p. \quad \square \tag{5}$$

In case the multiplier factor is larger than the pipeline latency, if $F > p - 1$, then the first thread will finish sooner with its production of cells and remains idle for some time. If $p \gg 1$, then having one thread out of many idle is not bad. The other case, if tracking all paths for one cell is smaller than the pipeline latency, if $F < p - 1$, is worse as many threads will be idle waiting for cells to process.

The above analysis applies to pipelined polyhedral homotopies to solve a random coefficient system. Consider the solving of the top dimensional system.

Corollary 1. *Let F be the multiplier factor in the cost of tracking the paths to solve the start system, relative to the cost of computing the cells. If the pipeline latency equals F, then the speedup to solve the top dimensional system with p threads will asymptotically converge to p, as the number of cells goes to infinity.*

Proof. Solving the top dimensional system consists in two stages. The first stage, solving a random coefficient system, is covered by Theorem 1. In the second stage, the solutions of the random coefficient system are the start solutions in a homotopy to solve the top dimensional system. This second stage is a pleasingly parallel computation as the paths can be tracked independently from each other and for which the speedup is close to optimal for sufficiently large problems. $\square$

3   Computing Lower Dimensional Solution Sets

The solution of the top dimensional system is an important first stage, which leads to the top dimensional solution set, provided the given dimension on input equals the top dimension. This section describes the second stage in a numerical irreducible decomposition: the computation of candidate generic points on the lower dimensional solution sets.

3.1   Cascades of Homotopies

The solutions of an embedded system with nonzero slack variables are regular solutions and serve as start solutions to compute sufficiently many generic points on the lower dimensional solution sets. The sufficiently many in the sentence above means that there will be at least as many generic points as the degrees of the lower dimensional solution sets.

Example 2. (a system with a 3-stage cascade of homotopies) Consider the following system:

$$\mathbf{f}(\mathbf{x}) = \begin{cases} (x_1 - 1)(x_1 - 2)(x_1 - 3)(x_1 - 4) = 0 \\ (x_1 - 1)(x_2 - 1)(x_2 - 2)(x_2 - 3) = 0 \\ (x_1 - 1)(x_1 - 2)(x_3 - 1)(x_3 - 2) = 0 \\ (x_1 - 1)(x_2 - 1)(x_3 - 1)(x_4 - 1) = 0. \end{cases} \tag{6}$$

In its factored form, the numerical irreducible decomposition is apparent. First, there is the three dimensional solution set defined by $x_1 = 1$. Second, for $x_1 = 2$, observe that $x_2 = 1$ defines a two dimensional solution set and four lines: $(2, 2, x_3, 1)$, $(2, 2, 1, x_4)$, $(2, 3, 1, x_4)$, and $(2, 3, x_3, 1)$. Third, for $x_1 = 3$, there are four lines: $(3, 1, 1, x_4)$, $(3, 1, 2, x_4)$, $(3, 2, 1, x_4)$, $(3, 3, 1, x_4)$, and two isolated points $(3, 2, 2, 1)$ and $(3, 3, 2, 1)$. Fourth, for $x_1 = 4$, there are four lines: $(4, 1, 1, x_4)$, $(4, 1, 2, x_4)$, $(4, 2, 1, x_4)$, $(4, 3, 1, x_4)$, and two additional isolated solutions $(4, 3, 2, 1)$ and $(4, 2, 2, 1)$.

Sorted then by dimension, there is one three dimensional solution set, one two dimensional solution set, twelve lines, and four isolated solutions.

The top dimensional system has three random linear equations and three slack variables z_1, z_2, and z_3. The mixed volume of the top dimensional system equals 61 and this is the number of paths tracked in its solution. Of those 61 paths, 6 diverge to infinity and the cascade of homotopies starts with 55 paths. The number of paths tracked in the cascade is summarized at the right in Fig. 3.

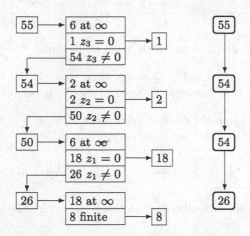

Fig. 3. At the left are the numbers of paths tracked in each stage of the computation of a numerical irreducible decomposition of $\mathbf{f}(\mathbf{x}) = \mathbf{0}$ in (6). The numbers at the right are the *candidate* generic points on each positive dimensional solution set, or in case of the rightmost 8 at the bottom, the number of *candidate* isolated solutions. Shown at the farthest right is the summary of the number of paths tracked in each stage of the cascade.

The number of solutions with nonzero slack variables remains constant in each run, because those solutions are regular. Except for the top dimensional system, the number of solutions with slack variables equal to zero fluctuates, each time different random constants are generated in the embedding, because such solutions are highly singular.

The right of Fig. 3 shows the order of computation of the path tracking jobs, in four stages, for each dimension of the solution set. The obvious parallel implementation is to have p threads collaborate to track all paths in that stage.

3.2 Speedup

The following analysis assumes that every path has the same difficulty and requires the same amount of time to track.

Theorem 2. *Let T_p be the time it takes to track n paths with p threads. Then, the optimal speedup S_p is*

$$S_p = p - \frac{p - r}{T_p}, \quad r = n \bmod p. \tag{7}$$

If $n < p$, then $S_p = n$.

Proof. Assume it takes one time unit to track one path. The time on one thread is then $T_1 = n = qp + r$, $q = \lfloor n/p \rfloor$ and $r = n \bmod p$. As $r < p$, the tracking of r paths with p threads takes one time unit, so $T_p = q + 1$. Then the speedup is

$$S_p = \frac{T_1}{T_p} = \frac{qp + r}{q + 1} = \frac{qp + p - p + r}{q + 1} = \frac{qp + p}{q + 1} - \frac{p - r}{q + 1} = p - \frac{p - r}{T_p}. \tag{8}$$

If $n < p$, then $q = 0$ and $r = n$, which leads to $S_p = n$. □

In the limit, as $n \to \infty$, also $T_p \to \infty$, then $(p - r)/T_p \to 0$ and so $S_p \to p$. For a cascade with $D + 1$ stages, Theorem 2 can be generalized as follows.

Corollary 2. *Let T_p be the time it takes to track with p threads a sequence of $n_0, n_1, \ldots, n_D$ paths. Then, the optimal speedup S_p is*

$$S_p = p - \frac{dp - r_0 - r_1 - \cdots - r_D}{T_p}, \quad r_k = n_k \bmod p, k = 0, 1, \ldots D. \tag{9}$$

Proof. Assume it takes one time unit to track one path. The time on one thread is then

$$T_1 = n_0 + n_1 + \cdots + n_D = q_0 p + r_0 + q_1 p + r_1 + \cdots + q_D p + r_D, \tag{10}$$

where $q_k = \lfloor n_k/p \rfloor$ and $r_k = n_k \bmod p$, for $k = 0, 1, \ldots, D$. As $r_k < p$, the tracking of r_k paths with p threads takes $D + 1$ time units, so the time on p threads is

$$T_p = q_0 + q_1 + \cdots + q_D + D + 1. \tag{11}$$

Then the speedup is

$$S_p = \frac{T_1}{T_p} = \frac{pT_p - dp + r_0 + r_1 + \cdots + r_D}{T_p} \tag{12}$$

$$= p - \frac{dp - r_0 - r_1 - \cdots - r_D}{T_p}. \quad □ \tag{13}$$

If the length $D + 1$ of the sequence of paths is long and the number of paths in each stage is less than p, then the speedup will be limited.

4   Filtering Lower Dimensional Solution Sets

Even if one is interested only in the isolated solutions of a polynomial system, one would need to be able to distinguish the isolated multiple solutions from solutions on a positive dimensional solution set. Without additional information, both an isolated multiple solution and a solution on a positive dimensional set appear numerically as singular solutions, that is: as solutions where the Jacobian matrix does not have full rank. A homotopy membership test makes this distinction.

4.1   Homotopy Membership Tests

Example 3. (homotopy membership test) Consider the following system:

$$\mathbf{f}(\mathbf{x}) = \begin{cases} (x_1 - 1)(x_1 - 2) = 0 \\ (x_1 - 1)x_2^2 = 0. \end{cases} \tag{14}$$

The solution consists of the line $x_1 = 1$ and the isolated point $(2,0)$ which occurs with multiplicity two. The line $x_1 = 1$ is represented by one generic point as the solution of the embedded system

$$E(\mathbf{f}(\mathbf{x}), z_1) = \begin{cases} (x_1 - 1)(x_1 - 2) + \gamma_1 z_1 = 0 \\ (x_1 - 1)x_2^2 + \gamma_2 z_1 = 0 \\ c_0 + c_1 x_1 + c_2 x_2 + z_1 = 0, \end{cases} \tag{15}$$

where the constants γ_1, γ_2, c_0, c_1, and c_2 are randomly generated complex numbers. Replacing the constant c_0 by $c_3 = -2c_1$ makes that the point $(2,0,0)$ satisfies the system $E(\mathbf{f}(\mathbf{x}), z_1) = \mathbf{0}$. Consider the homotopy

$$\mathbf{h}(\mathbf{x}, z_1, t) = \begin{cases} (x_1 - 1)(x_1 - 2) + \gamma_1 z_1 = 0 \\ (x_1 - 1)x_2^2 + \gamma_2 z_1 = 0 \\ (1 - t)c_0 + tc_3 + c_1 x_1 + c_2 x_2 + z_1 = 0. \end{cases} \tag{16}$$

For $t = 0$, there is the generic point on the line $x_1 = 1$ as a solution of the system (15). Tracking one path starting at the generic point to $t = 1$ moves the generic point to another generic point on $x_1 = 1$. If that other generic point at $t = 1$ coincides with the point $(2,0,0)$, then the point $(2,0)$ belongs to the line. Otherwise, as is the case in this example, it does not.

In running the homotopy membership test, a number of paths need to be tracked. To identify the bottlenecks in a parallel version, consider the output of Fig. 3 in the continuation of the example on the system in 6.

Example 4 (Example 2 continued). Assume the spurious points on the higher dimensional solution sets have already been removed so there is one generic point on the three dimensional solution set, one generic point on the two dimensional solution set, and twelve generic points on the one dimensional solution set.

At the end of the cascade, there are eight candidate isolated solutions. Four of those eight are regular solutions and are thus isolated. The other four solutions

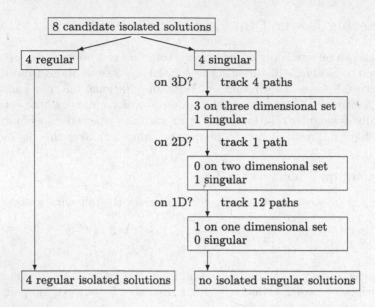

Fig. 4. Stages in testing whether the singular candidate isolated points belong to the higher dimensional solution sets.

are singular. Singular solutions may be isolated multiple solutions, but could also belong to the higher dimensional solution sets. Consider Fig. 4.

Executing the homotopy membership tests as in Fig. 4, first on 3D, then on 2D, and finally on 1D, the bottleneck occurs in the middle, where there is only one path to track.

Figure 5 is the continuation of Fig. 3: the output of the cascade shown in Fig. 3 is the input of the filtering in Fig. 5. Figure 4 explains the last stage in Fig. 5.

4.2   Speedup

The analysis of the speedup is another consequence of Theorem 2.

Corollary 3. *Let T_p be the time it takes to filter n_D, n_{D-1}, ..., $n_{\ell+1}$ singular points on components respectively of dimensions D, $D-1$, ..., $\ell+1$ and degrees d_D, d_{D-1}, ..., $d_{\ell+1}$. Then, the optimal speedup is*

$$S_p = p - \frac{(D-\ell)p - r_D - r_{D-1} - \cdots - r_{\ell+1}}{T_p}, \quad r_k = (n_k d_k) \bmod p, \quad (17)$$

for $k = \ell+1, \ldots, D-1, D$.

Proof. For a component of degree d_k, it takes $n_k d_k$ paths to filter n_k singular points. The statement in (17) follows from replacing n_k by $n_k d_k$ in the statement in (9) of Corollary 2.   □

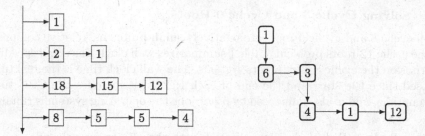

Fig. 5. On input are the candidate generic points shown as output in Fig. 3: 1 point at dimension three, 2 points at dimension two, 18 points at dimension one, and 8 candidate isolated points. Points on higher dimensional solution sets are removed by homotopy membership filters. The numbers at the right equal the number of paths in each stage of the filters. The sequence 4, 1, 12 at the bottom is explained in Fig. 4.

Although the example shown in Fig. 5 is too small for parallel computation, it illustrates the law of diminishing returns in introducing parallelisms. There are two reasons for a reduced parallelism:

1. The number of singular solutions and the degrees of the solution sets could be smaller than the number of available cores.
2. In a cascade of homotopies, there are as many steps as $D + 1$, where D is the expected top dimension. To filter the output of the cascade, there are $D(D + 1)/2$ stages, so longer sequences of homotopies are considered.

Singular solutions that do not lie on any higher positive dimensional solution set need to be processed further by deflation [11, 12], not available yet in a multi-threaded implementation. Parallel algorithms to factor the positive dimensional solutions into irreducible factors are described in [10].

5 Computational Experiments

The software was developed on a Mac OS X laptop and Linux workstations. The executable for Windows also supports multithreading. All times reported below are on a CentOS Linux 7 computer with two Intel Xeon E5-2699v4 Broadwell-EP 2.20 GHz processors, which each have 22 cores, 256 KB L2 cache and 55 MB L3 cache. The memory is 256 MB, in 8 banks of 32 MB at 2400 MHz. As the processors support hyperthreading, speedups of more than 44 are possible.

On Linux, the executable phc is compiled with the GNAT GPL 2016 edition of the gnu-ada compiler. The thread model is posix, in gcc version 4.9.4. The code in PHCpack contains an Ada translation of the MixedVol Algorithm [8], The source code for the software is at github, licensed under GNU GPL version 3. The blackbox solver for a numerical irreducible decomposition is called as phc -B and with p threads: as phc -B -tp. With phc -B2 and phc -B4, computations happen respectively in double double and quad double arithmetic [9].

5.1 Solving Cyclic 8 and Cyclic 9-Roots

Both cyclic 8 and cyclic 9-roots are relatively small problems, relative compared to the cyclic 12-roots problem. Table 1 summarizes wall clock times and speedups for runs on the cyclic 8 and 9-roots systems. The wall clock time is the real time, elapsed since the start and the end of each run. This includes the CPU time, system time, and is also influenced by other jobs the operating system is running.

Table 1. Wall clock times in seconds with `phc -B -tp` for p threads.

p	Cyclic 8-roots		Cyclic 9-roots	
	Seconds	Speedup	Seconds	Speedup
1	181.765	1.00	2598.435	1.00
2	167.871	1.08	1779.939	1.46
4	89.713	2.03	901.424	2.88
8	47.644	3.82	427.800	6.07
16	32.215	5.65	267.838	9.70
32	22.182	8.19	153.353	16.94
64	20.103	9.04	150.734	17.24

With 64 threads the time for cyclic 8-roots reduces from 3 min to 20 s and for cyclic 9-roots from 43 min to 2 min and 30 s. Table 2 summarizes the wall clock times with 64 threads in higher precision.

Table 2. Wall clock times with 64 threads in double and quad double precision.

	Cyclic 8-roots Seconds = hms format	Cyclic 9-roots Seconds = hms format
dd	53.042 = 53s	498.805 = 8 m 19 s
qd	916.020 = 15 m 16 s	4761.258 = 1 h 19 m 21 s

5.2 Solving Cyclic 12-Roots on One Thread

The classical Bézout bound for the system is 479,001,600. This is lowered to 342,875,319 with the application of a linear-product start system. In contrast, the mixed volume of the embedded cyclic 12-roots system equals 983,952.

The wall clock time on the blackbox solver on one thread is about 95 h (almost 4 days). This run includes the computation of the linear-product bound which takes about 3 h. This computation is excluded in the parallel version because the multithreaded version overlaps the mixed volume computation with polyhedral homotopies. While a speedup of about 30 is not optimal, the time reduces from 4 days to less than 3 h with 64 threads, see Table 3.

The blackbox solver does not exploit symmetry, see [1] for such exploitation.

Table 3. Times of the pipelined polyhedral homotopies versus the total time in the solver phc -B -tp, for increasing values 2, 4, 8, 16, 32, 64 of the tasks p.

p	Seconds = hms format	Speedup	Total seconds = hms format	Percentage
2	62812.764 = 17 h 26 m 52 s	1.00	157517.816 = 43 h 45 m 18 s	39.88%
4	21181.058 = 5 h 53 m 01 s	2.97	73088.635 = 20 h 18 m 09 s	28.98%
8	8932.512 = 2 h 28 m 53 s	7.03	38384.005 = 10 h 39 m 44 s	23.27%
16	4656.478 = 1 h 17 m 36 s	13.49	19657.329 = 5 h 27 m 37 s	23.69%
32	4200.362 = 1 h 10 m 01 s	14.95	12154.088 = 3 h 22 m 34 s	34.56%
64	4422.220 = 1 h 13 m 42 s	14.20	9808.424 = 2 h 43 m 28 s	45.08%

5.3 Pipelined Polyhedral Homotopies

This section concerns the computation of a random coefficient start system used in a homotopy to solve the top dimensional system, to start the cascade homotopies for the cyclic 12-roots system. Table 3 summarizes the wall clock times to solve a random coefficient start system to solve the top dimensional system.

For pipelining, we need at least 2 tasks: one to produce the mixed cells and another to track the paths. The speedup of p tasks is computed over 2 tasks. With 16 threads, the time to solve a random coefficient system is reduced from 17.43 h to 1.17 h. The second part of Table 3 lists the time of solving the random coefficient system relative to the total time of the solver. For 2 threads, solving the random coefficient system takes almost 40% of the total time and then decreases to less than 24% of the total time with 16 threads. Already for 16 threads, the speedup of 13.49 indicates that the production of mixed cells cannot keep up with the pace of tracking the paths.

Dynamic enumeration [15] applies a greedy algorithm to compute all mixed cells and its implementation in DEMiCs [14] produces the mixed cells at a faster pace than MixedVol [8]. Table 4 shows times for the mixed volume computation with DEMiCs [14] in a pipelined version of the polyhedral homotopies.

Table 4. Times of the pipelined polyhedral homotopies with DEMiCs, for increasing values 2, 4, 8, 16, 32, 64 of tasks p. The last time is an average over 13 runs. With 64 threads the times ranged between 23 min and 47 min.

p	Seconds = hms format	Speedup
2	56614 = 15 h 43 m 34 s	1.00
4	21224 = 5 h 53 m 44 s	2.67
8	9182 = 2 h 23 m 44 s	6.17
16	4627 = 1 h 17 m 07 s	12.24
32	2171 = 36 m 11 s	26.08
64	1989 = 33 m 09 s	28.46

J. Verschelde

5.4   Solving the Cyclic 12-Roots System in Parallel

As already shown in Table 3, the total time with 2 threads goes down from more than 43 h to less than 3 h , with 64 threads. Table 5 provides a detailed breakup of the wall clock times for each stage in the solver.

Table 5. Wall clock times in seconds for all stages of the solver on cyclic 12-roots. The solving of the top dimension system breaks up in two stages: the solving of a start system (start) and the continuation to the solutions of the top dimensional system (contin). Speedups are good in the cascade stage, but the filter stage contains also the factorization in irreducible components, which does not run in parallel.

p	Solving top system			Cascade and filter			Grand	Speedup
	Start	Contin	Total	Cascade	Filter	Total	Total	
2	62813	47667	110803	44383	2331	46714	157518	1.00
4	21181	25105	46617	24913	1558	26471	73089	2.16
8	8933	14632	23896	13542	946	14488	38384	4.10
16	4656	7178	12129	6853	676	7529	19657	8.01
32	4200	3663	8094	3415	645	4060	12154	12.96
64	4422	2240	7003	2228	557	2805	9808	16.06

A run in double precision with 64 threads ends after 7 h and 37 min. This time lies between the times in double precision with 8 threads, 10 h and 39 min, and with 16 threads, 5 h and 27 min (Table 3). Confusing quality with precision, from 8 to 64 threads, the working precision can be doubled with a reduction in time by 3 h, from 10.5 h to 7.5 h.

References

1. Adrovic, D., Verschelde, J.: Polyhedral methods for space curves exploiting symmetry applied to the cyclic n-roots problem. In: Gerdt, V.P., Koepf, W., Mayr, E.W., Vorozhtsov, E.V. (eds.) CASC 2013. LNCS, vol. 8136, pp. 10–29. Springer, Cham (2013). https://doi.org/10.1007/978-3-319-02297-0_2
2. Backelin, J.: Square multiples n give infinitely many cyclic n-roots. Reports, Matematiska Institutionen 8, Stockholms universitet (1989)
3. Bates, D.J., Hauenstein, J.D., Sommese, A.J., Wampler, C.W.: Software for numerical algebraic geometry: a paradigm and progress towards its implementation. In: Stillman, M.E., Takayama, N., Verschelde, J. (eds.) Software for Algebraic Geometry. IMA Volumes in Mathematics and its Applications, vol. 148, pp. 33–46. Springer, New York (2008). https://doi.org/10.1007/978-0-387-78133-4_1
4. Björck, G., Fröberg, R.: Methods to "divide out" certain solutions from systems of algebraic equations, applied to find all cyclic 8-roots. In: Gyllenberg, M., Persson, L.E. (eds.) Analysis, Algebra and Computers in Mathematical Research. LNM, vol. 564, pp. 57–70. Dekker, London (1994)

5. Chen, T., Lee, T.-L., Li, T.-Y.: Hom4PS-3: a parallel numerical solver for systems of polynomial equations based on polyhedral homotopy continuation methods. In: Hong, H., Yap, C. (eds.) ICMS 2014. LNCS, vol. 8592, pp. 183–190. Springer, Heidelberg (2014). https://doi.org/10.1007/978-3-662-44199-2_30
6. Chen, T., Lee, T.L., Li, T.Y.: Mixed volume computation in parallel. Taiwan. J. Math. **18**(1), 93–114 (2014)
7. Faugère, J.C.: Finding all the solutions of Cyclic 9 using Gröbner basis techniques. In: Computer Mathematics - Proceedings of the Fifth Asian Symposium (ASCM 2001). Lecture Notes Series on Computing, vol. 9, pp. 1–12. World Scientific (2001)
8. Gao, T., Li, T.Y., Wu, M.: Algorithm 846: MixedVol: a software package for mixed-volume computation. ACM Trans. Math. Softw. **31**(4), 555–560 (2005)
9. Hida, Y., Li, X.S., Bailey, D.H.: Algorithms for quad-double precision floating point arithmetic. In: 15th IEEE Symposium on Computer Arithmetic (Arith-15 2001), pp. 155–162. IEEE Computer Society (2001)
10. Leykin, A., Verschelde, J.: Decomposing solution sets of polynomial systems: a new parallel monodromy breakup algorithm. Int. J. Comput. Sci. Eng. **4**(2), 94–101 (2009)
11. Leykin, A., Verschelde, J., Zhao, A.: Newton's method with deflation for isolated singularities of polynomial systems. Theor. Comput. Sci. **359**(1–3), 111–122 (2006)
12. Leykin, A., Verschelde, J., Zhao, A.: Evaluation of Jacobian matrices for Newton's method with deflation to approximate isolated singular solutions of polynomial systems. In: Wang, D., Zhi, L. (eds.) Symbolic-Numeric Computation, Trends in Mathematics, pp. 269–278. Birkhauser (2007)
13. Malajovich, G.: Computing mixed volume and all mixed cells in quermassintegral time. Found. Comput. Math. **17**, 1293–1334 (2016)
14. Mizutani, T., Takeda, A.: DEMiCs: a software package for computing the mixed volume via dynamic enumeration of all mixed cells. In: Stillman, M.E., Takayama, N., Verschelde, J. (eds.) Software for Algebraic Geometry. IMA Volumes in Mathematics and Its Applications, vol. 148, pp. 59–79. Springer, New York (2008). https://doi.org/10.1007/978-0-387-78133-4_5
15. Mizutani, T., Takeda, A., Kojima, M.: Dynamic enumeration of all mixed cells. Discret. Comput. Geom. **37**(3), 351–367 (2007)
16. Sabeti, R.: Numerical-symbolic exact irreducible decomposition of cyclic-12. LMS J. Comput. Math. **14**, 155–172 (2011)
17. Sandén, B.I.: Design of Multithreaded Software. The Entity-Life Modeling Approach. IEEE Computer Society (2011)
18. Sommese, A.J., Verschelde, J., Wampler, C.W.: Numerical irreducible decomposition using PHCpack. In: Joswig, M., Takayama, N. (eds.) Algebra, Geometry, and Software Systems, pp. 109–130. Springer, Heidelberg (2003). https://doi.org/10.1007/978-3-662-05148-1_6
19. Sommese, A.J., Verschelde, J., Wampler, C.W.: Introduction to numerical algebraic geometry. In: Dickenstein, A., Emiris, I.Z. (eds.) Solving Polynomial Equations. Foundations, Algorithms and Applications. Algorithms and Computation in Mathematics, vol. 14, pp. 301–337. Springer, Heidelberg (2005). https://doi.org/10.1007/3-540-27357-3_8
20. Verschelde, J.: Algorithm 795: PHCpack: a general-purpose solver for polynomial systems by homotopy continuation. ACM Trans. Math. Softw. **25**(2):251–276 (1999). Software: http://www.phcpack.org
21. Verschelde, J., Yoffe, G.: Polynomial homotopies on multicore workstations. In: Maza, M.M., Roch, J.-L. (eds.) Proceedings of the 4th International Workshop on Parallel Symbolic Computation (PASCO 2010), pp. 131–140. ACM (2010)

Computing Limits with the RegularChains and PowerSeries Libraries: from Rational Functions to Topological Closures

(*Abstract of the Tutorial*)

Marc Moreno Maza[✉]

Department of Computer Science,
The University of Western Ontario, London, Canada
moreno@csd.uwo.ca

While computer algebra systems can perform highly sophisticated algebraic tasks, they are much less equipped for solving problems from mathematical analysis in a symbolic manner. Elementary problems in analysis, like the manipulation of Taylor series and the calculation of limits of univariate functions are supported, with some limitations, in general-purpose computer algebra systems such as MAPLE and MATHEMATICA. However, limits of multivariate functions and more advanced notions of limits, like topological closures, are almost absent from such systems. For instance, and quite surprisingly, MAPLE is not capable of computing limits of rational functions in more than two variables.

Many fundamental concepts in mathematics are defined in terms of limits and it is highly desirable for computer algebra to implement those concepts. However, limits are, by essence, hard to compute, or even not computable in an algorithmic fashion, say by doing finitely many rational operations on polynomials or matrices.

In this tutorial, we shall see how various types of limits can be computed by means of algebraic calculations. Examples will cover the Zariski closure of a constructible set, the tangent cone of an algebraic set at one of its singular points, and the limit of a real multivariate rational function at one of its poles.

The tutorial will include a presentation of the underlying mathematical concepts and algorithms as well as an extended software demonstration powered by the RegularChains and PowerSeries libraries. Both libraries are freely available in source from www.regularchains.org.

© Springer Nature Switzerland AG 2018
V. P. Gerdt et al. (Eds.): CASC 2018, LNCS 11077, p. 377, 2018.
https://doi.org/10.1007/978-3-319-99639-4

Author Index

Printed in the United States
By Bookmasters